RECENT ADVANCES IN COMPOSITES IN THE UNITED STATES AND JAPAN

A symposium
sponsored by
ASTM Committees
D-30 on High Modulus Fibers
and Their Composites and
E-9 on Fatigue
Hampton, VA, 6–8 June 1983

ASTM SPECIAL TECHNICAL PUBLICATION 864
Jack R. Vinson and Minoru Taya,
University of Delaware, editors

ASTM Publication Code Number (PCN)
04-864000-33

 1916 Race Street, Philadelphia, PA 19103

Library of Congress Cataloging in Publication Data

Recent advances in composites in the United States and
  Japan.

  (ASTM special technical publication; 864)
  "Papers presented at the United States/Japan Symposium
on Composite Materials"—Foreword.
  "ASTM publication code number (PCN) 04-864000-33."
  Includes bibliographies and index.
  1. Composite materials—United States—Congresses.
2. Composite materials—Japan—Congresses. I. Vinson,
Jack R., 1929-   . II. Taya, Minoru. III. ASTM
Committee D-30 on High Modulus Fibers and Their Composites.
IV. American Society for Testing and Materials.
Committee E-9 on Fatigue. V. United States/Japan
Symposium on Composite Materials (2nd: 1983:
Hampton, Va.) VI. Series.
TA418.9.C6R34   1985   620.1'18   85-6119
ISBN 0-8031-0436-7

03135942

NOTE

The Society is not responsible, as a body,
for the statements and opinions
advanced in this publication.

Printed in Baltimore, MD
July 1985

Foreword

This publication, *Recent Advances in Composites in the United States and Japan*, contains papers presented at the United States/Japan Symposium on Composite Materials which was held in Hampton, Virginia, 6–8 June 1983. The symposium was sponsored by ASTM Committees D-30 on High Modulus Fibers and Their Composites and E-9 on Fatigue in cooperation with the National Aeronautic and Space Administration. Jack R. Vinson and Minoru Taya, University of Delaware, served as symposium chairman and secretary, respectively. Jack R. Vinson and Minoru Taya are editors of this publication.

Related
ASTM Publications

Effects of Defects in Composite Materials, STP 836 (1984), 04-836000-33

Long Term Behavior of Composites, STP 813 (1983), 04-813000-33

Composite Materials: Testing and Design (6th Conference), STP 787 (1982), 04-787000-33

Damage in Composite Materials, STP 775 (1982), 04-775000-30

Test Methods and Design Allowables for Fibrous Composites, STP 734 (1981), 04-734000-33

A Note of Appreciation
to Reviewers

The quality of the papers that appear in this publication reflects not only the obvious efforts of the authors but also the unheralded, though essential, work of the reviewers. On behalf of ASTM we acknowledge with appreciation their dedication to high professional standards and their sacrifice of time and effort.

ASTM Committee on Publications

ASTM Editorial Staff

Contents

Design

Fabrication Methods

TESTING METHODS

ELEVATED TEMPERATURE AND ENVIRONMENTAL EFFECTS

THERMOMECHANICAL PROPERTIES

SUMMARY

Introduction

The Second United States-Japan Conference on Composite Materials was held on 6–8 June 1983 at the NASA Langley Research Center, Hampton, Virginia. It was sponsored by the American Society for Testing and Materials Committees D-30 on High-Modulus Fibers and Their Composites and E-9 on Fatigue in cooperation with the National Aeronautics and Space Administration. The conference presented, reviewed, and critiqued all the latest developments in composite materials occurring in both the United States and Japan.

The conference was the successor to the First Japan-United States Conference on Composite Materials, which was held in Tokyo in January 1981.

The chairman of American Organizing Committee for both conferences was Jack R. Vinson, University of Delaware, and the chairman of the Japanese Organizing Committee for both Conferences was Kozo Kawata, University of Tokyo. The American Organizing Committee for this conference included C. W. Bert, C. C. Chamis, A. Dhingra, K. Reifsnider, W. J. Renton, G. L. Roderick, R. Schapery, R. L. Sierakowski, R. Signorelli, M. Taya (secretary), W. J. Walker, S. S. Wang, and C. Zweben. Appreciation is hereby expressed to each committee member for his part in making the conference a great success, especially to G. L. Roderick for his help in providing the conference place. A third conference is being planned for 23–25 June 1986, Tokyo, Japan.

Appreciation is also expressed to the E. I. duPont de Nemours and Company, General Dynamics-Fort Worth, and the Vought Corporation for their financial support.

This volume provides the reader with a very complete set of timely papers that were presented at the conference. These represent the very latest findings in both countries in this rapidly developing area of science, engineering, and technology.

J. R. Vinson

University of Delaware, Newark, DE 19716;
conference chairman and coeditor.

M. Taya

University of Delaware, Newark, DE 19716;
conference secretary and coeditor.

1

Fracture

Hiroshi Fukuda[1]

Load Concentration Factors in a Chain-of-Bundles Probability Model

REFERENCE: Fukuda, H., "Load Concentration Factors in a Chain-of-Bundles Probability Model," *Recent Advances in Composites in the United States and Japan, ASTM STP 864,* J. R. Vinson and M. Taya, Eds., American Society for Testing and Materials, Philadelphia, 1985, pp. 5–15.

ABSTRACT: A chain-of-bundles probability model is often used to predict the strength of unidirectional composites. In this model, it becomes necessary first to calculate stress redistribution due to fiber breakage. A rather simple estimation of this stress redistribution has hitherto been conducted. This paper examines it in a more precise manner by adopting a shear-lag assumption. Both ordinary composites and hybrid composites are analyzed. Two idealized models are chosen in the analysis: (1) an infinite model in which a group of broken fibers is embedded in an infinite number of continuous fibers, and (2) a repeating model in which broken fibers appear repeatedly. Actual load concentration factors will fall in between the above two extreme cases. Results of load concentration factors are presented in terms of the number of fractured fibers. The present analysis provides data for a statistical calculation of the strength of composites.

KEY WORDS: load concentration factor, chain-of-bundles probability model, shear-lag, hybrid composites

The strength of a unidirectional composite is often predicted by a rule of mixtures. But this rule is only an approximate solution and, hence, more precise calculation becomes necessary. A statistical approach is an effective tool to understanding the failure of a composite in a more precise manner. This approach assumes that the strength of each fiber is not unique. When a tensile load is applied to the unidirectional composite, the weakest fiber will break first and a stress redistribution will take place. Next, failure must be considered under this redistributed stress field. Thus, it is first of all necessary to calculate the stress distributions around discontinuous fibers, in other words, to evaluate so called load sharing rules [1].

The first and simplest rule is an equal load sharing (ELS) rule. This rule

[1]Lecturer, Institute of Interdisciplinary Research, University of Tokyo, 4-6-1 Komaba, Meguro-ku, Tokyo 153, Japan.

5

assumes that when some of fibers are broken, the surviving fibers share the applied load equally. The work of Rosen [2], for example, is based on this ELS rule. Another limiting case is so called local load sharing (LLS) rule. This rule assumes that a load initially sustained by a broken fiber is shared by the nearest fibers only. A ratio of the maximum stress of the nearest fiber to the average fiber stress is called a load concentration factor, or, stress concentration factor (SCF). The ELS rule underestimates the SCF and the LLS rule overestimates it.

The SCFs affect directly the strength of composites in a statistical approach. For example, if a single fiber is broken, the probability that one of the two adjacent fibers will break due to the load concentration is [3]

$$p_{2/1} = 2[F(K_1\sigma) - F(\sigma)] - 2[F(K_1\sigma) - F(\sigma)]^2 \qquad (1)$$

where $F(\sigma)$ is a cumulative distribution function and K_1 is the SCF caused by one broken fiber. Thus precise calculation of SCF becomes necessary.

In this paper we calculate the SCF based upon a shear-lay assumption. Since an ordinary (nonhybrid) composite is a special case of hybrid composite as far as the analytical formulation is concerned, we use a model of a hybrid composite in the analysis. Results obtained here will become a basis of forthcoming statistical evaluation of the strength of composites.

Analysis

Infinite Model

At the first stage of failure of a composite, the weakest fibers will break individually as was shown by Rosen [2]. Figure 1a demonstrates this situation. What is the SCF of Point A? This will be primarily affected by discontinuous fibers, B. But another end of fiber, C, may affect a little the SCF of Point A. It

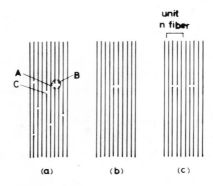

FIG. 1—*Actual fiber breakage and its idealizations.*

is not clever to calculate the SCF in such a complicated combination of broken fibers. Therefore, we choose two extreme models shown in Fig. 1b, c. The actual SCF of Fig. 1a will fall in between the above two idealized cases. The SCF of Fig. 1b is analyzed in this section.

Figure 2 shows the analytical model of an intermingled hybrid sheet where two kinds of fibers assume alternating positions. A bundle of fibers may be treated as if it were an individual fiber [4]. Here we use the terminology of low elongation (LE) and high elongation (HE) to distinguish two kinds of fibers in accordance with common usage [4,5], although the character of fiber elongation does not directly enter into the present mathematical derivations.

Hedgepeth [6] first applied the shear-lag assumption to calculate the SCF in a unidirectional sheet. Fukuda and Chou [7,8] modified this idea so as to fit to a hybrid sheet.

The force equilibrium equations of the mth LE and HE fibers are, respectively,

$$EA \ \frac{d^2u_m}{dx^2} + \frac{Gh}{d} \ (u_m^* + u_{m-1}^* - 2u_m) = 0$$

$$E^*A^* \ \frac{d^2u_m^*}{dx^2} + \frac{Gh}{d} \ (u_{m+1} + u_m - 2u_m^*) = 0 \tag{2}$$

where u_m is the displacement of the mth fiber and EA and G denote the fiber extensional stiffness and the shear modulus of the matrix. An asterisk is used to denote the quantity related to HE fibers.

The force-displacement relations are, under linear elasticity conditions,

$$p_m = EA \ \frac{du_m}{dx} \ , \qquad p_m^* = E^*A^* \ \frac{du_m^*}{dx} \tag{3}$$

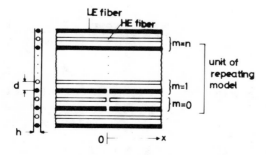

FIG. 2—Model of analysis.

where p_m is the axial load of the mth fiber. The boundary conditions are

$$
\begin{array}{lll}
p_m(0) = 0 & \text{and/or} & p_m^*(0) = 0 \quad \text{(broken fibers)} \\
u_m(0) = 0 & \text{and/or} & u_m^*(0) = 0 \quad \text{(unbroken fibers)} \\
p_m(\infty) = p, & & p_m^*(\infty) = (E^*A^*/EA)p
\end{array} \tag{4}
$$

The following dimensionless parameters are introduced here:

$$
\begin{array}{ll}
P_m = p_m/p & P_m^* = p_m^*/p \\
U_m = u_m/p\sqrt{d/EAGh} & U_m^* = u_m^*/p\sqrt{d/EAGh} \\
\xi = x/\sqrt{EAd/Gh} & R = E^*A^*/EA
\end{array} \tag{5}
$$

Thus Eqs 2 become

$$
U_m'' + U_m^* + U_{m-1}^* - 2U_m = 0
$$

$$
R\,U_m^{*\prime\prime} + U_{m+1} + U_m - 2U_m^* = 0 \tag{6}
$$

where $(\ )'' = d^2(\ )/d\xi^2$. Further, by introducing the following expressions:

$$
U_m(\xi) = \xi + \Sigma\, V_{m-k}(\xi)\, U_k(0) + \Sigma\, W_{m-k}(\xi)\, U_k^*(0)
$$

$$
U_m^*(\xi) = \xi + \Sigma\, V_{m-k}^*(\xi)\, U_k(0) + \Sigma\, W_{m-k}^*(\xi)\, U_k^*(0) \tag{7}
$$

Eqs 2 reduce to the equations concerning V, V^*, W, and W^* which are called influence functions.

The influence functions are solved as follows although the detail is not shown here:

$$
V_m = \frac{1}{\pi} \int_0^\pi (C_1 e^{-\lambda_1 \xi} + C_2 e^{-\lambda_2 \xi}) \cos m\theta\, d\theta
$$

$$
V_m^* = \frac{1}{\pi} \int_0^\pi C_3\, (e^{-\lambda_1 \xi} - e^{-\lambda_2 \xi})\, \frac{\cos m\theta + \cos(m+1)\theta}{2(1 + \cos\theta)}\, d\theta \tag{8}
$$

$$
W_m = \frac{1}{\pi} \int_0^\pi C_5\, (e^{-\lambda_1 \xi} - e^{-\lambda_2 \xi})\, \{\cos m\theta + \cos(m-1)\,\theta\}\, d\theta
$$

$$
W_m^* = \frac{1}{\pi} \int_0^\pi (C_7 e^{-\lambda_1 \xi} + C_8 e^{-\lambda_2 \xi}) \cos m\theta\, d\theta
$$

where

$$C_1 = C_8 = (2 - \lambda_2^2)K \qquad C_2 = C_7 = -(2 - \lambda_1^2)K$$

$$C_3 = (2 - \lambda_1^2)(2 - \lambda_2^2)K \qquad C_5 = -K$$

$$K = 1/(\lambda_1^2 - \lambda_2^2) \qquad \lambda_{1,2} = \sqrt{a \pm \sqrt{a^2 - b}}$$

$$a = 1 + 1/R \qquad b = (2/R)(1 - \cos\theta)$$

(9)

The SCFs of the mth group of fibers are defined as

$$P_m(0)/P_m(\infty) = 1 + \sum_k V'_{m-k}(0) U_k(0) + \sum_k W'_{m-k}(0) U_k^*(0)$$

$$P_m^*(0)/P_m^*(\infty) = 1 + \sum_k V_{m-k}^{*'}(0) U_k(0) + \sum_k W_{m-k}^{*'}(0) U_k^*(0)$$

(10)

where $U_k(0)$ and $U_k^*(0)$ can be determined from the first of boundary conditions, Eqs 4.

Repeating Model

The SCFs of Fig. 1c are calculated in this subsection. The SCFs in a nonhybrid composite of finite number of fibers have already been discussed in Ref 9, although some errata can be noticed there. This part is an expansion of the previous work to the hybrid composites. Suppose, for example, a unit as in Fig. 2. In this case, equilibrium equations of force becomes, in terms of dimensionless parameters,

$$U_0'' + U_0^* - U_0 = 0 \tag{11a}$$

$$U_m'' + U_m^* + U_{m-1}^* - 2U_m = 0 \qquad (0 < m \leqq n - 1) \tag{11b}$$

$$R U_m^{*''} + U_{m+1} + U_m - 2U_m^* = 0 \qquad (0 \leqq m \leqq n - 1) \tag{11c}$$

$$U_n'' + U_{n-1}^* - U_n = 0 \tag{11d}$$

From Eq 11a, U_0^* becomes

$$U_0^* = U_0 - U_0'' \tag{12}$$

Putting $m = 0$ in Eq 11c, U_1 can be calculated as follows:

$$U_1 = 2U_0^* - U_0 - R\,U_0^{*\prime\prime}$$

$$= U_0 - (2 + R)\,U_0'' + R\,U_0^{(4)}$$

(13)

where $U_0^{(4)} = d^4U_0/d\xi^4$. By repeating the above procedure, all U_m and U_m^* can be expressed by U_0 and its derivatives only. Substituting the above values into Eq 11d, we get a differential equation concerning U_0. Thus we can solve U_0 and hence all U_m and U_m^* also. If the displacement of each fiber is obtained, the axial load is calculated by Eqs 3 and finally we get the SCF.

Numerical Results and Discussion

We will start from an infinite model, the SCFs of which are calculated from Eqs 10. An ordinary composite, a Kevlar/graphite hybrid, and a glass/graphite hybrid are considered which can be represented by $R = 1$, $R = \frac{1}{2}$, and $R = \frac{1}{3}$, respectively. Although $1 + \cos\theta$, the denominator of Eqs 8, is zero at $\theta = \pi$, the value of $\{\cos m\theta + \cos(m + 1)\theta\}/(1 + \cos\theta)$ is finite. This was taken into consideration prior to the numerical integration of Eqs 8. Figure 3

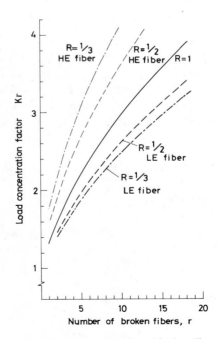

FIG. 3—*Load concentration factors versus number of broken fibers in an infinite model.*

shows the SCF caused by r successive broken fibers. Hedgepeth [6] calculated the SCF of nonhybrid composite as

$$K_r = \frac{4 \times 6 \times \ldots \times (2r + 2)}{3 \times 5 \times \ldots \times (2r + 1)} \qquad (14)$$

where K_r is the SCF of a continuous fiber adjacent to r successive broken fibers. The solid line of Fig. 3 exactly coincides with the above value, and therefore it can be concluded that the present analysis includes Hedgepeth's analysis as a special case of $R = 1$.

In a hybrid composite, the value of SCF depends on a model considered. Suppose, for example, fibers of HE, LE, and HE are broken (Fig. 4a). The adjacent fiber is LE fiber and the maximum load concentration will occur at Point A. In a case of Fig. 4b, Point B of the HE fiber is of most interest. The array of broken fibers is assumed to be symmetric. The bold lines (dashed and dot-dashed lines) of Fig. 3 show the SCFs of the LE fiber (Fig. 4a). With decreasing R, the SCFs of the LE fiber decreases. At the first stage of failure, LE fibers are expected to break because they have smaller failure strain. Therefore, the smaller SCF of the LE fiber is advantageous for hybrid composites. The SCF of the HE fiber, on the contrary, increases with decreasing R. But this demerit is secondary because HE fibers have relatively large failure strain.

Next we will consider the SCF of a finite-layer model where the formulation of the previous section is used. In this case the total number of fibers, n, also affects the value of SCF. Figure 5 shows the SCFs in a nonhybrid composite sheet where r is the number of broken fibers. Let us choose a case of $r = 5$. If n is seven, the number of surviving fibers is only two. These two fibers must sustain a total load $7p$ where p is the axial load of a fiber at infinity. Thus the SCF becomes 3.5. If n becomes nine, four survived fibers share $9p$ and the SCF is calculated as 2.725. At the limit of $n = \infty$, the SCF tends to Hedgepeth's result, $K_5 = 2.216$.

According to the LLS rule, the SCF is

$$K_r = 1 + r/2 \qquad (15)$$

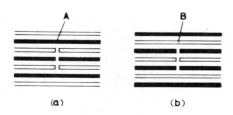

(a)   (b)

FIG. 4—Two patterns of fiber breakage.

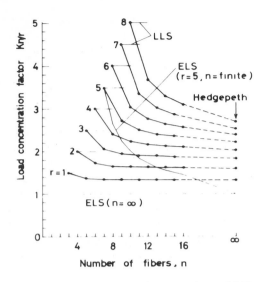

FIG. 5—*Load concentration factors in a repeating model (R = 1).*

which is independent of n. These values are shown in Fig. 5. The SCF under the ELS assumption is

$$K_{n/r} = r/(n - r) \tag{16}$$

These values are also shown in Fig. 5 for $r = 5$ only. For the case of $n = r + 2$ only, our results coincide with LLS and ELS, and with increasing n the difference between LLS and ELS becomes too large. This is the reason why more precise calculation becomes necessary.

Next we will discuss the SCFs in a hybrid composite. Figure 6 shows the SCFs for $R = \frac{1}{2}$, which approximately corresponds to the Kevlar/graphite hybrid. Again LE fibers and HE fibers must be distinguished. Let us consider a bold line of $r = 3$. This corresponds to Fig. 4a; that is, fibers of HE, LE, and HE are broken and the adjacent fiber is LE. The axial load sustained by a HE fiber is half that of the LE fiber. If n is five of three LE and two HE fibers, the total load is $3p + 2 \times \frac{1}{2} p = 4p$. Therefore, two surviving fibers must sustain $2p$ each. When n increases to seven, the SCF of the nearest fiber reduces to 1.80 and, at the limit of $n = \infty$, the SCF becomes $(K_{\infty/3})_{LE} = 1.61$. The fine line of $r = 3$ shows similar results at Point B of Fig. 4b.

Figure 7 shows the results for $R = \frac{1}{3}$. The SCFs of LE fibers for $R = \frac{1}{3}$ are smaller than the corresponding values for $R = \frac{1}{2}$. Therefore, it may be concluded that a glass/graphite hybrid is superior to a Kevlar/graphite hybrid as far as the failure of graphite fibers is concerned. But we must notice here that actual selection of materials depends on other factors such as the strength-to-

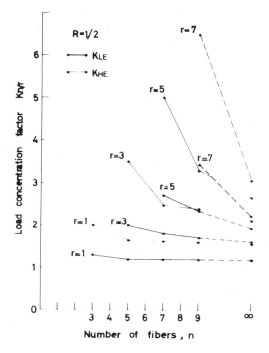

FIG. 6—*Load concentration factors in a repeating model (R = $\frac{1}{2}$).*

density ratio; in this sense a Kevlar/graphite is better than a glass/graphite. Further, we must pay attention to the compatibility in coefficients of thermal expansion of reinforcing fibers, which is one of the most important factors for hybridization. This factor is discussed by Bunsell and Harris [*10*].

Conclusions

The SCFs caused by broken fibers in an intermingled hybrid sheet have been calculated. Following are the conclusions:

1. For the special case of $R = 1$, the present results coincided with Hedgepeth's analysis. An expansion of his analysis to hybrid composites was done successfully.

2. Hedgepeth considered a limiting case of $n = \infty$. In this paper the SCFs of both the infinite model and the repeating model were calculated. The present analysis is also applicable to a small bundle. More precise SCFs than those derived from LLS and ELS rules were calculated here.

3. The SCFs in hybrid composites were also calculated. These have not been discussed hitherto in detail.

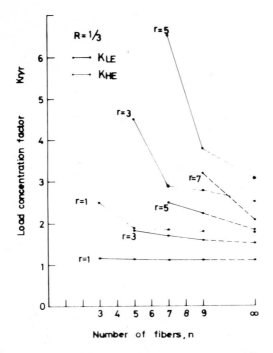

FIG. 7—*Load concentration factors in a repeating model (*R = 1/3).

4. The results obtained here will become a basis of forthcoming statistical prediction of the strength of composites.

Acknowledgment

This work is partly supported by the Ministry of Education, Japan, under Contract No. 57350003.

References

[1] Harlow, D. G. and Phoenix, S. L., *Journal of Composite Materials*, Vol. 12, No. 2, 1978, pp. 195–214.
[2] Rosen, B. W., *AIAA Journal*, American Institute of Aeronautics and Astronautics, Vol. 2, No. 11, 1964, pp. 1985–1991.
[3] Zweben, C., *AIAA Journal*, American Institute of Aeronautics and Astronautics, Vol. 6, No. 12, 1968, pp. 2325–2331.
[4] Zweben, C., *Journal of Materials Science*, Vol. 12, No. 7, 1977, pp. 1325–1337.
[5] Chou, T. W. and Kelly, A., *Annual Review of Materials Science*, Vol. 10, 1980, pp. 229–259.
[6] Hedgepeth, J. M., NASA Technical Note, D-882, National Aeronautics and Space Administration, 1961.

[7] Fukuda, H. and Chou, T. W., *Transactions*, Japan Society for Composite Materials, Vol. 7, No. 2, 1981, pp. 37-42.
[8] Fukuda, H. and Chou, T. W., *Journal of Applied Mechanics*, Vol. 50, No. 4, 1983, pp. 845-848.
[9] Fukuda, H. and Kawata, K., *Fibre Science and Technology*, Vol. 13, No. 4, 1980, pp. 255-267.
[10] Bunsell, A. R. and Harris, B., *Composites*, Vol. 5, No. 4, 1974, pp. 157-164.

Lawrence T. Drzal[1] and Michael J. Rich[2]

Effect of Graphite Fiber/Epoxy Matrix Adhesion on Composite Fracture Behavior

REFERENCE: Drzal, L. T. and Rich, M. J., **"Effect of Graphite Fiber/Epoxy Matrix Adhesion on Composite Fracture Behavior,"** *Research Advances in Composites in the United States and Japan, ASTM STP 864*, J. R. Vinson and M. Taya, Eds., American Society for Testing and Materials, Philadelphia, 1985, pp. 16–26.

ABSTRACT: Very little fundamental information is known about the mechanism by which surface treatments and surface finishes influence fiber-matrix adhesion and composite properties. This study is directed at elucidating the molecular mechanisms by which surface treatments affect fiber-matrix adhesion and composite properties.

A polyacrylonitrile based graphite fiber and epoxy matrix were chosen as a model system. Substantial changes were detected in the value of the interfacial shear strength attainable for this fiber-matrix system when only the interphase region was altered. Observation of the interphase under load and high resolution transmission electron microscope (TEM) of ultramicrotomed sections were able to discriminate between various failure mechanisms operating at the fiber-matrix interphase.

An untreated graphite fiber possesses a weak structural layer that can not support high shear loads. When this fiber is stressed in shear the fiber separates from the matrix. Failure analysis shows that the interfacial fracture path involves some fiber fracture in the outer layers as well as along the interface. Surface treatment removes this defect laden outer layer of the fiber and also adds surface chemical groups which increase the interaction with the matrix by a factor of two. The fracture path becomes purely interfacial with no fiber failure. Both mechanisms, that is, surface chemical interactions and morphological changes in the outer layers of the fiber contribute to the increased interfacial shear strength attainable with the surface treated fiber in the same matrix.

Application of a fiber surface finish increases the interfacial shear strength. The mechanism by which this can occur is through the creation of a brittle interphase around the fiber. The original finish layer devoid of crosslinking agent receives some through diffusion from the surrounding resin mixture. This results in a layer more brittle than the matrix. Stress analysis shows that this brittle layer promotes better shear transfer resulting in an increased interfacial shear strength. However, at the fiber breaks, matrix cracks grow perpendicular instead of parallel to the fiber surface under shear loading because of this brittle layer.

Small composite specimens containing the fibers with various surface treatments and possessing the various failure modes were fractured in order to determine if the interphase

[1]Materials research engineer, Air Force Wright Aeronautical Laboratories, Wright-Patterson Air Force Base, OH 45433.
[2]Research associate, University of Dayton Research Institute, Dayton, OH 45469.

effects had an influence on composite properties. Results indicate that the single filament behavior is observable in the composite.

KEY WORDS: fiber-matrix adhesion, interface, interfacial shear strength, interphase, composite fracture, graphite fiber adhesion, fiber surface treatment, fiber coatings, fiber finishes, fracture toughness

Graphite fiber reinforced epoxy composites provide structural materials which behave in a predictable manner obeying a rule of mixtures type of relationship. A key element of this predictability is the existence of an acceptable degree of adhesion between fiber and matrix in the composite. Variability of fiber-matrix adhesion has not been considered as either an adjustable parameter for optimizing composite properties or as a causative agent responsible for certain characteristics of composite material behavior.

The increasing potential uses for composite materials is necessitating substitution of different matrix materials and fibers into structural composite materials. These substitutions are driven by a desire to extend the operating environment to higher temperatures or moist conditions or both. Material substitution is not straight forward primarily because the fiber-matrix interface requires optimization for each new fiber-matrix combination. This effort has been undertaken in order to quantify the relationship between fiber-matrix interface adhesion and composite fracture properties so that ultimately a predictive model can be developed which will allow material substitution and interface optimization to be made *a priori* in composite materials.

Experimental

Fibers

A polyacrylonitrile based graphite fiber was chosen as the baseline material for this study. It is designated as a Type A-1 fiber (Hercules, Inc.). The fiber has a tensile modulus of about 241 GPa and a tensile strength of about 2.93 GPa at a 25 mm gage length. The fiber was supplied by the manufacturer untreated (AU-1), surface treated with an oxidative treatment (AS-1), and surface treated and coated with a finish of DER-330, a diglycidyl ether or Bisphenol-A (Dow Chemical Co.), to 100 to 200 nm in thickness (AS1C). Previous studies have measured surface areas of fibers produced in a similar manner and concluded that an increase in surface area does not accompany the fiber surface treatment used here [1,2]. The fibers were examined for uniformity of finish throughout the fiber tow. All fibers were uniformly coated with a thin layer of epoxy. Those on the outside of the tow, however, did have areas where the resin finish had formed droplets. Single filaments were selected from the center of the tow to avoid these outer fibers.

Matrix

The matrix used in this investigation was a stoichiometric mixture of a diglycidyl ether of bisphenol-A (Epon 828, Shell Chemical Co.) cured with 14.5 parts per hundred resin (phr) meta-phenylene diamine (m-PDA, Aldrich Chemical Co.). The m-PDA was of high purity and was kept in a cool, dark, inert environment to prevent deterioration. Published work has shown that a "darkened" m-PDA can affect the interfacial properties while leaving the bulk properties of the epoxy unchanged [3]. The Epon 828 was likewise kept in a cool, inert environment to prevent deterioration during the course of these experiments. The Epon 828 is chemically very similar to the DER-330 epoxy used as the finish.

The matrix system was weighed, vacuum melted, mixed, and debulked and then processed for 2 h at 75°C and for 2 h at 125°C followed by an overnight cooldown. Under these processing conditions, gelation occurs within the first 60 min at 75°C.

Interfacial Shear Strength

The adhesion between fiber and matrix was characterized through measurement of the interfacial shear strength. This is accomplished by embedding the fiber in a tensile coupon of matrix resin and subjecting the specimen to tensile loading. Since the tensile forces are transferred to the fiber through shear forces at the fiber-matrix interface and because the maximum strain of the fiber is much less than that of the matrix, the fiber will fracture into small segments within the matrix. As higher tensile loads are applied the fracture process continues until the interfacial forces no longer induce fracture in the fiber. At this point a minimum segment length is attained known as the critical transfer length which allows the interfacial shear strength to be determined.

The relationship between fiber tensile strength (σ_f), critical length to diameter ratio (l_c/d), and the interfacial shear strength (τ) is given by [4]

$$\tau = \frac{\sigma_f d}{2\,l_c} \tag{1}$$

Since a distribution of lengths is observed experimentally, this relationship has been altered to reflect Weibull statistics to the form

$$\tau = \frac{\sigma_f}{2\beta}\,\Gamma\left(1 - \frac{1}{\alpha}\right) \tag{2}$$

where (α) is the shape factor, (β) is the scale factor (that is, the mean value of l_c/d) and (Γ) is the gamma function. This relationship is used to evaluate the interfacial shear strength between fiber and matrix. Details of the technique and the data accumulation and reduction schemes have been published [5].

A microscopic technique was used to investigate the interfacial response between graphite fiber and matrix as a function of both shear loading and surface finish. Optical microscopy with transmitted polarized light was used to detect fiber fracture within the epoxy specimen and to document the load transfer from fiber to matrix near the region of the fiber break. In addition fiber fragment lengths could be measured directly within the specimen using a calibrated eyepiece.

Composite Fracture Surfaces

Unidirectional composite specimens with 50 to 60 volume percent fiber were prepared using the same fiber-matrix combinations investigated in the single fiber interfacial shear strength test. Ten centimeter sections of fiber tow were cut and stacked on a fluoropolymer coated section of screen. The weight of the fiber was adjusted to correspond to the volume fraction desired in the composite specimen based on the volume of the mold cavity. The reactive debulked epoxy mixture heated to 75°C was thoroughly pressed through this fiber mass prior to transfer to the mold cavity. The cavity was 25 mm wide by 100 mm long and contained spacers that fixed the final composite thickness at 2.5 mm. All components were coated with a permanent solid fluoropolymer to enhance release from the mold. Mold release was not used to preclude release agent migration to the fiber surface or transfer to the cured composite surface. After initial squeeze out of the excess epoxy, the composite was fabricated according to the same processing cycle used for the single filament specimens. The fully cured unidirectional composite specimen bar was then cut into specimens 12.5 by 12.5 mm, and an initial cut was made with a diamond saw either perpendicular or parallel to the fiber axis. Fixture holes were drilled into the specimen as shown in Fig. 1 for loading with a testing machine. Specimens were fractured at 20°C and 0.0025 cm/min. The fracture surfaces were examined uncoated in a scanning electron microscope (SEM).

Results and Discussion

Interfacial Shear Strength

The interfacial shear strength for each fiber-epoxy combination was measured using the single fiber specimen. The data were reduced according to Eq 2 and tabulated in Table 1. The alpha (scale factor) and beta (shape factor) for the computer generated Weibull fit are listed in columns 3 and 4. Column 5 contains the value of the fiber fracture strength measured at the gage length equal to the fiber critical length for that fiber. The coefficient of variation for the interfacial shear strength varied between 0.1 and 0.2.

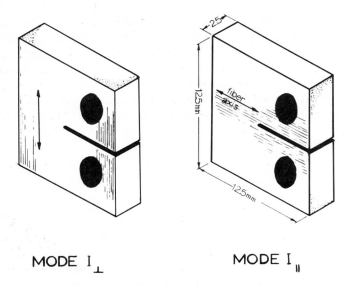

MODE I$_\perp$ MODE I$_\parallel$

FIG. 1—*Fracture specimens showing fiber orientation in relation to direction of the pre-crack.*

TABLE 1—*Interfacial shear strength for A-fiber/epoxy specimens.*

Fiber	τ, MPa	α	β	σ_f, GPa
AU-1	44.8	2.6	70.1	4.34
AS-1	74.4	3.5	38.8	4.52
AS-1C	93.6	3.8	30.3	4.58

The interfacial shear strength is very sensitive to changes in fiber-matrix adhesion. The untreated graphite fiber (AU-1) exhibits the lowest value of interfacial shear strength (44.8 MPa). The oxidative surface treatment given to this fiber to transform it into the AS-1 greatly improves the interfacial shear strength by almost a factor of two over the value obtained for the untreated fiber.

Application of the 100 to 200-nm coating of pure epoxy to the surface treated fiber (AS1C) also causes the interfacial shear strength to rise to still higher values. This finish layer increases the interfacial shear strength 25% over the value obtainable with an oxidatively treated fiber alone.

Locus of Failure

Microscopic observation of each of these specimens during the application of stress associated with the interfacial shear strength determination detected significant changes in interfacial behavior. Figure 2 contains photomicro-

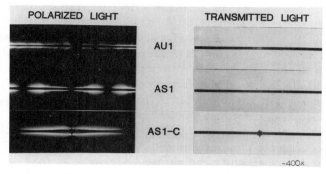

FIG. 2—*Transmitted and polarized light micrographs at ×400 of A-fibers under strain in the epoxy matrix.*

graphs of representative observations of each of the three fiber-matrix combinations in transmitted and transmitted polarized light.

The AU1 fiber which possesses the lowest degree of interfacial shear strength with this epoxy matrix completely separates from the matrix under load, and a cylindrical cavity can be observed forming at the fiber break. Under polarized light, growth of the photoelastic region around the fiber can be observed at very low stress levels until the entire fragment length has debonded.

Sections taken from this region show that the failure path occurs not only along the fiber-matrix interface but through the outer layers of the fiber as well. Graphite fiber fragments are present on the epoxy side of the fracture path.

The AS1 oxidatively treated fiber exhibits a different mode of behavior. At the same stress levels as the AU1 fiber, no detectable separation of the ends of the fiber fragments with increasing strain is observed. However, under polarized light conditions, a highly stressed area of matrix is visible at the fiber fragment ends. With increasing strain this elliptically shaped region moves further away from the fragment ends and leaves a narrow highly stressed region between this region and the fiber fragment end extending all around the fiber.

Recently published work [6] has shown that the elliptical region is associated with a growing interfacial crack tip. The narrow region between it and the fragment end is due to frictional stresses on the matrix where that crack has already separated the fiber from the matrix. Micromechanical stress analysis of this single fiber sample using properties characteristic of these materials shows that there is a net normal compressive component to the stresses existing at the interface responsible for the frictional response between fiber and matrix [7]. Locus of fracture examination of sectioned samples of this AS1 fiber specimen has indicated a change in failure path from a combination of interfacial and fiber to one that is strictly interfacial with no evidence of fiber failure near the surface.

Combining the values for the interfacial shear strength with measurements of the fiber surface oxygen content measured previously [8] leads to the conclusion that the oxidative surface treatment applied to these fibers operates by a two part mechanism. This effect can be visualized more readily by plotting the interfacial shear strength versus the surface oxygen content as shown in Fig. 3. The weak defect laden outer layer of the graphite fiber normally present after transformation of the polyacrylonitrile fiber into graphite fiber is removed during the surface treatment. At the same time surface chemical species, mostly oxygen with some nitrogen, are added to the fiber surface. The increase in interfacial shear strength is almost entirely attributable to the creation of a defect free surface layer and only slightly due to the addition of the surface chemical species.

This has been proven by carefully removing the surface chemical groups from the AS1 fiber to a level less than that which exists on the untreated AU1 fiber and detecting only a slight (10%) decrease in interfacial shear strength. If the surface chemical effect predominated, removal of these species would have reduced the interfacial shear strength to a level equal to or below that of the untreated AU1 fiber. This was not the case [6].

The surface treated fiber which had the 100 to 200-nm coating of pure epoxy applied to its surface after oxidative surface treatment (AS1C) behaves differently under load than either the AU1 or AS1 fiber. After a fiber breaks in the specimen, a matrix crack is observed to begin to propagate perpendicular to the fiber axis. Polarized light micrography highlights the stresses that

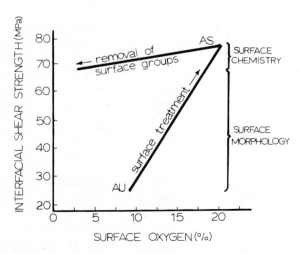

FIG. 3—*Interfacial shear strength versus surface oxygen content for the A-fibers (from Ref 6).*

develop at the fragment ends and also around the growing matrix crack. Interfacial crack growth does not occur (see Fig. 2).

Various explanations for the increase in interfacial shear strength have been proposed. Among them are the protection of the surface from flaws, improved wetting of the fiber, and enhancement of surface reactivity. These explanations have been proven to be unapplicable for this case and an alternate explanation proposed [9]. Namely, shear strength enhancement occurs through creation of a brittle interphase region around the fiber. The pure epoxy coating obtains amine curing agent by diffusion during the curing of the specimen until gelation occurs. Because of crosslinking reactions occurring at the same time, less that the stoichiometric amount of curing agent can diffuse into the layer adjacent to the fiber surface resulting in a gradient of epoxy composition formed in an interphase region near the fiber surface. Epoxy systems with less than the proper amount of curing agent increase in modulus compared to the stoichiometric mixture but decrease in fracture strength and fracture toughness [10,11]. This indicates that the coating behaves as a brittle interphase region promoting better stress transfer because of the higher modulus, but more susceptible to crack growth because of the lower fracture toughness. Various micromechanical models which include an interphase layer around the fiber show a direct correspondence between better stress transfer, in this case meaning higher interfacial shear strength, and a larger interphase modulus [12].

Composite Fracture Behavior

SEM photomicrographs of fracture surfaces of composites made with the same fibers (that is, AU1, AS1, and AS1C) and with the same Epon 828/mPDA matrix cured according to the same schedule show fracture behavior that parallels the differences in interfacial behavior observed with the single filament test.

Figure 4 shows that fracture surface of the composite with the crack being driven parallel to the fiber axis in an opening mode. The AU1 composite surface shows many bare fibers, fiber detachment and interfacial separation between fiber and matrix.

The AS1 fiber composite at approximately the same volume percent fiber content shows some bare fibers and some evidence of interfacial separation. However, the frequency of this observation has decreased and there is additional indication that the matrix is involved in the failure. Regions can be seen where the fracture is observed to occur through the matrix as well as interfacially.

The AS1C composite fracture surface made with the AS1C fiber which exhibited the highest value for the interfacial shear strength with this matrix also shows the greatest degree of matrix failure along with the least amount of

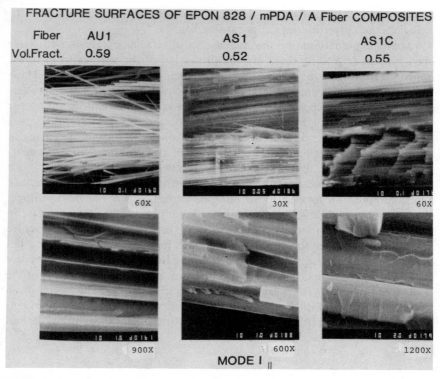

FIG. 4—*Scanning electron micrographs of the composite fracture surfaces parallel to the crack.*

bare fiber and interfacial separation. Most fibers appear to have some matrix adhering to their surface after fracture.

An attempt was made to calculate the stress intensity factor for each specimen. Calculations indicated that there was a continuous increase in fracture toughness with crack length. Some of this effect could be attributed to a slight degree of axial misalignment between the tows in the composite. In general however, the highest values were recorded for the AS1C composite followed by the AS1 composite with the lowest being obtained for the AU1 composite. These trends were valid for all values of crack length measured.

Fracture of the composite specimens in an opening mode with the crack perpendicular to the fiber axis coincided very closely with the interfacial shear strength failure observations. Figure 5 shows the AU1 fiber composite fracture surface displaying a large degree of fiber pullout.

The surface treated AS1 fiber composite fracture surface also exhibited a great degree of fiber pullout but the fragment lengths were much shorter than for the AU1 composite. This parallels the decrease in critical transfer length measured with the single fiber specimen for the AS1 fiber.

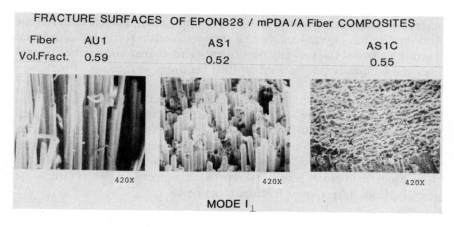

FIG. 5—*Scanning electron micrographs of the composite fracture surfaces perpendicular to the crack.*

The fracture surface of the AS1C fiber composite was dramatically differ-ent than for the other fibers. Planar regions were observed over areas of the fracture surface. Little if any pullout of the fiber from the matrix could be detected in these areas. This behavior is coincident with the observations from the single fiber tests. In that case a matrix crack grew at the fiber breaks which extended two or three fiber diameters into the matrix. The coalescence of these cracks appears to be responsible for the planar features of the com-posite fracture surfaces. The stress intensity factor for these specimens was not measured because of the difficulty in following the crack front.

Conclusions

A definite relationship exists between the level of fiber-matrix adhesion and the fracture behavior of composites. Low levels of adhesion characterized by interfacial separation between fiber and matrix at low levels of stress produce composite fracture surfaces with copious amounts of interfacial separation for opening mode crack growth parallel to the fiber axis and with a large de-gree of fiber pullout for opening mode fracture perpendicular to the fiber axis. Surface treatments which increase the fiber-matrix adhesion to the point where partial separation and interfacial crack growth occur produce compos-ite fracture surfaces with more cohesive matrix failure in the opening mode parallel to the fiber axis. For opening mode fracture perpendicular to the fi-ber axis, fiber pullout occurs although the pullout lengths are relatively short. Composites made from surface treated and coated fibers have the highest de-gree of interfacial shear strength. The opening mode fracture surfaces of these composites show almost complete cohesive matrix failure between fibers when the fracture is parallel to the fiber axis. The opening mode fracture surface of these composites with the crack propagating perpendicular to the fiber surface has planar regions with almost smooth features indicating that fracture progresses across adjacent fibers. This coated and surface treated fiber caused matrix crack growth at the fiber break in the single fiber specimen.

Although quantitative fracture toughness measurements were not conducted on these specimens some qualitative trends appear to exist. For opening mode fracture parallel to the fiber axis, increases in adhesion coincide with increased fracture toughness. For opening mode fracture perpendicular to the fiber axis, there may be an optimum level of adhesion. High interfacial shear strength promotes planar fracture of the composite which in turn would be expected to have a lower fracture toughness.

References

[1] Drzal, L. T., *Carbon*, Vol. 15, 1977, p. 129.
[2] Drzal, L. T., Mescher, J. A., and Hall, D. L., *Carbon*, Vol. 17, 1979, p. 375.
[3] Gutfreund, K. and Kutscha, D., "Interfacial Investigations in Advanced Fiber Reinforced Plastics," AFML-TR-67-275, Air Force Materials Laboratory, Dayton, Ohio, 1967.
[4] Kelly, A. and Tyson, W. R., *Journal of the Mechanics and Physics of Solids*, Vol. 13, 1965, p. 329.
[5] Drzal, L. T., Rich, M. J., Camping, J. P., and Park, W. J., *Proceedings*, 1980 Conference of RP/Composites Institute, Paper 20-C, 1980.
[6] Drzal, L. T., Rich, M. J., and Lloyd, P. F., *Journal of Adhesion*, Vol. 16, 1983, p. 1.
[7] Whitney, J. M., unpublished results.
[8] Hammer, G. E. and Drzal, L. T., *Applications of Surface Science*, Vol. 4, 1980, p. 340.
[9] Drzal, L. T., Rich, M. J., Koenig, M. F. and Lloyd, P. F., *Journal of Adhesion*, Vol. 16, 1983, p. 133.
[10] Selby, K. and Miller, L. E., *Journal of Materials Science*, Vol. 10, 1975, p. 12.
[11] Kim, S. L., Skibo, M. D., Manson, J. A., Hertzberg, R. W., and Janiszewski, J., *Polymer Engineering and Sciences*, Vol. 18, 1978, p. 1093.
[12] Rosen, B. W., *Fibre Composite Materials*, American Society for Metals, 1965.

Yutaka Kagawa,[1] Eiichi Nakata,[2] and Susumu Yoshida[2]

Fracture Behavior of SiC Matrix Composites Reinforced with Helical Tantalum Fiber

REFERENCE: Kagawa, Y., Nakata, E., and Yoshida, S., **"Fracture Behavior of SiC Matrix Composites Reinforced with Helical Tantalum Fiber,"** *Recent Advances in Composites in the United States and Japan, ASTM STP 864*, J. R. Vinson and M. Taya, Eds., American Society for Testing and Materials, Philadelphia, 1985, pp. 27–43.

ABSTRACT: The present study was started to examine the effect of ductile metallic fiber on the toughness and fracture behavior of engineering ceramics. Helical fibers were used for the purpose of better ensuring stress transfer from the matrix to the fiber than straight fibers.

A length of tantalum fiber 0.4 mm in diameter was formed into two types of helix. Helical or straight fibers were incorporated in a silicon carbide (SiC) matrix. The fibers were unidirectionally aligned to the longest axis of a specimen with uniform spacings. Specimen dimensions were 10 by 10 by 55 mm. Specimens were prepared by means of vacuum sintering. The volume fraction of the fiber was varied from 0.01 to 0.04. Three-point bending tests were carried out at room temperature in air. Measurement of the acoustic emission signal was made to investigate the fracture process of the composites. Results were compared with three kinds of specimens: helical fiber reinforced composites, straight fiber reinforced composites, and SiC matrix alone.

It was found that the fracture of the composites reinforced with tantalum fiber, whether straight or helical, was not so catastrophic as to be shattered to pieces like that of the SiC matrix alone. This arises because in fiber reinforced composites, fibers bridge the crack surfaces which have occurred in the matrix and carry the load which the matrix supported before fracture. But the fracture behavior of both types of composites differs in detail depending on their fiber geometry. The maximum load supported by the specimen after the matrix fracture was larger in the helical fiber composites than in the straight fiber composites. This seems to be due to the fact that the stress-transfer effect between the matrix and fiber was better for the helical fiber composites than for the straight fiber composites. Also, multiple fracture of the SiC matrix was observed in the helical fiber composites. Moreover, the total work done up to the complete fracture of a specimen, which was estimated from the load-displacement relation, was different by the fiber geometry and this was due to the differences of the energy absorption mechanism of the composites.

[1]Assistant, The Castings Research Laboratory, Waseda University, 8-26 Nishiwaseda 2-chome, Shinjuku-ku, Tokyo 160, Japan.

[2]Professor and visiting professor, respectively, School of Science and Engineering, Waseda University, 4-1 Ohokubo 3-chome, Shinjuku-ku, Tokyo 160, Japan.

KEY WORDS: composite materials, helical fiber, ceramic matrix, silicon carbide, fracture behavior, stress transfer, work of fracture

Some engineering ceramics such as silicon carbide (SiC) and silicon nitride (Si_3N_4) that have superior mechanical properties at elevated temperatures are expected to become important industrial materials for high-temperature structural use in the near future. However, as they are ceramics by nature, their brittleness and lack of reliability as structural materials may become a source of troubles.

On the other hand, it has been known since old times that small additions of fiber can often improve the strength and toughness of weak and brittle ceramic materials.

As for modern engineering ceramics, Lindley and Godfrey [1] produced a Si_3N_4 matrix composite reinforced by SiC fiber with a tungsten core and observed improvements of the work of fracture as well as of the strength. Brennan [2] showed that a tantalum fiber reinforced Si_3N_4 composite had high impact strength. Recently Guo et al [3] reported that a carbon fiber reinforced Si_3N_4 matrix composite possessed much higher toughness and thermal shock resistance than those of unreinforced Si_3N_4.

However, there have been no researches reported up to the present which try to see the effect of the geometrical shape of the reinforcing fiber on the strength and toughness of a ceramic matrix composite.

The present experiment has been performed as a preliminary one to control the fracture behavior of fiber reinforced ceramic matrix composites by changing the interaction between fiber and matrix using a fiber with modified geometrical shape. Helical tantalum fiber was used for reinforcement in this experiment. This fiber is expected to enable good stress transfer, even if the fiber-matrix interface is very weak, by deformation of the geometrical shape of the helix [4–9].

Experimental Procedure

Helical Tantalum Fiber

Tantalum fiber of 0.4 mm diameter was obtained commercially. It was closely wound into a helix on a mandrel of 0.5 mm diameter. The helix with a straight tail was elongated at a constant speed of 0.1 mm/min to a desired length.

To specify the geometrical shape of the helix, several parameters, namely, the helical angle (θ), the radius of helix (r_h), and the helical pitch (P_h), are defined as shown in Fig. 1. These parameters were measured by taking scanning electron micrographs (SEM). The measured values of the parameters for helices used in this experiment are given in Table 1. The appearances of the helical fiber are shown in Fig. 2.

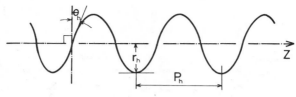

Z : HELICAL AXIS
θ_h : HELICAL ANGLE
r_h : HELICAL RADIUS
P_h : HELICAL PITCH

FIG. 1—*Definition of helical parameters.*

TABLE 1—*Geometrical constants in helical tantalum fiber used in the experiment (all tantalum fiber had a diameter of 0.4 mm).*

Fiber Type	Geometrical Constants of Helix		
	Angle, θ, deg	Radius, r_h, mm	Pitch, P_h, mm
Straight	...	...	...
Type I	74.1	0.155	3.42
Type II	48.9	0.345	2.49

Bare straight tantalum fiber and helical tantalum fibers were tension-tested in an Instron-type testing machine. All tests were performed with a crosshead movement of 1.0 mm/min with a specimen gage length of 50 mm. Change of the helical radius (r_h) during tension testing was also measured by an optical microscope.

Composite Preparation

Very pure α-SiC powder, prepared by Showa Denko K. K., was used as a starting material for the matrix. The average grain size of the powder was 0.88 μm. The α-SiC powder and polycarbosilane, whose average grain size was 30 μm, were mixed by milling and were moistened with a small amount of ethylalcohol. The amount of polycarbosilane in the mixed powder was always 13 weight%.

Tantalum fibers, both straight and helical, were aligned along the longest axis of a steel die with uniform spacings by the hand layup process. The arrangement of fibers in the composite is shown in Fig. 3. Then the powder in the die with embedded fibers was pressed for 60 s under a pressure of 89 MPa. The resulting green density of matrix was 70% of the theoretical value. The size of the green composite was 55 by 10 by 10 mm. The green composite was

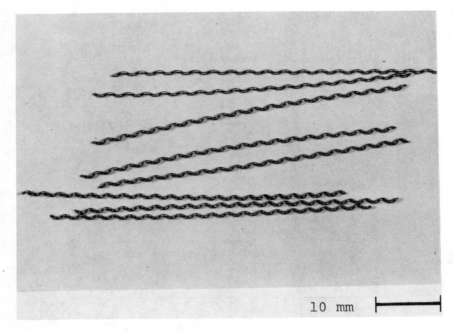

FIG. 2—*Appearance of Type II helical tantalum fibers.*

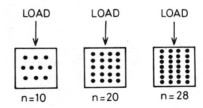

FIG. 3—*Arrangement of fibers in the composite.*

sintered in vacuum for 60 min at 1273 K. In this experiment, for convenience, vacuum sintering was used. The final average density of the matrix, which was taken over more than 30 pieces of specimen, was 60% of the theoretical one. The scatter of the final density was ranged from 58% to 63% of the theoretical one. The amount of shrinkage of the SiC matrix after sintering was less than 1%.

The volume fraction of the fiber in the composite was calculated by the following equation by measuring the width (w), and height (h) of the sintered specimen

$$V_f = n \pi r_f^2 / wh \sin \theta$$

where n was the number of fiber included in the composite. In this work the volume fraction of the fiber varied from 0.01 to 0.04, because it was known that small additions of ductile fiber could change the fracture behavior of the brittle materials.

Pullout Test

Pullout tests on a single tantalum fiber, both straight and helical, embedded in an SiC matrix were carried out.

The embedded length (l_e) of the fiber was held constant at 30 mm. As for the helical fiber, length l_e is defined as the axial length of the helix. The shape and dimension of the pullout test specimen is shown in Fig. 4a. The Instron-type testing machine was used under the same conditions as the tension testing for the bare fiber. To minimize influence of the chucks, the fiber was gripped as shown in Fig. 4b.

Three-Point Bend Test

A series of three-point bend tests was done on unnotched bars at room temperature (298 K) using an Instron-type testing machine. Specimen bars of 55-mm length, 10-mm height, and 10-mm width were used for three-point bend tests. The span length was 40 mm and the crosshead movement was held con-

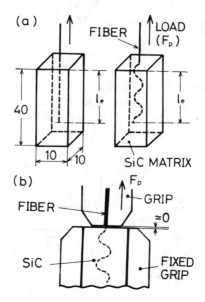

FIG. 4—*Shape and dimension of pullout test specimen (a), and schematic drawing of pullout test condition (b) (mm).*

stant at 1.0 mm/min throughout the experiments. Ten specimens were used for each condition. A schematic of the specimen geometry and the loading configuration is shown in Fig. 5.

Acoustic emission (AE) was monitored using an NAIS-type apparatus operating within a frequency range from 100 to 200 kHz, using a piezoelectric (PZT) transducer resonant at 140 kHz. The transducer was mounted at the upper part of the three-point bend test machine as shown in Fig. 5.

The fracture appearance of bend-tested specimens was also observed both macroscopically and microscopically.

Results and Discussion

Tensile Properties of Helical Tantalum Fiber

Figure 6 shows the load-elongation curves for bare straight fibers and helical ones. Change of the helical radius during tensile elongation is also shown in the same figure. The linear parts on the load-elongation curves of the helical fiber near the origin represent elastic or nearly elastic elongation of the helix. After the stage of elastic deformation, the fiber begins overall plastic deformation and the helix slackens, the helical angle increasing with decreasing helical radius. This stage is represented by the upward concave part of the load-elongation curve. The third stage corresponds to plastic deformation of the tantalum fiber itself and its fracture.

The ultimate tensile loads of these fibers are almost equal. But, naturally, elongations at ultimate tensile load are different for each fiber type.

Pullout Test

Figure 7 shows typical results on tensile load versus displacement relation in pullout tests. In the case of straight fiber, fiber pullout was observed. The pullout of the straight fiber shows that the embedded length of 30 mm is less

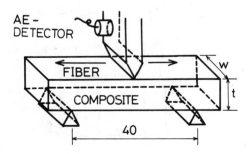

FIG. 5—*Schematic drawing of three-point bend test condition (mm).*

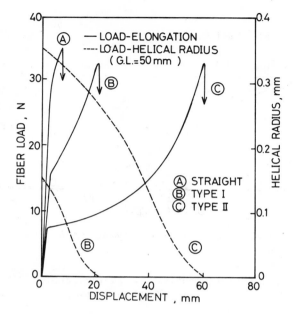

FIG. 6—*Typical load-elongation curves and load-helical radius curves for bare straight fiber and bare helical fiber.*

than half a critical length of fiber [*10*]. The shear stress τ_s which is required to debond the straight tantalum fiber from the matrix is calculated

$$\tau_s = F_p^{\max}/2\pi r_f l_e$$

where $F_p^{\max}$ is the critical load when pullout happens and l_e is the embedded length of the fiber. From this equation, the debonding stress, τ_s, is less than 7.7 MPa. This result is indicative of fairly weak bonding between tantalum fiber and SiC matrix.

When both types of helical tantalum fiber were tested instead of the straight one, fiber fracture occurred even when the embedded length was only of a pitch.

The interfacial bond strength must be controlled by the properties of tantalum fiber and the SiC matrix and is supposed to be the same both for straight and helical fibers. Therefore, the difference in the pullout property between them must be caused by the fiber geometry.

Load-Displacement Curves in Bend Tests

Figures 8*a*–8*d* show typical three-point bend test curves for SiC matrix alone (Fig. 8*a*), straight tantalum fiber reinforced composite (Fig. 8*b*), and

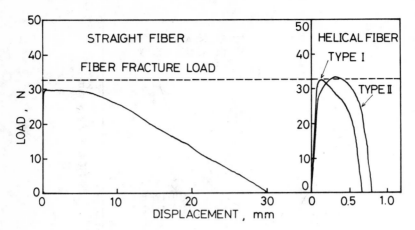

FIG. 7—*Load-displacement relations for pullout tests.*

helical tantalum fiber reinforced composites (Figs. 8c and 8d). The broken curve in Fig. 8 shows the acoustic emission ringdown count summation during the bend test.

For the composite reinforced with straight fibers, load increased with increasing displacement toward the maximum load. A sudden load drop in the load-displacement curve just after the maximum load corresponded to catastrophic fracture of the SiC matrix at the center of the specimen. When the matrix was fractured, the load that has been born by the matrix was immediately carried by the tantalum fiber and, therefore, the composite did not separate away as the unreinforced specimen. Then, fibers were pulled out from the matrix and bridged the fracture pieces of the SiC matrix by the frictional force between fiber and matrix. Figure 9 shows straight fibers pulled out from the fracture surface. The pullout of fibers was represented by the continuous and monotonous tail in the load-elongation curve. This phenomenon could be explained evidently from the pullout test result of straight fiber and matrix.

The first AE signal was observed just before fracture of the SiC matrix. After the matrix fracture, the AE signal increased proportionally to the fiber pullout from the fracture surfaces.

The bend load-displacement behavior of the SiC matrix reinforced with helical tantalum fiber was the same as that of the straight fiber reinforced one up to the maximum load. However, its behavior after the maximum load was completely different from the latter and, moreover, it varied depending upon the geometry of the helix.

As for the composite reinforced with Type I helical fiber, after the SiC matrix fractured at the maximum load, helical fibers that come out bridged the fractured surfaces similarly to the case of the straight fiber reinforced one. As shown clearly in the pullout tests, stress transfer from the matrix to Type I

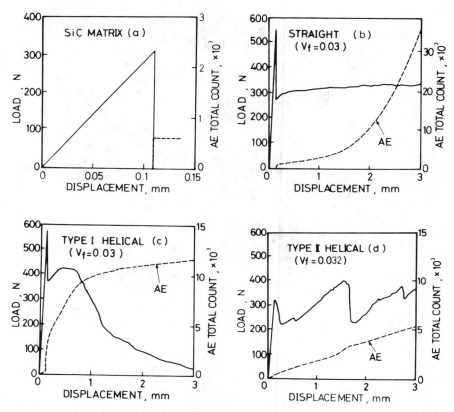

FIG. 8—*Typical three-point bend load-displacement curves. (a): SiC matrix, (b): straight fiber composite, (c): Type I fiber composite, (d): Type II fiber composite.*

fibers was made possible due to the geometry of the helix, even if the interface between the tantalum and SiC was very weak. Therefore, the helical tantalum fiber could carry the tensile load which had been borne by the matrix before fracture without further pulling out from the matrix. As a result, fibers were straightened and fractured one by one from the tension side. This was represented by the second peak on the load-displacement curve (Fig. 8c). The appearance of the fractured part is shown in Fig. 10, and an SEM of the fractured fibers is shown in Fig. 11. Almost all helical fibers (Type I) coming out from the fracture surface were straightened and fractured in ductile mode as shown in Fig. 11.

An AE signal was observed from the beginning of the three-point bend test. The origin of the AE signal may be the interface debonding caused by the change in curvature of the fiber due to the applied bending stress. These phenomena were unrelated to the fiber volume fraction.

5 mm

FIG. 9—*Fracture appearance of the straight fiber reinforced composite.*

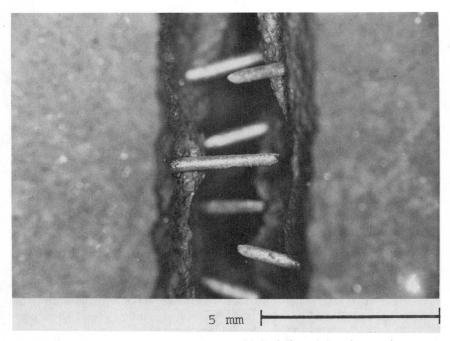

5 mm

FIG. 10—*Fracture appearance of the Type I helical fiber reinforced composite.*

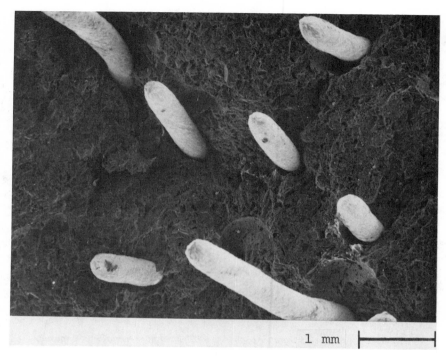

FIG. 11—*Fracture surface of Type I helical fiber reinforced composite.*

Two fracture modes were observed in Type II helical fiber reinforced composites. The first mode of fracture was observed for the composites with the fiber volume fraction of less than 0.02. This fracture mode is similar to that of Type I helical fiber reinforced composites (Fig. 8c). In the second type of fracture mode, after the initial matrix fracture, the helical fibers were pulled out from both sides of the fracture surface. The helical fibers pulled out from the matrix bridged the fracture surface and extended while the embedding matrix was broken to small pieces. This type of fracture appearance is shown in Fig. 12.

The extension of the helical fiber is shown in Fig. 8d as a saw-toothed wave on the load-displacement curve. This phenomenon was due to the interaction between the helical fiber and SiC matrix which occurred when the helical radius was decreased by tensile deformation. In this case, the AE signal increased linearly with the displacement.

Maximum Load and Volume Fraction

Figure 13 shows the plots of the maximum load in bend tests versus volume fraction of the fiber. As shown in this figure, a linear relation between the

FIG. 12—*Fracture appearance of the Type II helical fiber reinforced composite.*

maximum load and the volume fraction is obtained for the straight fiber reinforced composites.

In the case of helical fiber reinforced composites, the relation was different
according to the type of reinforcing fiber and fracture mode of the composites. As for the case of Type I helical fiber, the relation between the maximum
load and the volume fraction is linear, similar to the case of the straight fiber
reinforcing.

If the data of these two cases are compared at the same volume fraction, the
maximum loads of Type I helical fiber reinforced composite are slightly
greater than those of the straight fiber one. Type I helical fiber reinforced
composites showed the greatest maximum loads among the tested composites.

When the volume fraction of Type II helical fiber was less than 0.02, the
maximum load was nearly equal to that of the matrix alone. However, the
maximum load decreased with increasing volume fraction of fiber.

These results seem to be related closely to the tensile load carried by the
reinforcing fibers and the stress transfer ability between fiber and matrix.

Work of Fracture

Load-displacement curves in the bend tests of the composites are shown
schematically in Fig. 14. Generally, these curves can be divided into two re-

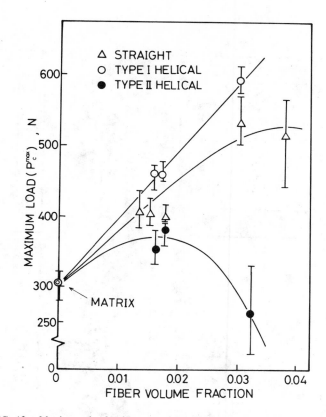

FIG. 13—*Maximum load-volume fraction plots for three types of composites.*

gions. Region I ranges from the beginning up to the maximum load, that is, the initial cracking of the SiC matrix. Region II is from just after the initial matrix cracking to the end of the bend test. The load-displacement behavior of composites during Region II varies depending on the types of the reinforcing fiber.

Figure 15 shows plots of work done versus volume fraction of the composites in Region I. The work is determined as the area between the load-displacement curve and the abscissa. Although no difference is observed the shape of load-displacement curves in Region I, there are seen small differences between fiber types in the work done for given volume fraction.

Work done at various arbitrary values of displacement for the straight and Type I helical fiber reinforced composites in Region II are plotted against the fiber volume fraction. The results are shown in Fig. 16. The work was determined by the same method as in Region I. The work increased with increasing displacement. The relation between the work done at the fiber volume fraction for straight and Type I reinforcement seems almost equal.

In the straight fiber reinforced composite, in Region II, fibers were pulled

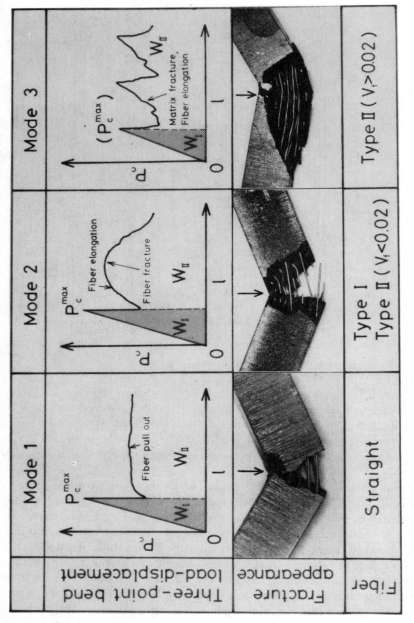

FIG. 14—Schematic representation of load-elongation curves and fracture appearance for three types of composites.

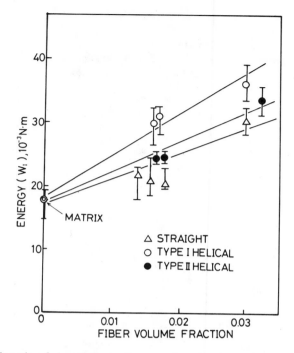

FIG. 15—*Plots of work done in Region I versus volume fraction of three types of composites.*

out from the both sides of the fracture surface as shown in Fig. 9. This process provides considerable amounts of absorbed energy [*11*]. However, in Type I helical fiber reinforced composites, ductile fracture of tantalum fiber occurred at the fracture surface of the matrix as shown in Fig. 11. The work was done when fibers were necking down [*12*] and this work contributed to the total absorbed energy in Region II. The differences of energy absorption mechanism in Region II basically depend on differences in interfacial shear strength.

Conclusions

From the results of this preliminary investigation of the SiC matrix composites containing tantalum fiber, it has been considered that it is possible to control the fracture process of ceramics by a small addition of helical tantalum fiber even when the interface is very weak and stress transfer is impossible in the straight fiber reinforced composite. The results of the three-point bend tests showed that deformation of the tested composites was independent of the type of the reinforcing fibers up to the maximum load. After the initial matrix fracture, the fracture process varied depending on the reinforcing fi-

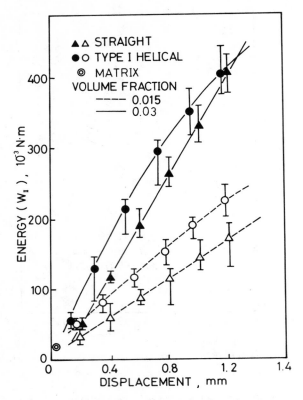

FIG. 16—*Plots of work done in Region II versus volume fraction for straight and helical fiber reinforced composites.*

ber type and its volume fraction. This is probably caused by the difference in the stress transfer mechanism and the magnitude of interaction between fiber and matrix. The work done in the composites after the initial matrix fracture was also influenced by the fracture process of the composites.

References

[1] Lindley, M. W. and Godfrey, D. J., *Nature*, Vol. 229, No. 5281, 1971, pp. 192–198.
[2] Brennan, J. J. in *Special Ceramics 6*, P. Popper, Ed., British Ceramic Research Association, 1975, pp. 123–134.
[3] Guo, J. K., Mao, Z. Q., Dibao, C., Wang, R. H., and Yan, D. S., *Journal of Material Science*, Vol. 17, 1982, pp. 3611–3616.
[4] Kagawa, Y., Okuhara, H., Watanabe, Y., Nakata, E., and Yoshida, S. in *Composite Materials: Mechanics, Mechanical Properties and Fabrication*, K. Kawata and T. Akasaka, Eds., Japan Society for Composite Materials, 1981, pp. 213–222.
[5] Kagawa, Y., Nakata, E., and Yoshida, S., *Journal of the Japan Institute of Metals*, Vol. 46, No. 6, 1982, pp. 632–639.

[6] Kagawa, Y., Watanabe, Y., Nakata, E. and Yoshida, S. in *Proceedings*, 1982 Joint Conference on Experimental Mechanics, Society for Experimental Stress Analysis, Part 1, 1982, pp. 230–235.

[7] Kagawa, Y., Watanabe, Y., Nakata, E., and Yoshida, S., *Journal of the Japan Institute of Metals*, Vol. 46, No. 9, 1982, pp. 900–907.

[8] Kagawa, Y., Nakata, E., and Yoshida, S., *Progress in Science and Engineering of Composites*, T. Hayashi, K. Kawata, and S. Umekawa, Eds., Japan Society for Composite Materials, 1982, Vol. 1, pp. 1457–1464.

[9] Kagawa, Y., Nakata, E., and Yoshida, S., *Journal of the Japan Institute of Metals*, Vol. 47, No. 9, 1983, pp. 760–767.

[10] Kelly, A. and Tyson, W. R. in *High Strength Materials*, V. Zackay, Ed., Wiley, New York, 1965, pp. 578.

[11] Cooper, G. A., *Journal of Material Science*, Vol. 5, 1970, pp. 645–654.

[12] Tetelman, A. S. in *Composite Materials: Testing and Design. ASTM STP 460*, American Society for Testing and Materials, Philadelphia, 1969, pp. 473–502.

Mark D. Kistner,[1] *James M. Whitney,*[1] *and*
Charles E. Browning[1]

First-Ply Failure of Graphite/Epoxy Laminates

REFERENCE: Kistner, M. D., Whitney, J. M., and Browning, C. E., **"First-Ply Failure of Graphite/Epoxy Laminates,"** *Recent Advances in Composites in the United States and Japan, ASTM STP 864*, J. R. Vinson and M. Taya, Eds., American Society for Testing and Materials, Philadelphia, 1985, pp. 44-61.

ABSTRACT: *In situ* transverse-ply failure in AS-4/3502 graphite/epoxy laminates is investigated by determining crack density as a function of laminate axial strain. These results are then related to laminate stress-strain response. Results show that plateaus in either the longitudinal or transverse stress-strain curve are associated with a significant increase in transverse-ply crack density. Stress analysis reveals a departure of inplane stresses from classical lamination theory near free edges. These stresses are found to be a function of laminate-ply orientations and stacking sequence. Thus, the onset of transverse-ply cracking may be influenced by the free edge in a standard tensile coupon.

KEY WORDS: composites, laminates, first-ply failure, transverse cracking, free-edge effects

The dominant method for determining transverse-ply failure is the standard 90° tension test. This test gives some useful information concerning the transverse strain capability of the material under consideration. However, a multi-directional laminate containing 90°-plies may provide a more realistic approach to both measuring and defining *in situ* transverse strength. In particular, a 90° tension test provides initial transverse ply failure data only, while a multi-directional laminate allows one to observe multiple cracks in transverse plies and assess their effect on laminate mechanical behavior. The plies adjacent to the 90°-plies provide constraint to the accumulation of damage in the 90°-plies. Such constraint plies have several important effects. In addition to offering resistance to crack growth in the transverse plies perpen-

[1] Materials research engineers, Materials Laboratory, Air Force Wright Aeronautical Laboratories, Wright-Patterson Air Force Base, OH 45433.

dicular to the load direction, they also reduce the effect of surface flaws on first-ply failure. The mismatch between ply thermal expansion coefficients induces residual stresses due to cool down from the cure temperature. As a result, the constraint-plies will affect the strain level at which transverse-ply failure occurs (that is, laminate strain is not equal to absolute ply strain).

Recently, a number of anomalies associated with *in situ* transverse-ply failure have been reported by Flaggs and Kural [1]. The *in situ* strength was found to depend strongly on the 90°-ply thickness and orientation of adjacent plies. No correlation was found between transverse unidirectional strength and *in situ* transverse-ply strength as predicted by lamination theory, which included the effects of residual thermal stresses. The authors concluded that these results suggest transverse strength should not be considered an intrinsic lamina property. Motivated by the works of Parvizi, Garrett, and Bailey [2] and Wang and Crossman [3,4], Flaggs and Kural pursued a fracture mechanics approach for predicting the *in situ* transverse-ply strength. Results were not, however, very promising.

In the present paper, *in situ* transverse-ply strength is examined with two objectives in mind. The first objective is to identify laminate parameters which influence *in situ* transverse-ply strength. The second objective involves assessing the potential of a laminate uniaxial tensile test for determining *in situ* transverse-ply strength.

Initial studies reported here are limited to graphite/epoxy material. Transverse-ply failure is investigated by examining the surface of a highly polished multi-directional tension specimen under magnification. The transverse-plies are examined as a function of laminate strain to determine crack density. These results are then related to the axial stress-strain curve. This will allow one to relate the knee in the curve to crack density for the laminate under consideration.

For certain filament dominated laminates, the stress-strain curve is essentially linear to failure. In such cases, transverse crack density will be correlated with the laminate transverse stress-strain curve (response perpendicular to the load), as a knee will usually appear in this curve indicating a stiffness change due to transverse-ply damage. Transverse response, as observed experimentally, will be also compared to analytical results obtained from lamination theory, including free-edge effects. It has been noted by Kriz and Stinchcomb [5] that the state of stress along the edge of a graphite/epoxy quasi-isotropic laminate during static tension and tension-tension cyclic loading can be changed significantly by the presence of moisture. Thus, it is appropriate to consider free-edge effects in the stress analysis performed in the present paper.

In addition to assessing the potential of a laminate test for determining *in situ* transverse strain capability of a particular composite material, this study should be helpful in determining strain levels at which significant damage occurs in composite laminates.

Experimental Procedure

The material used for this project is Hercules AS-4/3502 system consisting of continuous graphite fiber in an epoxy resin matrix. The following orientations were used:

$[90]_4$, $[90]_8$, $[90]_{12}$, $[90]_{16}$
$[0_4/90_4]_s$
$[0/90]_{4s}$
$[+\Theta/-\Theta/+\Theta/-\Theta/90_3/\overline{90}]_s$, $\Theta = 45°$, $60°$
$[+\Theta/90/-\Theta/90/+\Theta/90/-\Theta/\overline{90}]_s$, $\Theta = 45°$, $60°$

These orientations were chosen to show the effect of constraint-ply orientation and transverse-ply thickness upon the level at which transverse-ply cracking occurs. The transverse-ply cracking is observed by examination of a highly polished tension specimen under magnification as a function of laminate strain.

The panels, after fabrication, were examined by an ultrasonic C-scan and a photomicrograph. Also, a density and per-ply thickness was done to calculate a fiber volume on the void free panels. All of the laminates used appeared to be void free by examination of a polished cross section of the material from the center of each panel.

The tension specimens were prepared by end tabbing the panels then cutting 25.4 by 228.6 mm (1 by 9 in.) specimens out of the panels. A high speed diamond saw was used for the machining. The edges of the tension specimens were then sanded and polished using the following grits: (1) 60 grit silicon carbide paper; (2) 240 grit silicon carbide paper; (3) 400 grit silicon carbide paper; (4) 600 grit silicon carbide paper; and (5) 0.3 μm alumina on a polishing wheel. After polishing with the 0.3 μm alumina, the edge of the specimen should appear glossy and be free from large scratches.

Strain gages were mounted on the specimens. For the initial stress-strain curves, a large area of the specimen needed to be monitored. For these specimens, 25.4 mm (1 in.) strain gages were used. For tension specimens, which were used to measure the crack density and hysteresis effect as a function of elongation, 8.3 mm (0.25 in.) strain gages were used.

The tension testing was performed using a servohydraulic testing machine. Data were obtained by applying a constant loading rate rather than a constant strain rate. The loading rate was chosen by knowing the modulus, gage length, and dimensions of the specimens, and setting the rate of elongation of 0.25 mm (0.01 in.)/min across the specimen in the linear portion of the stress-strain curve. The rate of 0.25 mm (0.01 in.)/min across the length of the specimen was chosen so the knee in the stress-strain curve may be seen more easily than at faster rates.

For all of the panels, the acoustic emission from the specimen was monitored as the tensile load was applied. The acoustic emission sensor was attached to the specimen by applying a thin layer of high vacuum silicone grease to the surface of the sensor then using masking tape to hold it to the specimen's surface. This assured good contact between the sensor and the specimen. The sensitivity for the acoustic emission measurement for all of the tension tests is 1000 counts per 25.4 mm (1 in.), 40 decibels, and a 2 s time constant. With this sensitivity it was found that the first transverse crack appears as a jump of about 25.4 mm (1 in.) or more on the chart. Noise appears as jumps usually less than 12.7 mm ($\frac{1}{2}$ in.) in height.

For the crack density measurements, the acoustic emission is monitored until the first significant acoustic emission event occurs. The specimen is then taken out of the test machine and examined under magnification to look for transverse cracking. This process is repeated until the first transverse crack is detected. The specimen is then taken to the next strain level and examined. This process is repeated at approximately 0.1% strain intervals until ultimate failure of the specimen occurs. For each of the reloading cycles, the strain gage is zeroed. This leads to a lowering of the values of strain recorded. To correct for this, the stress is calculated for each strain level and the corresponding strain is taken from the stress-strain curve of the laminate.

A hysteresis study is also performed for the purpose of ascertaining the effect of load cycling on 90°-ply crack density. For this hysteresis study, the laminate is initially loaded to a strain level equal to approximately 75% of the maximum strain in the linear portion of the stress-strain curve. After each loading, the number of transverse cracks are counted. After the fifth loading, the strain level is raised to 125% of the maximum strain in the linear portion of the stress-strain curve. The specimen is loaded five times at this strain level, then taken to ultimate failure. For each strain level, the specimens are taken up to a constant value of recorded strain.

Laminate Stress Analysis

Laminate strains at which transverse-ply failure occurs can be experimentally determined. However, these data only represent mechanical strains and not the total strain. In particular, thermal residual strains due to laminate cure must be also considered when evaluating *in situ* transverse-ply failure. Classical lamination theory can be utilized for determining residual strains [6].

However, lamination theory neglects free-edge effects. Since the state of stress in a zone near the free edge of a laminate coupon specimen is three-dimensional in nature [7,8], the inplane stresses, as well as the interlaminar stresses, will be different then determined from classical lamination theory. Thus, before any conclusions are made concerning the effect of laminate stacking geometry on transverse-ply failure, one must consider a more complete stress analysis than provided by classical lamination theory.

Results are obtained from the global-local laminate model developed by Pagano and Soni [9]. Previous solutions generated in conjunction with free-edge models have concentrated on the interlaminar stresses due to the potential for delamination. For present purposes, however, attention is given to the inplane free-edge stresses. In particular, the normal stress parallel to the load direction in the 90°-plies are determined. These stresses are found to be a function of stacking sequence and, thus, differ from those determined from lamination theory where they are independent of stacking sequence for a particular set of ply orientations.

The global-local model is a self-consistent model which can define detailed response functions in a region of interest (local), while representing the remaining domain by effective properties (global). The local model represents each layer as a homogeneous anisotropic continuum. In the local region each layer or sublayer is represented by a higher order plate theory which includes transverse shear deformation and a linear strain through the thickness. The global region is based on a laminated plate theory which includes the same kinematic assumptions as in the local model. Continuity conditions across interface boundaries are enforced. The governing equations are based on principles of complimentary energy as illustrated in Fig. 1.

For the analytical results presented in this paper, the following unidirectional ply properties are utilized:

$$E_1 = 140 \text{ GPa (20 Msi)}, E_2 = 11 \text{ GPa (1.6 Msi)},$$
$$E_3 = E_2, G_{12} = 6 \text{ GPa (0.88 Msi)}, G_{13} = G_{12},$$
$$G_{23} = 3.8 \text{ GPa (0.55 Msi)}, \nu_{12} = \nu_{13} = 0.31,$$
$$\nu_{23} = 0.54, \alpha_1 = -0.9/°C \, (-0.5/°F) \times 10^{-6}$$
$$\alpha_2 = 23/°C \, (13/°F) \times 10^{-6}$$

Discussion

Results from the 90° tension data are shown in Table 1. The specimen used is 90_n where $n = 4, 8, 12, 16$. The specimens' edges are polished. A total of ten specimens from each group with four specimens strain gaged are tested to show the effect of thickness on 90° tensile strength. The results in Table 1 show no conclusive evidence of a variation in strain to failure with thickness.

For the multidirectional laminates, the crack density in the center 25.4 mm (1 in.) of each specimen is counted for each 90°-ply. Figure 2 shows the variation in crack density relative to stress-strain response for the $[0/90]_{4s}$ laminate. This laminate showed the largest variation in crack density across the thickness of the laminate. As shown in Fig. 2, the crack density is seen to increase in the outer 90° ply faster than the center 90°-ply. Also, it may be noted that the onset of cracking appears at about 0.49% elongation. The knee, as seen by the transverse gage, occurs at a low value of transverse crack density. Above the knee, the crack density is seen to increase rapidly to failure.

The $[0_4/90_4]_s$ laminate in Fig. 3 is seen to have transverse cracks initially due to fabrication. The crack density did not increase until about 0.5% elongation. At the knee in the transverse stress-strain curve, the crack density is

NEW APPROACH

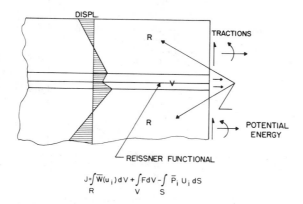

FIG. 1—*Global local model for laminate analyses.*

TABLE 1—*Thickness effect on transverse unidirectional tension tests.*

Thickness Effect on 90° Tension Tests			
Number of 90°-Plies	σ_0, MPa (ksi)	E_{11}, GPa (Msi)	E_f, %
4	47.8 (6.93)[a]	10.3 (1.49)	0.48
	48.4 (7.02)[b]		0.49[c]
8	61.2 (8.87)[a]	10.2 (1.48)	0.63
	60.6 (8.79)[b]		0.63[c]
12	53.2 (7.72)[a]	10.4 (1.51)	0.54
	56.6 (8.21)[b]		0.57[c]
16	60.7 (8.80)[a]	10.7 (1.55)	0.61
	57.3 (8.31)[b]		0.57[c]

[a] Four tests.
[b] Ten tests.
[c] Calculated.

seen to have a relatively low value. Above the knee, the crack density increases rapidly and appears to reach a saturation value at about 0.8% elongation.

The $[45/90/-45/90/45/90/-45/\overline{90}]_s$ laminate in Fig. 4 is shown to have the onset of transverse cracking at 0.49% elongation. At the stress-strain curve knee (about 0.66% elongation), the crack density is seen to be relatively low and rapidly increasing. Above the plateau in the stress-strain curves, the crack density is seen to increase slowly to failure.

The hysteresis effect of loading a specimen below and above the knee in the stress-strain curve is shown in Fig. 5. The upper curve in Fig. 5 is a stress-strain curve produced by loading the tensile coupon directly to failure. The lower curves are produced by loading another tensile coupon below and above

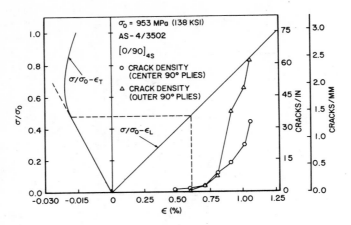

FIG. 2—*Variation in crack density relative to stress-strain response of [0/90]₄ₛ graphite/epoxy.*

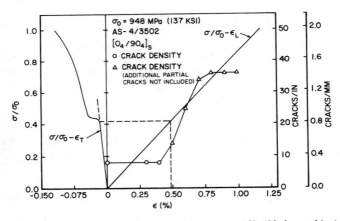

FIG. 3—*Crack density relative to stress-strain response of [0₄/90₄]ₛ graphite/epoxy.*

the knee in the stress-strain curve. This process is described in more detail in the experimental procedure section. Loading above the knee in the stress-strain curve is seen to have a significant effect upon modulus as previously shown by Whitney, Browning, and Grimes [10]. The increase in crack density upon reloading to the same strain level is shown in Table 2. The hysteresis curves for the first and second loading above the knee in the stress strain curve are shown in Fig. 6. The hysteresis effect is seen to dissipate upon reloading to the same strain level.

The $[+60/90/-60/90/+60/90/-60/\overline{90}]_s$ laminate in Fig. 7 is seen to

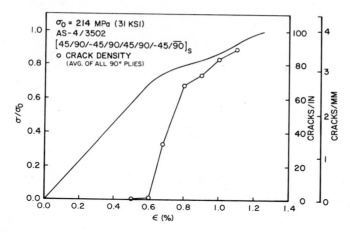

FIG. 4—*Average crack density relative to stress-strain response of [45/90/−45/90/45/90/−45/90]ₛ graphite/epoxy.*

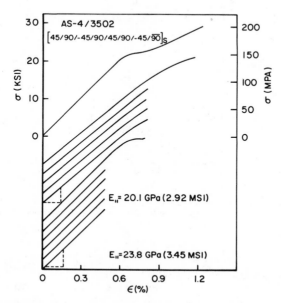

FIG. 5—*Hysteresis study on [45/90/−45/90/45/90/−45/90]ₛ graphite/epoxy.*

TABLE 2—*Hysteresis study along with crack density measurements AS-4/3502.*
[45/90/−45/90/45/90/−45/90]$_s$.

Strain Level, %	Loading Number	Avg Crack Density	
		Number/mm	(Number/in.)
0	0	0	(0)
0.48	1	0	(0)
	2	0.0008	(0.02)
	3	0.0008	(0.02)
	4	0.0008	(0.02)
	5	0.0008	(0.02)
0.80	6	2.677	(68.00)
	7	2.885	(73.29)
	8	3.071	(78.00)
	9	3.363	(85.43)
	10	3.493	(88.71)
Ultimate failure (1.25)	11	3.617	(91.86)

[a] 1 in. = 25.4 mm.

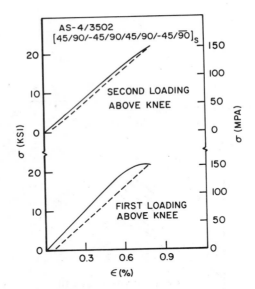

FIG. 6—*First and second loading above the knee in the stress-strain curve for [45/90/−45/ 90/45/90/−45/90]$_s$ graphite/epoxy.*

have the onset of transverse cracking at approximately 0.55% elongation. The crack density is seen to be a low value at the stress-strain curve knee. Above the plateau in the stress-strain curve, the crack density increases rapidly. This laminate showed a unique behavior compared to the other laminates examined. The first crack formed through all ±60 and 90° layers on

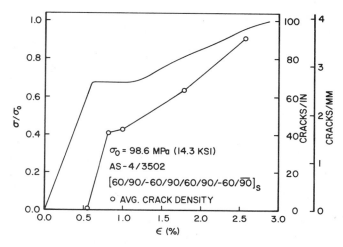

FIG. 7—*Average crack density relative to stress-strain response of [60/90/−60/90/60/90/−60/90]ₛ graphite/epoxy.*

one edge of the tension specimen then appeared to travel across the entire width. For the $[+60/-60/+60/-60/90_3/\overline{90}]_s$ laminate, ultimate failure occurs rapidly upon formation of the first crack. Similar ultimate strength behavior was observed for the $[+45/-45/+45/-45/90_3/\overline{90}]_s$ laminate, that is, failure occurs as the crack density begins to buildup. The stress-strain curve is linear to failure for both the $[+45/-45/+45/-45/90_3/\overline{90}]_s$ and $[+60/-60/+60/-60/90_3/\overline{90}]_s$ laminates.

A typical 90° crack is illustrated in Fig. 8. The partial or secondary cracks cited in Fig. 3 are illustrated in Fig. 9 where a photomicrograph of a $[0_4/90_4]_s$ laminate is shown. For each of the laminates investigated, no free-edge delamination is observed. Only some local delamination extending out of transverse cracks at the higher strain levels is observed.

In each of the laminates studied, a significant increase in the crack density is accompanied by a significant change in stress-strain behavior as observed from either a longitudinal or transverse strain gage. A summary of laminate results is given in Table 3. In assessing the value of each of the laminates under consideration for determining transverse ply strength, we must give some consideration to the laminate stress analysis. Results from the global-local model are illustrated in Figs. 10 to 13. In each of these figures σ_{XL} denotes values of σ_X as determined from classical laminated plate theory. Although these results are for residual stresses only, other solutions [11] indicate that mechanical loads induce the same trends.

For the $[0/90]_s$ class of laminates, a significant inplane stress concentration at the free-edge is noted. In the case of the $[0_4/90_4]_s$ laminate, this stress concentration is significantly higher than in the case of the $[0/90]_{4s}$ laminate.

FIG. 8—*A typical 90° crack in graphite/epoxy (×50 magnification).*

In addition, the stress concentration dissipates more rapidly away from the free edge for the laminate in which the 90°-plies are split. The large stress concentration factor for the $[0_4/90_4]_s$ laminate is likely to be the cause of the initial fabrication cracks observed. It is also interesting to note the stress gradient through the 90°-plies. The laminate with the 90°-plies split show the highest inplane stresses in the outer layers. This may be the reason that the crack density builds up more rapidly in the outer-ply of the $[0/90]_{4s}$ laminate

FIG. 9—*Partial or secondary cracks in $[0_4/90_4]_s$ of graphite/epoxy.*

as illustrated in Fig. 2. It is anticipated that these stress concentrations would induce crack initiation at the edge of the specimen rather than in the interior.

In the case of the angle-ply laminates, the inplane stresses at the edge actually drop off in the 90° plies. The $[+45/-45/+45/-45/90_3/\overline{90}]_s$ laminate displays the least change in inplane stress at the free edge. This behavior makes the $[\pm45/90]_s$ class of laminates an interesting candidate for determining *in situ* transverse-ply strength.

TABLE 3—*Summary of laminate results.*

Orientation	First Transverse Crack σ, MPa (ksi)	ε, %	Knee σ, MPa (ksi)	ε, %	Modulus E_{11}, GPa (Mis)	Ultimate σ, MPa ksi	ε, %
$[45/90/-45/90/45/90/-45/\overline{90}]_s$	125 (18.1)	0.49	156 (22.6)	0.66	23.4 (3.40)	202 (29.3)	1.24
$[45/-45/45/-45/90_3/\overline{90}]_s$	75.8 (11.0)	0.33	. . .	. . .	23.0 (3.34)	119 (17.2)	0.53
$[60/90/-60/90/60/90/-60/\overline{90}]_s$	61.7 (8.95)	0.55	66.5 (9.65)	0.60	11.2 (1.63)	99.3 (14.4)	2.90
$[60/-60/60/-60/90_3/\overline{90}]_s$	68.9 (10.0)	0.66	. . .	. . .	10.9 (1.58)	68.9 (10.0)	0.66
$[0/90]_{4s}$	379 (54.2)	0.49	419 (60.7)	0.62	73.8 (10.7)	978 (141.9)	1.23
$[0_4/90_4]_s$	. . .	. . .	363 (52.7)	0.48	73.8 (10.7)	948 (137.5)	1.16

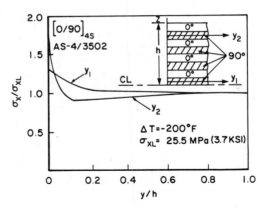

FIG. 10—*Inplane free-edge residual stress parallel to load direction using the global-local model.*

Summary and Conclusions

In situ transverse-ply strength has been examined by determining crack density as a function of laminate stress-strain behavior. A laminate stress analysis, which includes free-edge effects, has been performed to assess the effect of laminate geometry and stacking sequence on inplane edge stresses. The effect of thickness on 90° tensile strength has been also examined. Results obtained to date lead to the following conclusions:

1. Thickness appears to have little effect on unidirectional 90° tensile strength.

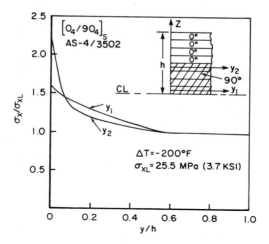

FIG. 11—*Inplane free-edge residual stress parallel to load direction using the global-local model.*

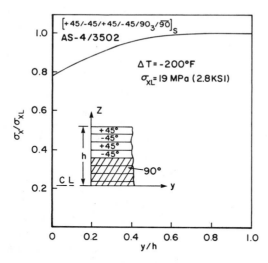

FIG. 12—*Inplane free-edge residual stress parallel to load direction using the global-local model.*

2. Laminate geometry does have an effect on *in situ* transverse-ply strength.

3. Stress analysis indicates that the inplane normal stress in the 90°-plies parallel to the load are a function of ply orientations and stacking sequence.

4. Data in conjunction with the stress analysis suggests that free edge stresses have an influence on *in situ* transverse-ply failure.

5. A uniaxial tension test in conjunction with a laminate of the $[\pm 45/90]_s$

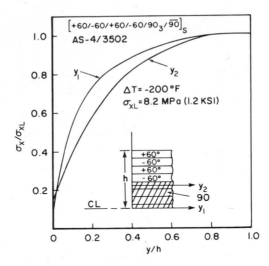

FIG. 13—*Inplane free-edge residual stress parallel to load direction using the global-local model.*

class appears to have merit in relating stress-strain behavior to the onset of significant transverse-ply damage.

Future Work

Future work should concentrate on two areas. The first involves performing a more complete stress analysis using the global-local model. In particular, the stress analysis should be performed at strain levels equivalent to the observed buildup strain levels of transverse-ply cracking. Residual stresses should be determined at more realistic values of ΔT [125 to 137°C (250 to 280°F)]. The edge stresses through the laminate thickness should be also determined.

In the second area, work should be concentrated on developing more generic conclusions concerning *in situ* transverse-ply failure. This requires an evaluation of composite laminates containing more ductile resins than the current state-of-the-art epoxy systems. Ductile resin systems are likely to introduce more complex behavior as illustrated in Figs. 14 and 15. These photomicrographs are taken from a $[+45/90/-45/90/+45/90/-45/\overline{90}]_s$ laminate fabricated from Imperial Chemical Industries' (ICI's) prepreg containing XA-S fibers in a polyetheretherketone (PEEK) thermoplastic matrix. In Fig. 14, only one transverse crack is observed at a laminate strain level of 2.25%. However, yielding between plies can be seen. This phenomenon is more clearly viewed in Fig. 15 where out-of-plane deformation (flow) between plies is observed. Thus, the concept of transverse-ply failure must be reconsidered for such materials.

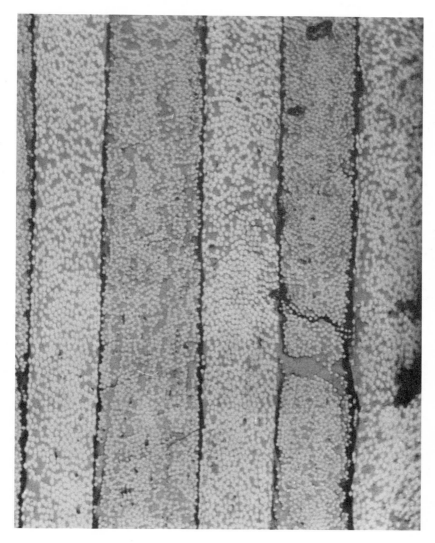

FIG. 14—*XA-S/PEEK graphite fiber-thermoplastic matrix taken to 2.25% elongation (×100 magnification).*

Acknowledgments

The authors gratefully acknowledge the support of the Materials Laboratory, Air Force Wright Aeronautical Laboratories, Wright-Patterson Air Force Base. The authors also acknowledge J. Camping, B. Ragland, and R. Cornwell for advice and technical help regarding test methods and test specimen preparation. A special acknowledgment is given to Dr. N. J.

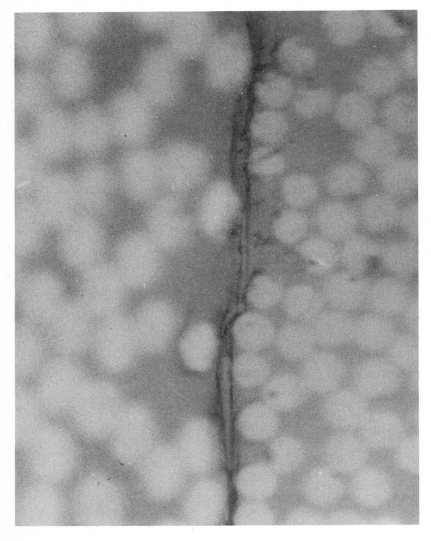

FIG. 15—*XA-S/PEEK graphite/thermoplastic taken to 2.25% elongation (×1000 magnification).*

Pagano and Dr. J. R. Soni for laminate stress analysis which enriched this paper. Also a special thanks goes out to Mrs. C. Waters and Miss C. Runyon for typing this manuscript.

References

[1] Flaggs, D. L. and Kural, M. H., *Journal of Composite Materials*, Vol. 16, March 1982, pp. 103–116.

[2] Parvizi, A., Garrett, K. W., Bailey, J. E., *Journal of Materials Science*, Vol. 13, 1978, pp. 195–201.

[3] Wang, A. S. D. and Crossman, F. W., *Journal of Composite Materials Supplement*, Vol. 14, 1980, pp. 71–87.

[4] Crossman, F. W. and Wang, A. S. D. in *Damage in Composite Materials, ASTM STP 775*, American Society for Testing and Materials, Philadelphia, 1982, pp. 118–139.

[5] Kritz, R. D. and Stinchcomb, W. W. in *Damage in Composite Materials, ASTM STP 775*, American Society for Testing and Materials, Philadelphia, 1982, pp. 63–80.

[6] Hahn, H. T. and Pagano, N. J., *Journal of Composite Materials*, Vol. 9, 1975, pp. 91–106.

[7] Pagano, N. J. and Pipes, R. B., *Journal of Composite Materials*, Vol. 5, 1971, pp. 50–57.

[8] Pagano, N. J. and Pipes, R. B., *International Journal of Mechanical Sciences*, Vol. 15, 1973, pp. 679–688.

[9] Pagano, N. J. and Soni, S. R., *International Journal of Solids and Structures*, Vol. 19, 1983, pp. 207–228.

[10] Whitney, J. M., Browning, C. E., and Grimes, S. C., *Composite Materials in Engineering Design*, American Society for Metals, 1973, pp. 441–447.

[11] Pagano, N. J. and Soni, S. R., "Free Edge Cracking of Composite Laminates Due to Curing Stresses," presented at the 7th ASTM Conference on Composite Materials: Testing and Design, American Society for Testing and Materials, Philadelphia, 2–5 April, 1984.

E. C. Klang[1] *and M. W. Hyer*[2]

Damage Initiation at Curved Free Edges: Application to Uniaxially Loaded Plates Containing Holes and Notches

REFERENCE: Klang, E. C. and Hyer, M. W., **"Damage Initiation at Curved Free Edges: Application to Uniaxially Loaded Plates Containing Holes and Notches,"** *Recent Advances in Composites in the United States and Japan, ASTM STP 864*, J. R. Vinson and M. Taya, Eds., American Society for Testing and Materials, Philadelphia, 1985, pp. 62–90.

ABSTRACT: This paper extends two ideas that have been used in the study of straight free edges to the case of curved free edges. The first idea is based on the notion that it is the differences in material properties of adjacent laminae that are responsible for initial free-edge interlaminar damage. To extend this to curved free edges the idea must be interpreted in terms of differences in shear and elongation strains of adjacent laminae. This interpretation involves both the material property differences and the spatial distribution of laminate strain at the hole edge. It is shown that a large strain level and a small difference in material properties can be as damaging as a small strain level and a large difference. From this it is evident interlaminar damage will probably begin at some circumferential location other than the net section. To explore these findings, the second extension was necessary. Specifically, the edge replication technique, using acetate film, was extended to record damage at a curved free edge. For a curved free edge the standard polishing technique was modified, as were the steps necessary to actually make a replica. These points are discussed in the paper. Finally, the damage recorded from uniaxially loaded plates containing central circular holes and semicircular edge notches are compared with the predictions of the difference theory.

KEY WORDS: composite laminates, notched laminates, free-edge effects, replication techniques, test methods

Delamination, transverse cracking, and other forms of damage along the free edges of composite laminates have been the subject of many, many, in-

[1]Department of Theoretical and Applied Mechanics, University of Illinois, Champaign-Urbana, IL 61801; formerly, Department of Engineering, Science and Mechanics, Virginia Polytechnic Institute and State University, Blacksburg, VA 24061.
[2]Department of Engineering Science and Mechanics, Virginia Polytechnic Institute and State University, Blacksburg, VA 24061.

vestigations. The vast majority of these studies have been conducted for the case of the straight free edge. These studies began due to the experimental observation that stacking sequence affected the failure of composites and the character of response at free edges. The work of Foye and Baker [1] is an example of these observations. The analytical investigation into these phenomena had its beginnings with the work of Puppo and Evensen [2] and the now-classic paper of Pipes and Pagano [3]. Analytically, the studies could be considered finished with the rigorous work of Wang and Choi [4,5]. Wang and Choi treated the straight free-edge problem as a generalized plane-strain elasticity problem. They showed that at the interface between adjacent laminae at the free edges the stresses are singular, that is, go to infinity. The power of the singularity at the interface is a function of the orientation of the two adjacent layers. The strength of the singularity is a function of the far-field boundary conditions. Thus, for example, the characteristics of the singularity as a function of θ can be determined for $[\pm\theta]_s$ laminates. In the decade between the work of Pipes and Pagano and Wang and Choi, hundreds of other investigations were conducted in efforts to determine stresses along straight free edges. The vast majority of these efforts used nonsingular finite elements and the answers obtained were a function of the fineness of the mesh employed near the edge. In some studies more than 50 elements were used through the thickness of one lamina and multiple-layer laminates taxed the capacities of many computers. Needless to say, without singular elements, care had to be used interpreting the results.

Experimentally, replication techniques using acetate films, X-rays, dye penetrants, and traveling microscopes have been used to investigate, record, and catalog damage along straight free edges.

For practical reasons, there is increasing interest in determining the behavior of laminates at curved free edges. Because of the desire to have access to as many portions of an airframe as possible, current conceptual designs of various composite aircraft components require cut outs and access holes in these components. These are essentially unloaded or open holes that can be quite large in diameter. The holes may be circular, square, or some other shape. Whatever the shape, they are geometric discontinuities. Because of this there will be stress concentrations associated with them. Couple these stress concentrations with the three-dimensional and singular behavior of stresses near free edges and the problem of determining the stresses at the hole boundary becomes very complex. Some work has been done, though, in attempts to study the problem [6–13]. Most of these studies were conducted before the singular nature of the stresses was fully appreciated. Except for Tang [11,12], who used a perturbation approach, all studies referenced used finite elements. Generally the meshes were quite coarse, owing to the fact that the discretization had to take place circumferentially, radially, and through the thickness. For these studies using coarse meshes and nonsingular elements, it is not clear what the results mean regarding the free-edge stresses.

Experimentally, investigation of curved free edges has received less attention than theoretical approaches to the problem, and certainly less than the experimental treatment of the straight free edge. The experimental investigations that have been done have been concerned more with the global behavior of the laminate than with the details of the behavior around the hole. Notched strength is one example of global behavior that has been of interest. One of the first investigations that did focus on a fundamental understanding of the curved free edge was by Daniel et al [14]. They used moiré grids, birefringent coatings, and strain gages to study specimens 254 mm (10.0 in.) wide with 25.4-mm-diameter (1.00-in.) holes. They investigated three material systems and six stacking arrangements. The strains and failure patterns at the hole were reported. It was found that initial failure occurred away from the net section and there was as much as a 17% difference in the strengths of two quasi-isotropic layups. This was attributed to stacking sequence. The authors also reported on an apparent relation between the lamina Poisson's ratio and the peel stress observed in some of the laminates. Rowlands et al [15] published a paper reporting on the performance of panels with cutouts other than circular. Their findings were similar to the findings for circular cutouts. Whitney and Kim [16] noted the differences and similarities between the curved free edge and the straight free edge. They found notched strength to be insensitive to stacking sequence and concluded that either damage at the hole edge did not affect ultimate strength, or, the effect of delamination was independent of stacking sequence. In a fatigue study, Whitcomb [9] sectioned several panels around the circumference of their central holes. He investigated peeling and interlaminar shear effects and concluded that peel stresses seemed to affect delamination less than shear stresses.

Recently, dye penetrants have been used to trace the initiation and propagation of damage near open holes. The method is quite powerful but there is the question of the dye affecting the material. The deplying technique has also seen usage near open holes.

This paper contributes further to the study of composite material behavior at curved free edges. An analytical approach and an experimental technique are discussed. Both are extensions of ideas that have been applied to the study of straight free edges. The analytical portion of the work is based on the concept of the difference in elastic properties of adjacent laminae. The concept assumes that it is the differences in material properties between adjacent laminae that are responsible for interlaminar failures. The larger the difference, the more likely it is that interlaminar failure will occur. Herakovich [17] applied this idea to straight free edges. The extension of the idea to curved edges is more involved than the straight free-edge counterpart. These extensions are discussed. The experimental portion of the study is based on the replication technique used on straight free edges. The polishing of the curved surface and the actual application with the acetate tape are the major problems in applying the technique to curved free edges. The steps necessary to overcome the problems are discussed.

Difference Considerations for Circular Boundaries

Like Poisson's ratio, the coefficient of mutual influence of the second kind, hereafter referred to simply as the coefficient of mutual influence, is defined as the ratio of strains. Specifically, the coefficient of mutual influence is defined as the ratio of a shearing strain to an elongation strain. Referring to Fig. 1, which shows a curved boundary and indicates the coordinate systems used in this discussion, the coefficient of mutual influence relates the in-plane x'-y' shear strain to the y' elongation strain, that is

$$\eta = \frac{\gamma_{x'y'}}{\epsilon_{y'}} \tag{1}$$

The Poisson's ratio under discussion is given by

$$\nu = - \frac{\epsilon_{x'}}{\epsilon_{y'}} \tag{2}$$

In Fig. 1 it is seen that the orientation of the circumferential x'-y' system relative to the laminate x-y system is θ. The fibers in the lamina make an angle β with respect to the x'-axis while they make an angle α with the x-axis. Furthermore

$$\alpha = \theta + \beta \tag{3}$$

In the convention here, x is the direction across the net section.

The definitions given by Eqs 1 and 2 apply under the condition that all stresses except $\sigma_{y'}$ are zero. At a traction-free curved edge this is certainly true. Using Hooke's Law, the two strains in each definition can be written in terms of this single nonzero stress. Since we are dealing with linear elasticity, this stress never actually appears. Using Hooke's Law, Eqs 1 and 2 become

$$\eta \equiv \frac{\gamma_{x'y'}}{\epsilon_{y'}} = \frac{\overline{S}_{26}}{\overline{S}_{22}} \tag{4}$$

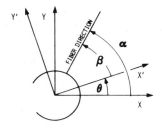

FIG. 1—*Coordinate systems, geometry, and nomenclature.*

$$\nu = -\frac{\epsilon_{x'}}{\epsilon_{y'}} = -\frac{\overline{S}_{12}}{\overline{S}_{22}} \tag{5}$$

where the $\overline{S}_{ij}$'s are the lamina compliances as measured in the x'-y' system. Using the definition of the off-axis compliances, Eqs 4 and 5 become

$$\frac{\gamma_{x'y'}}{\epsilon_{y'}} = \frac{\overline{S}_{26(\beta)}}{\overline{S}_{22(\beta)}}$$

$$= \frac{(2S_{11} - 2S_{12} - S_{66}) \sin^3 \beta \cos \beta - (2S_{22} - 2S_{12} - S_{66}) \sin \beta \cos^3 \beta}{S_{11} \sin^4 \beta + (2S_{12} + S_{66}) \sin^2 \beta \cos^2 \beta + S_{22} \cos^4 \beta} \tag{6}$$

$$\frac{\epsilon_{x'}}{\epsilon_{y'}} = -\frac{\overline{S}_{12(\beta)}}{\overline{S}_{22(\beta)}}$$

$$= -\frac{S_{12}(\sin^4 \beta + \cos^4 \beta) + (S_{11} + S_{22} - S_{66}) \sin^2 \beta \cos^2 \beta}{S_{11} \sin^4 \beta + (2S_{12} + S_{66}) \sin^2 \beta \cos^2 \beta + S_{22} \cos^4 \beta} \tag{7}$$

The unbarred S_{ij}'s are the compliances as measured in the lamina coordinate system and the subscript 1 refers to the fiber direction. The nomenclature here follows the nomenclature of Jones [18]. For a given lamina, α of Eq 3 is fixed. As the circumferential location, θ, on the circular boundary changes, $\beta = \alpha - \theta$ changes and so the strain ratios given by Eqs 6 and 7 also change. The essence of the elastic property difference idea is as follows: In a single unconstrained lamina, for a given $\epsilon_{y'}$, the $\gamma_{x'y'}$ shearing strain and the $\epsilon_{x'}$ contraction strain are given by Eqs 6 and 7. When this lamina is part of a laminate, its strains are dictated by the elastic properties of all the other laminae in the laminate. Classical lamination theory hypothesizes that in a laminate all laminae experience the same strains. Then at a particular point x'-y' in a laminate, in each lamina $\gamma_{x'y'}$, $\epsilon_{x'}$, and $\epsilon_{y'}$ are identical and are equal to the laminate strain. Call these laminate strains $\epsilon_{x'}^0$, $\epsilon_{y'}^0$, and $\gamma_{x'y'}^0$. However, Eqs 6 and 7 indicate that if a lamina experiences a strain $\epsilon_{y'} = \epsilon_{y'}^0$, then $\gamma_{x'y'}$ and $\epsilon_{x'}$ are governed only by the orientation and elastic properties of that lamina. Thus there is a difference between the shearing and contraction strains the individual laminae would have if they were free, and the strains they actually experience because they are in the laminate. In particular, for the ith lamina:

$$\{\gamma_{x'y'i}\}_{\text{from Eq 6 with } \epsilon_{y'} = \epsilon_{y'}^0} - \gamma_{x'y'}^0 = \text{difference in } \gamma_{x'y'i} = \Delta\gamma_{x'y'i} \tag{8}$$

$$\{\epsilon_{x'i}\}_{\text{from Eq 7 with } \epsilon_{y'} = \epsilon_{y'}^0} - \epsilon_{y'}^0 = \text{difference in } \epsilon_{y'i} = \Delta\epsilon_{y'i} \tag{9}$$

There is such a strain difference for every lamina. Herakovich [17] hypothesized that it was not these differences that were important but rather it was the difference in differences between adjacent layers that was the key parameter. Thus for the ith and $(i + 1)$st laminae the differences in the strain differences are

$$\Delta\gamma_{x'y'i+1} - \Delta\gamma_{x'y'i} = \{\gamma_{x'y'i+1}\}\text{from Eq 6 with } \epsilon_{y'} = \epsilon_{y'}^0,$$

$$- \{\gamma_{x'y'i}\}\text{from Eq 6 with } \epsilon_{y'} = \epsilon_{y'}^0, \quad (10)$$

$$\Delta\epsilon_{x'i+1} - \Delta\epsilon_{x'i} = \{\epsilon_{x'i+1}\}\text{from Eq 6 with } \epsilon_{y'} = \epsilon_{y'}^0,$$

$$- \{\epsilon_{x'i}\}\text{from Eq 6 with } \epsilon_{y'} = \epsilon_{y'}^0, \quad (11)$$

Henceforth the right-hand side of Eqs 10 and 11 will be referred to, respectively, as

$$\delta\gamma_{x'y'i,i+1} \text{ and } \delta\epsilon_{x'i,i+1}$$

Using Eqs 4 and 5, Eqs 10 and 11 become

$$\delta\gamma_{x'y'i,i+1} = \left\{ \frac{\bar{S}_{26}(\beta)_{i+1}}{\bar{S}_{22}(\beta)_{i+1}} - \frac{\bar{S}_{26}(\beta)_i}{\bar{S}_{22}(\beta)_i} \right\} \epsilon_{y'}^0 \quad (12)$$

$$\delta\epsilon_{x'i,i+1} = \left\{ \frac{\bar{S}_{12}(\beta)_i}{\bar{S}_{22}(\beta)_i} - \frac{\bar{S}_{12}(\beta)_{i+1}}{\bar{S}_{22}(\beta)_{i+1}} \right\} \epsilon_{y'}^0 \quad (13)$$

It is important to recall that Eqs 12 and 13 are written in the x'-y' system. The strain $\epsilon_{y'}^0$ is a function of circumferential location, θ, that is, $\epsilon_{y'}^0 = \epsilon_{y'}^0(\theta)$. The angle β for the $(i + 1)$st lamina will be given by $\beta_{i+1} = \theta - \alpha_{i+1}$ while for the ith lamina, $\beta_i = \theta - \alpha_i$. Thus the right-hand sides of Eqs 12 and 13 are functions of θ, that is, position along the free edge, and the notation becomes

$$\delta\gamma_{x'y'}(\theta)_{i,i+1} = \left\{ \frac{\bar{S}_{26}(\theta - \alpha)_{i+1}}{\bar{S}_{22}(\theta - \alpha)_{i+1}} - \frac{\bar{S}_{26}(\theta - \alpha)_i}{\bar{S}_{22}(\theta - \alpha)_i} \right\} \epsilon_{y'}^0(\theta) \quad (14)$$

$$\delta\epsilon_{x'}(\theta)_{i,i+1} = \left\{ \frac{\bar{S}_{12}(\theta - \alpha)_i}{\bar{S}_{22}(\theta - \alpha)_i} - \frac{\bar{S}_{12}(\theta - \alpha)_{i+1}}{\bar{S}_{22}(\theta - \alpha)_{i+1}} \right\} \epsilon_{y'}^0(\theta) \quad (15)$$

To shorten the notation, the bracketed term on the right side of Eq 14 will be denoted as $\delta\eta(\theta)_{i,i+1}$ while the bracketed right side of Eq 15 will be denoted as $\delta\nu(\theta)_{i,i+1}$. The equations then become

$$\delta\gamma_{x'y'}(\theta)_{i,i+1} = \delta\eta(\theta)_{i,i+1} \epsilon_{y'}^0(\theta) \quad (16)$$

$$\delta\epsilon_{x'}(\theta)_{i,i+1} = \delta\nu(\theta)_{i,i+1} \epsilon_{y'}^0(\theta) \quad (17)$$

The quantities $\delta\eta(\theta)_{i,i+1}$ and $\delta\nu(\theta)_{i,i+1}$ are strictly material properties and will be computed by using the definitions of the $\bar{S}_{ij}$. The $\epsilon_y^0{}'(\theta)$ will be computed using plane-stress finite-element solutions for particular curved free-edge geometries and laminate material properties. Since $\delta\gamma_{x'y'}(\theta)_{i,i+1}$ and $\delta\epsilon_{x'}(\theta)_{i,i+1}$ are related to both the difference of material properties and to the laminate strain level, a small difference of material properties and a large strain level could have the same effect as a large difference of material properties and a small strain level. For the straight free edge the strain is uniform with position along the edge and so it is the difference of the material properties that is the primary concern. For the curved free edge the material property differences are modulated by the laminate strain level and so both the differences in properties *and* the strain level must be considered.

Specific Problem Studies

The specific problem studied was the case of a laminated plate, either with a central circular hole or a semicircular edge notch, subjected to uniaxial tension. Figures 2*a* and 2*b* show the configurations of the two specimens. The three holes at each end of the specimens were for gripping the specimen. Five stacking arrangements were studied: $[-45/+45/0/90]_s$, $[0/-45/90/$

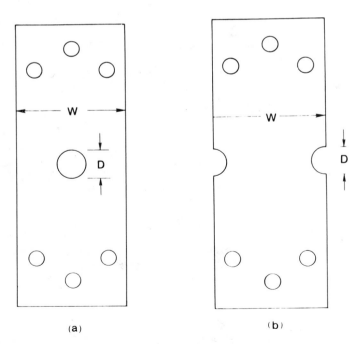

(a)   (b)

FIG. 2—*Configuration of specimens: (a), central-hole specimen; (b), edge-notch specimen.*

+45]$_s$, [(90/0)$_2$]$_s$, [($\pm$30)$_2$]$_s$, and [($\pm$10)$_2$]$_s$. These stacking arrangements are referred to the lengthwise axis of Fig. 2. The material used in the experiments was CELION-6000/PMR-15, a graphite polyimide, with the following material properties[3]:

$$E_1 = 137 \text{ GPa } (19.9 \times 10^6 \text{ psi})$$

$$E_2 = 7.58 \text{ GPa } (1.10 \times 10^6 \text{ psi})$$

$$G_{12} = 4.90 \text{ GPa } (0.71 \times 10^6 \text{ psi})$$

$$\nu_{12} = 0.35.$$

These data were determined from tests of on- and off-axis tension specimens. All material properties needed to obtain numerical results were determined by using classical lamination theory and the above basic material properties.

Table 1 indicates the geometry of the holes and notches. With two curved boundaries and five laminates, many numerical results were generated. To illustrate some of the findings, only the results from the two quasi-isotropic laminates will be reported here.

Finite-Element Analyses

The finite-element analyses to compute ϵ_y^0 (θ) were conducted using plane-stress hybrid quadrilateral elements. The plane-stress analysis was based on classical lamination theory. The finite-element meshes were similar to those of other investigators studying holes and notches in plates. The meshes for the edge notch were identical to the meshes for the central hole, boundary conditions being the only difference between the two analyses. Quarter-plane symmetry at the laminate level was exploited in the analyses. There were 20 elements around the quarter-hole circumference and 11 elements across the half net section. A displacement boundary condition, simulating the effect of the screw-type loading frame, was used at the grip end of the laminate. Laminate

TABLE 1—*Notch and hole geometry.*

Laminate	Hole/Notch Diameter, D, mm (in.)	Gross Width, W, mm (in.)
[−45/+45/0/90]$_s$	50.8 (2.00)	152 (6.00)
[0/−45/90/+45]$_s$	50.8 (2.00)	152 (6.00)
[(90/0)$_2$]$_s$	38.0 (1.50)	114 (4.50)
[($\pm$30)$_2$]$_s$	38.0 (1.50)	114 (4.50)
[($\pm$10)$_2$]$_s$	50.8 (2.00)	152 (6.00)

[3]Original measurements made in U.S. customary units.

strains at the edge of the curved boundary were computed directly from element response. This is opposed to an interpolation to the edge of the element by using information from within the element.

Numerical Results

Of course the difference in material properties from one layer to the next was not affected by whether the curved boundary was part of a central circular hole or part of a semicircular edge notch. Figures 3 and 4 show the quantities $\delta\eta(\theta)_{i,i+1}$ and $\delta\nu(\theta)_{i,i+1}$ as a function of circumferential location for the two quasi-isotropic stacking arrangements studied. For each stacking arrangement, differences in material properties between adjacent pairs of laminae are shown. It is interesting to note that for many pairs the maximum difference occurs at some point on the circumference other than at the net section, that is, at some location other than the $\theta = 0$ deg location. This observation led to the belief that one might not observe delamination at the net section for these cases. Transverse (through-the-thickness) cracks might initiate at the net section but it was felt that interlaminar failure would begin slightly away from the net section. As will be seen, experimental evidence indicated this to be the case. Also, due to the lack of symmetry of some of the differences with θ, it was felt that interlaminar failures, if they occurred, would not necessarily initiate concurrently at both $+\theta$ and $-\theta$ away from the net section. This is despite the fact that at a laminate level the specimens exhibited mirror image symmetry about the net section. It should be mentioned that the material property differences for the straight free edge correspond to the numerical results at the $\theta = 0$ deg location. It is well known that adjacent ±45-deg laminae produce a large mismatch in $\delta\eta$ at straight free edges. Thus for straight specimens adjacent ±45-deg laminae should be avoided and a $[0/-45/90/+45]_s$ laminate used instead of a $[-45/+45/0/90]_s$. However, examination of the magnitude of $\delta\eta(\theta)$ for all θ for the two laminates indicates neither stacking arrangement seems superior in terms of low values. The value of $\delta\eta(\theta)$ is not the whole story. The magnitude of the circumferential strain, $\epsilon_{y'}(\theta)$, is also important.

Figure 5a shows the finite-element predictions of $\epsilon_{y'}^0(\theta)$ as a function of θ for the quasi-isotropic specimens with a central hole. Superimposed on these figures are the strain values as measured in the actual experiments. For comparison purposes the strains have been normalized. For the finite-element solutions the specified displacement at the grip ends divided by the specimen length was used to normalize the strains. For the experimental data the strain from a far-field strain gage was used as a normalization factor. The strains were taken at several load levels so there is a spread to the data. This is indicated by the vertical lines. Figure 5b shows the finite-element predictions and experimental measures of circumferential strain around the curved edge of the quasi-isotropic notched specimens. The good correlation between experi-

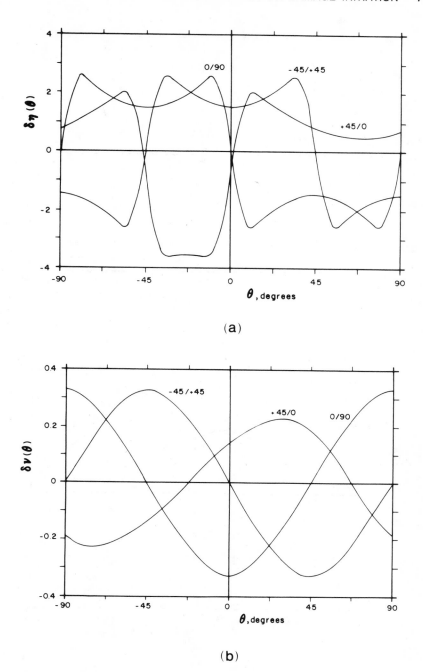

FIG. 3—*Differences in material properties for adjacent laminae in [−45/+45/0/90]ₛ lami-nate: (a), mismatch in coefficients of mutual influence; (b), mismatch in Poisson's ratios.*

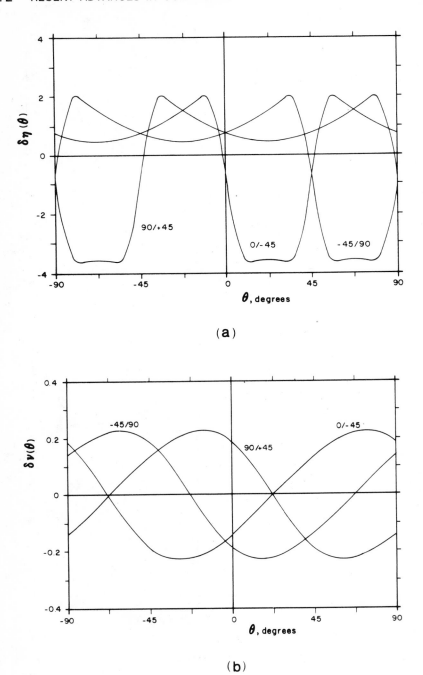

FIG. 4—*Differences in material properties for adjacent laminae in [0/−45/90/+45]ₛ lami-nate: (a), mismatch in coefficients of mutual influence; (b), mismatch in Poisson's ratios.*

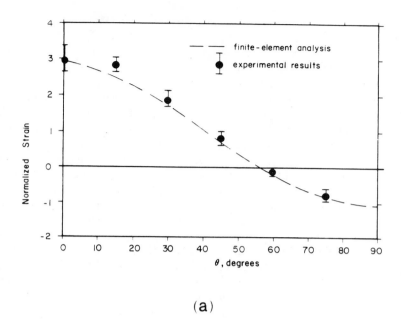

(a)

(b)

FIG. 5—*Circumferential strains in laminates:* (a), *around central hole;* (b), *around edge notch.*

ment and theory was similar to the good correlation for the central hole specimens.

Finally, Figs. 6–9 show the key parameters in this study. These figures show numerical results predicting the variation of the differences in the shearing strains, $\delta\gamma_{x'y'}(\theta)_{i,i+1}$ and the contracting strains, $\delta\epsilon_{x'}(\theta)_{i,i+1}$ for the two stacking sequences and for the two specimen geometries. These figures also show experimental data to be discussed later. The numerical results (solid lines) are computed from Eqs 16 and 17 and the results are a combination of the results in Figs. 3–5. For example, the numerical results in Fig. 6a are the product of the numbers in Fig. 3a and the numbers in Fig. 5a. As can be seen, the effect of the difference in material properties (Fig. 3a) is altered by the strain level (Fig. 5a), the alteration emphasizing some differences and de-emphasizing others. In essence, the hypothesis put forth in the paper states that it is from the numerical results of Figs. 6–9 that the location of the first interlaminar damage can be predicted. For a given laminate, initial damage could be expected to begin at the circumferential location and lamina interface at which the largest values of $\delta\gamma_{x'y'}$ or $\delta\epsilon_{x'}$ or both occur. For example, in a $[0/-45/90/+45]_s$ laminate with a hole, Fig. 8a, interlaminar shear damage could be expected to begin at the $0/-45$ interface at $\theta = 10$ deg. As will be seen, the situation appears to be more complex than that.

Experimental Setup

Figure 10 shows a typical specimen in the loading fixtures. The strain gaging is visible in the figure. The specimens were mounted in a displacement-controlled tension machine using friction grips. On each end of the specimen the two halves of the grips were clamped together with three bolts which passed through the specimens. The hole diameters in the specimens, seen in Fig. 2, were larger than the bolt diameters so that there would be no bearing of the bolts on the specimen. This method of gripping the specimen was used to minimize the risk of failing the specimen in the grip region. The strain gages were mounted circumferentially every 15 deg around the hole edges. The gages were Type CEA-13-032UW-120 manufactured by the Micro-measurements Group [19]. The gage backing was trimmed so that the active gage was at the hole edge. There were also strain gages mounted in the far field. These were larger gages, Type CEA-13-062UW-120.

The specimens were loaded slowly, 0.51 mm/min (0.02 in./min), in 445 N (100 lb) increments. For the narrower specimens this corresponded to a 550-psi (383-MPa) increment in far-field stress. At each increment the load was held and an acetate film replica of the hole edge was taken. Generally 6 to 12 replicas, at increasing load levels, were taken. After the specimen had failed the acetate film could be reviewed to catalog the types, locations, and load levels of damage initiation and progression. With this approach to loading and replicating, the exact strain level producing damage could not be deter-

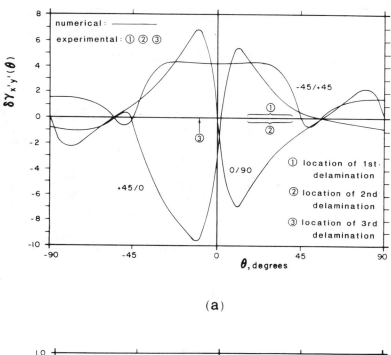

(a)

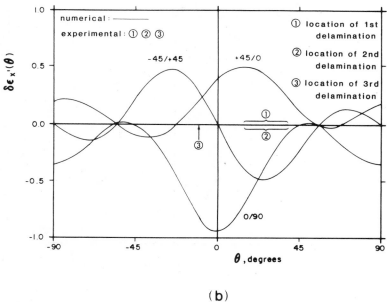

(b)

FIG. 6—*Differences in strains for adjacent laminae in* $[-45/+45/0/90]_s$ *laminate with hole: (a), radial-circumferential shear strains; (b), radial elongations strains.*

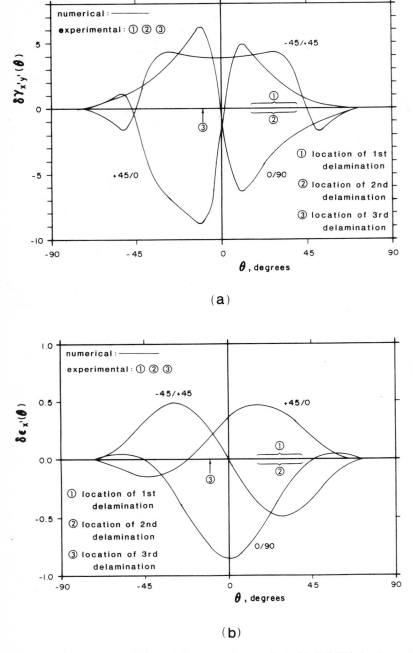

FIG. 7—*Differences in strains for adjacent laminae in* [−45/+45/0/90]ₛ *laminate with notch:* (a), *radial-circumferential shear strains;* (b), *radial elongation strains.*

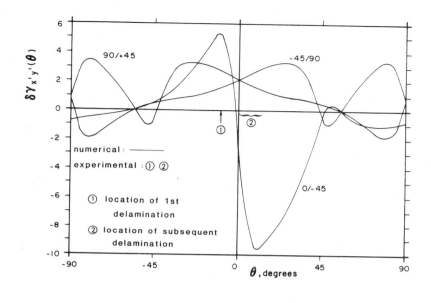

(a)

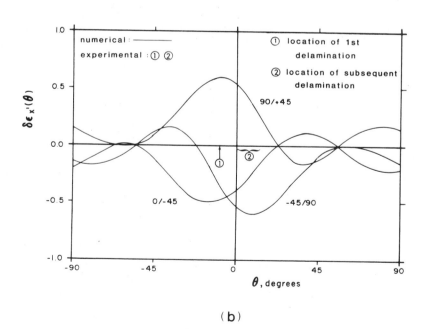

(b)

FIG. 8—*Differences in strains for adjacent laminae in [0/−45/90/+45]ₛ laminate with hole:* (a), *radial-circumferential shear strains;* (b), *radial elongation strains.*

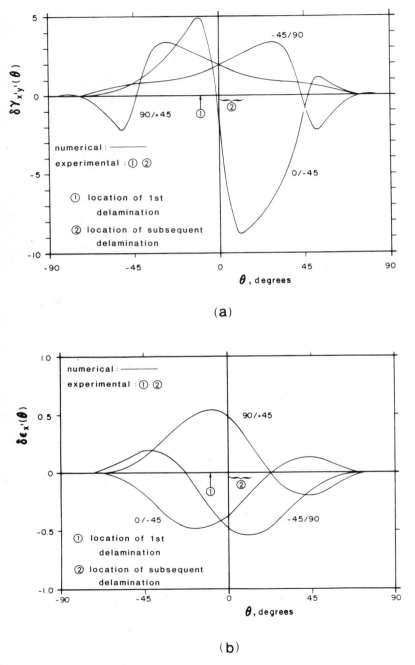

FIG. 9—*Differences in strains for adjacent laminae in* $[0/-45/90/+45]_s$ *laminate with notch:* (a), *radial-circumferential shear strains;* (b), *radial elongation strains.*

FIG. 10—*Edge-notch specimen in loading frame.*

mined. It could be determined only to within a certain interval. This is inherent in the replication technique since continuous recording is not possible. In the section on experimental results, the stress level at the lower end of the increment interval is reported. For the load levels here this leads to a 5% uncertainty in the exact stress level at which a damage event occurred.

Before discussing the results of the experiment and the comparisons with the theory presented, the technique for replicating a curved fee edge is discussed.

Replicating of Curved Surfaces

Surface replication was originally developed for studying the microstructure of metals. The technique has been successfully extended to the replication of straight free edges of composite laminates. A replica of the edges of laminates reveals the very fine detail of the individual layers as well as the delamination and cracking of these layers when the laminate is subjected to a load. One key to successful surface replication is a highly polished surface. The other key to successful replication is to insure the replicating film does not move relative to the surface being replicated and, in fact, to have a slight pressure holding the film to the surface. These requirements make replication of curved surfaces more difficult than replication of flat surfaces. These aspects are discussed next.

To polish the curved-edge surface of the central holes and the notches, a felt bob and a vertical drill press were used. A felt bob is nothing more than a felt grinding wheel which is stiff enough to remain cylindrical under working forces but absorbent enough to hold polishing compound. The felt bob used here was 12.7 mm (0.5 in.) in diameter. The specimens to be polished were oriented in a plane parallel to the horizontal drill press table. To keep the axis of the felt bob perpendicular to the plane of the hole, and to keep the specimen approximately 25.4 mm (1.00 in.) off the table surface and parallel to it, horizontal spacers were used. These spacers were made so that the specimens could slide around parallel to the table.

The holes and notches in the specimens were drilled with an ultrasonic core drill. Thus the hole and notch surfaces were in good condition. However, if there are nicks and gouges in the surface to be replicated, these should be smoothed before the surface is polished. Once the large defects are eliminated, the polishing process can begin. Two grits (or sizes) of polishing compound were used, 5 μm and 3 μm. Many polishing compounds come as powders and need to be combined with water to form a dilute slurry solution. This slurry was applied to the surface and it was found that the easiest method of application was a squeeze bottle with a nozzle and a relatively small opening for squirting the compound. The coarse (5μm) grit was applied first and the felt bob was rotated at a low speed, about 500 rpm. This portion of the operation continued for 5 min with constant movement of the specimen on the table. After 5 min the felt bob was removed and either replaced or cleaned thoroughly. This was to insure that none of the 5-μm grit remained in the bob when the fine (3μm) grit was used. Once the bob was cleaned, or replaced, the polishing procedure was repeated for the 3-μm grit. (After both grits have been used, the surface should have a mirror-like gloss. If this is not the case, more polishing with the fine grit is required. A good check on the quality of the surface is to make an actual replica and examine it. Individual fibers are easy to identify if the surface is properly prepared.) Figure 11 illustrates the polishing process. In the figure the slurry is being applied. The felt bob is visible. The bars across the specimen keep the horizontal spacers in place.

The first step in the actual replication was the choice of replicating film, or tape. The primary differences in the various replicating materials available are the thickness and the form [20]. Replicating film is made of acetate and comes in strip or in sheet form. Various thicknesses are available, and for the

FIG. 11—*Polishing edge notch with drill press and felt bob.*

application here a thin 0.076-mm (0.003 in.) film was preferred due to its low stiffness. The low stiffness allowed the film to conform to the contour of the curved surface. The film was cut to be about 12.7 mm (0.5 in.) wider than the width of the curved surface to be replicated. The length of the film was about one-half the hole circumference. [If the hole is extremely large, the replicating process can be repeated several times around the circumference, in incremental fashion. However, for holes less than 50.4 mm (2.00 in.) in diameter this should not be necessary.] After the film was cut to the proper length and width, the surface to be replicated was cleaned. Despite the high polish on the surface, dust and other particles accumulated on the surface. These particles led to imperfect replicas. The particles ultimately became sites for air bubbles between the surface and the film. A clean cloth soaked in acetone was used to remove all foreign matter from the surface. Then a light coating of acetone was applied to the surface and allowed to evaporate. This step also helped keep air bubbles to a minimum. Once the surface was cleaned and a thin coating of acetone applied, the replicating film was brought into contact with the surface. Finally, acetone was injected between the film and the polished surface using a hypodermic injection needle. The acetone softened the acetate film and allowed it to conform to the cracks and other three-dimensional features in the surface.

When replicating a straight free edge, to insure good contact and to force all air bubbles out, the film is pushed against the polished surface with a finger. If any significant length of straight edge is to be replicated, both ends of the film are held in place with scotch-tape and a finger is run the length of the film, pushing it against the surface with a slight pressure. Extreme care must be taken at this point to insure that there is no shear-type movement of the film relative to the surface. Any movement will result in a blurred replica. For the curved free edge, one end of the strip of film was taped to the hole edge. The taped end of the film was at the top of the hole. The replicator's finger held the other end (bottom) of the film strip. This was the delicate part of the replicating operation. With care, the bottom end of the strip could be held such that the strip conformed to the circular hole. Putting a slight amount of compressive force on the bottom end of the film strip caused the strip to snap-buckle to the shape of the circular edge. Once this buckling had taken place the lower end of the film strip could be fastened to the specimen using another piece of tape or a spring-type clip. The replicator then had two hands free. The acetone was then injected between the film and the polished surface. Figure 12 illustrates this step. In this figure spring clips are shown at each end of the strip. Capillary action and gravity caused the acetone to spread uniformly over the contact surface. Unlike the straight-edge procedure, the replicator's finger was not used to push the softened film against the surface. This was attempted; however, due to the geometry of the hole and the forces in the buckled strip, the strip pulled away from the surface when finger pressure was applied, forming chords of the circle between the fastened ends and the finger. This caused blurred replicas because there was circumferential movement of the strip relative to the surface. However, it was discovered that the slight compressive force used to buckle the film strip into the form of the circular hole boundary could be made large enough to actually eliminate the need for additional finger pressure. This was quite fortunate. However, it was alluded to earlier that attaining the correct compressive force was a delicate part of the operation. Too much compressive force caused the film to snap away from the curved surface when the injection needle was inserted between it and the surface. Practice was required to obtain the proper pressure on the film strip.

With the proper pressure, for the hole sizes here, injection of the acetone was required only at one point. If it is necessary, injection can be done at several locations. However, the replication may be blurred due to the slight movement of the film which has already softened at other injection locations. Once the acetone was injected, and all surfaces wetted, the film was not moved until all the acetone had evaporated and the replica film had hardened again. After the film had hardened, it was carefully peeled from the surface and mounted between two glass plates. The two plates were clamped or taped together to keep the replica film from curling and thereby distorting the image. The replica could be viewed with a microscope, or in a microfiche reader,

FIG. 12—*Acetate film in place in an edge notch and injecting acetone between film and sur-face.*

while between the glass plates. A $\times 54$ magnification of the replica has been found to be satisfactory for examining the results.

Several other comments on this procedure should be made. First, if in the polishing process the felt bob is not kept perpendicular to the plane of the specimen edges, the edges of the surface to be replicated will round off. When this happens it is difficult to replicate the full specimen thickness. Second, any specimen instrumentation, such as strain gaging, should be mounted after polishing. The water and powder mixture used for polishing can have a detrimental effect on the instrumentation. The presence of the instrumentation, with wires, etc., can make it difficult to polish properly. The polished surface can easily be protected during the installation of instrumentation.

It should be noted that several aspects of the replicating process as described here might be considerably more difficult in the field than they were in the laboratory environment.

Experimental Results

Edge replicas corresponding to various load levels for the specimens were analyzed upon completion of each test. The replicas yielded a damage history record that could easily be followed. The damage history for each laminate is described below. It should be noted that in each case only one half of the hole circumference was replicated.

$[-45/+45/0/90]_s$ Quasi-Isotropic Laminate

Center Hole Specimen—The first transverse crack occurred at a far-field stress of 67.6 MPa (9.8 ksi). This crack was located in the adjacent 90-deg plies at $\theta \approx 0$. Several transverse cracks appeared in this region and the cracking seemed to be symmetric with respect to θ. Soon after transverse cracking occurred in the 90-deg plies, it was followed by transverse cracking in the +45-deg plies at angles +15 deg $\leq \theta \leq$ +40 deg. At still higher loads the transverse cracks in the 90-deg plies caused cracking parallel to the plane of the laminate near $\theta = 0$. Some cracks were interlaminar and some cracks were within an individual lamina (intralaminar). Such cracks indicate high peel stresses at the net section.

Further loading resulted in delamination at the $-45/+45$ interface in the region +15 deg $\leq \theta \leq$ +40 deg. An example of this is shown by the replicate in Fig. 13a. The delaminations shifted to the +45/0 interface, as shown in Fig. 13b, as the loading was increased. Delamination was observed at the +45/0 interface in the region of $\theta = -10$ deg prior to failure. This is shown in Fig. 13c. Final failure occurred when the 0-deg ply fractured. Figures 6a and 6b indicate the locations and sequence of the observed delaminations.

A significant observation is that nearly all of the damage occurred for $\theta > 0$, thus indicating that the cracking was not symmetric with respect to $\theta = 0$. Furthermore, the final failure surface was not parallel with the net section.

Edge Notch Specimen—The first crack appeared in the adjacent 90-deg plies at a stress of 80.7 MPa (11.7 ksi). The rest of the damage was quite similar to the damage accumulation in the center-hole specimen and will not be repeated here. Figures 7a and 7b illustrate the locations and sequence of the observed delaminations. It is worthwhile to note that this specimen also exhibited nonsymmetric cracking with respect to $\theta = 0$.

$[0/-45/90/+45]_s$ Quasi-Isotropic Laminate

Center-Hole Specimen—Transverse cracks appeared first in the 90-deg plies at 80.7 MPa (11.7 ksi) and $\theta \approx -15$ deg. The cracking then seemed to shift to the -45-deg plies at higher loads at the same circumferential location. Final transverse cracking appeared in the adjacent 45-deg plies at a stress of 121 MPa (17.6 ksi) and at angles of +30 deg $\leq \theta \leq$ +45 deg.

Delamination was first observed between the 0 deg and -45-deg plies at a stress of 110 MPa (15.9 ksi). The circumferential location of the delamination was $\theta \approx -10$ deg. At still higher loads the delaminations shifted from the $0/-45$ interface to the $-45/90$ interface. This is shown in Fig. 14a. Prior to failure, delaminations occurred in the region 0 deg $\leq \theta \leq +10$ deg at the $0/-45$ interface, as shown in Fig. 14b. (In Fig. 14b part of the replica is missing because the outer lamina had moved away from the surface.) Figures 8a and 8b indicate the locations and sequence of the delaminations.

Ultimate failure occurred again when the 0-deg plies fractured. As in the case of the $[-45/+45/0/90]_s$ laminate, the cracking was not symmetric with respect to $\theta = 0$.

Edge-Notch Specimen—The initiation of transverse cracking was found to be nearly opposite that found in the center-hole specimen. The first transverse cracks occurred in the adjacent 45-deg plies at a lower stress, 67.6 MPa (9.8 ksi), than that of the center-hole specimen. Cracking was then found in the 90-deg plies at the same load, 80.7 MPa (11.7 ksi), as that for the center-hole specimen. After this cracking had occurred the rest of the damage history was quite similar to that of the center-hole specimen. Figures 9a and 9b show the delamination locations. Again the cracking was not symmetric with respect to $\theta = 0$.

Comparison of Experimental and Theoretical Results

The correlation between the predictions of the theory and the observations of the experiment was not particularly good. Despite this, it is felt there is a strong case to be made for the idea of using the difference idea to predict qualitative, and even quantitative, information regarding curved free-edge interlaminar damage. For the $[-45/+45/0/90]_s$ laminate with a hole, it would appear from Fig. 6a that initial damage in the form of a shear-induced delamination would occur at the $+45/0$ interface at $\theta = -10$ deg. Such a delamination was observed. However, it was discovered only after delamination damage had been observed at the $-45/+45$ and $+45/0$ interfaces in the region of 15 deg $\leq \theta \leq 40$ deg. In the region 15 deg $\leq \theta \leq 40$ deg the shear differences for the $-45/+45$ and the $+45/0$ interfaces are significant and so this may not be surprising. What could be important is that the differences in Poisson strains, Fig. 6b, are peaking for the $-45/+45$ and $+45/0$ interfaces in the range 15 deg $\leq \theta \leq 40$ deg while the difference in Poisson effects is not as large for the $+45/0$ interface at $\theta \approx -10$ deg. Thus an interaction between the difference in the coefficients of mutual influence and the difference in Poisson's ratios could be much more important than one difference or the other. This is like a mixed-mode failure in fracture mechanics. If this logic is followed, however, there should have been interlaminar damage between the 0-deg and 90-deg lamina at -15 deg $\leq \theta \leq -10$ deg. Both the coefficient of mutual influence and the Poisson's ratio differences are high in this region. No interlaminar cracking occurred.

For the notch in the $[-45/+45/0/90]_s$ laminate, the damage initiation did not exactly follow the trend of Fig. 7a as regards shear delamination. However, a delamination at the $+45/0$ interface at $\theta = 10$ deg was evident prior to final failure. Again, interaction between the two types of differences could affect the order of damage initiation and accumulation.

For the $[0/-45/90/+45]_s$ laminate with a hole, Fig. 8a indicates an overall shear delamination tendency at $\theta = 10$ deg at the $0/-45$ interface. Also, at $\theta = -10$ deg, at the same interface, delamination could begin. However, at $\theta = 10$ deg, the Poisson difference for that laminae pair is not as large as it is for $\theta = -10$ deg. Interaction could well favor the -10-deg location. Ex-

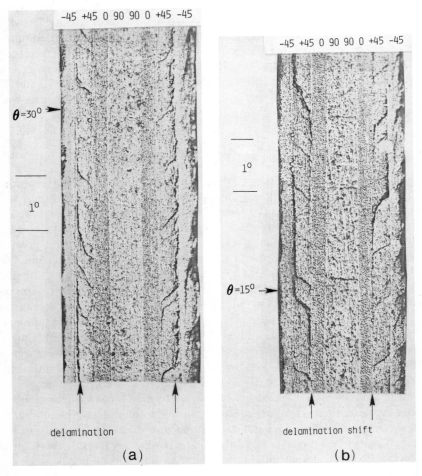

FIG. 13—*Replicas of hole in $[-45/+45/0/90]_s$ laminate at various circumferential locations and gross-section stress levels: (a), $\theta \approx 30$ deg, 164 MPa (23.8 ksi); (b), $\theta \approx 15$ deg, 179 MPa (25.8 ksi); (c), $\theta \approx -10$ deg, 192 MPa, 27.8 ksi.*

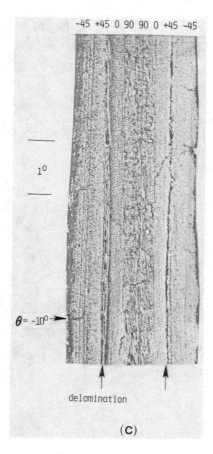

-45 +45 0 90 90 0 +45 -45

1^0

$\theta = -10^0 \rightarrow$

delamination

(C)

FIG. 13—*Continued.*

perimentally, damage started at $\theta = -10$ deg between the 0 deg and -45-deg laminae. Thus in this case there is good correlation.

Figure 9a indicates the $[0/-45/90/+45]_s$ laminate with an edge notch should show some interlaminar effects at $\theta = 10$ deg at the $0/-45$ interface. This happened prior to failure but only after other interlaminar damage occurred at other locations. From Fig. 9b, it is seen that both the Poisson and the shear differences are high near $\theta = -10$ deg for the $0/-45$ interface. That interlaminar damage was observed here as well as at $\theta = +10$ deg is thus not surprising. Therefore, it seems interaction needs to be considered.

Discussion and Conclusion

Two ideas have been presented in this paper. The first idea extends the notion of difference, or mismatch, in the elastic response of adjacent laminae

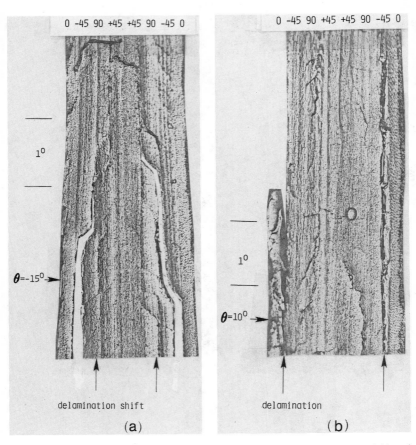

FIG. 14—*Replicas of hole in [0/−45/90/+45]$_s$ laminate at various circumferential locations and gross-section stress levels: (a), $\theta \approx -15$ deg, 151 MPa (21.9 ksi); (b), $\theta \approx 10$ deg, 164 MPa (23.8 ksi).*

to explain interlaminar effects at curved free edges. The second idea extends the replication technique to the case of curved free edges. The latter extension seemed successful. The former extension could not be considered completely successful. However, there was enough correlation between the mismatch predictions and experimental observations to warrant further examination into the idea. This examination should include interaction of the Poisson mismatch and the mismatch of the coefficient of mutual influence. One aspect of laminate response that the mismatch theory did reveal was the lack of symmetric failures at the net section. For some time investigators have observed that failure of globally symmetric laminates was not symmetric with respect to the net section. It is seen here that, analytically, the mismatches in strains are not symmetric with respect to the net section. Thus there is a strong correlation with observation.

Finally, it could be argued that after the first damage has occurred, the stress state around the hole is altered, and using the stress distribution associated with an undamaged laminate is wrong. To a degree this is true. However, the difference technique is designed to predict the location of the first damage and, it is hoped, the load level which causes first damage. It is felt that the stresses are so intense in the presence of a large difference in material properties that a simple-to-use idea, as opposed to a complicated and expensive three-dimensional finite-element model, should be able to predict the basic mechanics of the problem.

Acknowledgments

The work reported on here was supported by the National Aeronautics and Space Administration (NASA)-Virginia Tech Composites Program, Cooperative Agreement NCC1-15.

References

[*1*] Foye, R. L. and Baker, D. J., "Design of Orthotropic Laminates," *Proceedings*, 18th American Institute of Aeronautics and Astronautics SDM Conference, Denver, CO, April 1980.

[*2*] Puppo, A. H. and Evensen, H. A., "Interlaminar Shear in Laminated Composites Under Generalized Plane Stress," *Journal of Composite Materials*, Vol. 4, 1970, pp. 204–210.

[*3*] Pipes, R. B. and Pagano, N. J., "Interlaminar Stresses in Composite Laminates Under Uniform Axial Extension," *Journal of Composite Materials*, Vol. 4, 1970, pp. 538–548.

[*4*] Wang, S. S. and Choi, I., "Boundary-Layer Edge Effects in Composite Laminates: Part 1—Free-Edge Stress Singularities," *Journal of Applied Mechanics*, Vol. 49, No. 3, Sept. 1982, pp. 541–548.

[*5*] Wang, S. S. and Choi, I., "Boundary-Layer Edge Effects in Composite Laminates: Part 2—Free-Edge Solutions and Basic Characteristics," *Journal of Applied Mechanics*, Vol. 49, No. 3, Sept. 1982, pp. 549–560.

[*6*] Dana, J. R. and Barker, R. M., "Three-Dimensional Analysis for the Stress Distribution Near Circular Holes in Laminated Composites," College of Engineering Report VPI-E-74-18, Virginia Polytechnic Institute and State University, Blacksburg, VA, 1974.

[*7*] Rybicki, E. F. and Schmueser, D. W., "Three-Dimensional Finite-Element Stress Analysis of Laminated Plates Containing a Circular Hole," AFML-TR-76-92, Air Force Materials Laboratory, 1976; available National Technical Information Science, Springfield, VA, (Document No. ADA-032-428).

[*8*] Rybicki, E. F. and Schmueser, D. W., "Effect of Stacking Sequence and Lay-up Angle on the Free Edge Stresses Around a Hole in a Laminated Plate Under Tension," *Journal of Composite Materials*, Vol. 12, 1978, pp. 300–313.

[*9*] Whitcomb, J. D., "Experimental and Analytical Study of Fatigue Damage in Notched Graphite/Epoxy Laminates," National Aeronautics and Space Administration, Technical Memorandum 80121, June 1978; available National Technical Information Science, Springfield, VA.

[*10*] Griffin, O. H., Kamat, M. P., and Herakovich, C. T., "Three-Dimensional Inelastic Finite Element Analysis of Laminated Composites," College of Engineering Report VPI-E-80-78, Virginia Polytechnic Institute and State University, Blacksburg, VA, 1980.

[*11*] Tang, S., "Interlaminar Stresses Around Circular Cutouts in Composite Plates Under Tension," *AIAA Journal*, American Institute of Aeronautics and Astronautics, Vol. 15, No. 11, 1977, pp. 1631–1637.

[*12*] Tang, S., "A Variational Approach to Edge Stresses of Circular Cutouts in Composites," *Proceedings*, 20th AIAA/ASME/ASCE/AHS SDM Conference, American Institute of Aeronautics and Astronautics, St. Louis, MO, 4–6 April 1979.

[*13*] Raju, I. S. and Crews, J. H., Jr., "Three-Dimensional Analysis of $[0/90]_s$ and $[90/0]_s$ Laminates with Central Circular Hole," *Composites Technology Review*, Vol. 4, No. 4, 1982, pp. 116–124.

[*14*] Daniel, I. M., Rowlands, R. E., and Whiteside, J. B., "Effects of Material and Stacking Sequence on Behavior of Composite Plates with Holes," *Experimental Mechanics*, Jan. 1974, pp. 1–9.

[*15*] Rowlands, R. E., Daniel, I. M., and Whiteside, J. B., "Geometric and Loading Effects on Strength of Composite Plates with Cutouts," *Composite Materials: Testing and Design (Third Conference), ASTM STP 546*, American Society for Testing and Materials, Philadelphia, 1974, pp. 361–375.

[*16*] Whitney, J. M. and Nuismer, R. J., "Stress Fracture Criteria for Laminated Composites Containing Stress Concentrations," *Journal of Composite Materials*, Vol. 8, 1974, pp. 253–265.

[*17*] Herakovich, C. T., "On the Relationship Between Engineering Properties and Delamination of Composite Materials," *Journal of Composite Materials*, Vol. 15, 1981, pp. 336–348.

[*18*] Jones, R. M., *Mechanics of Composite Materials*, McGraw-Hill, New York, 1975.

[*19*] Micromeasurements Group, Raleigh, NC.

[*20*] Ernest F. Fullam, Inc., Schenectady, NY.

Takashi Horiguchi[1]

Fracture Toughness of Fiber-Polymer Cement Concrete System

REFERENCE: Horiguchi, T., **"Fracture Toughness of Fiber-Polymer Cement Concrete System,"** *Recent Advances in Composites in the United States and Japan, ASTM STP 864*, J. R. Vinson and M. Taya, Eds., American Society for Testing and Materials, Philadelphia, 1985, pp. 91–109.

ABSTRACT: Fracture mechanics parameter and energy absorption capacity of fiber-polymer cement concrete are experimentally evaluated and then K_{Ic}, J_{Ic}, toughness index, modulus of toughness, impact toughness, and bending toughness are examined by means of an analysis of factorial experiments.

Five factors—namely, steel fiber geometry, volume fraction of fiber, unit polymer content, water-cement ratio, and relative notch depth—are assumed to affect these fracture mechanics parameters and toughness parameters.

Experimental scatter of the test method of these fracture parameters is discussed in order to determine the testing method of toughness property of the fiber-polymer cement concrete composite. The confidence interval and significant factors effect are also calculated. Finally, the usefulness and the limitation of fracture mechanics for the application of the cementitious concrete design are discussed.

KEY WORDS: composite materials, fiber reinforced polymer cement concrete, metal fibers, polymer dispersions, fracture (materials), toughness, statistical analysis, design of experiments, orthogonal arrays

Two of the inherent weaknesses of cement concrete materials are their low tensile strength and poor deformation capacity. Embedding fibers into a cement matrix, for example, steel fibers, has proved successful in improving its deformation characteristics [1,2]. The addition of polymer dispersion in the cement matrix, on the other hand, has been recognized to improve the flowability (workability) of the matrix, its ductility, cracking strain, and the strength of the interfacial bond [3,4]. This implies that polymer dispersion plays a role in overcoming the difficulty of low elongation at break and possibly of the low fracture toughness of the cement matrix [5].

Kaplan [6] was the first to investigate the fracture toughness of concrete,

[1]Associate professor, Department of Civil Engineering, Hokkaido Institute of Technology, Sapporo, Japan.

using a flexural beam notched on the tension face, and confirmed the applicability of linear fracture mechanics to cement concrete specimens. Since his work, many studies on this subject have been conducted [7–10]. Recently, the J-integral formulated by Rice [11] has been proposed as a failure criterion for elastic-plastic behavior of fiber reinforced concrete by Mindess et al [12]. They concluded that J_{Ic} is a promising fracture criterion for the fiber reinforced concrete, while K_{Ic} (or G_{Ic}) is not adequate except for the cement paste. On the other hand, the energy absorption capacities calculated from the load-deflection curve have been proposed as a toughness parameter. Toughness index (TI), which is one of the tests recommended by American Concrete Institute (ACI) Committee 544 on fiber reinforced concrete, has been proposed to determine the energy absorption capacity of a fibrous concrete [13]. It is calculated by dividing the area under the load-deflection curve out to a center deflection of 1.9 mm (0.075 in.) by the area under the curve up to the first crack strength. This first crack strength is, however, difficult to measure accurately and thus significant experimental scatter arises [14]. In addition to the above, toughness parameters such as flexure toughness [15] and modulus of toughness [16] have been proposed to describe the characteristics of the fracture toughness.

These parameters concerned with fracture toughness or toughness properties are well known for describing the effectiveness of fiber addition to concrete, and the current literature contains a considerable number of reports on this subject. However, with regard to the statistical approach to the problem, little work has been done in the field of cementitious concrete materials.

According to Mindess et al [12], the value of J_{Ic} ranges from 109.3 to 524.5 N/m (0.624 to 2.995 lb/in.) for identical mixtures. This experimental scatter is due mainly to the fiber orientation, since it is very difficult to achieve a truly random fiber dispersal in the substance. Although the effects of short fiber orientation on fracture behavior have been investigated [17–19], it is still not clear how large the scatter is.

In the present study, an experimental design theory developed by Fisher [20] is applied to evaluate the fiber efficiency, and then to verify the applicability of the fracture parameters together with bending toughness (T_b) and impact toughness (T_p) on steel fiber reinforced polymer cement concrete (SFRPCC).

Design of Experiment

The response of the SFRPCC is affected by many factors, that is, fiber orientation, volume fraction of fiber, and unit polymer content, and the interaction among these factors. Where a number of factors are variables, in the traditional method, one factor is taken as a variable under the condition where the other factors are constant and the response due to this factor is read for several levels. Then next factor is taken as a variable. This process is repeated until all factors of interest have been dealt with. In spite of the large amount of testing that may be required with this approach, it may be quite inadequate if interactions among these factors occur—that is, if the response of a dependent variable to one factor is affected by the levels which other factors influence.

To overcome this difficulty, factorial experiments are conducted to investigate the effects of various factors simultaneously rather than conducting a series of single-factor experiments. The interpretation of the results of a factorial experiment depends upon the statistical variance ratio test of significance (analysis of variance). The statistical variance ratio F_A for the factor A is

$$F_A = \left(\frac{S_A}{\phi_A}\right)\Bigg/\left(\frac{S_e}{\phi_e}\right)$$
$$= V_A/V_e \tag{1}$$

where S, ϕ, and V are the sums of squares, the degrees of freedom, and the mean square sums, respectively, and the suffixes A and e represent factor A and error source, respectively.

If F_A is less than $100 - \alpha\%$ point of the F distribution [that is, $F_A < F(\phi_A, \phi_e, \alpha)$], the effect of the factor A on the experimental results is not significant. However, if F_A exceeds the $100 - \alpha\%$ point of the F distribution, the effect of the factor A is judged significant. F-values for the 5% and 1% right tail points are used from the table of Snedecor.

When there exists a significant factor, the confidence interval (confidence limit) CL and the factor effects ρ are calculated using Taguchi's formulas [21]

$$\mathrm{CL} = \pm \sqrt{\frac{F(\phi_A, \phi_e, \alpha)V_e}{ne}} \tag{2}$$

$$\rho = \frac{S_A - \phi_A V_e}{S_T} \tag{3}$$

where ne is the repetition times of the experiment, and S_T is the total sum of squares.

The variance ratio of the interaction between the two factors A and B is also tested by using F_{AB} and $F(\phi_{AB}, \phi_e, \alpha)$

$$F_{AB} = \left(\frac{S_{AB}}{\phi_{AB}}\right)\Bigg/\left(\frac{S_e}{\phi_e}\right)$$

$$\phi_{AB} = (\phi_A - 1)(\phi_B - 1) \tag{4}$$

In this study, basically 16 units of experiment were conducted for four factors each at two levels, which is described by L_{16} orthogonal arrays (2^4) [22]. As for the fracture toughness, an additional factor with three levels was incorporated in the experiment (that is, L_{48} orthogonal arrays $2^4 \times 3$). These factors are steel fiber geometry (form of fiber), volume fraction of fiber, unit polymer content, water-cement ratio, and relative notch depth. The levels used for each factor are given in Table 1 and the corresponding complete orthogonal arrays of the factorial design in Table 2.

Experimental Details

Materials

Two different types of steel fiber, a round straight steel and a crimped steel 30 mm (1.2 in.) in length and 0.5 mm (0.02 in.) in diameter, were used in 1% and 1.5% by volume. The polymer in the Ethyl acrylate emulsion was dispersed in water in the form of very small spherical particles about 0.1 μm in diameter. Ordinary portland cement was used in all the specimens. Washed and dried river sand and coarse aggregate gravel, mostly round-shaped of 10 mm (0.4 in.) maximum size, were used.

Preparation of Specimens

A mixer with a flexible rubber bowl was used in order to achieve optimum homogeneity and to minimize the amount of the void. The polymer dispersion was mixed with water and added after all other materials had been mixed. The batching procedure is as follows:

1. Aggregates and cement are mixed for 30 s.
2. Fibers are dumped in and then mixed for 60 s.
3. Polymer dispersion with added water is dumped in and then mixed for 90 s.
4. Setting 30 s.
5. Mixing 30 s.

TABLE 1—Code for factors and levels.

Factor Source	Factor Code	Level Code Number		
		1	2	3
Form of steel fiber	Kf	crimped	straight	...
Volume fraction of fiber	Vf	1%	1.5%	...
Polymer content	P	0	50 kg/m³	...
Water-cement content	W	40%	45%	...
Relative notch depth	Nd	0.15	0.24	0.33

TABLE 2—L_{16} factorial design.

Test Series	Factor Code				
	K_f	V_f	W	p	N_d
1	1	1	1	1	123
2	1	1	1	2	123
3	1	1	2	1	123
4	1	1	2	2	123
5	1	2	1	1	123
6	1	2	1	2	123
7	1	2	2	1	123
8	1	2	2	2	123
9	2	1	1	1	123
10	2	1	1	2	123
11	2	1	2	1	123
12	2	1	2	2	123
13	2	2	1	1	123
14	2	2	1	2	123
15	2	2	2	1	123
16	2	2	2	2	123
pcc[a]	...	...	1	2	123
cc[b]	...	...	1	1	123

[a]Polymer modified cement concrete.
[b]Conventional cement concrete.

Eighteen types of mixtures (units) were prepared in this study. For each mixture, 15 specimens of size 75 by 75 by 400 mm (3 by 3 by 16 in.) were cast, and for the fracture toughness specimens a groove was cut at midspan with a 0.4-mm-wide (0.016 in.) diamond saw, one day prior to testing. The notch depth ranged from 11 to 25 mm (0.43 to 1.0 in.). Three specimens were prepared for each notch depth. The flexure specimens were tested in third-point loading [ASTM Test Method for Flexural Strength of Concrete (Using Simple Beam with Third-Point Loading) (C 78-75)] as shown in Fig. 1.

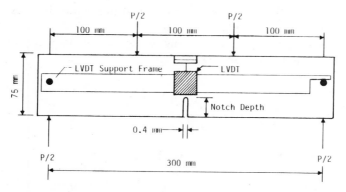

FIG. 1—Center notched beam tested in four-point bending.

Testing Procedure

All tests were conducted at a specimen age of 28 days. The experiment was designed to investigate the effects of the factors on the following test results:

fiber efficiency (average fiber spacing S),
fracture toughness (K_{Ic}),
J-integral (J_{Ic}),
toughness index (TI),
modulus of toughness (MOT),
bending toughness (T_b), and
impact toughness (T_p).

First, the fiber efficiency was examined in order to estimate the degree of fiber dispersion, and then compared with theoretical calculations. The total number of fibers in the area of tensile cross section N_f was counted at the fracture surface of the test specimens. The total number of fibers passing the unit area, n_f, is

$$n_f = N_f/(100 \times 50) \qquad \text{(fib/mm}^2\text{)} \qquad (5)$$

If mutual placing of the fiber faces in the section is supposed to be (1) square pattern, (2) triangular pattern, and (3) hexagonal pattern, the average fiber spacings S_{sq}, S_{tri}, and S_{hex}, respectively, are

$$S_{sq} = \sqrt{\pi d_f^2/(4\beta V_f)}$$
$$= \sqrt{1/n_f} \qquad (6)$$

$$S_{tri} = 1.07 \, S_{sq} \qquad (7)$$

$$S_{hex} = 0.88 \, S_{sq} \qquad (8)$$

where d_f and β are the diameter and efficiency factor of the fiber, respectively, and V_f is the volume fraction of the fiber.

For a specimen in third-point loading, K_{Ic} was calculated using the expression of Brown and Srawley [23].

$$K_{Ic} = \frac{6M}{Bd} \sqrt{a} \, F(a/d)$$
$$M = Pl/2 \qquad (9)$$

where $F(a/d)$ is the finite width correction factor given by

$$F(a/d) = 1.99 - 2.47 \, (a/d) + 12.97 \, (a/d)^2 - 23.17 \, (a/d)^3 + 24.80 \, (a/d)^4 \qquad (10)$$

where

a = crack length,
d = specimen depth,
M = ultimate bending moment,
P = applied load,
l = shear span, and
B = specimen width.

J_{Ic} was calculated using the expression of Mindess et al [12]

$$J_{Ic} = \frac{2}{B(d - a)} \int_0^{\delta_c} P d\delta_c \qquad (11)$$

where δ_c is the vertical displacement of the point of maximum load.

The definition of the toughness parameters is graphically demonstrated in Fig. 2. Bending toughness (T_b) was measured from load-deflection curves of both unnotched and notched specimens.

The impact toughness (T_p) testing procedure [5] is shown in Fig. 3. As

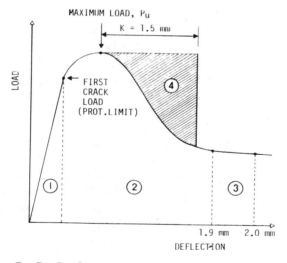

①,②,③,④: Areas under the curve up to each limit

$$TI = \frac{① + ②}{①}$$

$$MOT = -④ \frac{100}{Pu \cdot K} + 200$$

$$Tb = ① + ② + ③$$

FIG. 2—*Calculation of toughness parameter.*

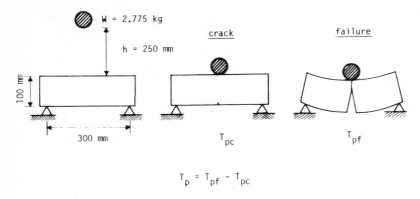

$$T_p = T_{pf} - T_{pc}$$

FIG. 3—*Impact toughness test procedure.*

shown in the figure, the test was performed on a smooth rigid floor to mini-mize the energy loss due to bouncing. A hardened steel ball 87.5 mm (3.4 in.) in diameter was dropped repeatedly, and the number of blows to cause the first visible crack was counted. Ultimate failure occurs when the cracks have opened sufficiently to make the specimen touch the baseplate. Impact tough-ness was calculated from the number of blows at the ultimate failure stage minus the number of blows at the first visible crack stage (that is, $T_p = T_{pf} - T_{pc}$).

Test Results and Discussion

Fiber Efficiency

The analysis of variance for N_f data is shown in Table 3. The effect of vol-ume fraction of fiber is highly significant and the factor effect is 55%. The $N_f =$ values increase about 50% in addition to 0.5% fiber content as shown in Fig. 4a. Theoretical calculations based on geometric theory [24–26], the ste-reological approach [27], and X-ray analysis [28] were compared with the ex-perimental average fiber spacings S_{sq}, S_{tri}, and S_{hex} shown in Fig. 4b. If the triangular pattern is used for the fiber spacing model, the experimental results show good accordance with the geometric theory. On the contrary, the experimental results and the stereological fiber spacing show good agreement when the model is supposed to be the hexagonal placing pattern. In most practical cases, however, it will be very difficult to predict the fiber pattern in the section, and consequently it will be impossible to determine the exact av-erage fiber spacing. Statistically, the volume fraction of fiber affects the fiber efficiency strongly (significant at 99% confidence level), and the factor effect reaches 55%; that is, 45% of the fiber efficiency is affected by unknown fac-tors or experimental error. The confidence interval of the fiber efficiency af-

TABLE 3—*Analysis of variance for counting of the fiber efficiency.*

Factor Code (Source)	Degree of Freedom	Mean Sum of Squares	Variance Ratio, F	Percentage of Factor Effect
K_f	1	. . .	. . .	. . .
V_f	1	6456.12	19.41[a]	55.1
p	1	. . .	. . .	. . .
W	1	. . .	. . .	. . .
$K_f \times V_f$	1	. . .	. . .	. . .
$V_f \times P$	1	. . .	. . .	. . .
$K_f \times P$	1	. . .	. . .	. . .
$K_f \times W$	1	. . .	. . .	. . .
Error	7	4656.68	. . .	44.9
Total	15	11112.80		100.0

[a]Significant at 1% level.

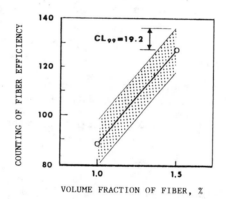

FIG. 4a—*Counting of fiber efficiency versus volume fraction of fiber.*

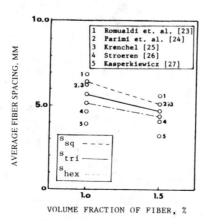

FIG. 4b—*Average fiber spacing versus volume fraction of fiber.*

fected by volume fraction of fiber was $\pm 22\%$ at 1.0% volume fraction of fiber and $\pm 15\%$ at 1.5%.

Fracture Toughness

The analysis of variance for fracture toughness (K_{Ic}) data is shown in Table 4. Three factors were highly significant at the 1% level and there was an interaction between the relative notch depth and the fiber volume fraction. The K_{Ic}-values increased in proportion to the relative notch depth and to the volume fraction of fiber, and decreased in proportion to the water-cement ratio (see Fig. 5). Figure 6 shows the interaction between the two factors. As the volume fraction of fiber increases, the rate of K_{Ic} increases quite rapidly according to the relative notch depth increase. This means that these two factors, volume fraction of fiber and relative notch depth, are not *independent* factors which affect the fracture toughness. The more the volume fraction of fiber increases, the more the fracture toughness of the SFRPCC is sensitive to the notch depth.

Toughness Parameters

A summary of the variance ratio for the toughness parameter results is given in Table 5. The statistical analysis yields the following results:

1. As for J_{Ic} data, the volume fraction of fiber is the only factor that is significant at the 5% level, and the 95% confidence interval is ± 362 N/m

TABLE 4—*Analysis of variance for* K_{Ic} *data.*

Factor Code (Source)	Degree of Freedom	Mean Sum of Squares	Variance Ratio, F	Percentage of Factor Effect
K_f	1	...	...	...
V_f	1	1.02	20.4^a	24.7
p	1	0.10	2.0	...
W	1	0.42	8.4^a	9.9
N_d	2	0.39	8.0^a	16.5
$N_d \times V_f$	2	0.30	6.0^b	12.3
$N_d \times P$	2	0.18	3.6	...
$V_f \times K_f$	1	0.02	0.4	...
$V_f \times P$	1	0.01	0.2	...
$K_f \times P$	1	0.02	0.4	...
$K_f \times W$	1	...	...	...
Error	33	0.05	...	36.6
Total	47	4.05		100.0

^{a}Significant at 1% level.
^{b}Significant at 5% level.

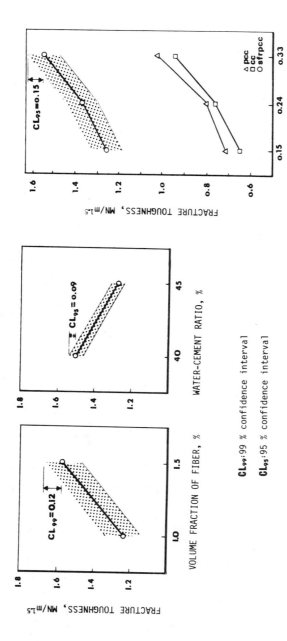

FIG. 5—*Fracture toughness main effects.*

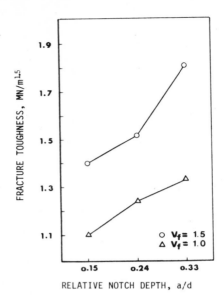

FIG. 6—*Volume fraction of fiber and relative notch depth interaction for fracture toughness.*

(2.07 lb/in.), which is about ±80% of the J_{Ic} mean value at 1% fiber volume fraction (Fig. 7). This interval explains why the aforementioned remarkable experimental scatter of J_{Ic}, reported by Mindess et al, occurs. The result implies that the *J*-integral is not an adequate test method for estimating the fracture toughness properties of SFRPCC.

2. There is neither significant factor nor interaction in the toughness index. This means that the effect of the factors on the toughness index is more feeble than that of experimental scatter.

3. Unit polymer content effect is significant at 5% level in modulus of toughness and the confidence interval is ±16.3 (Fig. 7). This toughness parameter is for estimating a debonding characteristics of SFRPCC after the maximum load (see Fig. 2). The experimental results show the improvement of interfacial bond strength by adding polymer dispersion. However, this test method shows remarkable experimental scatter as in Table 5.

4. Although there is neither significant factor nor interaction in bending toughness of unnotched specimens, for the specimens with the 11-mm (0.44 in.) notch, three significant factors and an interaction are recognized for bending toughness response. These factors are volume fraction of fiber, unit polymer content, and water-cement ratio (Fig. 8). The interaction is analyzed between the form of fiber and the volume fraction of fiber. Results similar to the 11-mm (0.44 in.) notched bending toughness case are also obtained for 18-mm (0.72 in.) notched bending toughness (Fig. 9). (See also Fig. 10.)

TABLE 5—*Summary of testing the variance ratio, F.*

Factor Code	Toughness Parameters							
	J_{Ic}	TI	MOT	$T_b(0)^a$	$T_b(1)^b$	$T_b(2)^c$	$T_b(3)^d$	T_p
K_f	...	...	...	...	...	...	2.76	...
V_f	6.56[e]	...	3.56	2.85	14.44[f]	31.01[f]	42.26[f]	30.23[f]
P	...	3.01	5.07[e]	2.01	4.88[e]	8.94[e]	...	11.41[f]
W	1.77	...	...	1.52	10.94[f]	7.49[e]	3.08	1.94
$K_f \times V_f$	...	...	...	...	7.63[e]	5.81[e]	...	...
$V_f \times P$	2.34	...	...	2.79	...	...	1.69	9.69[f]
$K_f \times P$	...	...	...	1.55	...	1.66	...	...
$K_f \times W$	...	...	...	...	...	...	...	...
Error	66.16	88.19	69.34	72.41	30.68	23.11	24.68	23.34

[a]Unnotched specimen.
[b]11-mm (0.44 in.) notched specimen.
[c]18-mm (0.72 in.) notched specimen.
[d]25-mm (1 in.) notched specimen.
[e]Significant at 5% level.
[f]Significant at 1% level.

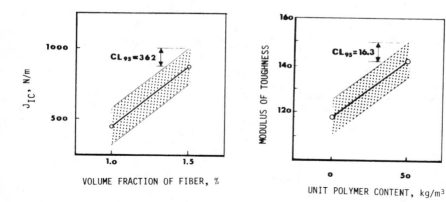

FIG. 7—J_{Ic} *and MOT main effects.*

The volume fraction of fiber is highly significant and the unit polymer content is significant as well as the water-cement ratio. When the notch depth is up to 25 mm (1 in.), bending toughness is affected strongly by the volume fraction of fiber (Fig. 11).

5. Both the volume fraction of fiber and the unit polymer content are significant in impact toughness (Fig. 12 and Table 5). There exists an interaction between the volume fraction of fiber and the unit polymer content (Table 5). It seems that impact toughness is a promising toughness parameter for estimating the dynamic toughness property of SFRPCC not only because of

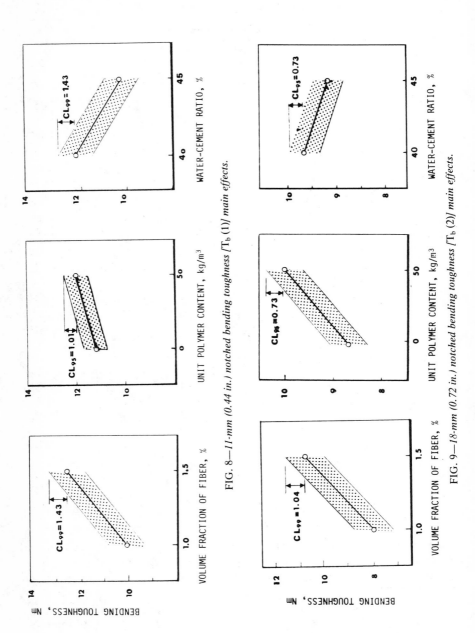

FIG. 8—*11-mm (0.44 in.) notched bending toughness [T_b (1)] main effects.*

FIG. 9—*18-mm (0.72 in.) notched bending toughness [T_b (2)] main effects.*

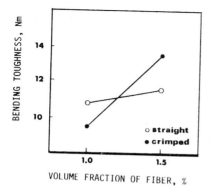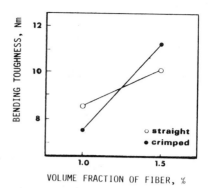

FIG. 10—*Interactions for 11- and 18-mm (0.44 and 0.72 in.) notched bending toughness* [T_b (1), T_b (2)].

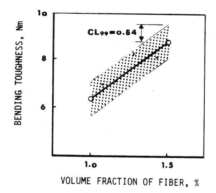

FIG. 11—*Effect of volume fraction of fiber on 25-mm (1 in.) notched bending toughness* [T_b (3)].

the smaller experimental scatter but also because of the significant effect of the two aforementioned factors.

6. Experimental error is calculated and is given in Table 5. Figure 13 shows the relationship between the experimental error and toughness parameters diagrammatically. This error varies between 22% and 88%. It is quite evident that these toughness parameters are affected strongly by the experimental scatter of fiber efficiency, which has been discussed previously, since the toughness of the SFRPCC depends on the debonding and post-debonding friction characteristics between the matrix and the fiber at the post-cracking stage. However, from Fig. 13 it is clear that these methods of toughness testing are classified by two groups which are concerned in the experimental error. The methods of J_{Ic}, TI, MOT, and unnotched bending toughness are

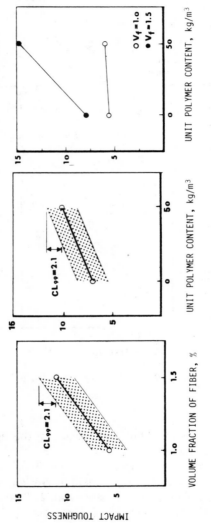

FIG. 12—*Impact toughness main effects and interaction.*

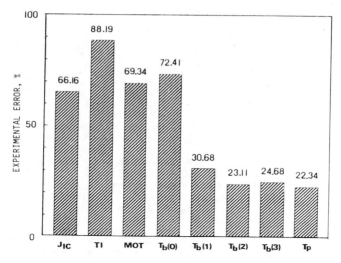

FIG. 13—*Evaluation of statistical scatter versus toughness parameters.*

enumerated as one group whose experimental errors are directly affected by the experimental scatter of the fiber efficiency. On the contrary, the other group, such as the method of notched bending toughness and impact toughness, is mitigated by the experimental scatter due to the effect of fiber efficiency. These findings support the view that the applicability of fracture mechanics to the SFRPCC is quite limited.

Conclusions

Based on the analysis of variance, the following conclusions may be drawn:

1. It is quite evident that fiber efficiency is strongly affected by the volume fraction of the fiber. It should be noted, however, that fiber efficiency has considerable statistical scatter (see Fig. 4).

2. The fracture toughness of SFRPCC is strongly notch-sensitive. This implies that linear elastic fracture mechanics is useless for this type of material.

3. The test result of the toughness parameter indicates that the statistical interval of each parameter should be considered not only for estimation of toughness property but also for using the value for the purpose of design.

4. Impact toughness seems to be a promising toughness parameter for estimating the dynamic toughness property of SFRPCC.

5. The presence of interaction among the factors implies that the toughness of SFRPCC is not always the linear function of the volume fraction of fiber and of the unit polymer content.

References

[1] Hughes, B. P. and Hattuhi, N. I., "Load-Deflection Curves for Fibre-Reinforced Concrete Beams in Flexure," *Magazine of Concrete Research*, Vol. 30, No. 105, Dec. 1978, pp. 199–228.

[2] Swamy, R. N. and Al-Ta'an, S. A., "Deformation and Ultimate Strength in Flexure of Reinforced Concrete Beams Made with Steel Fiber Concrete," ACI *American Concrete Institute Journal, Proceedings*, Vol. 78, No. 36, Sept. 1981, pp. 395–405.

[3] Mangat, P. S., "Strength and Deformation Characteristics of an Acrylic Polymer-Cement Composite," *Materials and Structures*, Vol. 11, No. 66, Nov.-Dec. 1978, pp. 435–443.

[4] Okada, K. and Ohama, Y., "Status of Concrete-Polymer Composites in Japan" in *Proceedings*, Third International Congress on Polymers in Concrete, Koriyama, Japan, May 1981, pp. 3–19.

[5] Horiguchi, T. and Paillère, A. M., "Amelioration des bétons armés de fibre d'acier par l'addition des polymer divers" *Proceedings*, Réunion International des Laboratoires d'Essais et de Recherches sur les Materiaux et les Constructions Symposium on Future for Plastics in Building and Civil Engineering, Liège, Belgium, June 1984.

[6] Kaplan, M. F., "Crack Propagation and the Fracture of Concrete," *American Concrete Institute Journal, Proceedings*, Vol. 58, No. 5, Nov. 1961, pp. 591–610.

[7] Naus, D. J. and Lott, J. L., "Fracture Toughness of Portland Cement Concrete," *American Concrete Institute Journal, Proceedings*, Vol. 66, No. 39, June 1969, pp. 481–489.

[8] Gjørv, O. E., Sørensen, S. I., and Arnesen, A., "Notch Sensitivity and Fracture Toughness of Concrete," *Cement and Concrete Research*, Vol. 7, No. 3, May 1977, pp. 333–344.

[9] Cook, D. J. and Crookham, G. D., "Fracture Toughness Measurements of Polymer Concrete," *Magazine of Concrete Research*, Vol. 30, No. 105, Dec. 1978, pp. 205–214.

[10] Petersson, P. E., "Crack Growth and Development of Fracture Zones in Plain Concrete and Similar Materials," Report TVBM-1006, Lund Institute of Technology, Sweden, 1981.

[11] Rice, J. R., "A Path Independent Integral and the Approximate Analysis of Strain Concentration by Notches and Cracks," *Journal of Applied Mechanics, Transactions*, American Society of Mechanical Engineers, June 1968, pp. 379–386.

[12] Mindess, S., Lawrence, F. V., and Kesler, C. E., "The J-Integral as a Fracture Criterion for Fiber Reinforced Concrete," *Cement and Concrete Research*, Vol. 7, No. 6, Nov. 1977, pp. 731–742.

[13] Henager, C. H., "A Toughness Index of Fiber Concrete" in *Proceedings*, Réunion International des Laboratoires d'Essais et de Recherches sur les Materiaux et les Constructions Symposium, 1978, pp. 79–86.

[14] Horiguchi, T., Ando, T., and Kurose, T. in "Toughness Property of Fiber-Polymer Cement Concrete" in *Proceedings*, Japan Society of Civil Engineers, Hokkaido, No. 39, 1983, pp. 347–352 (in Japanese).

[15] Shah, S. P. and Rangan, B. V., "Fiber Reinforced Concrete Properties," *American Concrete Institute Journal, Proceedings*, Vol. 68, No. 14, Feb. 1971, pp. 126–135.

[16] Umeyama, K., Okamura, Y., and Kobayashi, K. in "Method of Evaluating Flexural Toughness of Steel Fiber Reinforced Concrete" in *Proceedings*, Japan Concrete Institute Second Meeting, May 1981, pp. 201–204 (in Japanese).

[17] Hannant, D. J. and Spring, N., "Steel-Fibre-Reinforced Mortar: A Technique For Producing Composites with Uniaxial Fibre Alignment," *Magazine of Concrete Research*, Vol. 26, No. 86, Mar. 1974, pp. 47–48.

[18] Strøeven, P., "Morphometry of Fibre Reinforced Cementitious Materials: Part II," *Materials and Structures*, Vol. 12, No. 67, Jan.-Feb. 1979, pp. 9–20.

[19] Johnston, C. D., "Steel Fibre Reinforced and Plain Concrete: Factors Influencing Flexural Strength Measurement," *American Concrete Institute Journal, Proceedings*, Vol. 79, No. 14, Mar.-April 1982, pp. 131–138.

[20] Fisher, R., *The Design of Experiments*, Oliver and Boyd, London, 8th ed., 1960.

[21] Taguchi, G., *The Design of Experiments*, Maruzen, Japan, 4th Ed., 1962.

[22] Isobe, K., *How to Use the Orthogonal Arrays*, Japan Standards Association, 1979.

[23] Brown, W. F. and Srawley, J. E. in *Plane Strain Crack Toughness Testing of High-Strength Metallic Materials, ASTM STP 410*, American Society for Testing and Materials, Philadelphia, 1966, p. 11.

[24] Romualdi, J. P. and Mandel, J. A., "Tensile Strength of Concrete Affected by Uniformly Distributed and Closely Spaced Short Lengths of Wire Reinforcement," *American Concrete Institute Journal, Proceedings*, Vol. 61, No. 6, June 1964, pp. 657–670.

[25] Parimi, S. R. and Shridhar Rao, J. K. in "Effectiveness of Random Fibres in Fibre Reinforced Concrete," *Proceedings*, International Conference on Mechanical Behaviour of Materials, Kyoto, Japan, 1971, p. 813.

[26] Krenchel, H., "Fibre Spacing and Specific Fibre Surface" in *Proceedings*, Réunion International des Laboratoires d'Essais et de Recherches sur les Materiaux et les Constructions Symposium on Fibre Reinforced Cement and Concrete, 1975, pp. 69–79.

[27] Strøeven, P., "Morphometry of Fibre Reinforced Cementitious Materials: Part I," *Materials and Structures*, Vol. 11, No. 61, 1978, pp. 31–38.

[28] Kasperkiewics, J., "Fibre Spacing in Steel Fibre Reinforced Composites," *Materials and Structures*, Vol. 10, No. 55, 1977, p. 25.

Hiroichi Ohira[1] and Nobuhide Uda[1]

Knee Point and Post-Failure Behavior of Some Laminated Composites Subjected to Uniaxial Tension

REFERENCE: Ohira, H. and Uda, N., **"Knee Point and Post-Failure Behavior of Some Laminated Composites Subjected to Uniaxial Tension,"** *Recent Advances in Composites in the United States and Japan, ASTM STP 864*, J. R. Vinson and M. Taya, Eds., American Society for Testing and Materials, Philadelphia, 1985, pp. 110–127.

ABSTRACT: To investigate the phenomenon of initial failure, the post-failure behavior and the behavior under unloading and reloading, uniaxial tension tests were conducted on three kinds of laminated carbon fiber reinforced plastic coupon specimens, that is, ±30-deg, 90-deg laminates, ±45-deg, 90-deg laminates and 0-deg, 90-deg laminates. The test results are compared with the authors' theory. In the experimental stress-strain diagram, the notable incremental strain at a constant stress level was observed at the knee point, that is, the point of initial failure. This indicates that the assumption of the brittle matrix of the composite is adequate for the specimens tested, and the assumption of the ductile matrix is invalid. An inverse method is presented which makes it possible to estimate the failure stresses of constituent layers and also the effective curing temperature of an individual laminate from the experimental stress-strain diagram. The experiment showed that the unloading path is in agreement with one of two possible theoretical predictions. The reloading path coincided with the unloading path.

KEY WORDS: composite materials, fiber reinforced composites, carbon fiber reinforced plastics, tension tests, stress-strain diagrams, stress analysis, fracture mechanics, fracture strength, unloading, brittleness

The object of the present investigation is to attain a better understanding of the phenomenon of the initial failure and the post-failure behavior of laminated composite plates subjected to in-plane loads.

Tsai et al, in their pioneering work [1], showed theoretically and experimentally that the stress-strain diagram of a cross-ply composite under tension

---

[1] Professor and research assistant, respectively, Department of Aeronautical Engineering, Kyushu University, Fukuoka 812, Japan.

is represented by a bilinear line. The intersection of the two straight segments implies the initial failure, and the intersection was called the knee point. Their composite was glass fiber reinforced plastics (GFRP).

The present authors, in previous papers [2,3], have presented a theory concerning the strength of the composite plates of general lamination subjected to in-plane loads. A possible approach has been shown to analyze the post-failure state and also the case of unloading.

The present work has been undertaken to verify experimentally the authors' theory. The assumptions of the ductile matrix or the brittle matrix and the assumptions involved in estimating the unloading paths are examined by comparing with the results of uniaxial tension tests conducted on three kinds of laminated coupon specimens made of carbon fiber reinforced plastics (CFRP).

Theoretical Predictions

A brief account of the theory is given in the Appendix.

The x,y are the reference axes. Associated with each constituent layer of the laminate, the axes of orthotropy L, T are taken: L along the fiber and T transverse to it. When referred to L-T axes, the stresses in layer i are $\sigma_L^{(i)}$, $\sigma_T^{(i)}$, and $\tau_{LT}^{(i)}$. Their critical values corresponding to failure are denoted by $F_L^{(i)}$, $F_T^{(i)}$ and $F_{LT}^{(i)}$, respectively. Furthermore, $F_{Lt}^{(i)}$ means F_L in tension and $F_{Lc}^{(i)}$ in compression, so do $F_{Tt}^{(i)}$ and $F_{Tc}^{(i)}$. The theory considers two cases for the property of the matrix of the composite, that is, the ductile matrix and the brittle matrix. The ductile matrix assumes an elastic and perfectly plastic property for $\sigma_T^{(i)}$ versus $\epsilon_T^{(i)}$ or $\tau_{LT}^{(i)}$ versus $\gamma_{LT}^{(i)}$ with the critical stresses $F_T^{(i)}$ or $F_{LT}^{(i)}$, respectively. The brittle matrix assumes that, after reaching the critical value, the stress drops to zero.

Figure 1 shows the theoretical results for a CFRP cross-ply subjected to uniaxial tension σ_x. This is the same problem as that of Tsai [1], except that the GFRP of Tsai is now replaced by CFRP. As regards the abscissa, ϵ_x is the strain measured from the stress-free hot state in the autoclave. In the load-free initial state at room temperature, the strain is ϵ_{x0} where the strain in the ordinary sense ($\epsilon_x - \epsilon_{x0}$) is zero. The lamination angle $\theta^{(i)}$ and the thickness ratios $h^{(i)}$ of constituent layers are as indicated in the figure. The material constants used are the ones to be described later in the section "Material Constants." The theory yields two branches of loading path, one for the ductile matrix and the other for the brittle matrix. The initial failure occurs at the knee point in the $F_{Tt}^{(2)}$ mode, that is, the tensile F_T failure in the 90-deg layer. Along the ductile branch, $F_{Tt}^{(1)}$ failure occurs on the way. Finally, the fracture occurs in the $F_{Lt}^{(1)}$ mode, that is, the fibers break. This ductile branch is approximately equivalent to the diagram of Tsai for GFRP. In the case of the brittle branch, we note the considerable increase in strain at a constant stress immediately after the knee. We call this incremental strain the "slip." Due to

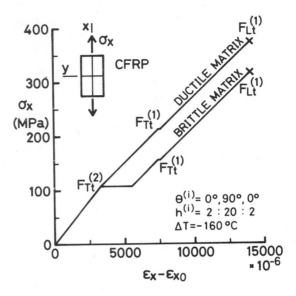

FIG. 1—*CFRP cross-ply (theory)*.

$F_{Tt}^{(2)}$ failure, the 90-deg layer cannot carry load any more and a redistribution of stresses occurs which changes the overall strains of the composite. This is the cause of the slip. When the slip is terminated, the load can be increased again. Two branches are parallel to each other. The abscissas of the failure points of $F_{Tt}^{(1)}$ on both branches are identical, and the same applies also for the fracture points of $F_{Lt}^{(1)}$. Figure 2 shows the theoretical results for a CFRP coupon specimen of a $\pm$30-deg, 90-deg laminate. This laminate was chosen to examine the effect of replacing the 0-deg layers of the cross-ply by angle plys. The diagram is sensitive to the magnitudes of temperature and the failure stresses. Figure 2 illustrates the case when the temperature difference ΔT is one half of the nominal value. For this coupon specimen, the final fracture is in the $F_{Tt}^{(1)}$ mode. For simplicity's sake, only $F_{Tt}^{(1)}$ and $F_{Lt}^{(1)}$ failures are indicated in Figs. 1 and 2. It is to be understood that what occurs in layer 1 occurs also in other layers of the same absolute lamination angles. Figure 3 shows schematically the unloading paths after the load exceeded the initial failure load. Two unloading paths designated by U1 are parallel to the initial elastic line. Paths U1 have been explained in the previous paper [2], and are obtained by assuming that the failed layer regains the elastic property. The path designated by U2 associated with the brittle branch is the new one obtained in the present investigation by assuming the failed layer does not recover from damage. Path U2 is parallel to the second loading line of the brittle branch, as shown.

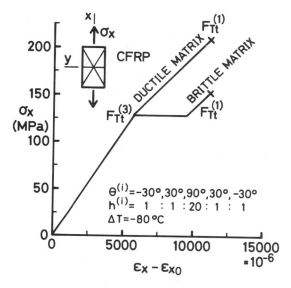

FIG. 2—*CFRP ±30-deg, 90-deg laminate (theory).*

Experimental Procedure

The CFRP coupon specimens used were 0-deg specimens, 90-deg specimens, ±45-deg specimens, ±30-deg specimens, ±30-deg, 90-deg laminates, ±45-deg, 90-deg laminates, and 0-deg, 90-deg laminates. The specimens were manufactured by the Toray Co. They were made by laminating graphite/epoxy prepregs. The prepreg (Toray P3060-15) was made of carbon fibers (Torayca T300, high-strength type) and epoxy resin (Toray 3601, for aircraft use). The nominal thickness of one prepreg was 0.15 mm. Specimens were cured in an autoclave. Tension tests were performed on a 10-ton testing machine (Instron 1125). Several strain gages were bonded on each specimen. The gages were connected to the data acquisition system consisting of a 20-channel scanner, an analog/digital converter and a microcomputer. Details of the experimental procedure and specimen dimensions have been reported previously [4–6].

Material Constants

The material constants for our CFRP were obtained by conducting tension tests on 0-deg specimens, 90-deg specimens, and ±45-deg specimens. Tests were made in accord with ASTM Test Method for Tensile Properties of Fiber-Resin Composites (D 3039) and ASTM Recommended Practice for In-Plane

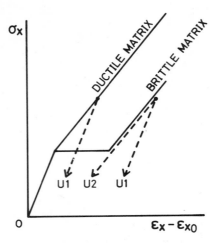

FIG. 3—*Unloading paths (theory).*

Shear Stress-Strain Response of Unidirectional Reinforced Plastics (D 3518-76), and the results have been reported previously [4–6]. The mean values of the test results were rounded off to give the following basic values: elasticity constants $E_L = 150$ GPa, $E_T = 9.10$ GPa, $G_{LT} = 4.60$ GPa, $\nu_L = 0.320$, $\nu_T = \nu_L E_T/E_L$; failure stresses $F_{Lt} = 1.90$ GPa, $F_{Tt} = 68.0$ MPa, $F_{LT} = 90.0$ MPa, $F_{Lc} = -1.90$ GPa(**), $F_{Tc} = -490$ MPa(**); thermal constants $\alpha_L = 0.2 \times 10^{-6}$ 1/°C(*), $\alpha_T = 35 \times 10^{-6}$ 1/°C(*), $\Delta T = -160$°C(*); fiber volume fraction $V_f = 60\%$(*), where E_L, E_T = Young's moduli, G_{LT} = shear modulus, ν_L, ν_T = Poisson's ratios, α_L, α_T = coefficients of linear expansion, and ΔT = temperature difference between room temperature and curing temperature. The values attached with * are taken from data of the Toray Co., and those with ** are estimated values. The latter values are not critical for the present case.

Experimental Results for Laminates

Tension tests were performed on three kinds of symmetric laminates. They were made in the shape of coupons that were 25.4 mm wide and 290 mm long (test section length 160 mm). The thickness ratio of the 90-deg layer of each set of laminates was designed to be relatively large, so that the initial failures would occur in the F_{Tt} mode and such failure phenomena would be experimentally observable. Several strain gages were bonded on each specimen. Short and long gages were used. The gage lengths were 5 and 10 mm for short gages, and 30, 50, 60, or 68 mm for long gages. The layout of strain gages was changed from one specimen to another hoping to get desirable readings from the gages. Figure 4 shows an example of strain gage layout.

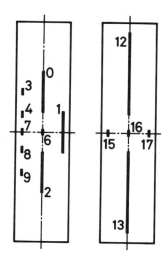

FIG. 4—*Layout of short and long strain gages, No. 8 specimen.*

±30-Deg, 90-Deg Laminates

Ten specimens of this kind were tested. Multiple initial $F_{Tt}^{(3)}$ failures occurred in narrow zones that run across the full width of each specimen. Such
narrow zones appeared here and there along the length of the specimen in
completely random fashion. Unless such failure zones fortunately hit gages,
the readings from those gages were nothing but the linear elastic deformations. Figure 5 shows the stress-strain diagrams obtained experimentally
from 5-mm short gages. In the diagrams, the description No. 9, CH.12, for
instance, indicates the specimen number and the channel number of the
scanner to which the gage was connected. In Fig. 5a, we observe the first
rising line, then the horizontal straight line which is the slip, and the final
rising line. This, at least qualitatively, coincides with the theoretical branch
of the brittle matrix of Fig. 2. This indicates that the assumption of the brittle
matrix is more adequate than the assumption of the ductile matrix. A total of
57 short gages and a total of 23 long gages were attached on ten specimens.
Twelve short gages picked up the readings similar to Fig. 5a. When two short
gages on an identical specimen were placed at different longitudinal locations, the failure stresses at the slips were generally not of the same magnitudes, and neither were the slip lengths. This means that two independent
failure zones occurred at the corresponding gage portions. Figure 5b shows
the two-step type of diagram. It is obvious that two failure zones of different
failure stresses occurred within its gage length. The failure of the shorter slip
length would have occupied a narrower portion of the gage length. Thirty-
three short gage readings were of this type. Figures 5c and 5d show the three-

step and multistep types, respectively. Seven short gages picked up readings similar to Fig. 5c, and three short gages measured readings similar to Fig. 5d. Figure 6 shows the stress-strain diagram obtained from a 30-mm long gage. Twenty out of a total of 23 long gages attached on five specimens showed a similar trend. This diagram should not be compared with the duc-

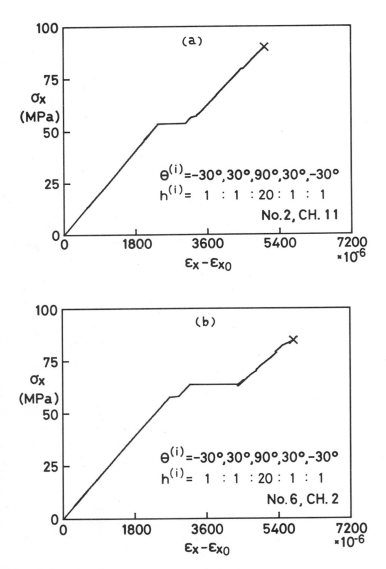

FIG. 5—*Readings of 5-mm short strain gages:* (a) *one-step type,* (b) *two-step type,* (c) *three-step type, and* (d) *multistep type.*

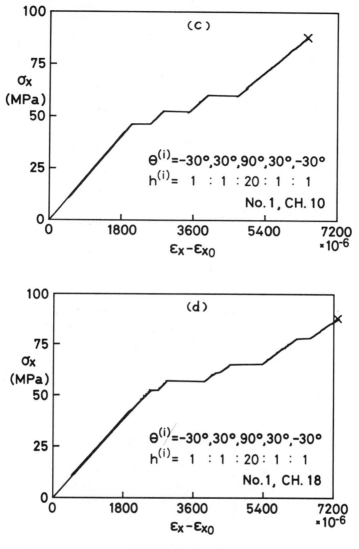

FIG. 5—*Continued.*

tile branch of Fig. 2. This is one feature of the multistep type brittle failure where the scattering of the failure stresses of local zones covered up to the final fracture stress. Figure 7 shows the results of the unloading and reloading tests that were performed on two specimens. The unloading was initiated after the applied stress exceeded the initial failure stress. From Figs. 7a and 7b, we see the unloading path is straight and the reloading path exactly coincides

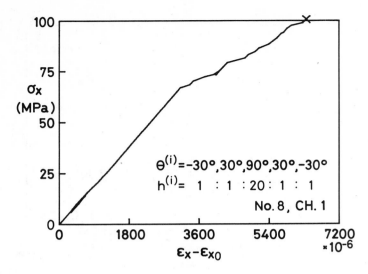

FIG. 6—*Readings of 30-mm long strain gage.*

with the unloading path. The reloading path goes farther up beyond the initial unloading stress, keeping the same inclination. When the unloading path of Fig. 7(a) is extended to intercept the abscissa, we note that there exists some amount of residual strain. This indicates that the initial thermal strains are released by the failure of the 90-deg layer.

±45-Deg, 90-Deg Laminates

Ten specimens of this kind were tested. Figure 8 shows the stress-strain diagram picked up from a 5-mm short gage. Unlike Fig. 7a, the slipline is now horizontal. This diagram is, at least qualitatively, an ideal one which is expected from the theory. A total of 54 short gages were bonded on ten specimens. Six of them picked up readings similar to Fig. 8, and nine of them picked up readings similar to Fig. 5a.

0-Deg, 90-Deg Laminates (Cross-Plys)

Six specimens of this kind were tested. Figure 9 shows the stress-strain diagram picked up from a 5-mm short gage. A total of 65 short gages and a total of six long gages were bonded on six specimens. Twenty-six short gages picked up the readings similar to Fig. 9. Figures 8 and 9 are the evidences that the inclined slipline of Fig. 7a is only a rare case. Figure 10 shows the readings of a 5-mm short gage when the unloading-reloadings were applied five times. Two unloading lines seen in the diagram are the fourth and the fifth of the

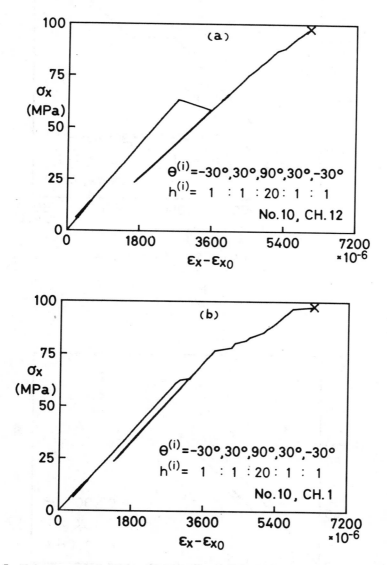

FIG. 7—*Unloading and reloading paths:* (a) *readings of 5-mm short strain gage and* (b) *readings of 30-mm long strain gage.*

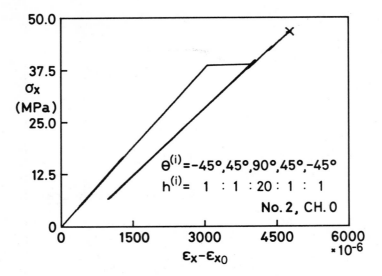

FIG. 8—*Readings of 5-mm short strain gage.*

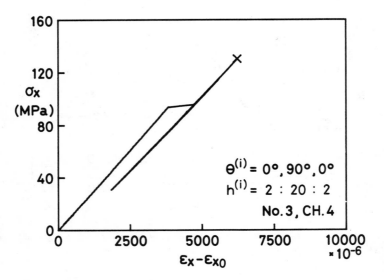

FIG. 9—*Readings of 5-mm short strain gage.*

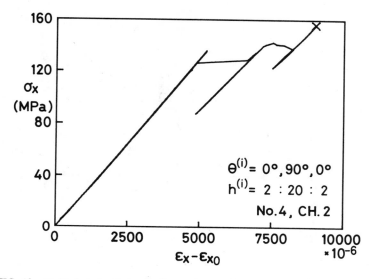

FIG. 10—*Multiple unloading-reloading paths, readings of 5-mm short strain gage.*

unloadings. The first three unloadings were overlapped on the initial elastic line of the diagram. The initial failures corresponding to the first three un- loadings took place at different portions of the specimen.

Discussion

The experimental stress-strain diagrams presented so far strongly indicate that the assumption of the brittle matrix employed in the theory is adequate for the specimens tested and the assumption of the ductile matrix is invalid. This is in great contrast to the conclusion of Tsai [1], who explained his ex- perimental results for GFRP cross-ply by an equivalent assumption of the ductile matrix.

Figure 11 is another illustration of the theoretical prediction for the stress- strain diagram of the ±30-deg, 90-deg laminate. Only the case of the brittle matrix is shown. The numerical values show in the diagram are the values computed by using the elastic constants and the coefficients of linear expan- sion α_L and α_T given before. The temperature difference ΔT and the failure stresses $F_{Tt}^{(i)}$ and $F_{LT}^{(i)}$ are left to be variables. F_{Tt} failure in the 90-deg layer occurs at point a. In the ±30-deg layers, the failure occurs either in the F_{Tt} mode at point c or in the F_{LT} mode at point d. If $(\sigma_x)_c > (\sigma_x)_a$ and $(\sigma_x)_c < (\sigma_x)_d$, then point c is the final fracture point, and point d is to be discarded. If $(\sigma_x)_d > (\sigma_x)_a$ and $(\sigma_x)_d < (\sigma_x)_c$, then point d is the final fracture point, and point c is to be discarded. If $(\sigma_x)_c < (\sigma_x)_a$ and $(\sigma_x)_d < (\sigma_x)_a$, then point a becomes the final fracture point, and points c and d must be discarded. The

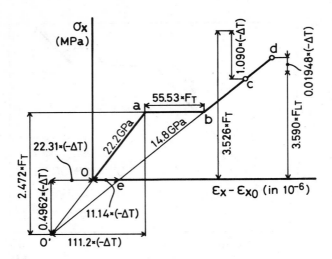

FIG. 11—*Theoretical stress-strain diagram for ±30-deg, 90-deg specimens.*

important implications of Fig. 11 are the following: The inclinations of line $0a$ and line bc are independent of the temperature and the failure stresses; the slip length ab is dependent only on F_{T_t} in the 90-deg layer and is directly proportional to it; the magnitude of the residual strain $0e$ is dependent only on ΔT and is directly proportional to it; when ΔT is reduced to zero, the origin of the diagram comes down along the extension of line $0a$ to coincide with the intersection point $0'$ of lines $0a$ and bc.

Figure 12 illustrates how the stress-strain diagram becomes when local failures occur within the gage length of a strain gage. Figure 12a defines the notations for the case of uniform failure, where the residual strain $\bar{\epsilon}_T$ represents the release of thermal strains due to the failure of the 90-deg layer. Figure 12b shows the case when one local failure zone occupies $c\%$ of the gage length l. The slip length is contracted to $c\%$ of Fig. 12a. The inclination of the second rising line $2'3'$ assumes a value E_a whose magnitude is in between E_e and E_p where $1/E_a = (c/E_p) + (1 - c)/E_e$. The intersection of lines 01 and $2'3'$ coincides with that of Fig. 12a. Figure 12c shows the case when two successive local failures of the different failure stresses σ_{K1} and σ_{K2} occur within the full gage length. The corresponding slip lengths in the case of uniform failure are denoted by $\Delta\epsilon_{K1}$ and $\Delta\epsilon_{K2}$, respectively. Then, in Fig. 12c, the first slip length is $c\%$ of $\Delta\epsilon_{K1}$ and the second slip length is $(1 - c)\%$ of $\Delta\epsilon_{K2}$. The expression for E_a of the second rising line is the same as the one given for Fig. 12b. The inclination of the final rising line is E_p. Again, the three rising lines meet at point $0'$ of Fig. 12a. The magnitudes of the released thermal strains are as shown in Figs. 12a to 12c. Other cases of local failures will be dealt with similarly.

The general views thus given by means of Figs. 11 and 12 enable us to grasp the characteristics hidden in the experimental individual stress-strain diagrams. Upon this basis, a quantitative inspection is now made, as an example, of the experimental stress-strain diagram of Fig. 13. When all the material constants that have been given before are used in the theory, the resulting stress-strain diagram does not agree with Fig. 13 except for the inclination $E_e = 22.2$ GPa of the initial elastic line. The initial failure stress derived from the theory is $(\sigma_x)_a = 88.7$ MPa, which is too large compared with the experi-

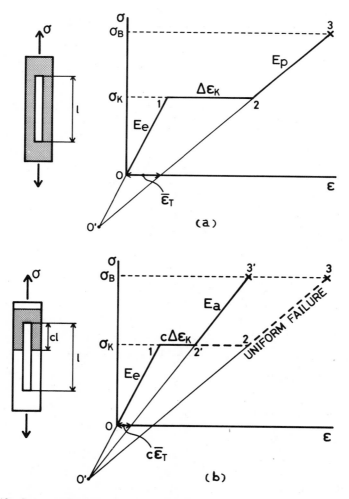

FIG. 12—*Stress-strain diagrams due to local failures within the gage lengths:* (a) *uniform failure,* (b) *one local failure covering only a fraction of gage length, and* (c) *two local failures covering full gage length.*

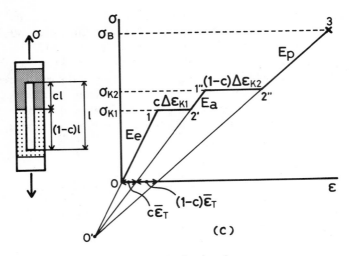

FIG. 12—*Continued.*

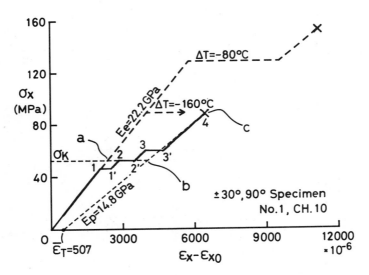

FIG. 13—*Quantitative comparison of theory and experiment for ±30-deg, 90-deg specimen.*

mental value $(\sigma_x)_2 = 52.4$ MPa. Even when ±10% variation of each parameter is permitted, no essential improvement is attained. Hence, this ordinary approach is abandoned, and an inverse method of the following is employed. Line *bc* of the predicted inclination $E_p = 14.8$ GPa is drawn in the vicinity of line 3′4. Its exact lateral location is not known, but, since line 3′4 is nearly parallel to line bc, point c is taken to coincide with point 4. Since the diagram is of three-step type, its implication is obvious from Fig. 12c. Firstly, from the intersection of line bc with the abscissa, we obtain the temperature difference $\Delta T = -46°C$. Secondly, the mean initial failure stress is taken at point 2, then $\sigma_K = 52.4$ MPa; when this magnitude is equated with the theoretical expression of the respective point of Fig. 11, we obtain for the failure stress of $\theta^{(3)} = 90$-deg layer the value $F_{Tt}^{(3)} = 30$ MPa; when a similar manipulation is made on point c of Fig. 13, we obtain for the failure stress of $\theta^{(1)} = 30$-deg layer the value $F_{Tt}^{(1)} = 38$ MPa. Summarizing, the values estimated from Fig. 13 are $F_{Tt}^{(3)} = 30$ MPa, $F_{Tt}^{(1)} = 38$ MPa and $\Delta T = -46°C$, while the values given before in the section "Material Constants," are $F_{Tt}^{(3)} = F_{Tt}^{(1)} = 68$ MPa and $\Delta T = -160°C$. It may be reasonable to consider that the full nominal value of the temperature difference is not effective in developing the thermal stresses, because the continuity between layers could not be perfect throughout the curing period. The difference of the failure stresses may be attributed to the fact that the mechanical properties of the unidirectional basic test specimens are not quite reproduced in the constituent layers of the builtup laminate.

Conclusions

1. The theory predicts two possible paths on the stress-strain diagram depending on the assumption of the ductile matrix or the brittle matrix. The experimental results strongly revealed the property of the brittle matrix. That is, the experimental stress-strain diagram showed the notable slip after the initial failure. Hence the deformation behavior of the CFRP tested can be explained only by the assumption of the brittle matrix, and the assumption of the ductile matrix is invalid.

2. The theoretical stress-strain diagram for the laminate was derived quantitatively by using the material constants obtained from basic tests. When such a theoretical diagram was compared with that of experiment, the agreement of the initial inclination was satisfactory. However, the magnitude of the initial failure stress or that of the final fracture stress disagreed. Various causes of the discrepancy can be thought of. Before conducting complex investigations, a simple means, called the inverse method, has been attempted within the accuracy of the present theory. The method gives estimated effective values of the curing temperature of the individual specimen and also the failure stresses of the constituent layers of the specimen.

3. The theory predicts two possible unloading paths on the stress-strain diagram. One is parallel to the initial elastic line, and the other is parallel to the loading path at the terminating point of the slip. The experiment showed the trend which agrees with the latter. The reloading path coincided with the unloading path.

Acknowledgments

The present work was supported by a Grant-in-Aid for Scientific Research, Ministry of Education, Japan, during the period from 1981 to 1982. We wish to thank Associate Professor K. Kunoo for his discussions. The computation was performed at the Computer Center, Kyushu University.

APPENDIX

Theoretical Background

The theoretical background of Figs. 1, 2, 3, and 11 is explained briefly. Details were published in the authors' previous papers [2,3].

The stresses $\{\sigma_L^{(i)}\} \equiv [\sigma_L^{(i)}, \sigma_T^{(i)}, \tau_{LT}^{(i)}]^T$ of layer i as functions of the applied loads $\{\sigma_x\} \equiv [\sigma_x, \sigma_y, \tau_{xy}]^T$ are written

$$\{\sigma_L^{(i)}\} = [K^{(i)}]\{\sigma_x\} + ([K^{(i)}]\{J\} - \{G^{(i)}\})\Delta T + (\{p^{(i)}\} - [K^{(i)}]\{Q\}) \quad (1)$$

The first term is due to the applied loads. The second term is due to the temperature difference ΔT. The third term is the plastic stresses due to failure of the ductile matrix. In the present case, $\sigma_y = \tau_{xy} = 0$. In the case of the brittle matrix, the third term drops out.

The failure criteria according to the maximum stress theory are

$$\text{If } \sigma_T^{(i)} \geq F_{LT}^{(i)} \text{ then } F_{Tt}^{(i)} \text{ failure}$$

$$\text{If } \tau_{LT}^{(i)} \geq F_{LT}^{(i)} \text{ then } +F_{LT}^{(i)} \text{ failure} \quad (2)$$

$$\text{If } \tau_{LT}^{(i)} \leq F_{LT}^{(i)} \text{ then } -F_{LT}^{(i)} \text{ failure}$$

with three more similar criteria for $F_{Lt}^{(i)}$, $F_{Lc}^{(i)}$, and $F_{Tc}^{(i)}$ failures. Substituting Eq 1 into Eq 2, the failure load $\{\sigma_x\}$ and the specific mode of failure are determined.

The method of analyzing the post-failure state is explained by taking the case of $F_{Tt}^{(f)}$ failure. The stress-strain relations are written as $\{\sigma_L^{(i)}\} = [D^{(i)}]\{\epsilon_L^{(i)}\}$, where

$$[D^{(i)}] = \begin{bmatrix} D_{11}^{(i)} & D_{12}^{(i)} & 0 \\ D_{21}^{(i)} & D_{22}^{(i)} & 0 \\ 0 & 0 & D_{33}^{(i)} \end{bmatrix} = \frac{1}{1 - \nu_L^{(i)}\nu_T^{(i)}} \begin{bmatrix} E_L^{(i)} & \nu_L^{(i)}E_T^{(i)} & 0 \\ \nu_T^{(i)}E_L^{(i)} & E_T^{(i)} & 0 \\ 0 & 0 & (1 - \nu_L^{(i)}\nu_T^{(i)})G_{LT}^{(i)} \end{bmatrix}$$

$$(3)$$

Immediately after $F_{Tt}^{(f)}$ failure, the stiffness components of that layer are replaced by $D_{22}^{(f)} = D_{12}^{(f)} = D_{21}^{(f)} = D_{33}^{(f)} = 0$ and $D_{11}^{(f)} = E_L^{(f)}$. This change of $[D^{(f)}]$ induces a variation of the overall strain $\{\epsilon_x\}$. This is the cause of the slip. When the slip deformation is terminated, the load can be increased again. And, consulting with the criteria (2), the succeeding failures can be analyzed similarly.

In case of unloading from a load level in the post-failure state, two extreme postulates are made. The first postulate is that $[D^{(f)}]$ regains the elastic property. The second postulate is that $[D^{(f)}]$ does not recover from damage.

References

[1] Tsai, S. W. and Azzi, V. D., "Strength of Laminated Composite Materials," *AIAA Journal*, Vol. 4, No. 2, 1966, p. 296.

[2] Ohira, H. and Uda, N., "On the Knee-point of Cross-Ply Composite" in *Progress in Science and Engineering of Composites*, T. Hayashi et al., Eds., *Proceedings*, Fourth International Conference on Composite Materials, (ICCM-IV), Tokyo, 1982, p. 473.

[3] Ohira, H. and Uda, N., "Strength of Laminated Composite Plates Subjected to Combined Loads" in *Proceedings*, 24th Conference on the Strength of Structure, Kanazawa, Japan, July 1982, p. 34 (in Japanese).

[4] Ohira, H., "Investigation on the Non-Linear Behavior of Composite Materials under Loading and Unloading," Report of the Research Project Supported by Grant-in-Aid for Scientific Research, Ministry of Education, Japan, March 1983 (in Japanese).

[5] Ohira, H., Kunoo, K., Ono, K., and Uda, N., "Measurement of Material Properties of Carbon/Epoxy Composites—Case of Tension Applied Parallel to Fibers," *Technology Reports of the Kyushu University*, Fukuoka, Japan, Vol. 56, No. 4, 1983, p. 497 (in Japanese).

[6] Ohira, H., Kunoo, K., Ono, K., and Uda, N., "Measurement of Material Properties of Carbon/Epoxy Composites—Case of Tension Applied Perpendicular to Fibers," *Technology Reports of the Kyushu University*, Fukuaka, Japan, Vol. 56, No. 4, 1983, p. 505 (in Japanese).

Fatigue

Nikolaos Tsangarakis,[1] John M. Slepetz,[1] and John Nunes[1]

Fatigue Behavior of Alumina Fiber Reinforced Aluminum Composites

REFERENCE: Tsangarakis, N., Slepetz, J. M., and Nunes, J., **"Fatigue Behavior of Alumina Fiber Reinforced Aluminum Composites,"** *Recent Advances in Composites in the United States and Japan, ASTM STP 864,* J. R. Vinson and M. Taya, Eds., American Society for Testing and Materials, Philadelphia, 1985, pp. 131–152.

ABSTRACT: The fatigue behavior of two different batches of an alumina fiber reinforced aluminum composite (FP/Al) was investigated. Both batches had a nominal 55% fiber volume fraction with fibers uniaxially oriented in the test direction. Tension-tension fatigue tests were conducted on flat, untabbed, contoured specimens in a 90 KN MTS axial fatigue test machine at $R = 0.1$. The cyclic frequency of the fatigue tests was 30 Hz except for a limited number conducted at various frequencies to determine if fatigue behavior was frequency dependent. Some specimens were strain gaged so that load-strain response could be monitored during fatigue tests. Metallographic and fractographic examination of specimens was conducted to evaluate failure modes and damage mechanisms.

Preliminary static and fatigue tests showed a significant difference in mechanical properties of the two batches of FP/Al investigated. The first batch had an endurance limit of 410 MPa compared to 330 MPa for the second batch. Static strength and modulus were correspondingly higher for the first batch than the second. Microstructural differences were also observed, such as a significantly higher number of debonded fibers in the second batch of FP/Al, that could account for the difference in mechanical properties. There was no decrease in secant modulus with fatigue cycling for either batch of material, contrary to reported behavior for boron/aluminum composites. In spite of observed fabrication defects, the ratio of endurance limit and residual strength to ultimate tensile strength was greater than 60% for both batches tested. Fiber failure was found to dominate the fatigue life of FP/Al, and failure of the composite generally occurred after a sufficient number of fibers fractured at a given cross section.

KEY WORDS: composite materials, metal matrix, aluminum, fatigue, tests, mechanical properties

The fatigue behavior of metal matrix composites has been studied extensively in recent years. Composites such as boron/aluminum and graphite/aluminum have received particular attention; however, little work has been reported on fatigue of alumina fiber reinforced aluminum composites. Ear-

[1]Materials engineers, Army Materials and Mechanics Research Center, Watertown, MA 02172.

lier studies have shown fatigue behavior of metal matrix composites to be influenced by fiber breakage which may occur during fabrication or during the first few fatigue cycles [1] and lead to premature failure of the composite. Premature failures may be also associated with phenomena such as a fiber pull-out resulting from inadequate fiber to matrix bonding. Stinchcomb et al [2] found cyclic frequency to be a factor in fatigue behavior of boron/aluminum. Dvorak and Johnson [3] have shown that the endurance limit of unidirectional boron/aluminum is strongly dependent on the stress amplitude and nearly independent of the mean stress level. Dvorak devised a fatigue failure model based on plastic straining of the composite matrix. Gouda et al [4] constructed a different model based on fiber debonding and the development of H-shaped cracks in the matrix.

In the present study, the fatigue behavior of polycrystalline alumina (Al_2O_3) fiber reinforced aluminum composite (FP/Al) was investigated. Fatigue tests were conducted at various stress amplitudes over a range of cycles to failure. The minimum stress level was kept constant in these tests. The endurance limit of the composite was determined based on survival of 10^7 load cycles. Fatigue damage and changes in the load-strain behavior were monitored during the tests. Metallographic and fractographic analyses were conducted to assist in characterizing damage mechanisms. The behavior of FP/Al was compared to that of boron/aluminum, and the applicability of proposed fatigue models was investigated. The effect of cyclic frequency on fatigue life was also examined.

Description of Materials and Specimen Configuration

Polycrystalline alumina fiber (FP)[2] reinforced aluminum alloy composite test coupons, 1.27 by 0.254 by 15.24 cm were fabricated from liquid-infiltrated as-cast plates by DuPont [5]. To ensure good bonding between aluminum and alumina an Al-2.5Li alloy is used by the manufacturer. Lithium promotes wetting of the relatively inert Al_2O_3 filaments with aluminum. The presence of such small amounts of lithium effectively raises the elastic modulus and decreases the density of the aluminum [6]. Careful control of the lithium concentration and casting process parameters is required to achieve a well bonded, low porosity composite without degrading the fiber properties. Typical filament properties [7] are given in Table 1.

Two different batches of FP/Al composite were obtained for the present study. Both batches contained a 55% nominal volume fraction of fibers oriented parallel to the longitudinal test direction. Typical static properties [5] for this fiber volume fraction and orientation are: 600 MPa ultimate tensile strength, 234 GPa primary elastic modulus, and 0.24 Poisson's ratio.

In preliminary tests it was found that rectangular specimens with bonded

[2]Trade name, E. I. Dupont DeNemours and Company, Wilmington, DE.

TABLE 1—*Fiber filament properties.*

Ultimate tensile strength	1380, MPa
Elastic modulus	380, GPa
Density	3.9, g/cm³
Diameter	20, μm

tabs, commonly employed in fatigue studies of metal matrix composites, were inadequate for FP/Al because failure usually occurred in the tab. Satisfactory fatigue test results were obtained using an untabbed streamline contoured specimen previously developed for testing organic matrix composites [8]. The specimen configuration and dimensions are shown in Fig. 1. The streamline contour is designed to minimize the transverse tensile and shear stresses in the transition region and to produce a uniform maximum axial stress in the minimum width section. Streamline specimens were machined from the FP/Al coupons [8] and were used to determine both tensile and fatigue properties. Machining damage to the edges of the specimen was observed from photomicrographs and found to consist of a zone of broken filaments extending to a depth of 0.05 mm beneath the surface.

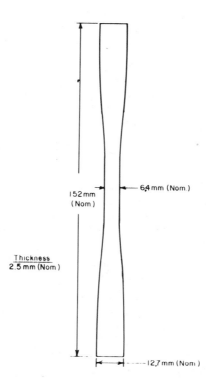

FIG. 1—*Contoured specimen used in static and fatigue tests.*

A few static tension tests were performed on small rectangular specimens of the Al–2.5Li alloy. These specimens were machined from samples of the melt used in casting the composite plates and were limited to a maximum size of 38 by 5 by 1.5 mm, due to the size and shape of the melt samples obtained from the supplier.

Test Procedures

Tension tests were conducted in a 90 KN capacity Instron test machine at a cross-head speed of 0.5 mm/min. Biaxial strain gages were mounted on both sides of the specimens for measuring longitudinal and transverse strain. Load and transverse strain were recorded versus longitudinal strain. Tension tests were run on virgin and fatigued specimens. Before loading to failure, each virgin specimen was subjected to several loading cycles between zero and approximately 80% of the ultimate tensile strength. This procedure enables any cyclic changes in the hysteresis loop to be recorded and provided additional information about the fatigue damage mechanisms. Fatigue runout specimens were cycled several times, duplicating the fatigue loading range, before loading to failure. This was done to observe any change in the elastic modulus resulting from the fatigue loading. Initial tangent and secant moduli, ultimate tensile strength, strain to fracture, and Poisson's ratio were determined from the load versus strain records. The small, matrix alloy specimens were loaded monotonically to failure. Grip failures occurred in all tests of these specimens so that the information obtained was limited to the elastic modulus, and in some cases, the 0.2% offset yield strength.

Fatigue tests were conducted in a 90 KN MTS axial fatigue test machine equipped with serrated grips, similar to those used by Stinchcomb [2]. Some specimens were strain-gaged to obtain (during fatigue) a dynamic load-strain response signal which was displayed on an oscilloscope. When it was desired to obtain a static load-strain curve, fatigue cycling was stopped; and a static test was executed under manual control. Fatigue cycling was then continued. In this way, the secant modulus was monitored continuously under both static and dynamic conditions. In all fatigue tests, the minimum cyclic stress was kept constant at 38 MPa. Except as otherwise noted, the fatigue frequency was 30 Hz.

Metallographic and fractographic analyses were conducted on selected virgin and fatigue tested specimens from both batches of FP/Al. Analysis of the fracture surfaces was conducted, employing both optical and scanning electron microscopy. Metallographic analysis of longitudinal and transverse sections was carried out, both near and away (approximately 3 mm) from the fatigue fracture surface. Microstructural features examined included porosity, excessive fiber-matrix reaction, atypical fiber morphology, and broken fibers. Measurements also were made on representative specimens for fiber diameter, fiber volume fraction, and unbonded fiber areas. Four randomly selected, typical areas were measured at ×500 magnification to determine the fiber volume fraction, V_f, from

$$V_f = \frac{\pi \cdot n_t \cdot \overline{d}_f^2}{4A_c}$$

where

A_c = total area measured,
n_t = total number of fibers, and
$\overline{d}_f$ = average fiber diameter.

A sample size of 80 filaments, 20 from each randomly selected area, was used to obtain the average fiber diameter. The percentage of unbonded fibers (N_{fd}) was determined by dividing the total number of unbonded fibers in A_c by n_t.

X-ray diffraction analysis was conducted on the composite, as well as on fibers removed from the composite, to identify any reaction products present. Some chemical analyses of the matrix and Auger analyses of unbonded fiber areas were also performed.

Results and Discussion

Mechanical Testing

The average tensile properties of the two batches of the FP/Al composite specimens employed in the fatigue study are given in Table 2. The average values are based on a minimum of four tests for each batch. A significant difference in the tensile properties exists between the two batches. The composite material of the first batch had a higher mean tensile strength and secant modulus, approximately 7 and 8%, respectively, above that of the second batch. However, both batches exhibited a wide spread (10% for the first batch and 12% for the second batch) in tensile strength value. The modulus

TABLE 2—*Tensile and fatigue properties of FP/AL.*

Batch	UTS,[a] MPa	UTS*,[b] MPa	E,[c] GPa	EL,[d] MPa	e_f,[e] %	ν[f]
1	589	554	215	410	0.29	0.26
2	543	525	204	330	0.31	0.31

[a]UTS = ultimate tensile strength in the fiber direction.
[b]UTS* = residual tensile strength of fatigue runout specimens.
[c]E = secant modulus.
[d]EL = endurance limit.
[e]e_f = strain to fracture in the fiber direction.
[f]ν = Poisson's ratio.

variation within a given batch was generally smaller than the strength variation. The strength and modulus values obtained for the first batch of FP/Al are in reasonable agreement with published results [5]; however, the values for the second batch are somewhat lower. Reasons for this discrepancy are discussed in detail later.

Figure 2 shows typical load versus axial strain curves for a virgin specimen and a fatigue runout specimen from the first batch of FP/Al. In Fig. 3, S-N curves for the two batches of FP/Al composite are depicted along with a S-N curve for 6061-T6 aluminum alloy [9]. The endurance limit of the first batch was 410 MPa, which corresponds to 69% of the ultimate tensile strength (UTS). As with the static properties, this value is in agreement with previously published data. However, the endurance limit of the second batch of FP/Al was only 330 MPa, or 60% of the UTS, a significantly lower value than obtained for the first batch. The residual tensile strength of fatigue runout specimens of both batches was only 7% less than the strength of respective virgin specimens (see Table 2). The secant modulus of runout specimens did not show any measurable decrease due to cyclic loading. This differs from behavior observed for boron/aluminum [3]. In the latter case a decrease of approximately 6% in the static secant modulus was observed for specimens fatigued to 2×10^6 cycles at the same R-ratio (0.1) and relative maximum stress as in the present study. Another observation in the present study was that the static secant modulus and the dynamic secant modulus were identical.

The load-strain history in Fig. 2 shows that the virgin specimen underwent

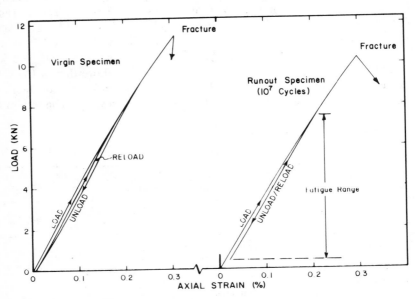

FIG. 2—*Static tests on a virgin and a fatigue runout specimen.*

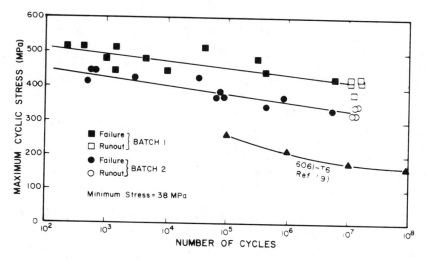

FIG. 3—S-N *curves for FP/Al, 55 volume %.*

cyclic plastic deformation in the unload-reload sequence. The characteristic hysteresis loop which developed during that sequence persisted with continued cyclic loading and unloading to the same maximum load. In contrast, the load-strain curve of the runout specimen in Fig. 2 did not exhibit such a hysteresis loop during static load-unload cycles within the fatigue range of the specimen. The runout specimen was apparently able to shakedown elastically in this range (about 75% of the range for the virgin specimen). Dvorak [*10*] demonstrated the important role played by cyclic plastic deformation and shakedown in the fatigue behavior of boron/aluminum. It seems reasonable to consider whether these phenomena likewise influence the behavior of FP/Al composites. To do this it is necessary to determine the shakedown range of the FP/Al specimens. The properties which define the shakedown range for a given composite are the yield strength of the matrix and the elastic modulus of the composite [*10*].

The shakedown range can be estimated from the linear region of the unloading segment in the cyclic hysteresis loop of a composite specimen. It should be noted that this region is difficult to define with accuracy. The stress-strain characteristics of the matrix material exhibit a gradual rather than sharp transition from elastic to plastic behavior, which makes the definition of the linear unloading region somewhat arbitrary. The as-cast matrix alloy properties are not likely to be useful in estimating the shakedown range because of significant differences from *in situ* matrix material. Microstructural analysis of the two revealed a much finer grain size in the *in situ* material, a feature which normally indicates a higher yield strength. Using the specimen load-strain curve in Fig. 2, the linear strain range of the FP/Al composite unloading curve was estimated to be 0.22%. Together with the average composite modulus of 215 GPa, this value gives a shakedown range of about 470 MPa for the first batch of FP/Al specimens. The maximum cyclic stress (with a minimum stress of 38 MPa) which would exceed the shakedown range is about 510 MPa. From Fig. 3 it can be seen that four specimens of the first batch were loaded above the estimated shakedown range, and failure occurred in these cases at less than 10^5 cycles. On the other hand, a number of specimens were loaded below the shakedown range, but above the observed endurance limit, with failures ranging from about 10^3 to 10^6 cycles.

As an unrelated observation, it was found that part of the permanent deformation incurred during the first few cycles of fatigue was recovered after the termination of fatigue testing. This is evident in the first part of the curve for the runout specimen in Fig. 2. The time interval between static testing and termination of the fatigue testing was 120 to 250 h. Such time dependent recovery may be attributed to the motion and rearrangement of dislocations in the metal matrix under the influence of internally acting residual stresses. This phenomenon is perhaps worthy of further investigation.

The results of tests to determine effect of cyclic frequency on the fatigue behavior of specimens from the first batch of FP/Al composite showed that there is no measurable frequency effect on the fatigue limit. This is contrary to the observations made by Stinchcomb [2] on boron/aluminum.

Microscopic/Metallographic Analysis

Typical microstructures observed on Batches 1 and 2 are shown in Figs. 4 to 7. Fiber diameter, fiber volume fraction, and unbonded fiber fraction measurements for both batches are summarized in Table 3. The standard deviation of fiber diameter did not vary significantly between specimens in each of the batches examined and, in most cases, was less than 10%. Between batches, the variation can be seen to be even less. The mean volume fraction of fibers was essentially the same in both batches.

The most significant difference observed between the two batches was a greater degree of fiber-matrix reaction and a higher percentage of unbonded fibers in Batch 2. A typical microstructure for Batch 1 is shown in Fig. 4 in which the fibers appear to be well bonded with little evidence of fiber-matrix reaction. The unbonded fiber areas (Fig. 5) and the higher degree of fiber-matrix reaction (Fig. 6) found in Batch 2 can be attributed to less than optimal processing conditions [5]. For example, careful control of the lithium concentration, as well as casting time and temperature, are required to avoid excessive fiber reaction. In addition, wetting of the fibers can be inhibited by incomplete burnout of the fugitive organic binder, resulting in unbonded fiber areas. Auger analysis of several unbonded areas in Batch 2 revealed a carbon residue, indicating that there was some entrapment of the organic

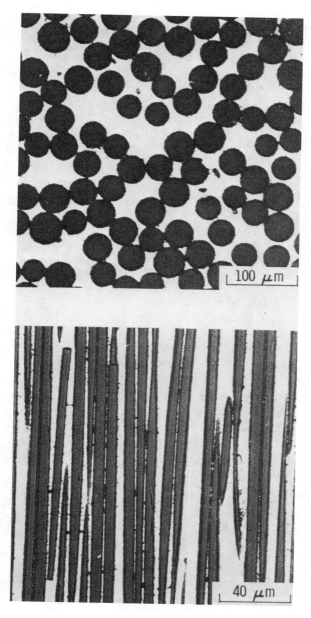

FIG. 4—*Typical microstructure for FP/Al, Batch 1.*

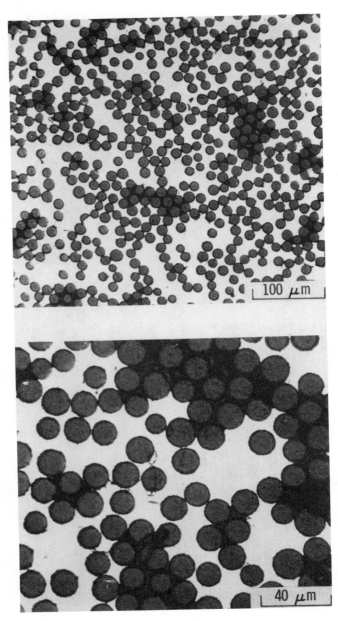

FIG. 5—*Typical unbonded fiber areas observed in FP/Al, Batch 2.*

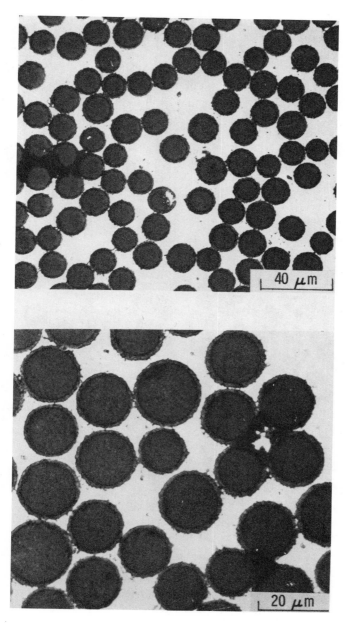

FIG. 6—*Example of excessive fiber-matrix reaction seen in FP/Al. Batch 2.*

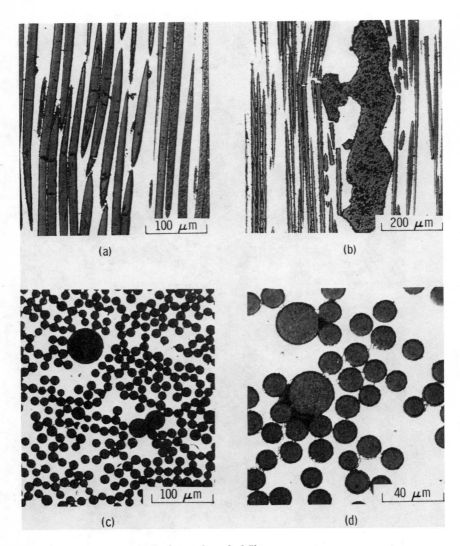

(a) Broken and crooked filaments.
(b) Porous, coarse grain, oversized fibers.
(c) Oversized, split and bridged filaments.
(d) Sintered filaments.

FIG. 7—Typical imperfections found in FP/Al composites.

TABLE 3—*Analysis of fiber parameters in the two batches.*

Specimen	$\bar{d}_f$, μm	V_f, %	n_t	N_{fd}, %
DK	20.1	56.8	542	3.8
DK-7	20.3	56.1	576	4.9
DK-10	20.1	55.1	572	3.7
DK-29	20.6	52.4	522	4.8
first batch averages	20.3	55.1		4.3
1	20.0	57.2	606	11.8
1-1	19.8	52.7	564	15.0
1-2	20.3	54.4	551	24.9
1-4	19.6	57.0	626	14.2
1-5	19.7	55.4	602	31.0
1-6	20.4	52.6	542	12.1
1-7	19.5	53.2	588	8.4
1-8	19.8	55.2	590	16.2
1-12	19.6	53.5	583	4.4
1-17	19.5	55.3	617	5.3
1-20	20.1	58.8	612	9.2
second batch averages	19.8	55.0		13.9

binder. The matrix alloy used for infiltration contained about 2.5 weight percent lithium. Analysis of the matrix chemistry from the as-cast composites revealed a decrease in lithium content to approximately 2.2 weight percent. There is, apparently, some depletion of lithium in the process of forming the fiber-matrix bond. It has been proposed that the reaction products may contain lithium aluminate [5] or a spinel [11]. X-ray diffraction analysis of several composite specimens did not reveal either product. However, fibers that were removed by acid digestion of the matrix (50% HCL-50% H_2O) did show evidence of the lithium aluminate ($LiAlO_2$). Similar observations have been made by other investigators [5,12]. There was some evidence of small amounts of a lithium-aluminum intermetallic phase in the matrix of both batches. This phase always appeared acicular and usually formed a bridge between fibers. The two batches of FP/Al investigated are believed to be representative of the quality level currently obtainable from the manufacturer.

Imperfections associated with the fiber manufacturing process and found in all specimens examined are shown in Fig. 7. These imperfections occupy less than 1% of the composite volume and do not appear to adversely affect the mechanical properties of the composites. They can be classified as (*a*) oversize filaments, (*b*) porous, coarse grain, and oversize fibers, (*c*) bridged filaments, (*d*) crooked filaments, and (*e*) broken filament bundles. The oversize fibers are nominally 10 times the average filament diameter (20 μm). Similarly, the coarse grain size of the porous fibers is 10 times the normal filament grain size (0.5 μm). Although not shown, matrix rich areas were also observed occasionally in both batches.

Examination of fatigued specimens from both batches of FP/Al revealed features that could explain the observed differences in their mechanical behavior. Two essentially different types of fatigue fracture surfaces were observed (Fig. 8). Batch 1 (Fig. 8a) exhibited a relatively smooth surface containing several plateaus separated by steps that were approximately 60 μm in depth. In contrast, Batch 2 (Fig. 8b) exhibited a rougher fracture surface containing many fiber pullouts and plateaus with steps that were twice the depth seen in the first batch. These topographical differences were more pronounced for the higher stress level, low cycle failures. Examples of fiber pullout and unbonded fiber areas were typically observed in the fatigue fracture surfaces of specimens from the second batch. A side view and a metallographic section of the fracture surface of specimen 1-6 are shown in Fig. 9. This specimen failed after 7.7×10^4 cycles at 386 MPa. The fracture surface profile and fiber pullouts are clearly evident in Fig. 9a. Also shown are unbonded areas that are interconnected with broken fibers to form secondary cracks (Fig. 9b).

A typical SEM view (Fig. 10a) of specimen 1-4 shows several arrested cracks found on the fracture surface. This specimen failed after 3.6×10^4 cycles at 431 MPa. Figure 10b illustrates the matrix ductility at a well-bonded fiber for a specimen of the second batch and also shows localized necking of the matrix. Ligaments of the matrix are still attached to the broken fiber even though most of the matrix has pulled away from the fiber. Specimens that had failed after 10^6 cycles also exhibited localized necked down matrix areas. SEM analysis also showed the occasional presence of two types of oversized fiber (see Fig. 7) on fracture surfaces. It was not possible to correlate their presence with anomalous static or fatigue failure behavior or to identify such imperfections as fracture origin sites.

Oxide particles of unknown origin were found on several Batch 2 fracture surfaces. The particles appeared to be primarily alumina; however, energy dispersive analysis of X-rays revealed evidence of significant levels of titanium and much less significant levels of calcium and silicon in the particles.

Examination of accumulated fatigue damage on failed and runout specimens of both batches of FP/Al showed fiber fracture to be the primary damage mechanism. There was also evidence of formation of H-shaped cracks due to the linking up of unbonded filaments with broken filaments. Examples of broken fibers and H-cracks are shown in Fig. 11 and 12. The transverse cross sections clearly reveal split filaments and broken filament debris. The longitudinal sections illustrate several examples of H-cracks which were frequently observed in fatigue tested specimens of both batches. On the other hand, there was no evidence of widespread arrested matrix cracking in either batch. Occasionally, however, secondary transverse cracks developed near the primary fracture surface (as seen in Fig. 13 & 14) and propagated through both matrix and fibers. This type of damage was found to occur more frequently in specimens of the second batch.

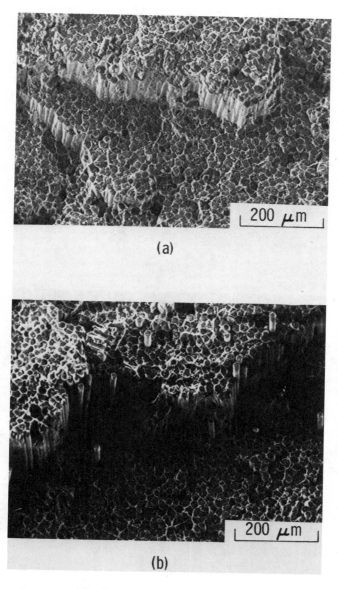

(a) Batch 1,
(b) Batch 2.

FIG. 8—Typical SEM fracture surface topography for FP/Al.

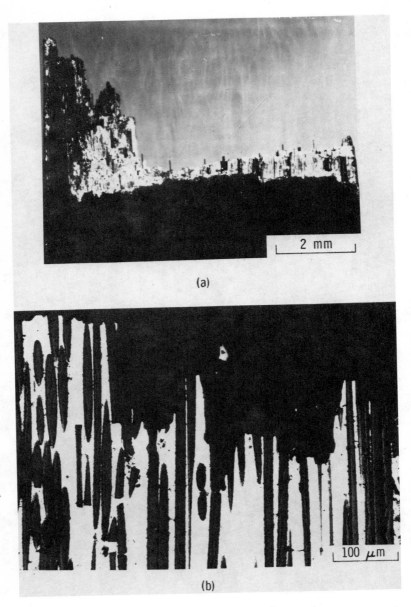

(a) Fiber pullouts and fracture surface roughness.
(b) Microstructure near the fracture surface.

FIG. 9—FP/Al, Batch 2, fracture surface sections.

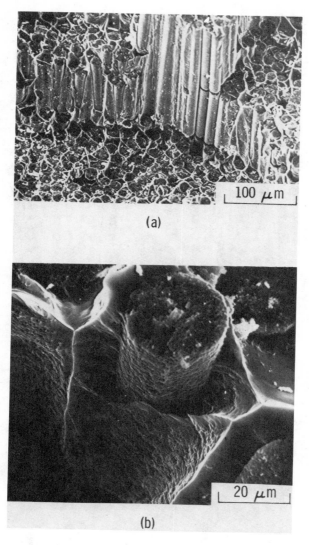

(a) Arrested secondary cracks.
(b) Matrix ligaments attached to a broken filament.

FIG. 10—FP/Al, Batch 2, SEM fracture surfaces.

FIG. 11—*FP/Al, Batch 1, runout specimen at 431 MPa.*

FIG. 12—*FP/Al, Batch 1, runout specimen at 431 MPa.*

Discussion

It is useful to consider the observations just discussed within the context of fatigue failure theories for metal matrix composites. The shakedown model proposed by Dvorak [10] implies that fatigue behavior is governed by plastic cyclic straining of the matrix. If the fatigue stress range does not exceed the elastic shakedown range of the matrix and the fibers are stressed below their endurance limit, then fatigue damage does not occur; and the specimen never fails. For loading above the shakedown range, matrix cracking occurs due to cyclic plastic straining, and further damage can develop progressively. Perfect bonding between fiber and matrix is assumed. The shakedown model does not address the mechanism of fatigue crack propagation. On the other hand, crack propagation is considered in the model proposed by Gouda et al [4]. In this model, it is assumed that fatigue behavior is controlled by a combination of fiber debonding and fracture, along with the formation of H-shaped cracks through the matrix. The H-cracks progressively link up to form a primary fracture region across the specimen. It is interesting to note that both models were tested experimentally on boron/aluminum. Depending upon the strength of the fiber-matrix bond in the particular batch of material investigated, either model could predict behavior which agreed with experiments. As seen in the present study, the bond characteristics can vary considerably from batch to batch of material. The failure region in a composite which had undergone cyclic plastic straining would exhibit extensive matrix cracking. On the other hand, the failure region of a composite in which H-crack behavior dominated would show widespread fiber debonding and pullout. The H-crack model predicts a horizontal *S-N* curve for defect-free fibers and weak fiber/matrix bonding; whereas, for perfectly bonded fibers, the model would predict a characteristic S-shaped *S-N* curve. Experimentally obtained *S-N* curves for boron/aluminum have tended to lie somewhere between the two extreme cases.

Neither of the fatigue models proposed for boron/aluminum adequately describes the behavior of the FP/Al composite investigated in the present study. Contrary to expected behavior in Dvorak's model, there was no evidence of extensive matrix cracking in failure regions of FP/Al specimens; nor was there any decrease in secant modulus which would have resulted from such cracking. Shakedown was not observed to be an important factor in fatigue behavior in as much as a significant number of specimens failed at maximum cyclic stress levels below the estimated shakedown range. Only a few specimens were cycled at stress levels exceeding this range, and, even in those cases, the amount of cyclic plastic deformation which occurred was probably not extensive. Gouda's model likewise does not seem to apply to FP/Al.. Although there was some evidence of H-cracks in fatigue specimens of both batches of material tested, there was no indication of progressive linking of H-cracks to form the primary fracture path.

The most significant observation in both failed and runout specimens of

FP/Al was extensive fiber fracture, including multiple fractures of individual fibers. The fatigue life of FP/Al, therefore, appears to be fiber dominated to the extent that failure occurs only when an insufficient number of fibers remains intact to transfer load at the critical section. Prior to failure, load transfer through fibers is still effective as the secant modulus does not show a decrease despite considerable fiber fracture. This would imply that the fiber and matrix remain well bonded. The difference in behavior between FP/Al and boron/aluminum is most likely due to the large difference in size and tensile strength of the alumina and boron fibers. The strain to fracture of boron fibers is over three times that of silica fibers, and their respective fatigue strengths have perhaps the same ratio. On the other hand, the shakedown range of FP/Al is nearly equal to that of boron/aluminum because the matrix yield strength and composite modulus are the same for both composites. Conceivably, fatigue failure of silica fibers can occur in FP/Al at stress levels below the shakedown range which defines the threshold level for matrix fatigue. The fatigue strength of boron fibers is probably well above the shakedown range of boron/aluminum, and matrix rather than fiber failure would be expected for cyclic loading just above this range. The difference in fiber characteristics undoubtedly affects the sequence of events in the two composites after a fatigue crack is initiated. The larger and stronger boron fibers have greater resistance to crack propagation and deflect the crack along the weaker fiber/matrix interface in boron/aluminum. This would result in a very erratic crack path, featuring debonding and fiber pullout prior to fatigue failure. In the FP/Al composite, however, the combination of weaker, smaller fibers and stronger interface would make it easier for a fatigue crack to propagate across fibers on a relatively flat plane.

It is interesting to note that when fatigue strength is normalized by composite tensile strength, the S-N curves for FP/Al and boron/aluminum are nearly identical [5] despite fundamental differences observed in behavior prior to fatigue failure. This indicates that, regardless of the path taken by fatigue cracks or the nature of damage to matrix and interface preceding fracture, the final failure event is fiber dominated in both composites.

Conclusions

1. Progressive fiber fracture was found to be the dominant damage mechanism controlling the fatigue behavior of FP/Al. Once a sufficient number of broken filaments developed at a cross section, composite failure occurred. Fatigue failure models proposed by Dvorak and Gouda do not explain this behavior.

2. Two batches of FP/Al were examined. The endurance limit of the first batch was 410 MPa (69% of the UTS) and that of the second batch, 330 MPa (60% of the UTS). The residual tensile strength of runout specimens of both batches was 93% of the UTS of the virgin material.

3. No measurable changes in the secant modulus occurred in either batch of FP/Al with fatigue cycling.

4. Fiber imperfections associated with the manufacturing process were not found to significantly influence the tensile and fatigue properties of FP/Al. On the other hand, fabrication defects, such as unbonded fibers, did affect these properties. The second batch of FP/Al, which had a significantly greater number of unbonded filaments, exhibited a lower endurance limit, secant modulus, and tensile strength than the first batch.

5. The fatigue limit was found to be independent of the cyclic frequency.

References

[1] Chamis, C. C. and Sullivan, T. L. in *Fatigue of Composite Materials*, ASTM STP 569, American Society for Testing and Materials, Philadelphia, 1975, pp. 95–114.

[2] Stinchcomb, W. W. et al in *Fatigue of Composite Materials*, ASTM STP 569, American Society for Testing and Materials, Philadelphia, 1975, pp. 115–129.

[3] Dvorak, G. J. and Johnson, W. S., *International Journal of Fracture*, Vol. 16, No. 6, Dec. 1980, pp. 582–602.

[4] Gouda, M. et al in *Fatigue of Fibrous Composite Materials*, ASTM STP 723, American Society for Testing and Materials, Philadelphia, 1981, pp. 101–115.

[5] Champion, A. R. et al, *Proceedings*, International Conference on Composite Materials, B. Norton et al, Eds., *Metallurgical Society of the American Institute of Mining, Metallurgical and Petroleum Engineers*, 1978, pp 883–904.

[6] Starke, E. A. and Sanders, T. H., *Journal of Metals*, Aug. 1981, pp. 24–33.

[7] Krueger, W. H. and Dhingra, A. K., "Alumina Fiber Reinforced Metal Composites for Potential Automotive Engine Applications," presented at the American Institute of Chemical Engineers, Detroit, Michigan, Aug. 1981.

[8] Oplinger, D. W., Gandhi, K., and Parker, B., "Studies of Tension Test Specimens for Composite Materials," AMMRC TR82-27, Army Materials and Mechanics Research Center, Watertown, Massachusetts, April 1982.

[9] *Aerospace Structural Metals Handbook*, Vol. II, Nonferrous Alloys, March 1966.

[10] Tarn, J. A., Dvorak, G. J., and Rao, M. S. M., *International Journal of Solids and Structures*, Vol. 11, No. 6, 1975, pp. 751–764.

[11] Prewo, K. M., "Fabrication and Evaluation of Low Cost Alumina Fiber Reinforced Metal Matrices," Interim Technical Report R77-912245-3, United Technologies Research Center, East Hartford, Contract No. N00014-76-C-0035, May 1977.

[12] Olsen, G. C., "Effects of Temperature, Thermal Exposure and Fatigue on an Alumina/Aluminum Composite," TP-1975, NASA Langley Research Center, Dec. 1980.

Charles E. Harris[1] and Don H. Morris[2]

An Evaluation of the Effects of Stacking Sequence and Thickness on the Fatigue Life of Quasi-Isotropic Graphite/Epoxy Laminates

REFERENCE: Harris, C. E. and Morris, D. H., **"An Evaluation of the Effects of Stacking Sequence and Thickness on the Fatigue Life of Quasi-Isotropic Graphite/Epoxy Laminates,"** *Recent Advances in Composites in the United States and Japan, ASTM STP 864,* J. R. Vinson and M. Taya, Eds., American Society for Testing and Materials, Philadelphia, 1985, pp. 153–172.

ABSTRACT: A test program has been conducted in which the effect of specimen thickness on the fatigue lives of quasi-isotropic, graphite/epoxy lamintes was investigated. Three replicate specimens for each notched and unnotched geometry at 16, 32, and 64-ply thicknesses of a $[90/45/0/-45]_{ns}$ and $[45/0/-45/90]_{ns}$ laminate were tested in compression-compression fatigue. The fatigue lives of the notched specimens did not appear to be a strong function of laminate stacking sequence or specimen thickness. The stress concentration at the hole dominated over the interlaminar stresses at the straight free edge. The unnotched specimens of the $[90/45/0/-45]_{ns}$ laminate with tensile interlaminar normal stresses delaminated more readily than did the $[45/0/-45/90]_{ns}$ laminate with compressive interlaminar normal stress. The life of the 16-ply unnotched specimens was lower than was the life of the 32 and 64-ply specimens. Delaminations were located at the interface where the maximum τ_{xz} shear stress occurred regardless of the sense of magnitude of the interlaminar normal stress. Finally, a specially designed antibuckling fixture was found to be effective in preventing out-of-plane motion without overconstraining the specimen.

KEY WORDS: composite material, graphite/epoxy, fatigue, thickness effect, interlaminar stresses

The role of the interlaminar stresses at the free edges of laminated composites in the delamination of laminates has been previously explored in a number of investigations. The magnitude of these interlaminar stresses have been

[1]Assistant professor, Department of Aerospace Engineering, Texas A&M University, College Station, TX 77840 (formerly at UPI&SU).

[2]Professor, Department of Engineering Science and Mechanics, Virginia Polytechnic Institute and State University, Blacksburg, VA 24061.

shown to be a function of laminate thickness [1]. Furthermore, the boundary layer at the laminate edge over which these stresses exist has also been shown to be approximately equal to the laminate thickness [1]. It is therefore of interest to explore the role of laminate thickness on the fatigue life and mode of damage formation in composite laminates.

The laminates studied were T300/5208 graphite/epoxy with $[90/45/0/-45]_{ns}$ and $[45/0/-45/90]_{ns}$ stacking sequences, where ns means multiple layers of the stacking sequence symmetric about the midplane. The two laminate types were selected because the $[90/45/0/-45]_{ns}$ laminate has high tensile interlaminar normal stresses under compressive loading and the other laminate has high compressive interlaminar normal stresses.

All specimens of a specific layup were tested at the same stress range. The stress ranges, determined from a series of preliminary tests were selected such that failure occurred in the 10^4 to 10^6 cycle range. The fatigue specimens were loaded with direct compressive bearing loads on the specimen ends. Specially designed compression test fixtures included antibuckling guides to minimize out-of-plane motion.

Nondestructive examinations were performed in order to locate the initiation of damage and to monitor and record the progression of damage. These examinations consisted of a combination of enhanced X-ray radiography, edge replication, and visual observations.

Interlaminar stress distributions at the straight free edge of the unnotched laminates were utilized to interpret the test results. The Pipes-Pagano approximation [1] was used to determine the interlaminar normal stress, σ_{zz}. The interlaminar shear stress distributions, τ_{xz} and τ_{yz}, for the unnotched laminates were generated by a finite-element analysis developed by Rooney and Herakovich [2].

Description of Test Program, Specimens and Test Fixtures

Description of Test Program

A series of preliminary tests were conducted with the objectives of refining test fixtures, addressing the problem of specimen alignment, and establishing the stress levels for the "production" tests. Refinements to test fixtures are discussed in the following paragraphs. Strain gages attached on opposite faces and across the specimen width and along the length confirmed that satisfactory specimen alignment could be achieved [3]. Other preliminary test results are reflected in the test conditions described next for the production tests.

The production tests consisted of three replicate tests at each test condition for both notched and unnotched geometries. All notched specimens were tested at a maximum compressive stress of 38 ksi (262 MPa). The unnotched specimens of the $[45/0/-45/90]_{ns}$ laminate were tested at a maximum compressive stress of 60 ksi (414 MPa), and the unnotched specimens of the $[90/45/0/-45]_{ns}$

laminate were tested at a maximum compressive stress of 48 ksi (331 MPa). All tests were conducted at an R-ratio of 10 and a frequency of 10 Hz.

Description of Specimens

Laminate panels of T300/5208 were prepared by a tape layup and autoclave curing process. All specimens of a specific thickness were cut from the same panel. The $[90/45/0/-45]_{ns}$ specimens were cut at a 90° orientation to the 0° outer-ply fiber direction. The $[45/0/-45/90]_{ns}$ specimens were cut at a 45° orientation to the 0° outer-ply fiber direction.

Fatigue tests were conducted using two basic specimen types. The unnotched specimen was a straight coupon with a 25.4 mm (1.0 in.) width and 203 mm (8.0 in.) length. The notched specimen, was a straight coupon 50.8 mm (2.0 in.) wide and 203 mm (8.0 in.) in length with a 6.35 mm (0.25 in.) diameter center, circular hole. Specimen thicknesses were 16-plies [typically 2.11 mm (0.083 in.)], 32-plies [4.19 mm (0.165 in.)] and 64-plies [8.46 mm (0.333 in.)].

Description of Notched Specimen Test Fixture

The notched specimens were loaded by direct compressive bearing on the specimen ends. Flat plates with end supports constrained the specimen ends and contacted the flat surfaces of the test machine crossheads (Fig. 1). Antibuckling supports prevented out-of-plane motion of the specimens by contact-supporting both faces of the specimen along the full length and width of the specimen with the exception of the "test section." Dry lubrication was used to minimize friction between the specimen and the supports. A test fixture "test section" was formed by cutting 41.3 mm by 41.3 mm (1.625 in. by 1.625 in.) windows out of the antibuckling supports surrounding the central region of the specimen where the hole was located. The size of this window was optimized by a previous test program conducted by Phillips [4] at the NASA Langley Research Center. The contacting surfaces of the antibuckling supports were machined to insure a close fit along the specimen length. Further details of the test fixture may be found in Ref 3.

Description of Unnotched Specimen Test Fixture

The unnotched test fixture is shown in Fig. 2. Load was applied by direct compressive bearing at the specimen ends. As with the notched specimen fixture, the specimen ends were constrained by flat plates with specimen end constraints. These flat plates contacted the flat surfaces of the test fixture crossheads. Antibuckling supports contacted the full length of the specimen faces to prevent out-of-plane motion. Dry lubrication was used to minimize friction. The contact surfaces of the antibuckling supports were machined to maximize contact with the specimen along its length as well as to provide a vertical alignment for the specimen. The width of the specimen contacted by the antibuckling support was 6.35 mm (0.25 in.) and the fixture had positioning guides to ensure that the specimen centerline and support contact centerline coincided. Further details may be found in Ref 3.

Experimental Procedures

Fatigue Test Procedures

The fatigue tests were conducted on a closed-loop electrohydraulic MTS testing machine in the load control mode. The R-ratio was 10, and the alternating load was sinusoidal at 10 Hz. Upon installation, the alignment of a specimen was checked by visual means and by strain gage readings. Detailed alignment procedures are given in Ref 3.

Enhanced X-Ray Procedure

A Hewlett-Packard Faxitron X-ray system was used for all X-ray examinations. Damaged areas of the specimen were enhanced by the use of zinc iodide. The specimens were typically subjected to the enhancer under the static mean load for a period of not less than 1 min, removed from the testing machine, and X-rayed. Specimens were then realigned in the test machine and the test continued. Removing and realigning a specimen did not appear to affect the results [3].

Edge Replica Procedure

The edge replica technique provides very detailed information on damage at the edge of a specimen [5]. An edge replica is obtained by loosely attaching to the specimen a piece of specially prepared acetate replicating tape. Acetone is then injected between the tape and specimen edge, and the tape is pressed against the edge. After about a minute the tape is peeled from the specimen resulting in a replicate of the damaged edge of a specimen. This technique was used to locate the regions of delamination and the progression of delamination.

Test Results

Notched Specimen Test Results

The notched specimen test results for both laminates are displayed in Fig. 3.

The progression of damage, in general, was similar and the failure modes were the same for all specimens, regardless of thickness and laminate type, except for one specimen. This specimen was a 16-ply specimen of the [45/0/−45/90] laminate which cracked while under the maximum static load when the testing

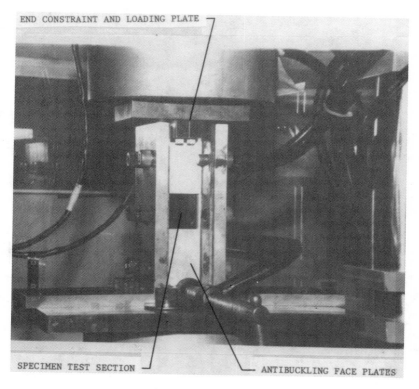

FIG. 1—*Notched specimen test fixture.*

machine was being set. Since the specimen did not fail, that is, it still supported the maximum load, it was fatigue loaded and failed at 1040 cycles.

The ranges of the fatigue life for each test condition (specific thickness and laminate type) overlap, as shown in Fig. 3. With the exception of the previously described deviate test, the 16-ply and 32-ply data for both laminates are essentially the same. However, the highest fatigue lives belonged to the 64-ply specimens. There is little difference between the lives of the two 64-ply laminates.

Damage in the typical specimen became visible in the 3000 to 8000 cycle range and was directly adjacent to the hole. X-ray examinations taken early in life (3500, 4000, or 5000 cycles) typically revealed small damage adjacent to the hole prior to the damage being visually observed. The damage then progressed at a moderate rate until it was on both faces and on both sides of the hole. The damage was usually about one hole diameter wide. Damage then slowed considerably as if a damage plateau had been reached. Towards the end of life (85 to 90%) damage began to progress rapidly. The X-ray photograph in Fig. 4 illustrates typical damage late in life. Frequently the damage "jumped" (with an associated audible sound) perhaps to half the ligament on

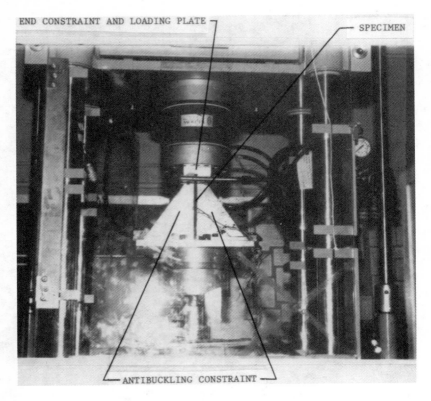

FIG. 2—*Unnotched specimen test fixture.*

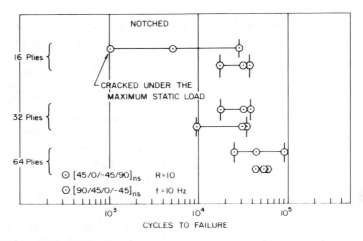

FIG. 3—*Notched specimen fatigue data, at a maximum compressive stress of 38 ksi (262 MPa).*

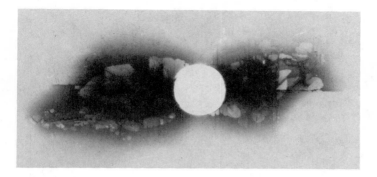

FIG. 4—*Typical damage around hole late in life.*

either side of the hole. Near the end of life the damage would "jump" completely across the ligament to one side of the hole, then do the same on the other side. This usually occurred on both faces. This was accompanied by considerable audible cracking sounds and almost immediate failure. (It should be noted that since the damage was localized around the hole and across the width adjacent to the hole, there does not appear to be any influence from the antibuckling guides on the formation of damage.)

There was no appreciable difference in the appearance of failed specimens of either laminate. The damage was generally confined to the region immediately adjacent to the hole. However, long axial delaminations did typically accompany the catastrophic failure in the test section. These delaminations were never visible in X-rays so they were thought to occur at catastrophic failure as opposed to preceding or precipitating failure.

Unnotched Specimen Test Results

The unnotched data for both laminates are shown in Fig. 5. The specimens of the [45/0/−45/90] laminate were tested at a maximum compressive stress of 414 MPa (60 ksi), whereas the specimens of the [90/45/0/−45] laminate were tested at a maximum compressive stress of 331 MPa (48 ksi).

The progression of damage and mode of failure was similar for all tests except for one 16-ply specimen which failed in a deviate manner. This specimen cracked while loaded statically at the maximum load. Since it would still support load it was cyclically loaded and failed at 1320 cycles. An edge replica was taken after the initial crack at static load. The location of the crack was identical to the location of the first delamination for the other 16-ply specimens.

The test data shown in Fig. 5 points up the differences in fatigue lives of the two laminates. The interlaminar normal stress, σ_{zz}, is tensile in the [90/45/ 0/−45] laminate and compressive in the [45/0/−45/90]$_{ns}$ laminate, with the magnitude being about the same. (Refer to the next section for the interlaminar

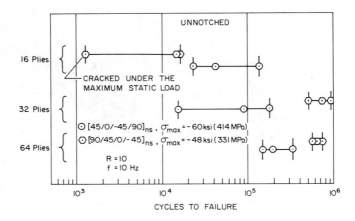

FIG. 5—*Unnotched specimen fatigue data.*

stress distributions). The fatigue lives of the [90/45/0/−45] laminate (except at 16-plies) are somewhat higher than those of the [45/0/−45/90] laminate. However, the maximum stress level was 20% lower. Any thickness effect trends brought about by differences in the magnitude of the interlaminar normal stress are not so obvious. The range of fatigue lives for the [45/0/−45/90] laminates overlap, and there is very little difference between the 16 and 32-ply data. The 64-ply data is slightly higher but overlaps the 32-ply data. This is somewhat opposite to the trend shown by the other laminate. The [90/45/0/−45] laminate, with tensile interlaminar normal stress, shows little difference in the lives of the 32 and 64-ply specimens. However, the lives of the 16-ply specimens are an order of magnitude lower than those of the 32 and 64-ply specimens. In fact there is a tendency for the 16-ply [90/45/0/−45] specimens to delaminate in the vicinity of the maximum test load during test setup. The [90/45/0/−45] specimens exhibit less data scatter than do the [45/0/−45/90]$_{ns}$ specimens, if one excludes the specimen which failed at 1320 cycles.

In spite of the differences in the fatigue lives of the unnotched specimens there was virtually no difference in the initiation, type, and progression of damage, as well as the type of final catastrophic failure. Damage initiated at some random location along the specimen length in the form of a small delamination at the free edge. Tables 1 and 2 give the ply interface location of these delaminations for typical specimens as determined from edge replicas. (A typical edge replica is shown in Fig. 6.) In general the delaminations occurred at the 0/−45 interface, plies 2 and 3 for [45/0/−45/90] and plies 3 and 4 for [90/45/0/−45]. From this initiation the delaminations extended along the specimen length until it was more or less full length. The delamination then worked through the width and finally along the opposite edge lengthwise. The test fixture end constraints prevented the cracks from extending to the spec-

TABLE 1—*Delamination locations[a] for the $[45/0/-45/90]_{ns}$ laminates.*

Specimen Identification	Number of Cycles (1000s)	Location of Replica Along Specimen	Location of Delaminations
FU7-16	19.7[b]		6/7 [−45/0]
	23.0		3/4 [0/45], 6/7 [−45/0]
	failure		2/3 [−45/0], 3/4 [0/45], 6/7 [−45,0]
FU9-32	10^2	top	11/12 [0/45], 14/15 [−45/0]
	10^2	middle	14/15 [−45/0]
FU1-32	12^2		14/15 [−45/0]
	30		14/15 [−45/0]
	35.8	top	11/12 [0/45], 14/15 [−45/0]
	52	bottom	11/12 [0/45], 14/15 [−45/0]
	75	middle	10/11 [−45/0]
	81		10/11 [−45/0]

[a]Interface locations are from the laminate midplane and are designated by giving the number of plies from the midplane followed by the ply fiber angle orientation.
[b]The first replica was taken immediately after the first delamination was visible, that is, at the initiation of macroscopic damage.

imen ends. The antibuckling constraint did not prevent damage formation across the specimen width. The process is illustrated by the X-ray photographs shown Fig. 7. Frequently while the first delamination was progressing a delamination at the same vertical location measured from the opposite face would initiate. These delaminations are also listed in Tables 1 and 2. As these delaminations extended, the outer plies that were split from the laminate buckled under the compressive load (Fig. 6) and caused the formation of the matrix cracks that are visible Fig. 7.

The next major damage event was the initiation of a second delamination. Sometimes but not always this delamination occurred at the same elevation location as the initiation of the first delamination. The second delaminations opened at a $0/-45$ interface but 4-plies inward from the initial $0/-45$ delamination. The location of the initiation of the second delamination was usually the location of the final catastrophic failure. However, catastrophic failure did not occur until the second delamination extended lengthwise and through the width. Usually a third and sometimes a fourth delamination would initiate. These multiple delaminations normally initiated at the elevation location of second delamination and may have formed relative to both faces before final failure. The third and fourth delaminations did not extend along the specimen length nor across the width. One distinction in thickness is that only two delaminations opened in the 16-ply specimens before failure and multiple delaminations (3 or 4) may open in the 32 and 64-ply specimens. In addition, delaminations in the 16-ply specimens did not extend lengthwise and across the width as did the delaminations in the 32 and 64-ply specimens.

TABLE 2—*Delamination locations[a] for the [90/45/0/−45]$_{ns}$ laminates.*

Specimen Identification	Number of Cycles (1000s)	Location of Replica Along Specimen	Location of Delaminations
FU17-64	179[b]	front, middle	29/30 [−45/0], 30/31 [0/45]
			29/30 [−45/0] on opposite face
	227	top, front	30/31 [0/45]
	385	front, middle	29/30 [−45/0], 30/31 [0/45]
			29/30 [−45/0] on opposite face
	535	top, front	29/30 [−45/0], 30/31 [0/45]
			29/30 [−45/0] on opposite face
	535	bottom, front	28/29 [90/−45] changes interface to 29/30 [−45/0]
	593	middle, back	27/28 [45/90], 29/30 [−45/0]
	593	top, front	25/26 [−45/0] other splits indistinguishable
	618	top, front	20/21 [90/−45] changes interface to 21/22 [−45/0]
			24/25 [90/−45] changes interface to 25/26 [−45/0]
			29/30 [−45/0]
			29/30 [−45/0] on opposite face
FU18-64	185[b]		30/31 [0/45]
FU19-64	130[b]	front, top	30/31 [0/45] (right face)
	135	front, middle	29/30 [−45/0] and 30/31 [0/45] (right face)
	205	front, middle	same as above
	205	back middle	same as above
	226	back, middle	25/26 [−45/0], 29/30 [−45/0] (right face)
			29/30 [−45/0] on opposite face
	226	back, top	29/30 [−45/0]
			29/30 [−45/0] on opposite face, 31/32 [45/90]
	342	front, middle	28/29 [90/−45] changes interface to 29/30 [−45/0]
	Failure		major delamination between 1/2 [−45/0] which changes interface to the other 1/2 [−45/0]

[a]Interface locations are from the laminate midplane and are designated by giving the number of plies from the midplane followed by the ply fiber angle orientation.
[b]The first replica was taken immediately after the first delamination was visible, that is, at the initiation of macroscopic damage.

Failed specimens exhibited a localized region of massive damage where the catastrophic failure occurred. This location normally coincided with the formation of multiple delaminations prior to failure. The 32 and 64-ply specimens exhibited the extensive delaminations that were typical of all specimens except the 16-ply specimens. Load redistribution may be an exploration for the difference between the 16-ply and the 32 and 64-ply specimens. When a delamination forms between the second and third plies, inward from a face,

FIG. 6—*Typical edge replica of an unnotched specimen.*

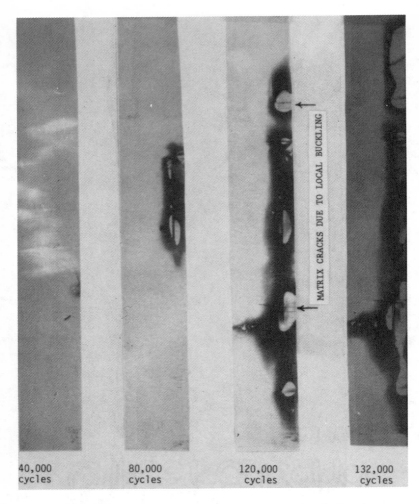

FIG. 7—*Typical progression of delamination damage in an unnotched specimen.*

the redistributed load to the undamaged laminate is much higher for a 16-ply specimen than is the case for the 32 and 64-ply laminates with the same delamination. The formation of only two delaminations was typical of the 16-ply specimens and catastrophic failure closely followed the formation of the second delamination.

Interlaminar Stress Analysis of the Unnotched Geometry

As a result of the mismatch of Poisson's ratio between adjacent layers and the mismatch of the coefficient of mutual influence, interlaminar stresses are necessary at the free edges of laminates for equilibrium [6]. There have been many studies concerning the computation of interlaminar stresses. Finite difference schemes, the finite element methods, and other approximations have been developed. Herein, the interlaminar normal stress distributions were calculated using the approximation method proposed by Pipes and Pagano [1]. The interlaminar shear stress distributions were obtained from a finite-element analysis by Rooney and Herakovich [2]. Because of the thickness effect the interlaminar normal stress distributions must be generated for the specific laminates and thicknesses of this study. Since the shear stress analyses are not a function of thickness, the existing analyses [2] for the $[45/0/-45/90]_s$ and $[90/45/0/-45]_s$ laminates are valid for the 16, 32, and 64-ply cases studied herein.

Interlaminar normal stresses were calculated for the unnotched $[45/0/-45/90]_{ns}$ laminate at 16, 32, and 64-plies for a static load corresponding to an applied laminate stress, σ_{xx}, of -414 MPa (-60 ksi). The stresses are plotted for each thickness in Fig. 8. Stresses in the $[90/45/0/-45]_{ns}$ laminate at 16, 32, and 64-plies were computed for a static load corresponding to an applied laminate stress -331 MPa (-48 ksi). These stresses are plotted in Fig. 9. The applied laminate stresses correspond to the peak of the alternating fatigue load for the $[45/0/-45/90]_{ns}$ and $[90/45/0/-45]_{ns}$ laminates. Note that the σ_{zz} stresses in the $[45/0/-45/90]_{ns}$ laminate are compressive while those in the $[90/45/0/-45]_{ns}$ laminate are tensile under the applied compressive load.

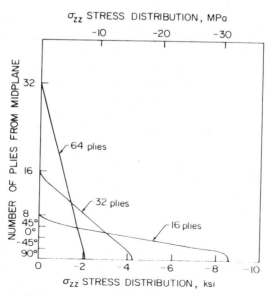

FIG. 8—*Interlaminar normal stress distributions for the $[45/0/-45/90]_{ns}$ laminates under a compressive stress of 60 ksi (414 MPa).*

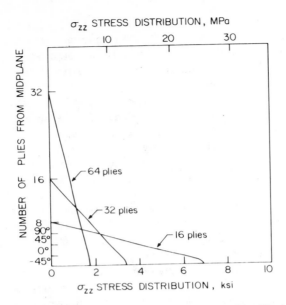

FIG. 9—*Interlaminar normal stress distributions for the [90/45/0/−45]$_{ns}$ laminates under a compressive stress of 48 ksi (331 MPa).*

It is interesting to note that as the laminate thickness increases the magnitude of σ_{zz} decreases. In fact, doubling the thickness halves the σ_{zz} stress. A review of the Pipes and Pagano model [1] reveals why this is the case. When a stack of [90/45/0/−45]$_s$ plies are added to the laminate the moment is increased. Since the forces due to the individual ply stresses σ_{yy}, add to zero, a pure couple, independent of location from the midplane, is added. Therefore, doubling the thickness and doubling the applied load results in a doubled moment. However, doubling the laminate thickness doubles the boundary layer region over which the interlaminar stress, σ_{zz}, exists. Since σ_{zz} in the Pipes and Pagano model is an inverse function of the thickness squared, the stress is then halved.

The shear stresses, τ_{yz} and τ_{xz}, were generated by a finite-element analysis [2]. The analysis was quasi three-dimensional, generalized plane strain. The elements were constant strain elements with a high concentration of small elements in the boundary layer along the free edge. The laminate cross section was represented by a two-dimensional grid where individual plies were treated as homogeneous orthotropic with properties of unidirectional (which can be off axis) fibrous composites. The nondimensional shear stress distributions for the [90/45/0/−45]$_s$ laminate are shown in Fig. 10, and those for the [45/0/−45/90]$_s$ laminate are shown in Fig. 11. The stress distributions shown are equally applicable to the 16, 32, and 64-ply cases of this study since τ_{yz} and τ_{xz} are not functions of the laminate thickness. The distribution shown is

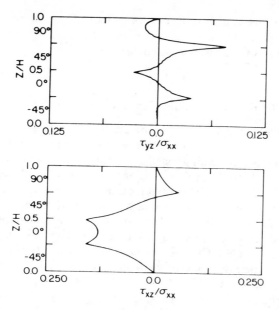

FIG. 10—*Shear stress distributions for the [90/45/0/−45]_s laminate.*

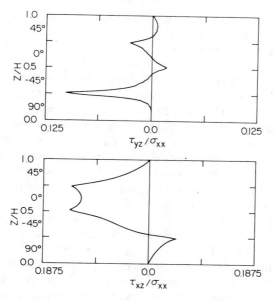

FIG. 11—*Shear stress distributions for the [45/0/−45/90]_s laminate.*

simply the same for each repeated stack of four-plies. The finite-element results show the shear stress, τ_{xz}, to be the largest in magnitude of the three interlaminar stresses for these two laminates.

Relationship Between Interlaminar Stresses and the Fatigue Test Results

The most likely type of damage to be produced by the interlaminar stresses is a delamination. When the normal stress, σ_{zz}, is tensile all three interlaminar stress components are working together to produce a delamination. When σ_{zz} is compressive the two interlaminar shear stresses still work together to produce a delamination. The interfacial location of the delamination may be determined by the peak stress location of one dominant stress such as σ_{zz} or τ_{xz}, or it may be determined by a combination of the stresses working together.

The notched geometry fatigue tests of both laminate types were all conducted at the same load ranges. There is very little difference in the fatigue lives when comparing specimen thickness or laminate type (see Fig. 3). Furthermore, the damage begins at the hold and progresses across the width, adjacent to the hole, with substantial matrix cracking accompanying the delaminations. In spite of the dramatic differences in the interlaminar normal stresses of these two laminates, the dominant factor affecting damage and failure is the stress concentration at the hole and not the interlaminar stresses at the straight free edges.

Recalling Fig. 5 we see that the unnotched specimens do exhibit different fatigue lives when comparing the two laminates. This is evident by the fact that the $[45/0/-45/90]_{ns}$ laminate specimens, with compressive interlaminar normal stresses, had fatigue lives in the same decade at 414 MPa (-60 ksi) as did the $[90/45/0/-45]_{ns}$ laminate, with tensile interlaminar normal stresses, and a stress of 331 MPa (-48 ksi). However, differences in the fatigue lives as a function of thickness are not so evident. Since the interlaminar normal stress varies by a factor of two between 16 and 32-plies and another factor of two between 32 and 64-plies, it seems unlikely that the interlaminar normal stress is the sole or dominant contributor toward damage and failure. It is more likely that damage is brought about by the combination of the tensile or compressive normal stress and the relatively high interlaminar shear stress τ_{xz}. This will be more fully explored in the following paragraphs by examining the location, initiation, and progression of damage.

It is evident from the nondestructive examinations that the interlaminar stresses are the dominant factor in the damage development in the unnotched specimens. Comparing the X-rays of the two geometries (Figs. 4 and 7), it is evident that matrix cracking is much more prevalent in the notched specimens. In fact the damage around the holes may begin as matrix cracks. On the other hand the damage in the unnotched specimens is predominantly delaminations. There are far fewer matrix cracks, and those that are present are associated with the large out-of-plane buckling of the plies that are split from the remaining undamaged laminate. This large, localized out-of-plane displacement and associated broken plies is illustrated by the edge replica shown in Fig. 6.

Finally, the sequence of X-rays shown in Figure 7 demonstrates that the initiation and progression of damage is an edge effect. The delamination begins at the edge and runs along the length of the specimen before penetrating through the specimen width.

Now that delaminations have been established as the primary type of damage, the relationship between the interlaminar stress distribution and the interfacial location of the delaminations will be examined. The location of delaminations have been determined from edge replicas and are tabulated in Table 1 for the $[45/0/-45/90]_{ns}$ laminate and Table 2 for the $[90/45/0/-45]_{ns}$ laminate for several typical specimens. (A complete listing is given in Ref 3.) A cursive review of these tables quickly establishes the $-45/0$ interface (plies located from laminate midplane) as the most common location of delaminations regardless of laminate type. Also the adjacent interface, $0/45$, is the location of almost all other delaminations. A review of the interlaminar stress distributions, Figs. 8 through 11, provides an explanation for the preference of these two interfaces for delaminations. The interlaminar shear stress, τ_{xz}, is a maximum value at these two interfaces for both laminate types. It is also of comparable magnitude and algebraic sign for both laminate types and thicknesses. In the case of the $[90/45/0/-45]_{ns}$ laminate the interlaminar normal stress is slightly more tensile at the $-45/0$ interface than at the $0/45$ interface. This would make the $-45/0$ interface slightly more preferential than the $0/45$ as is supported by the data in Table 2. One might expect just the opposite to be the case for the $[45/0/-45/90]_{ns}$ laminate because the $-45/0$ interface has a higher compressive normal stress. However, the data in Table 1 still indicates the $-45/0$ interface to be preferential. Perhaps this is because at the first, or outermost delaminations the magnitude of the interlaminar normal stress is small relative to the τ_{xz} shear stress so it may have little to do with the formation of the delamination. Moving inward to the second or third interior delamination, the interlaminar normal stress is higher, substantially so for the 16 and 32-ply laminates. Again one sees from the tables that the preference is still at the $-45/0$ or $0/45$ interfaces, with the $-45/0$ being more preferential than $0/45$ for both laminate types. The τ_{xz} shear stress has the same distribution and magnitude in interior stacks of $[90/45/0/-45]$ plies as it does in the outerplies. Therefore, since the delaminations are forming at the interfaces of maximum shear stress regardless of laminate type, that is, without regard to $+\sigma_{zz}$ or $-\sigma_{zz}$, it follows that the location of maximum interlaminar shear stress, τ_{xz}, determines the preference for delamination locations.

Delamination interface changes also appear to be consistent with the interlaminar shear stress distribution. In some instances a delamination at the $90/-45$ interface changed to the more preferential $-45/0$ interface. Also in one instance an interface change between the $-45/0$ and $0/45$ occurred. This might be expected since they have equal τ_{xz} shear stresses. Furthermore in those instances where a delamination occurred at the $90/-45$ or $45/90$ interface, they always appeared to be associated with an interface change to the more preferential $-45/0$ interface.

Discussion of Test Results

There have been a number of fatigue studies in which quasi-isotropic graphite/epoxy laminates were tested and studies in which the role of interlaminar stresses at the free edges in the damage development process have been investigated. However, none of these studies examine the laminate thickness effect, and none were for the specific laminate stacking sequences studied herein. In order to put the results of this study in perspective some comparisons are nonetheless available and will be cited below.

Whitney and Kim [7] studied the static fracture of $[90/0/\pm45]_s$ and $[\pm45/0/90]_s$ laminates which have high tensile and high compressive interlaminar normal stresses at straight free edges. They found the tensile strength of unnotched specimens to be reduced by the tensile interlaminar normal stress, but found no difference in the strength of notched specimens of the two different laminates provided the geometry was such that damage began at the notch rather than the straight free edge. This is consistent with the results herein where there was no appreciable difference in the fatigue lives of the two notched laminates.

The damage development in the unnotched specimens are strongly related to the τ_{xz} interlaminar shear stress distribution. Delaminations occurred where the τ_{xz} stress is maximum regardless of the magnitude and algebraic sign of the interlaminar normal stress. There are similar results in the literature. Stalnaker and Stinchcomb [8] studied $[0/\pm45/90]_s$ and $[0/90/\pm45]_s$ laminates and found that the first laminate formed delaminations primarily produced by tensile interlaminar normal stresses. In both laminates they reported branch delaminations at the interfaces where high τ_{xz} shear stresses existed even when the interlaminar normal stress is compressive. Herakovich et al [6] performed an analytical study using a tensor polynomial failure criterion in which quasi-isotropic graphite/epoxy laminates were included and found that the interlaminar normal stress was never the dominant stress for initiation of failure. Failure was either due to τ_{xz} or a combination of τ_{xz} and tensile σ_{zz}. Finally, Whitcomb [9] studied graphite/epoxy quasi-isotropic laminates experimentally and with a finite-element analysis. His analysis results and experimental results compared well, and he reported that delaminations were more likely to occur in areas where the interlaminar normal stress is high tensile and the τ_{xz} shear stress is high. Whitcomb [9] further reported that delaminations were found and predicted to occur in regions where the τ_{xz} stress is high but the interlaminar normal stress is compressive.

Conclusions

The notched specimens exhibited essentially no difference in fatigue lives between the $[45/0/-45/90]_{ns}$ and $[90/45/0/-45]_{ns}$ laminates (based on three replicate tests). Furthermore, there was no strong thickness effect although the highest lives belonged to the 64-ply specimens, and the lowest lives belonged to the 16-ply specimens. It was concluded that the stress concentration effect at the hole dominates over the interlaminar stresses in precipitating damage and failure.

The unnotched specimens did exhibit a difference in the fatigue lives between the two laminates. The $[90/45/0/-45]_{ns}$ laminate with tensile interlaminar normal stresses delaminated more readily than did the other laminate. The fatigue lives of both laminate types were somewhat affected by specimen thickness as evidenced by the fact that the greater lives belonged to the thicker specimens. The damage in both laminates was predominantly delaminations which began at the edge, progressed along the specimen length, then through the width and along the adjacent edge. There was a strong correlation between the interlaminar stress distributions and the formation of delaminations. The τ_{xz} shear stress was found to be the dominant interlaminar stress because delaminations formed at the interface where τ_{xz} was a maximum regardless of the normal stress, σ_{zz}, being tension or compression.

The antibuckling fixtures for both specimen types were effective in preventing specimen buckling. In neither case did the fixture appear to overconstrain the specimen and interfere or influence the formation and progression of damage.

Enhanced X-ray and edge replication form a good nondestructive examination team for defining fatigue damage when edge effects dominate. Specimen sectioning is probably necessary to define through-the-thickness damage of notched specimens.

It should be noted that the conclusions stated previously are for the specific laminates included in this study. The results and conclusions herein may not be generally applicable to other laminate types.

Acknowledgment

The financial support provided for this work by NASA Grant NAG-1-264 from the Fatigue and Fracture Branch of NASA-Langley is gratefully acknowledged. Further, sincere appreciation is extended to E. P. Phillips of NASA-Langley for designing the compression test fixture and for his encouragement and helpful discussions. The authors wish to thank Dr. C. T. Herakovich for providing discussion, consultation, and information concerning the computation and interpretation of interlaminar stresses.

References

[1] Pagano, N. J. and Pipes, R. B., *Journal of Composite Materials*, Vol. 4, 1970, pp. 538–548.

[2] Rooney, M. and Herakovich, C. T., "An Approximate Method for Optimizing Stacking Sequence in the Presence of Edge Effects," (to be published as a Virginia Polytechnic Institute and State University report), Blacksburg, VA.

[3] Harris, C. E. and Morris, D. H., "An Evaluation of the Effects of Stacking Sequence and Thickness on the Fatigue Life of Quasi-Isotropic Graphite/Epoxy Laminates," VPI-E-83-16, Virginia Polytechnic Institute and State University, Blacksburg, VA, 1983.

[4] Phillips, E. P., "Effects of Truncation of a Predominantly Compression Loan Spectrum on the Life of a Notched Graphite/Epoxy Laminate," NASA Technical Memorandum 80114, National Aeronautics and Space Administration, Hampton, VA, 1979.

[5] Stalnaker, D. O. and Stinchcomb, W. W., "An Investigation of Edge Damage Development in Quasi-Isotropic Graphite Epoxy Laminates," Technical Report VPI-E-77-24, Virginia Polytechnic Institute and State University, Blacksburg, VA, Sept. 1977.

[6] Herakovich, C. T., Nagarkar, A, and O'Brien, D. A., *Modern Developments in Composite Materials and Structures*, J. R. Vinson, Ed., The American Society of Mechanical Engineers, pp. 53–66.

[7] Whitney, J. M. and Kim, R. Y., *Composite Materials: Testing and Design (Fourth Conference), ASTM STP 617*, American Society for Testing and Materials, Philadelphia, 1977, pp. 229–242.

[8] Stalnaker, D. O. and Stinchcomb, W. W., *Composite Materials: Testing and Design (Fifth Conference, ASTM STP 647*, American Society for Testing and Materials, Philadelphia, 1979, pp. 620–641.

[9] Whitcomb, J. D., "Experimental and Analytical Study of Fatigue Damage in Notched Graphite/Epoxy Laminates," NASA Technical Memorandum 80121, National Aeronautics and Space Administration, Hampton, VA, 1979.

G. R. Kress[1] and W. W. Stinchcomb[1]

Fatigue Response of Notched Graphite/Epoxy Laminates

REFERENCE: Kress, G. R. and Stinchcomb, W. W., "Fatigue Response of Notched Graphite/Epoxy Laminates," *Recent Advances in Composites in the United States and Japan, ASTM STP 864*, J. R. Vinson and M. Taya, Eds., American Society for Testing and Materials, Philadelphia, 1985, pp. 173–196.

ABSTRACT: Although composites have gained widespread use, we do not, as yet, have a precise and complete understanding of the mechanisms of damage development in composite materials. Recent research results have pointed out the need to treat damage as a collective condition—that is, a damage state, rather than as an assembly of discrete and independent damage modes. This paper presents the results of an investigation to determine the damage states which develop in two quasi-isotropic graphite/epoxy laminates with center holes due to tension-tension cyclic loads, to determine the influence of stacking sequence on the initiation and interaction of damage modes and the process of damage development, and to establish the relationships between the damage states and the strength, stiffness, and life of the laminates. Damage was monitored nondestructively throughout the loading history using X-ray radiography, moiré interferometry, and stiffness change. Some specimens were deplied after specific numbers of cycles to determine the distribution of damage in each ply around the hole and to confirm the components and size of the damage state observed nondestructively. Fatigue life and residual strength tests were also performed.

Fatigue damage in the two laminates included matrix cracks in all plies followed by delaminations. The density of matrix cracks and the distribution of the damage zone (matrix cracks plus delaminations) in laminates cycled at the same percent of notched tensile strength were strongly dependent on the local constraint and distribution of interlaminar stresses as governed by the stacking sequence. The distinctly different damage states which developed in the two initially quasi-isotropic laminates due to similar load histories produced stiffness changes of 15% to 20%, different rates of change in residual strength (an initial increase followed by a decrease), and a factor of two to four difference in fatigue life.

KEY WORDS: composite materials, notched laminates, fatigue, damage stiffness, strength, life, nondestructive evaluation

[1]Engineering Science and Mechanics Department, Virginia Polytechnic Institute and State University, Blacksburg, VA 24061-4899. The first author is presently a researcher at the Institute for Structural Mechanics, DFVLR, Braunschweig, West Germany.

The response of composite materials and structures to time-varying loads must be understood to insure the reliability and survivability of composite structures. One particularly important aspect of the fatigue of composite materials is the behavior of laminates with geometric discontinuities such as holes, joints, ply terminations, and inherent defects. This paper addresses the specific subject of fatigue response of graphite/epoxy laminates with a center hole subjected to tension-tension cyclic loads. The objectives of the investigation are to:

1. determine the nature of damage induced in graphite/epoxy laminates with center holes by cyclic tensile loading, and

2. establish the influence of damage on strength, stiffness, and life of the laminates.

Two quasi-isotropic laminates were selected to give different distributions of interlaminar stresses around the hole. The laminates were tested under cyclic stresses ranging between 65% and 95% of the tensile strength of notched laminates. Damage was monitored nondestructively using X-ray radiography and stiffness change. Some specimens were deplied to confirm the components and size of the damage zone observed nondestructively. Fatigue life and residual strength data were recorded.

Materials and Specimens

T300/5208 graphite/epoxy prepreg tape 305 mm (12-in.) wide was layed up at NASA-Langley to form 406- by 305-mm (16 by 12-in.) panels with two stacking sequences: $[0,90,\pm45]_s$ (Type A) and $[45,90,-45,0]_s$ (Type B). Specimens 254 mm long by 38 mm wide (10 by 1.5 in.) were cut from each panel and a 9.5-mm-diameter (0.375-in.) center hole was ultrasonically drilled in the center of each specimen. The two stacking sequences give different interlaminar stresses at the hole and at the free edge. The Type A laminate has high interlaminar shear stresses at the $+45/-45$ interfaces whereas these interfaces are not present in the Type B laminate. The interface between 45- and 90-deg plies is common to both laminates and is a region of tensile interlaminar normal stress and high interlaminar shear stress in both laminates [1].

Experimental Procedure

Five specimens of each type of laminate were selected for monotonic tension tests to determine tensile strength and Young's modulus. Two aluminum pads, approximately 6.4 mm square by 1.5 mm thick (0.25 in. square by 0.06 in. thick) were bonded to one side of the specimen using a silicone adhesive. The tabs had "V" notches machined in them and were positioned above and below the hole so that the knife edges of a 25.4-mm (1.0 in.) extensometer

located in the notches centered the extensometer over the hole. The specimens were hydraulically gripped and loaded to fracture in an 89-kN (20-kip)
MTS loading frame.

Tensile strength and modulus data for the two laminates are given in Table
1. The mean value and standard deviation of tensile strength is slightly
greater for Type A notched laminates than for Type B notched laminates. The
first indication of damage, as detected by acoustic emission and load-strain
data, occurred at a lower load in the Type B laminate than in the Type A
laminate, although the load at "first damage" corresponded to approximately 60% of the failure load in each case.

Fatigue tests were run on the Types A and B laminates in load control at 10
Hz with a stress ratio $R = 0.1$. S-N curves for the two laminates are shown in
Fig. 1. The fatigue strength of Type A laminates is greater than that of Type B
laminates. The Type A laminates that were cycled at a maximum stress of
80% of their mean tensile strength and the Type B laminates that were cycled

TABLE 1—*Tensile strength and modulus data.*

	Mean Tensile Strength, MPa	Standard Deviation, MPa	Mean Young's Modulus, GPa[a]	Standard Deviation, GPa
Type A	273	20	34.8	0.86
Type B	251	7	33.9	0.78

[a]Modulus values computed using loading direction stress and strain data.

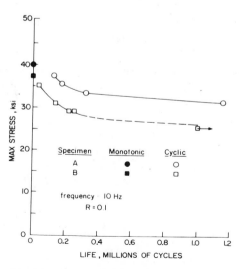

FIG. 1—S-N *curves for Types A and B laminates. (1 ksi = 6.895 MPa).*

at a maximum stress of 70% of their mean tensile strength survived one million cycles.

S-N curves of the type shown in Fig. 1 are useful for estimating expected life when laminates are cycled at a constant stress amplitude. However, the curves do not provide information on damage initiation and the rate of damage growth during the fatigue lifetime. A record of stiffness change throughout the fatigue life provides important information on the continuous damage process. Figure 2 shows the degradation of stiffness due to damage in Types A and B laminates cycled at maximum stresses corresponding to 90% of their mean tensile strength. The tests were terminated near the end of the fatigue life, as estimated from *S-N* data. The cycle axis is logarithmically scaled, and the stiffness data are not recorded by the data acquisition system during the first 100 cycles (10 s) of the test. From these curves it is clear that the damage process begins early in the fatigue life of both laminates, with greater stiffness reduction in the Type A laminate. However, the rate of damage development can be misinterpreted because of the logarithmic cycles axis. For example, the sharp decrease in stiffness near the end of the fatigue life is not a "sudden death" as it might appear using a logarithmic scale. These stiffness changes actually occur between 40 000 and 180 000 cycles (during approximately 78% of the fatigue life) for the Type A specimen and between 125 000 and 160 000 cycles (during approximately 22% of the fatigue life) for the Type B specimen. To illustrate this point, the stiffness data are replotted in Fig. 3 using a linear cycles scale. Both specimens suffer an initial stiffness decrease during the early portion of the fatigue life. The first stage of stiffness reduction is followed by a second stage during which the stiffness degradation rate is lower. A third stage of stiffness reduction is marked by an increase in the stiffness degradation rate. The third stage of stiffness reduction begins after approximately 22% of the life of the Type A laminate and after approximately 78% of the life of the Type B laminate.

At selected times in the loading history, nondestructive inspection of the damage was made using zinc iodide enhanced X-radiography. The deply technique [2] was also used to identify the interfaces on which delamination occurred.

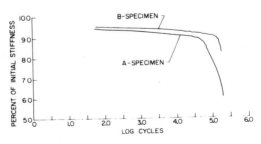

FIG. 2—*Stiffness as a function of log cycles for Types A and B specimens, $\sigma_{max} = 0.9\ \sigma_{ult}$.*

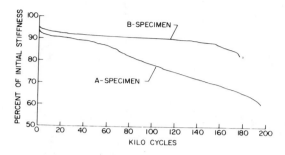

FIG. 3—*Stiffness as a function of linear cycles for Types A and B specimens,* $\sigma_{max} = 0.9\ \sigma_{ult}$.

Fatigue Damage

Data on the early stage of damage development were obtained by sequentially loading a Type A and a Type B specimen to introduce damage, unloading the specimens and making zinc-iodide enhanced X-ray radiographs, reloading the specimens to a higher load, and repeating the process until the specimens failed. The first damage in the Type A specimen was detected after loading to 60% of the mean tensile strength for this laminate. Short matrix cracks approximately 5 mm (0.2 in.) long formed in the zero-deg plies tangent to each side of the hole. Several shorter cracks in the 90-deg plies also formed along the sides of the hole. Unloading and reloading to a 70% stress level extended the cracks in the zero- and 90-deg plies, increased the number of cracks in the 90-deg plies, and introduced cracks in the +45-deg and −45-deg plies. Repeated unloading and loading to higher stress levels (80%, 90%, 95%, 100%, 105%, and 110% of mean tensile strength for Type A laminates) produced an increase in density and length of cracks in the non-zero-deg plies and a slight increase in the length of the two zero-deg ply cracks. The first delaminations were detected in the 90/45 interface after loading to the 80% stress level and are more clearly observed in the radiograph made after the 90% stress level, Fig. 4a. The radiograph after the 105% stress level shows the progressive development of matrix cracks and delamination, Fig. 4b. The Type A specimen failed at a stress of 113% of the mean tensile strength upon reloading after the 110% stress level.

The first matrix cracks in the Type B specimen were observed on radiographs made after loading to 50% of the mean tensile strength of Type B laminates. Cracks formed in the +45-deg plies on lines passing through or near the center of the hole. A single, short crack in the 90-deg plies was also detected. Unloading and reloading to the 60% stress level introduced matrix cracks in all plies with cracks in the zero-deg plies forming along tangents to the hole. Subsequent unloading and reloading to higher stress levels (80, 90, 95, 100, 105, 110, 115, 120, 125, and 130%) increased the density and extent of the cracks in the nonzero-deg plies. The two tangent zero-deg ply cracks

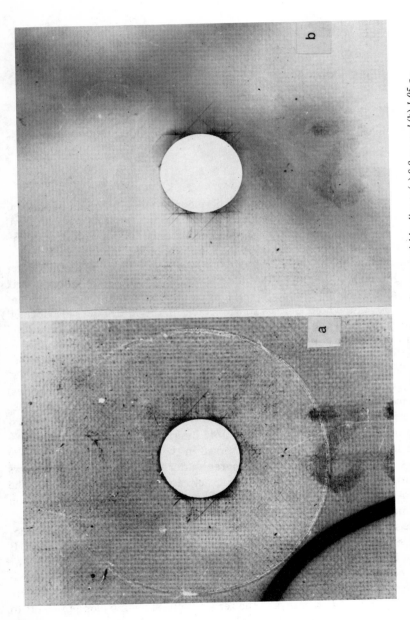

FIG. 4—*Radiographs of damage in a Type A specimen after sequential loading to* (a) *0.9* σ_{ult} *and* (b) *1.05* σ_{ult}.

grew to a length of one hole diameter after loading to the 100% stress level. Continued sequential loading increased the rate of growth of these two cracks. Edge cracks in the +45-deg ply were first noted in the 90% radiograph, Fig. 5a. Delaminations were first detected on the 60% radiograph but are more apparent in the 45/90 and 90/−45 interfaces on the 80% radiograph. The radiograph shown in Fig. 5b, made after loading to the 130% stress level, shows extensive matrix cracking and delamination. The apparent damage in the Type B laminate is much more widespread than in the Type A laminate, Fig. 4b, even though the Type B laminate failed at a higher stress level (135% of the Type B mean tensile strength). The apparent increase in strength will be discussed later in the paper. It is important to remember that although the two specimens were loaded quasi-statically, the specimens were cyclically loaded to successively higher loads each time a radiograph was made. Therefore, the damage must be considered as progressive development, or propagation, of the damage modes which initiated at a low stress level.

The observations on the locations of initial delaminations in Types A and B laminates agree with the analysis of O'Brien and Raju [3]. However, the stress levels at which the delaminations initiated were lower than those predicted by the strain energy release rate analysis, which considered delamination as the only damage mode. Experimental data for unnotched T300/5208 quasi-isotropic laminates by O'Brien [4] show that delaminations initiate in the laminates with matrix cracks at lower critical strain energy release rates than in the laminates without matrix cracks. This observation is supported by the results of Jamison and Reifsnider [5] and Wang [6]. The fact that the measured strains at initiation of delamination in Types A and B laminates, in which matrix cracking was the first damage mode, were lower than the initiation strains predicted by the analysis considering only delamination is consistent with other reported results. Matrix cracking is an important phase of the damage development process. Although matrix cracks, by themselves, may not significantly affect the strength and life of structural laminates, they do act as catalysts to initiate other strength and life-limiting damage modes such as delamination and fiber fracture [5].

Series of constant amplitude tension-tension fatigue tests were conducted on Types A and B specimens to monitor damage development and measure the corresponding residual strength. Radiographs of damage in Type A specimens cycled at the 80% stress level (ratio of the maximum cyclic stress to the mean tensile strength) and Type B specimens cycled at the 85% level are shown in Figs. 6 and 7. Data for other stress levels are given in Ref 7.

Figure 6a shows matrix cracks in all plies of a Type A specimen after only ten cycles. The density and length of the cracks in the nonzero-deg plies increase throughout the loading history (10^3, 10^4, 10^5, and 6×10^5 cycles). The two tangent zero-deg ply cracks increase in length throughout the cyclic loading and a few additional cracks in the zero-deg plies are shown in the 6×10^5

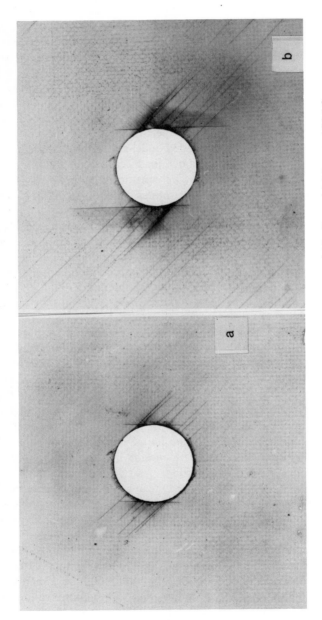

FIG. 5—*Radiographs of damage in a Type B specimen after sequential loading to* (a) *0.9* σ_{ult} *and* (b) *1.3* σ_{ult}.

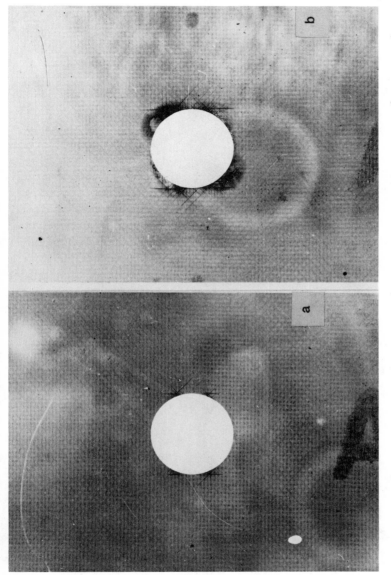

FIG. 6—*Radiographs of damage in Type A specimens cycled at $\sigma_{max} = 0.8\ \sigma_{ult}$ for (a) 10 cycles, (b) 10^3 cycles, (c) 10^4 cycles, (d) 10^5 cycles, and (e) 6×10^5 cycles.*

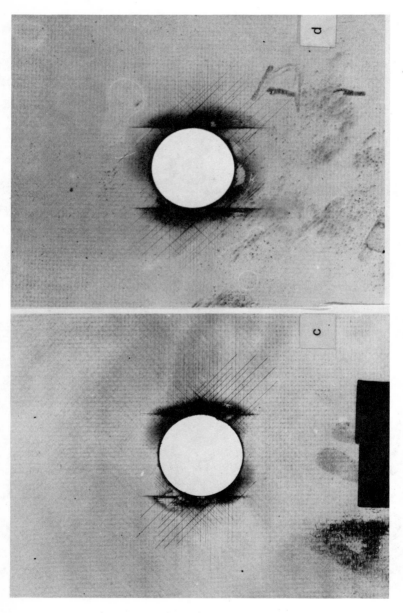

FIG. 6—Continued.

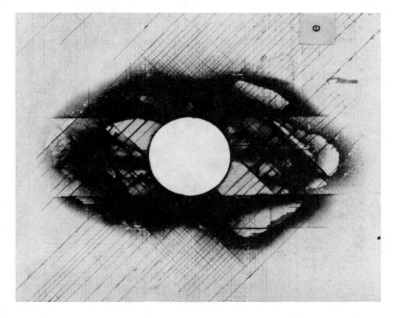

FIG. 6—*Continued.*

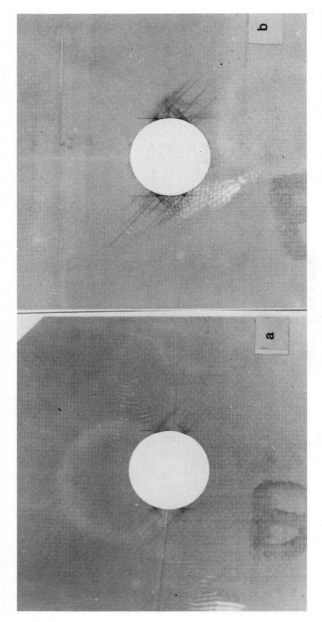

FIG. 7—Radiographs of damage in Type B laminates cycled at $\sigma_{max} = 0.85\ \sigma_{ult.}$ for (a) 10 cycles, (b) 10^3 cycles, (c) 10^4 cycles, and (d) 10^5 cycles.

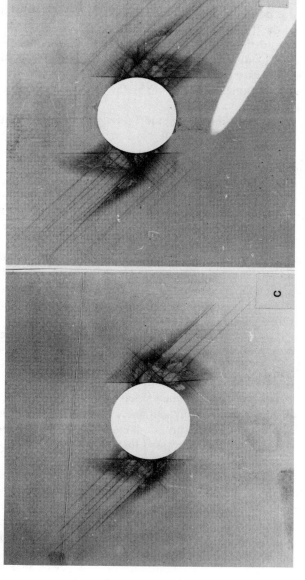

FIG. 7—*Continued.*

cycles radiograph. In the same radiograph, edge-initiated cracks in the nonzero-deg plies can be seen. Delamination at the hole has developed by 10^3 cycles and continues to grow as shown in the subsequent radiographs. More information on the size and location of the delaminated regions will be given in the discussion of the results of deplying.

The progressive development of matrix cracks in the Type B laminate is similar to that in the Type A laminate. After only ten cycles, cracks are evident in all plies, and the density and length of cracks in nonzero-deg plies increase during the loading history. Only two zero-deg ply cracks, tangent to the hole, can be seen after 10^5 cycles. Delaminations, which are present by 10^3 cycles, appear to be bounded on one side by the zero- and 45-deg ply matrix cracks which are tangent to the hole.

The radiographs shown in Figs. 6 and 7 are projections of damage in a finite volume of material onto a plane. The location of the damage in the thickness direction cannot be determined from these radiographs without some degree of speculation. Furthermore, the shape and size of the delaminated region on a given interface cannot be determined from these radiographs. The stero-radiography technique [8] and the deply technique [2] can be used to resolve damage in the thickness direction. The deply technique was selected in this investigation because it gives details of delamination zone size and shape for each deplied layer.

Prior to deplying, the specimens were loaded to a low, constant-tension load while the damaged region was painted with gold chloride, which infiltrates and stains the damaged material. The stained damaged region can be easily identified using a light or scanning electron microscope. The specimens were deplied using a thermal soak at 418°C (784°F) for 30 to 50 min in an electrically heated oven with controlled air exchange. This process causes the destruction of the resin-rich interfaces, which can then be separated. Layers to be deplied are covered with masking tape and lifted from the laminate using a razor blade to assist the separation.

The Type A and Type B specimens cycled at the 90% stress level to the third stage of stiffness reduction (Figs. 3 and 4) were selected for deply. Figures 8 and 9 show the essential features of damage on interfaces between plies whose orientation is shown by the crossed lines. Major delaminations on the 0/90 interface of the Type A specimen are bounded by the tangent cracks in the zero-deg plies and extend into the ligaments on each side of the hole (see Figs. 6c and 6d). Smaller delaminations are associated with small cracks in the zero-deg ply above and below the hole (see Fig. 6e). These cracks and delaminations appear in the final stage of fatigue life. Delaminations on the 90/45 interface are bounded by cracks in these plies. Delaminations on the 45/−45 interface completely surround the hole.

The delaminated regions in the B laminate are generally smaller than in the A laminate; however, direct comparisons of the delaminations in the two laminates are not possible because the different orientations of the adjacent plies

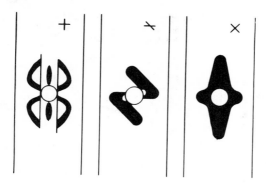

FIG. 8—*Damage on 0/90, 90/45, and 45/−45 interfaces of a Type A laminate after 180 000 cycles at $\sigma_{max} = 0.9\ \sigma_{ult}$.*

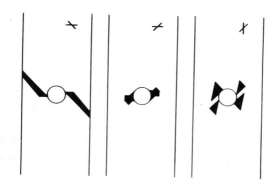

FIG. 9—*Damage on 45/90, 90/−45, and −45/0 interfaces of a Type B laminate after 160 000 cycles at $\sigma_{max} = 0.9\ \sigma_{ult}$.*

produce different constraints and different stresses on the interfaces. The constraints on the 45/90 interface in the Type B laminate are similar to those on the 90/45 interface in the Type A laminate, and the delaminations on both interfaces are bounded by cracks in the 45- and 90-deg plies.

Residual Strength and Stiffness

Figure 10 shows residual strength and stiffness data for Type B laminates cycled at the 85% level. The corresponding damage patterns are shown in Fig. 7. The two stiffness curves show the effect of gage length on the stiffness values determined from quasi-static stress-strain data recorded at the prescribed number of cycles. A clip-type strain gage extensometer centered over the hole was used to measure displacement over a 25.4-mm (1.0 in.) gage length. A direct-current differential transformer (DCDT) extensometer cen-

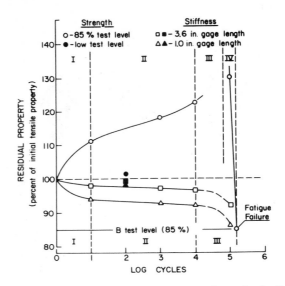

FIG. 10—*Stiffness and residual strength as a function of log cycles for Type B laminates,* $\sigma_{max} = 0.85 \ \sigma_{ult}.$

tered over the hole measured displacement over a 92-mm (3.625 in.) gage length. The stiffness change values based on the short extensometer data are larger. However, if the damage zone spreads outside the 1.0 gage length, the stiffness change values determined from the short extensometer are inaccurate.

The effect of damage development around the hole on the response of a Type B laminate is indicated by the stiffness curves shown in Fig. 10. The limits of Regions I, II, and III for stiffness change are somewhat arbitrary, but are chosen to reflect transitions in the damage growth process, Fig. 7, as represented by stiffness change. In Region I, the stiffness decreases at a decreasing rate as small matrix cracks develop near the hole. In Region II, the stiffness degrades linearly with log cycles and the matrix cracks in the zero- and 45-deg plies extend away from the hole and delaminations form in the region damaged by matrix cracks. Region III begins as the stiffness degradation rate increases. In this region, the sharp decrease in stiffness is associated with further extension of the cracks in the zero- and 45-deg plies, an increase in the density of the cracks in the 45-deg plies, and growth of delamination along the 45-deg matrix cracks. Early in the life of the laminate (Region I) the damage is confined to the approximate zone of stress concentration around the hole. In later stages of life, the stresses around the hole are redistributed and the stress concentration around the hole is changed due to the growth of the damage zone away from the hole.

The change in stress concentration at the hole due to damage is reflected by a change in the tensile strength of the notched laminates. During the first few (ten) cycles, the damage, primarily matrix cracks, reduces the effect of the stress concentration and the residual strength increase is on the order of 10%. The actual strength increase does depend on the cyclic stress level, with the higher stress levels corresponding to slightly higher strength increases than the lower stress levels during the early part of fatigue life. Stiffness change and residual strength data for a Type B laminate after 100 cycles at a low cyclic stress amplitude are also shown in Fig. 10. Both the increase in residual strength and the decrease in stiffness are less than those for the same type of specimen cycled at the 85% stress level.

The second region of the residual strength curve, approximately 10% of the fatigue life, is characterized by a slow but constant increase in residual strength over a logarithmic cycles scale. During this time, the density and length of matrix cracks continue to increase and delaminations form. At the end of the second region, the increase in the residual strength is 115 to 120%. The limits of Regions I and II for residual strength are approximately the same as observed for stiffness change.

The third region of the residual strength curve is marked by a further increase in residual strength. Although the data in this region are incomplete, it appears that the maximum residual strength is reached between 50 and 80% of the lifetime. The largest increase in residual strength was 42% for the Type B laminates. During the final region of the fatigue life, the residual strength decreases until the strength equals the level of maximum cyclic stress. The last two regions of the residual strength-log cycles curve correspond to the third region of the stiffness reduction curve where the stiffness change shows a sharp decrease on the log cycles plot. Figure 11 shows the residual strength data for the damaged Type A laminates shown in Fig. 6. During the first 20%

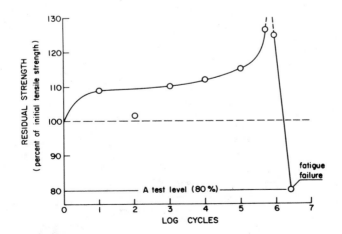

FIG. 11—*Residual strength as a function of log cycles for Type A laminates.* $\sigma_{max} = 0.80 \, \sigma_{ult}$.

of fatigue life, matrix cracks and delaminations form around the hole and the tensile strength of the laminate increases. As damage continues to develop, the strength of the laminate decreases to the level of the maximum applied cyclic stress, and fracture occurs. The logarithmic scale used to represent the lifetime may cause some confusion in interpreting these results. The residual strength data have been replotted in Fig. 12 using a linear scale to represent cycles.

The appearance of fracture surfaces depends on the loading history of the specimens. A typical static failure fracture surface of a Type A specimen is located along a line passing through the hole, Fig. 13a. Specimens with fatigue damage resulting in an increased residual strength and loaded monotonically to failure have a less localized fracture surface, as recorded in Fig. 13b for a Type A laminate after 500 000 cycles at the 80% stress level. Figure 13c shows the widely distributed damage and fracture surface of a highly damaged specimen which failed in fatigue after 1.3 million cycles at the 80% stress level.

A speculative explanation of the paradox of fatigue damage causing an increase in strength of notched laminates has been proposed by several investigators [9–13]. Some of the damage which develops at the hole reduces the local concentration of stress (wear-in), thereby increasing the apparent tensile strength of the notched laminate. Theoretically, the tensile strength should approach the tensile strength of the unnotched ligaments of material adjacent to the hole as the stress concentration factor approaches unity. Moiré interferometry data for the damaged, notched laminates provide some justification for this argument. The moiré technique is an accurate method for determining the two-dimensional displacement field on the surface of a material

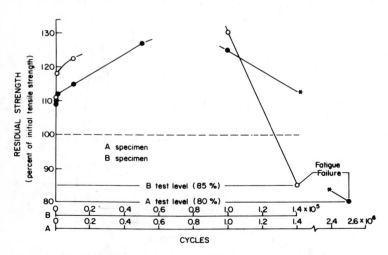

FIG. 12—*Residual strength as a function of linear cycles for Types A and B laminates.*

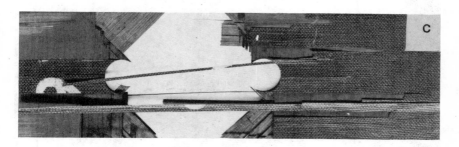

FIG. 13—*Fracture surfaces of Type A laminates:* (a) *monotonic tension, no prior damage;* (b) *monotonic tension after 500 000 cycles at* $\sigma_{max} = 0.8\ \sigma_{ult}$; *and* (c) *fatigue failure after 1.3 million cycles at* $\sigma_{max} = 0.8\ \sigma_{ult}$.

[14]. As damage initiates and grows within a laminate, for example, the changes in the surface displacement field affected by that damage can be observed and measured using the moiré technique [15]. In this investigation, moiré interferometry was used to measure the damage-related reduction in the local, load direction strain on the surface of A and B laminates. The details of the experimental procedure are given in Ref 7. Briefly, the method employs two gratings with pitches on the order of the wavelength of light. As the active grating attached to a specimen undergoes inplane deformation relative to the reference grating, wave trains of light passing through the gratings combine to produce steady-state constructive and destructive interfer-

ence. If a photographic plate is placed in the path of the light interference zones, the plate will be exposed in regions of constructive interference to produce a fringe pattern where the moiré fringes are the locus of points of constant displacement. The 60 000-lines-per-inch grating on the specimen was produced by a photoetching technique. A mechanical loading fixture was used in recording moiré fringe patterns of the specimens under load.

Figure 14 is the moiré fringe pattern for load direction displacements in a Type A laminate after ten cycles at the 80% stress level. The matrix cracks, of length ℓ_1, in the zero-deg plies and the associated fringes can be easily identified. With the exception of the crack tips, the strain along the crack and along the edge of the laminate is nearly uniform. The details of the fringe pattern at the crack tip are difficult to resolve. Recalling that a fringe is the

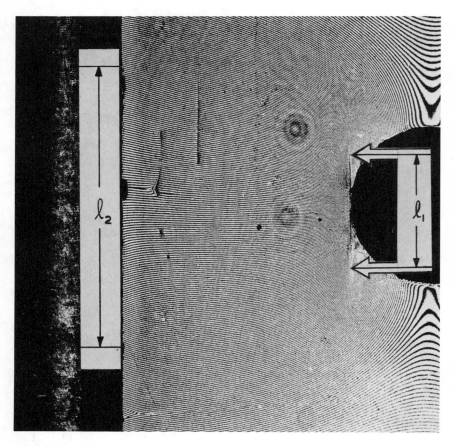

FIG. 14—*Moiré fringes in a Type A laminate with fatigue damage after ten cycles at $\sigma_{max} = 0.8\ \sigma_{ult}$.*

locus of points of constant displacement, the fringes emerging from the crack tips are traced to the edge of the specimen where the distance between them, ℓ_2, is measured. Since the displacements over the distances ℓ_1 and ℓ_2 are equal, the ratio of ℓ_2 to ℓ_1 is a measure of the nominal strain concentration associated with the cracks and the hole.

Figure 15 shows that the strain concentration decreases as the average length of the two zero-deg ply cracks increases. Noting that the residual strength of the damaged laminates increased as the length of the matrix cracks increased during the early part of the fatigue life, Fig. 16 shows the

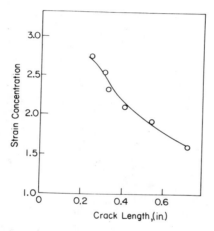

FIG. 15—*Strain concentration as a function of length of cracks in the zero-deg plies of Type A laminates (1 in. = 25.4 mm).*

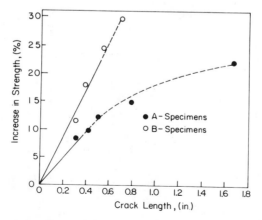

FIG. 16—*Increase in tensile strength of Types A and B laminates as a function of length of cracks in the zero-deg plies (1 in. = 25.4 mm).*

correlation between residual strength and the average length of the two tangent zero-deg ply cracks. The residual strength data are plotted against the corresponding measured values of strain concentration in Fig. 17 to show that the residual strength increase corresponds to the reduced strain concentration caused by the matrix cracks in the zero-deg plies.

The apparent increase in tensile strength due to fatigue damage and the reduction in tensile strength which leads to fatigue failure are due to the way in which the laminate stresses are redistributed due to damage. Early in the fatigue life, the matrix cracks and small delaminations effectively increase the radius of curvature of the hole to reduce the strain concentration and isolate ligaments on each side of the hole. As the ligaments form, the tensile strength of the laminate increases. As damage continues to develop and grows into the ligaments, the rate of strength increase diminishes until finally the damage in the ligaments reduces the strength. Continued cyclic loading leads to fatigue failure.

Conclusions

Damage

1. X-radiography (with zinc iodide penetrant), the deply technique, the moiré method, and stiffness measurements provide information on damage modes (matrix cracking and delamination), distribution of damage, and the consequences of damage on laminate response.

2. The modes and distribution of damage and attendant fatigue response of the notched laminates are stacking sequence dependent.

3. Chronology of damage consists of matrix cracks at the hole followed by

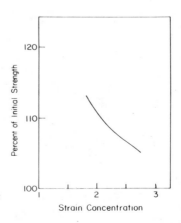

FIG. 17—*Increasing residual strength with decreasing strain concentration for Type A laminates.*

delaminations. The process of ply cracking is a function of stacking sequence. The initiation and growth of delaminations are influenced by the existing matrix cracks. Both damage modes are dependent on *local* stress fields.

Response

1. The increase in residual strength during the early portion of life is due to the redistribution of the stress field around the hole caused by "damage." The reduction in strain concentration in Type A and Type B laminates can be related, in part, to zero-deg ply matrix cracks tangent to the hole.
2. The reduction in laminate strength leading to fatigue failure is due to damage modes (cracks and delaminations) which develop in the ligaments adjacent to the hole during the middle and final stages of fatigue life.

Acknowledgment

The authors gratefully acknowledge the support of the National Aeronautics and Space Administration (NASA) Langley Research Center under Grant NAG-1-232, monitored by Dr. T. K. O'Brien. Special appreciation is extended to Dr. Daniel Post, Dr. Anand Asundi, and Mr. Robert Czarnek for their advice in planning and conducting the moiré experiments. We also extend our sincere appreciation to Mrs. Barbara Wengert and Mrs. Charlene Christie for typing the manuscript.

References

[1] Kress, G. R., "Fatigue Response of Notched Graphite-Epoxy Laminates," M.S. Thesis in Engineering Mechanics, Virginia Polytechnic Institute and State University, Blacksburg, VA, May 1983.
[2] Freeman, S. M., "Damage Progression in Graphite-Epoxy by a Deplying Technique," AFWAL-TR-81-3157, Wright Patterson Air Force Base, Dayton, OH, Dec. 1981.
[3] O'Brien, T. K. and Raju, I. S., "Strain Energy Release Rate Analysis of Delamination Around an Open Hole in Composite Laminates," presented at the 25th AIAA/ASME/ASCE/AHS Structures, Structural Dynamics, and Materials Conference, American Institute of Aeronautics and Astronautics, Palm Springs, CA, May 1984.
[4] O'Brien, T. K. in *Effects of Defects in Composite Materials, ASTM STP 836*, D. Wilkins, Ed., American Society for Testing and Materials, Philadelphia, 1984, pp. 125-142.
[5] Jamison, R. D. and Reifsnider, K. L., "Advanced Fatigue Damage Development in Graphite Epoxy Laminates," Interim Report to AF Wright Aeronautical Laboratories, AFWAL TR-82-3103, Wright Patterson Air Force Base, Dayton, OH, Dec. 1982.
[6] Wang, A. S. D., Kishore, N. N., and Li, C. A., "On Crack Development in Graphite-Epoxy $[O_2/90_n]_s$ Laminates Under Uniaxial Tension," presented at the International Symposium on Composites: Materials and Engineering, University of Delaware, Newark, DE, 1984.
[7] Stinchcomb, W. W., Kress, G. R., Reifsnider, K. L., and Henneke, E. G., "Fatigue Damage in Notched Composite Laminates Under Tension-Tension Cyclic Loads," Final Report, NASA Grant NAG-1-232, National Aeronautics and Space Administration, July 1983.
[8] Sendeckyj, G. P., Maddux, G. E. and Porter, E. in *Damage in Composite Materials, ASTM STP 775*, K. L. Reifsnider, Ed., American Society for Testing and Materials, Philadelphia, 1982, pp. 16-26.

[9] Chang, F. H., Gordon, D. E., and Gardner, A. H. in *Fatigue of Filamentary Composite Materials, ASTM STP 636*, K. L. Reifsnider and K. N. Lauraitis, Eds., American Society for Testing and Materials, Philadelphia, 1977, pp. 57-72.

[10] Roderick, G. L. and Whitcomb, J. D. in *Fatigue of Filamentary Composite Materials, ASTM STP 636*, K. L. Reifsnider and K. N. Lauraitis, Eds., American Society for Testing and Materials, Philadelphia, 1977, pp. 73-88.

[11] Sendeckyj, G. P., Stalnaker, H. D., and Kleismit, R. A. in *Fatigue of Filamentary Composite Materials, ASTM STP 636*, K. L. Reifsnider and K. N. Lauraitis, Eds., American Society for Testing and Materials, Philadelphia, 1977, pp. 123-140.

[12] Reifsnider, K. L., Stinchcomb, W. W., and O'Brien, T. K. in *Fatigue of Filamentary Composite Materials, ASTM STP 636*, K. L. Reifsnider and K. N. Lauraitis, Eds., American Society for Testing and Materials, Philadelphia, 1977, pp. 171-184.

[13] Whitcomb, J. D. in *Fatigue of Fibrous Composite Materials, ASTM STP 723*, American Society for Testing and Materials, Philadelphia, 1981, pp. 48-63.

[14] Post, Daniel, "Optical Interference for Deformation Measurements—Classical Holographic and Moiré Interferometry," *Mechanics of Nondestructive Testing*, W. W. Stinchcomb, Ed., Plenum Press, New York, 1980.

[15] Highsmith, A. L. and Reifsnider, K. L., "Measurement on Nonuniform Microstrain in Composite Laminates," *Composites Technology Review*, Vol. 4, No. 1, 1982.

John H. Cantrell, Jr.,[1] *William P. Winfree,*[1]
and Joseph S. Heyman[1]

Profiles of Fatigue Damage in Graphite/Epoxy Composites from Ultrasonic Transmission Power Spectra

REFERENCE: Cantrell, J. H., Jr., Winfree, W. P., and Heyman, J. S., **"Profiles of Fatigue Damage in Graphite/Epoxy Composites from Ultrasonic Transmission Power Spectra,"** *Recent Advances in Composites in the United States and Japan, ASTM STP 864,* J. R. Vinson and M. Taya, Eds., American Society for Testing and Materials, Philadelphia, 1985, pp. 197–206.

ABSTRACT: Early fatigue damage in non-unidirectional multi-ply graphite/epoxy composites is manifested by a distribution of cracks and disbonds through the bulk of the material. Such damage is subtle and is difficult to detect with conventional ultrasonic technology. Consequently, a new ultrasonic measurement technique called phase-insensitive tone-burst spectroscopy has been developed. The new technique eliminates problems associated with phase cancellation and pulse shape artifacts inherent to conventional broadband ultrasonic spectral measurement systems and produces clean spectral information irrespective of specimen inhomogeneity or irregularities in surface geometry. Application of the new technique to measurements of graphite/epoxy composites has yielded frequency-domain profiles that show distinct differences in ultrasonic attenuation, attenuation slope, and velocity for each specimen experiencing a different level of fatigue damage.

KEY WORDS: graphite/epoxy composites, fatigue, ultrasonics, damage mechanisms, nondestructive evaluation, acoustoelectric transducer, tone-burst spectroscopy, ultrasonic frequency analysis, ultrasonic attenuation, ultrasonic velocity, phase cancellation, pulse shape artifacts

Although many papers have been written on various aspects of fatigue in composite materials, relatively few have been published on the ultrasonic characterization of fatigue damage in such materials. One reason may be that composites are inhomogeneous, often geometrically irregular structures

[1]Research physicists and NDE program manager, respectively, NASA Langley Research Center, Hampton, VA 23665.

197

which give rise to measurement artifacts such as phase cancellation when conventional ultrasonic measurement techniques are employed. Such artifacts often lead to results which are difficult to interpret or analyze in terms of the true material state. In this paper we introduce a new ultrasonic measurement technology, called phase-insensitive tone-burst spectroscopy, which yields frequency-domain information (spectrum analysis) that is free of problems associated with ultrasonic phase cancellation and pulse shape artifacts inherent to conventional ultrasonic frequency analysis systems. The application of the new measurement technique is shown to provide a significant and sensitive measure of early fatigue damage in those graphite/epoxy composite specimens tested.

We first discuss fatigue damage mechanisms in graphite/epoxy composites. We then introduce phase-insensitive tone-burst spectroscopy and give the preliminary results of applying the new technique to spectral measurements of graphite/epoxy composites.

Damage Mechanisms in Graphite/Epoxy Composites

Unlike fatigue damage in metals, which is generally manifested by a single self-same crack nucleation site, fatigue of composite materials is manifested by several failure mechanisms. Some of these failures mechanisms are illustrated in Fig. 1, which shows a photomicrograph of typical fatigue damage in a graphite/epoxy composite having a $(90/\pm45/0)_s$ fiber ply sequence. Sections through the damage zone clearly reveal through-the-ply and interlaminar cracks distributed through the bulk of the material. More advanced failure mechanisms, not shown in Fig. 1, include fiber disbonding from the epoxy matrix and fiber breakage.

The acoustic analog of the optical photomicrograph is the ultrasonic C-scan. The C-scan is commonly used for acoustically imaging macroscopic damage in materials such as gross delaminations and large flaws in composites, and residual stress buildup around crack sites in metals. The imaging parameter is generally the ultrasonic attenuation or phase velocity through a volume of the material defined by the cross-sectional area of the ultrasonic beam and the thickness of the material. The image is a mapping of the set of values of the imaging parameter obtained from a surface scan of the material onto a two-dimensional representation of that surface. A modified isometric C-scan attenuation image of an unfatigued composite specimen of the type in Fig. 1 is shown in the top picture of Fig. 2 and that of a fatigued composite sample is shown beneath it. It is perhaps surprising that there is such little difference in the two images. The reason, however, is apparent. Fatigue damage in laminated composites is distributed through the bulk of the material as a network of stress-relieving transverse cracks having a somewhat predictable spacing [1]. The distribution in damage together with the resolution that is obtainable at the frequencies amenable to ultrasonic propagation in compos-

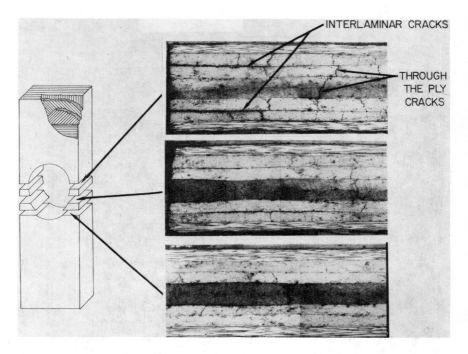

FIG. 1—*Typical fatigue damage in a graphite/epoxy composite.*

ites produces an image that is rather insensitive to macrosopic spatial variations or scans.

An alternative to spatial variation is frequency variation. Rather than map a "d-c" level of the image parameter onto a two-dimensional representation of the material surface as is commonly done in C-scans, we may, instead, consider measuring variations in attenuation or velocity as a function of the frequency of the ultrasonic wave (that is, ultrasonic frequency analysis). Further, the ultrasonic frequency analysis performed in any part of the damage zone may provide significant information on fatigue damage in the graphite/epoxy structure.

Phase-Insensitive Tone-Burst Spectroscopy

Our approach to the problem of obtaining ultrasonic spectral information from composite materials is to combine the phase-insensitive acoustoelectric transducer [2,3] with the newly developed frequency-tracked tone-burst spectroscopy technique [4]. The new measurement technique is called phase-insenstive tone-burst spectroscopy. We shall discuss the acousteoelectric

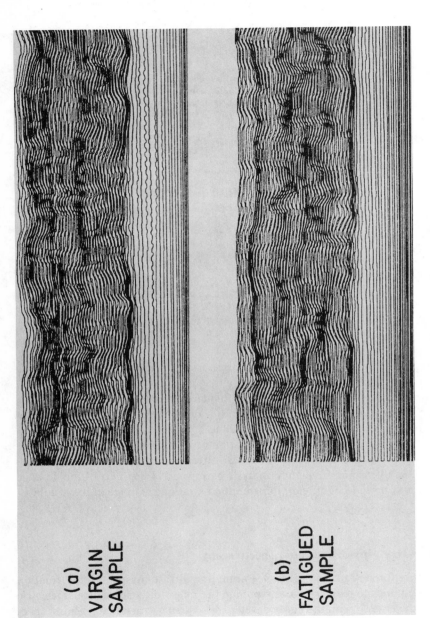

FIG. 2—*C-scan of graphite/epoxy composites.*

(a)
VIRGIN
SAMPLE

(b)
FATIGUED
SAMPLE

transducer and the frequency-tracked tone-burst spectroscopy technique separately.

The Acoustoelectric Transducer

Conventional piezoelectric transducers generate an electrical output signal whose amplitude is proportional to the integral of the transducer surface pressure. For normal incidence of plane monochromatic waves, the electrical signal is dependent only on the acoustic amplitude. However, for multifrequency or multiphase waves, such as those encountered in irregular or inhomogeneous materials (for example, composites), the electrical signal is no longer related only to the amplitude of the incident acoustic waves. These effects can be severe at ultrasonic frequencies [5,6], and can result in total phase cancellation [7].

Transducer interference artifacts are largely eliminated through the use of phase-insensitive acoustic power sensors. A practical ultrasonic power transducer called an acoustoelectric transducer (AET) has recently been developed for nondestructive evaluation. The device, based on the acoustoelectric effect occurring in piezoelectric semiconductors, couples energy from the acoustic wave to free charge carriers in semiconducting cadmium sulfide and produces an ultrasonically phase-insensitive electrical output signal. Experiments with inhomogeneous and geometrically irregular materials have shown that the AET gives more reliable and more accurate measurements of ultrasonic attenuation than do measurements with conventional piezoelectric transducers [8]. Details of the acoustoelectric effect and the acoustoelectric transducer are given elsewhere [2,3,5,9–11].

Frequency-Tracked Tone-Burst Spectroscopy

Frequency-tracked tone-burst spectroscopy is a technique that combines many of the attributes of time-domain broadband pulse methods and frequency-domain narrowband continuous-wave methods. A block diagram of the measurement system is shown in Fig. 3. A continuous-wave radio frequency (RF) signal of frequency, f, from a sweep generator (tracking generator) is passed through a transmitter gate which is turned on for time, T, with an on-to-off ratio of approximately 90 dB. The gated RF pulse from the transmitter is amplifed and excites a conventional broadband piezoelectric (PZT) transducer. The piezoelectric transducer generates a gated ultrasonic signal of time duration, T, and frequency, f, called an ultrasonic tone-burst. The ultrasonic tone-burst propagates through the specimen under test and impinges on a second transducer which may be either a conventional piezoelectric transducer or an AET. We have used the AET because of its insensitivity to phase variations in the incident ultrasonic wave. The signal from the AET is amplified, sampled and held, and stored in the y-channel of a data normalizer/recorder. The x-channel of the normalizer/recorder records the fre-

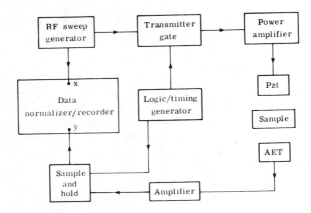

FIG. 3—*Measurement system block diagram.*

quency, f, of the ultrasonic tone-burst. If the frequency, f, is made to change continuously through a selected range of frequencies, the normalizer/recorder generates an ultrasonic power transmission frequency spectrum of the specimen under test. A detailed mathematical analysis of the frequency-tracked tone-burst spectroscopy technique and a comparison with conventional pulse spectrum analysis methods have been published elsewhere [4].

Tone-burst spectroscopy is the only method available for obtaining ultrasonic spectral information using the AET. The electrical output of the AET is a d-c signal level which is proportional to the total acoustic flux (acoustic energy per unit area per unit time) incident on the device. Since the d-c output does not distinguish between acoustic fluxes of different frequencies, broadband shock pulses of conventional pulse spectrum analysis methods cannot be used with the AET to extract spectral information. However, when the AET is used with the tone-burst spectroscopy technique (phase-insensitive tone-burst spectroscopy), the d-c level is defined at the tracking frequency, f, and therefore, does permit a power spectrum to be generated.

Application to Graphite/Epoxy Composites

We have used the phase-insensitive tone-burst spectroscopy technique to obtain ultrasonic transmission power spectra of graphite/epoxy composites having a $(0_2/90_2)_s$ fiber ply sequence. All specimens were rectangular plates of nominal dimensions 2 mm by 4 cm by 25 cm and were cut from the same lot.

The specimens were immersed in a water bath with a water delay line between the specimen and the transducers on either side of the specimen. The transmitter gate time T was set such that T was long compared with an ultrasonic transit time in the specimen but short compared with ultrasonic transit times in the water delay lines. Hence, a standing-wave equilibrium condition was established in the specimen but not in the delay lines. In the equilibrium condition the wave energy introduced into the specimen balanced the wave attenuation exactly. This arrangement gave results equivalent to continuous-wave (standing-wave) measurements but did not suffer from RF cross-talk problems inherent to such measurements [12]. In addition, the standing-wave pattern of the frequency spectrum obtained was that of the isolated specimen only.

The results of measurements on one virgin and two fatigued specimens are shown in Fig. 4. The spectra of Fig. 4 are those of the specimens only, since the spectral characteristics of the tranducers and delay lines have been deconvolved from the total spectral signal by use of a digital data storage/normalizer. The solid curve is the frequency spectrum obtained from an unfatigued specimen and is representative of the standing-wave resonance pattern expected from such a specimen. The decrease in signal amplitude with increasing frequency results from the increase in ultrasonic attenuation with increasing frequency. An almost identical curve was obtained from a second virgin specimen cut from the same lot. The reproducibility of the profiles for a given specimen was also excellent.

The dashed curve is the spectrum obtained from a specimen (hereafter called Specimen 2) which was fatigued for 11 000 cycles using a constant-amplitude 10-Hz sinusoidal axial load along the zero-degree plies at a maximum stress of 50% static ultimate tensile strength. The spectrum shows a significant shift toward lower frequencies in the resonance pattern relative to

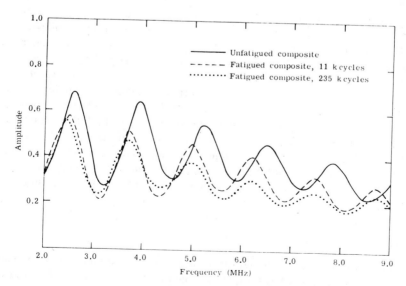

FIG. 4—*Measurement results for three specimens.*

the unfatigued specimen pattern. According to the propagating wave model [12], such a shift can result either from an increase in specimen thickness or a decrease in the ultrasonic phase velocity. Measurements of the specimen thicknesses reveal that to within experimental error (approximately ±2%) the shift can be accounted for entirely by variations in the specimen thickness. Hence, the specimen phase velocity appears to be unchanged by the present level of fatigue damage. Figure 4 shows that the ultrasonic amplitude at all frequencies has decreased substantially in Specimen 2 in concert with the appearance of microcracks in the bulk of the material. It is also important to note that the frequency dependence of the ultrasonic amplitude in Specimen 2 is different from that of the unfatigued specimen. Radiographic analysis reveals the presence of an average of 7.5 microcracks per centimetre along the axial direction in the bulk of the specimen. No cracks were observed in the unfatigued specimen.

The dotted curve of Fig. 4 is the frequency spectrum of a sample (hereafter called Specimen 3) which was fatigued for 235 000 cycles under the same conditions as Specimen 2. Radiographic analysis reveals an average 33.5 microcracks per centimetre along the axial direction through the bulk of the specimen. The frequency shift of the resonance pattern of Specimen 3 relative to the unfatigued specimen is approximately the same as that of Specimen 2. As with Specimen 2, the present shift is accounted for to within measurement error (±2%) by variations in the specimen thickness. The measurements of the three specimens indicated that the ultrasonic phase velocity of $(0_2/90_2)_s$ graphite/epoxy composites in the range 2 to 9 MHz is constant irrespective of fatigue damage at the levels used in the present experiments. These results are in agreement with the measurements of Ringermacher [13] (also for $(0_2/90_2)_s$ ply sequence) and of Williams and Doll [14] (for uniaxial plies).

The ultrasonic transmission amplitude at low frequencies in Specimen 3 does not differ greatly from that of Specimen 2 but at higher frequencies the difference becomes appreciable. Examination of the three spectra of Fig. 4 indicates that an increase in specimen fatigue damage not only produces an increase in the ultrasonic attenuation at all frequencies, but also produces changes in the frequency dependence of the attenuation. Although one may expect the increase in crack density with increased fatigue cycling to contribute significantly to the profile changes in the frequency spectra, the extent of the crack density contribution is not clear at present. In order to answer such questions it is necessary to quantify the spectral profiles. Research in this direction is currently in progress.

Conclusion

The inability of conventional ultrasonic C-scan technology to provide adequate information on early fatigue damage in graphite/epoxy composites has led to the development of a new ultrasonic measurement technology called phase-insensitive tone-burst spectroscopy. The new spectrometer is obtained by combining the phase-insensistive acoustoelectric transducer with a measurement technique called frequency-tracked tone-burst spectroscopy.

The new technique has been applied in a preliminary study of fatigue damage in composite materials and has resulted in ultrasonic frequency-domain profiles that provide a substantial amount of acoustic information in a single measurement. The profiles indicate that, in addition to an increase in ultrasonic attenuation at each frequency in the range 2 to 9 MHz, measurable changes in the attenuation as a function of frequency occur at each new level of fatigue in the specimens tested. The profiles also indicate that the ultrasonic phase velocity in the range 2 to 9 MHz is (to within experimental error of $\pm 2\%$) independent of the early state of fatigue and is in agreement with the results of other researchers. The sensitivity of the phase-insensitive tone-burst spectroscopy technique to changes in the material acoustic attenuation and attenuation versus frequency profile early in the fatigue process is encouraging. The new technique has great promise as a measurement methodology to obtain a reliable quantitative characterization of fatigue damage in composite materials. More comprehensive studies are in progress.

Acknowledgment

We wish to thank Dr. H. I. Ringermacher and Mr. John D. Whitcomb for their valuable comments.

References

[1] Reifsnider, K. L., in *Recent Advances in Engineering Science: Proceedings of the Fourteenth Annual Meeting*, Bethlehem, PA, Nov. 14-16, 1977, pp. 373-384, (A78/40301 17-31).

[2] Miller, J. G., Heyman, J. S., Weiss, A. N. , and Yuhas, S. D., "Power Sensitive Detector for Echocardiography and Other Medical Ultrasonic Applications," presented at the American Institute of Ultrasound in Medicine, Seattle, WA, Oct. 8-10, 1974.

[3] Heyman, J. S., "Phase Insensitive Acoustoelectric Transducer," *Journal of the Acoustical Society of America*, Vol. 64, July 1978, pp. 243-249.

[4] Cantrell, J. H., Jr., and Heyman, J. S., "Ultrasonic Spectrum Analysis Using Frequency-Tracked Gated rf Pulses," *Journal of the Acoustical Society of America*, Vol. 67, May 1980, pp. 1623-1628.

[5] Heyman, J. S. and Cantrell, J. H., Jr., "Application of an Ultrasonic Phase Insensitive Receiver to Material Measurements" in *Proceedings*, 1977 Ultrasonic Symposium, Institute of Electrical and Electronics Engineers, New York, IEEE Catalog No. 77CH1264-1SU, 1977, pp. 124-128.

[6] Marcus, P. W. and Carstensen, E. L., "Problems with Absorption Measurements of Inhomogeneous Solids," *Journal of the Acoustical Society of America*, Vol. 58, Dec. 1975, pp. 1334-1335.

[7] Fuller, E. R., Jr., Granato, A. V., Holder, J., and Naimon, E. R., "Ultrasonic Studies of the Properties of Solids," *Methods of Experimental Physics*, Vol. 11, pp. 371-442, E. C. Coleman, Ed., Academic Press, New York, 1974.

[8] Heyman, J. S. and Cantrell, J. H., Jr., "Effects of Material Inhomogeneities on Ultrasonic Measurements: The Problem and a Solution" in *Nondestructive Evaluation and Flaw Criti-*

cally for Composite Materials, ASTM STP 696, R. V. Pipes, Ed., American Society for Testing and Materials, Philadelphia, 1979, pp. 45–56.

[9] Weinreich, G., "Ultrasonic Attenuation by Free Carriers in Germanium," *Physical Review*, July 1957, pp. 317–318.

[10] Hutson, A. R. and White, D. L., "Elastic Wave Propagation in Piezoelectric Semiconductors," *Journal of Applied Physics*, Vol. 33, Jan. 1962, pp. 40–47.

[11] Southgate, P. D., "Use of a Power Sensitive Detectors in Pulse-Attenuation Measurements," *Journal of the Acoustical Society of America*, Vol. 39, March 1966, pp. 480–483.

[12] Bolef, D. I. and Miller, J. G., "High-Frequency Continuous Wave Ultrasonics," *Physical Acoustics*, Vol. 8, W. P. Mason and R. N. Thurston, Eds., 1971, pp. 95–201.

[13] Ringermacher, H. I. in *Proceedings*, 1980 Ultrasonics Symposium, Institute of Electrical and Electronics Engineers, New York, IEEE Catalog No. 80CH1602-2, 1980, pp. 957–960.

[14] Williams, J. H., Jr., and Doll, B., "Ultrasonic Attenuation as an Indicator of Fatigue Life of Graphite Fiber Epoxy Composite," *Materials Evaluation*, Vol. 38, May 1980, pp. 33–37.

Stress Analysis

Toshio Mura[1] and Minoru Taya[2]

Residual Stresses In and Around a Short Fiber in Metal Matrix Composites Due To Temperature Change

REFERENCE: Mura, T. and Taya, M., "**Residual Stresses In and Around a Short Fiber in Metal Matrix Composites Due To Temperature Change,**" *Recent Advances in Composites in the United States and Japan, ASTM STP 864*, J. R. Vinson and M. Taya, Eds., American Society for Testing and Materials, Philadelphia, 1985, pp. 209-224.

ABSTRACT: Residual stresses in and around short fibers in metal matrix composites due to temperature change are evaluated. An analytical result for the residual stress in the fiber is obtained by use of Somigliana dislocations at the matrix-fiber interface. These dislocations are introduced to accommodate the misfit of the thermal expansion difference between fibers and the matrix. These dislocations are responsible for the residual stresses after the composite material is cooled down to room temperature.

KEY WORDS: metal matrix composite, residual stress, short fiber, dislocations

Metal matrix composites (MMCs) are advantageous over polymer based composites in several environments, particularly at elevated temperature. However, the performance of MMCs must be carefully evaluated from various viewpoints. One of the most important points is to assess the amount of thermal residual stresses. There are a number of thermal environments which induce the residual stresses. Here we consider the case of a complete temperature excursion, for example, changing from room temperature T_0 to high temperature T_1, and returning to T_0. If T_1 is less than the creep relaxation temperature T_r, then only elastic/plastic analysis is required to compute the residual stresses, though the analysis is elaborate for a complicated geometry of the constituent in the composite. On the other hand, when T_1 is larger than

[1] Professor, Department of Civil Engineering, Material Research Center, Northwestern University, Evanston, IL 60201.

[2] Associate professor, Department of Mechanical and Aerospace Engineering, University of Delaware, Newark, DE 19716.

T_r, then the stress relaxation can take place, which is enhanced as the time elapses. Thus the mechanism which is operative above T_r involves elastic, plastic, and creep deformation. A comprehensive treatment in this direction was made by Garmong [1], but it is limited to one dimensional stress state, that is, unidirectional composite system.

On the other hand one emphasis in this paper is on a short fiber metal matrix composite (SFMMC) which requires a three-dimensional analysis. To our knowledge, there have been no studies aimed at the three-dimensional analysis of thermal residual stresses induced in SFMMC except for a purely elastic analysis, for example [2]. Therefore, we develop here a simple model to predict the thermal residual stresses induced in SFMMC after a complete temperature excursion ($T_0 - T_1 - T_0$).

We first state the formulation and give the results and discussions. Finally the conclusion is given.

Formulation

To simulate SFMMC, we consider an infinite isotropic body containing an ellipsoidal inhomogeneity (fiber) as shown in Fig. 1. The infinite body (composite) experiences a temperature excursion, $T_0 - T_1 - T_0$, where T_1 is assumed to be larger than the relaxation temperature of the matrix T_r and of

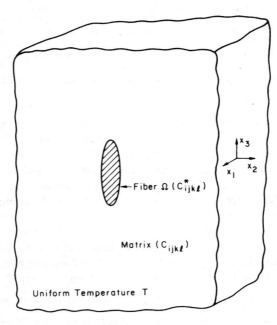

FIG. 1—*A theoretical model.*

course, $T_r > T_0$. At T_1 the composite is kept for a reasonably long period so that the stresses that were caused by the mismatch of the thermal expansion of the matrix and fiber can be relaxed fully. This full relaxation of stress will result in the plastic deformation at the matrix-fiber interface which may be simulated by Somigliana dislocations. Upon cooling from T_1 to T_0, the Somigliana dislocations will cause residual stresses in the composite. To make the problem simpler, the case of one fiber embedded in the infinite body is considered here. We note, however, that the case of multiple fibers can be treated without difficulty by the concept of "average stress approach" [2-4].

Let the elastic moduli of the matrix be C_{ijkl} and those of the ellipsoidal inhomogeneity be C^*_{ijkl}. This composite material is at an elevated temperature and the difference of the thermal expansion strain is denoted by e^T_{ij} defined in the inhomogeneity Ω. The material is also subjected to a uniform stress σ^0_{ij}, at infinity and the corresponding elastic strain is denoted by e^0_{ij}

$$\sigma^0_{ij} = C_{ijkl} e^0_{kl} \tag{1}$$

We assume that the matrix and fiber are completely elastic except at the matrix-fiber interface. According to Eshelby's theory [5], the thermal strain e^T_{ij} and the inhomogeneity of Ω cause an internal stress field (eigenstress). This stress field of Eshelby, however, is caused by the perfect bonding of the interface. The bonding will be destroyed due to the high temperature. This debonding is interpreted as a nonelastic deformation which is modeled by Somigliana dislocations b_i distributed on the interface of Ω. We will evaluate b_i since these dislocations remain at the interface even after cooling.

Somigliana Dislocations

The stress field at an elevated temperature T_1 under applied stress σ^0_{ij} is $\sigma^0_{ij} + \sigma^E_{ij} + \sigma^V_{ij}$, where σ^E_{ij} is the stress field caused by e^T_{ij} in Ω and the stress disturbance due to the inhomogeneity and σ^V_{ij} is the stress field caused by Somigliano dislocations b_i on the boundary S of Ω. The equivalent inclusion method of Eshelby is now employed. The ellipsoidal inhomogeneity Ω is replaced by an ellipsoidal inclusion Ω with a uniform eigenstrain e^*_{ij}. The inclusion has the same elastic moduli as those of the matrix but it has a uniform distribution of e^*_{ij} which is determined from the equivalency conditions for stresses and strains in the inhomogeneity and the inclusions.

The inclusion has the same geometry as the inhomogeneity, which is ellipsoidal

$$\frac{x_1^2}{a_1^2} + \frac{x_2^2}{a_2^2} + \frac{x_3^2}{a_3^2} = 1 \tag{2}$$

The displacement field caused by e_{ij}^T and e_{ij}^* is [6]

$$u_i^E(\underline{x}) = -\int_\Omega G_{ip,q}(\underline{x} - \underline{x}')C_{pqmn}\{e_{mn}^T(\underline{x}') + e_{mn}^*(\underline{x}')\}d\underline{x}' \qquad (3)$$

where

$dx' = dx_1'dx_1'dx_3',$
$G_{ip,q} = \partial G_{ip}/\partial x_q$, and
$\quad G_{ip}$ = Green's function in elasticity.

The displacement field caused by b_i is [7]

$$u_i^V(\underline{x}) = -\int_S G_{ip,q}(\underline{x} - \underline{x}')C_{pqmn}b_m(\underline{x}')n_n dS(\underline{x}') \qquad (4)$$

where

dS = surface element of S and
n_i = normal vector to dS.

The total displacement field is $u_i^E + u_i^V + u_i^0$, where u_i^0 is caused by σ_{ij}^0 in the homogeneous material before the introduction of e_{ij}^T and e_{ij}^*.

The compatible strains corresponding to u_i^E, u_i^V, and u_i^0 are denoted by e_{ij}^E, e_{ij}^V, and e_{ij}^0. The corresponding stresses in Ω are written as

$$\sigma_{ij}^E = C_{ijkl}(e_{kl}^E - e_{kl}^T - e_{kl}^*)$$

$$\sigma_{ij}^V = C_{ijkl}e_{kl}^V \qquad (5)$$

$$\sigma_{ij}^0 = C_{ijkl}e_{kl}^0$$

On the other hand, we can write

$$\sigma_{ij} = \sigma_{ij}^E + \sigma_{ij}^V + \sigma_{ij}^0 = C_{ijkl}^*(e_{kl}^E - e_{kl}^T + e_{kl}^V + e_{kl}^0) \qquad (6)$$

since the stress and strain in the inhomogeneity are equivalent to those in the inclusion. The equations to determine e_{kl}^* are obtained when Eq 5 is substituted into Eq 6.

Since

$$e_{kl}^E = (1/2)(u_{k,l}^E + u_{l,k}^E)$$

we have from Eq 3

$$e_{kl}^E = -\int_\Omega G_{kp,ql}C_{pqmn}(e_{mn}^T + e_{mn}^*)d\underline{x}' \qquad (7)$$

Strictly speaking, the right hand in Eq 7 should take a symmetric part with respect to k, l, but Eq 7 does not influence further calculations, since C_{ijkl} is multiplied eventually. The elastic strain e_{kl}^V is obtained from $(1/2)(u_{k,l}^V + u_{l,k}^V)$. Before this calculation, Eq 4 is rewritten as

$$u_i^V(\underset{\sim}{x}) = \int_\Omega C_{pqmn}(b_{m,n}G_{ip,q} - b_m G_{ip,qn})d\underset{\sim}{x}'$$

$$= -\int_\Omega G_{ip,q}C_{pqmn}e_{mn}^{**}d\underset{\sim}{x}' + b_i(\underset{\sim}{x}) \tag{8}$$

where

$$e_{ij}^{**}(\underset{\sim}{x}) = -(1/2)\{b_{i,j}(\underset{\sim}{x}) + b_{j,i}(\underset{\sim}{x})\} \tag{9}$$

In the derivation of Eq 8, the property of Green's function

$$C_{pqmn}G_{ip,qn}(\underset{\sim}{x} - \underset{\sim}{x}') = -\delta_{mi}\delta(\underset{\sim}{x} - \underset{\sim}{x}') \tag{10}$$

is used and the function b_i, which is originally defined on S, is analytical by extended into domain Ω as Asaro [8] did. The strain corresponding to u_i^V is

$$e_{kl}^V = -\int_\Omega G_{kp,ql}C_{pqmn}e_{mn}^{**}d\underset{\sim}{x}' - e_{kl}^{**} \tag{11}$$

It should be emphasized that e_{kl}^V is the elastic strain, but e_{ij}^{**} is a fictitious strain, formally defined by Eq 9.

The integrals in Eqs 7 and 11 are explicitly performed by Eshelby for $\underset{\sim}{x} \in \Omega$. That is

$$e_{kl}^E = S_{klmn}(e_{mn}^T + e_{mn}^*)$$
$$\tag{12}$$
$$e_{kl}^V = S_{klmn}e_{mn}^{**} - e_{kl}^{**}$$

where S_{klmn} is Eshelby's tensor [9]. From Eqs 5, 6, and 12, we have

$$\sigma_{ij} \equiv C_{ijkl}\{S_{klmn}(e_{mn}^T + e_{mn}^*) - e_{kl}^T - e_{kl}^* + S_{klmn}e_{mn}^{**} - e_{kl}^{**} + e_{kl}^0\}$$

$$= C_{ijkl}^*\{S_{klmn}(e_{mn}^T + e_{mn}^*) - e_{kl}^T + S_{klmn}e_{mn}^{**} - e_{kl}^{**} + e_{kl}^0\} \tag{13}$$

The six equations in Eq 13 are used for solving six unknowns of e_{ij}^* as linear functions of e_{ij}^T, e_{ij}^{**}, and e_{ij}^0. In this paper we consider only when $e_{ij}^0 = 0$; hence, Eq 13 is reduced to

$$\sigma_{ij} = C_{ijkl}\{S_{klmn}(e_{mn}^T + e_{mn}^*) - e_{kl}^T - e_{kl}^* + S_{klmn}e_{mn}^{**} - e_{kl}^{**}\}$$

$$= C_{ijkl}^*\{S_{klmn}(e_{mn}^T + e_{mn}^*) - e_{kl}^T + S_{klmn}e_{mn}^{**} - e_{kl}^{**}\} \qquad (14)$$

To make the computation even simpler, we assume here that the inhomogeneity Ω is isotropic in stiffness but transversely isotropic in thermal expansion. Thus, C_{ijkl} and C_{ijkl}^* are given by

$$C_{ijkl} = \lambda\delta_{ij}\delta_{kl} + \mu(\delta_{ik}\delta_{jl} + \delta_{il}\delta_{kj}) \qquad (15)$$

$$C_{ijkl}^* = \lambda^*\delta_{ij}\delta_{kl} + \mu^*(\delta_{ik}\delta_{jl} + \delta_{il}\delta_{kj}) \qquad (16)$$

where λ, λ^* and μ, μ^* are Lamé's constants of the matrix and inhomogeneity. At this stage, e_{ij}^{**} is still unknown. We will show next the method of determining e_{ij}^{**}.

Residual Stress Field

The stress field at the elevated temperature T_1 considered in the previous section is relaxed by a proper choice of Somigliana dislocations or, equivalently, a proper choice of e_{ij}^{**}. The amount of the stress relaxation depends upon the creep rate of the matrix material and the time elapsed at T_1. Garmong showed [1] that most of the stress is relaxed after a short period when the matrix is a typical monolithic material. In this paper we assume that the stress will be relaxed completely so that $\sigma_{ij} = \sigma_{ij}^E + \sigma_{ij}^V = 0$ everywhere. The Somigliana dislocations represented by b_i or e_{ij}^{**} are caused by the interface sliding or debonding, generally by the plastic deformation localized on the interface. Such b_i remains on the matrix-fiber after cooling, and it gives the maximum residual stress. Therefore, the maximum residual stress is σ_{ij}^V defined by Eq 5 when b_i is determined from $\sigma_{ij}^E + \sigma_{ij}^V = 0$.

Thus, the condition to determine e_{ij}^{**} is simply that σ_{ij} in Eq 14 at temperature T_1 equals to zero. We assume in the following computation that a short fiber is a prolate ellipsoid with the major axis being the x_3-axis as shown in Fig. 1. Hence, this composite system gives rise to a transversely isotropic material. Therefore, the number of the nonvanishing components of the stress and strain is two, one along the x_3-axis and the other along the x_1-axis (or x_2-axis).

Also the nonvanishing Eshelby's tensors are S_{1111} ($= S_{2222}$), S_{1122} ($= S_{2211}$), S_{3311} ($= S_{3322}$), S_{1133} ($= S_{2233}$), and ($= S_{3333}$), and they are given explicitly in the Appendix I. By setting $ij = 11$ and 33 in Eq 14, we obtain

$$C_{11}(e_{11}^T + e_{11}^* + e_{11}^{**}) + C_{12}(e_{33}^T + e_{33}^* + e_{33}^{**})$$

$$= \frac{(\lambda^* + \mu^*)}{\mu} e_{11}^* + \frac{\lambda^*}{2\mu} e_{33}^* \qquad (17)$$

$$C_{21}(e_{11}^T + e_{11}^* + e_{11}^{**}) + C_{22}(e_{33}^T + e_{33}^* + e_{33}^{**})$$

$$= \frac{\lambda^*}{\mu} e_{11}^* + \frac{(\lambda^* + 2\mu^*)}{2\mu} e_{33}^* \quad (18)$$

where C_{ij} are given by

$$C_{11} = \frac{(\lambda - \lambda^*)}{\mu} (S_{1111} + S_{1122} + S_{3311} - 1)$$

$$+ \left(1 - \frac{\mu^*}{\mu}\right)(S_{1111} + S_{1122} - 1),$$

$$C_{12} = \left(\frac{\lambda - \lambda^*}{2\mu}\right)(2S_{1133} + S_{3333} - 1) + \left(1 - \frac{\mu^*}{\mu}\right)S_{1133},$$

$$C_{21} = \frac{(\lambda - \lambda^*)}{\mu} (S_{1111} + S_{1122} + S_{3311} - 1) + 2\left(1 - \frac{\mu^*}{\mu}\right)S_{3311}, \text{ and}$$

$$C_{22} = \frac{(\lambda - \lambda^*)}{2\mu} (2S_{1133} + S_{3333} - 1) + \left(1 - \frac{\mu^*}{\mu}\right)(S_{3333} - 1).$$

We solve for $e_{11}^T + e_{11}^* + e_{11}^{**}$ and $e_{33}^T + e_{33}^* + e_{33}^{**}$ in Eqs 17 and 18 to obtain

$$e_{11}^T + e_{11}^* + e_{11}^{**} = F_2(e_{11}^T + e_{11}^{**}) + F_3(e_{33}^T + e_{33}^{**}) \quad (20)$$

$$e_{33}^T + e_{33}^* + e_{33}^{**} = F_4(e_{11}^T + e_{11}^{**}) + F_1(e_{33}^T + e_{33}^{**}) \quad (21)$$

where F_i ($i = 1, 2, 3,$ and 4) are given in the Appendix II. Upon substitution of Eqs 20 and 21 into Eq 14, the stresses σ_{11} and σ_{33} are expressed as

$$\frac{\sigma_{11}}{2\mu} = H_1\{F_2(e_{11}^T + e_{11}^{**}) + F_3(e_{33}^T + e_{33}^{**})\}$$

$$+ H_2\{F_4(e_{11}^T + e_{11}^{**}) + F_1(e_{33}^T + e_{33}^{**})\} \quad (22)$$

$$\frac{\sigma_{33}}{2\mu} = H_3\{F_2(e_{11}^T + e_{11}^{**}) + F_3(e_{33}^T + e_{33}^{**})\}$$

$$+ H_4\{F_4(e_{11}^T + e_{11}^{**}) + F_1(e_{33}^T + e_{33}^{**})\} \quad (23)$$

where H_i ($i = 1, 2, 3$, and 4) are given in the Appendix II. By setting $\sigma_{11} = \sigma_{33} = 0$ in Eqs 20 and 21, we obtain

$$e_{11}^{**} = -e_{11}^T, \qquad e_{33}^{**} = -e_{33}^T \qquad (24)$$

These simple results indicate that the residual stress can be calculated as the eigenstress in the perfectly elastic body caused by eigenstrain $-e_{ij}^T$ defined in Ω.

Therefore, the residual stresses at temperature T_0 are obtained by setting $e_{11}^T = e_{33}^T = 0$ in Eqs 22 and 23 and then substituting Eq 24 into Eqs 22 and 23. The result is

$$\frac{\sigma_{11}}{2\mu} = -(F_2 H_1 + F_4 H_2)e_{11}^T - (F_3 H_1 + F_1 H_2)e_{33}^T \qquad (25)$$

$$\frac{\sigma_{33}}{2\mu} = -(F_2 H_3 + F_4 H_4)e_{11}^T - (F_3 H_3 + F_4 H_4)e_{33}^T \qquad (26)$$

where e_{11}^T and e_{33}^T are given by

$$e_{11}^T = (\alpha_{fT} - \alpha_m)(T_1 - T_0)$$
$$e_{33}^T = (\alpha_{fL} - \alpha_m)(T_1 - T_0) \qquad (27)$$

and where α_m is the thermal expansion coefficient of the matrix, α_{fT} and α_{fL} are that of the fiber along the transverse and longitudinal direction, respectively. It is noted that the stress field inside the fiber is uniform and given by Eqs 25 and 26.

The stresses just outside the fiber $\sigma_{pq}^{(out)}$ is related to those inside the fiber, $\sigma_{pq}^{(in)}$ and $e_{ij}^* + e_{ij}^{**}$ as [10].

$$\sigma_{pq}^{(out)} = \sigma_{pq}^{(in)} + C_{pqmn}\left\{-C_{klij}(e_{ij}^* + e_{ij}^{**})n_l n_n \right.$$

$$\left. \times \frac{(\lambda + 2\mu)\delta_{km} - (\lambda + \mu)n_k n_m}{\mu(\lambda + 2\mu)} + e_{mn}^* + e_{mn}^{**}\right\} \qquad (28)$$

where $\sigma_{pq}^{(in)}$ are given by Eqs 25 and 26, and n_i is the x_i^{th} component of the unit vector outward normal to the surface of the fiber, Ω. Basically we are concerned with two points at which stresses are computed. The two points are shown in Fig. 2. Point A is just outside the end of the fiber, at $x_1 = x_2 = 0$ and $x_3 = l/2$ where l is the fiber length. Point B is located just outside the side face of the fiber at $x_2 = x_3 = 0$ and $x_1 = d/2$ where d is the fiber diameter.

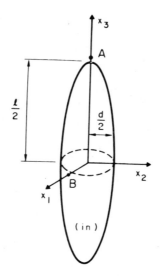

FIG. 2—*The calculation points for the stresses in and just outside the fiber, A and B.*

The components of the vectors $\underline{n}$ at A and B are given by $(0, 0, 1)$ and $(1, 0, 0)$, respectively. Thus the stresses at A and B are computed as

$$\frac{\sigma_r^A}{2\mu} = \frac{\sigma_{11}^{(in)}}{2\mu} + \left(\frac{1+\nu}{1-\nu}\right)(e_{11}^* + e_{11}^{**})$$

$$\frac{\sigma_\theta^B}{2\mu} = \frac{\sigma_{11}^{(in)}}{2\mu} + \frac{1}{(1-\nu)}(e_{11}^* + e_{11}^{**}) + \frac{\nu}{(1-\nu)}(e_{33}^* + e_{33}^{**}) \qquad (29)$$

$$\frac{\sigma_z^B}{2\mu} = \frac{\sigma_{33}^{(in)}}{2\mu} + \frac{\nu}{(1-\nu)}(e_{11}^* + e_{11}^{**}) + \frac{1}{(1-\nu)}(e_{33}^* + e_{33}^{**})$$

where the superscripts A and B denote the points of stress computed, and the subscripts r, θ, z are cylindrical coordinates with z being along the x_3-axis. It is noted that $\sigma_z^A = \sigma_{33}^{(in)}$ and $\sigma_r^B = \sigma_{11}^{(in)}$ due to the continuity of the traction across the matrix-fiber interface.

Results and Discussions

Since the target SFMMC in this study is carbon/aluminum system, we have used the following thermomechanical properties [2,11,12]

$$E_f/E_m = 8.2$$

$$\nu_m = 0.3$$

$$\nu_f = 0.3$$

$$\alpha_m = 2.3 \times 10^{-5}/°C$$

$$\alpha_{fL} = 0.$$

$$\alpha_{fT} = 2.3 \times 10^{-5}/°C$$

$$l/d = 10.$$

$$T_1 - T_0 = 400°C$$

(30)

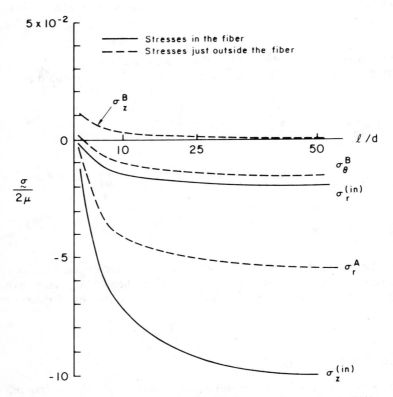

FIG. 3—*The residual stresses normalized by 2μ versus the aspect ratio (l/d).*

It should be noted here that these data are designed to simulate a typical carbon fiber and aluminum with the restriction that the fiber is isotropic in stiffness and transverse isotropic in thermal expansion. In order to investigate the effect of the stiffness ratio of the fiber to matrix (E_f/E_m) and the fiber aspect ratio (l/d), we have taken several values of E_f/E_m and l/d below and above their original values listed previously. The values of stresses normalized by 2μ (μ is the matrix shear modulus) are computed inside the fiber $\{\sigma_z^{(in)}, \sigma_r^{(in)}\}$ and just outside the fiber at Points A and B (σ_r^A, σ_θ^B and σ_z^B) by use of Eqs 25, 26, and 29. They are plotted as a function of l/d (Fig. 3) and E_f/E_m

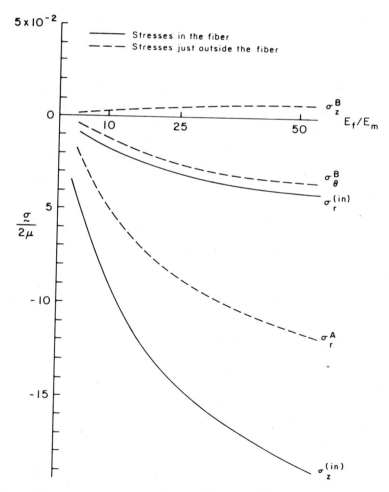

FIG. 4—*The residual stresses normalized by 2μ versus the stiffness ratio (fiber to matrix),* E_f/E_m.

(Fig. 4) where the solid curves denote the stresses inside the fiber and the dash curves denote those just outside the fiber. It follows from these figures that the maximum stress occurs in compression in the fiber along the fiber axis, which is enhanced as the values of l/d and E_f/E_m increase. Thus, one can observe that a short fiber will pop out of the matrix as the composite is sectioned due to this large compressive stress in the fiber.

This is illustrated in Fig. 5 where the residual stress σ_z is plotted along the traverse axes (x_1 and x_2). It is noted in Fig. 5 that the magnitude of σ_z in the matrix is very small due to the infinite matrix, but it will increase as the volume fraction of fibers f increases at the cost of the compressive stress in the fiber [2].

Next we compute the Somigliana dislocations b_i which are related to e_{ij}^{**} by Eq 9. By integrating Eq 9, we obtain

$$b_i(\underline{x}) = -e_{ij}^{**}x_j \qquad (31)$$

In this derivation, we have used that the antisymmetric components of $b_{i,j}$ after multiplied by e_{ij}^{**} vanish. The value of b_i along the matrix-fiber interface is computed for the data given by Eq 30 and the value normalized by ($l/2$) is plotted in Fig. 6. It is obvious from Fig. 6 and Eq 26 that the value of b_i is equal in magnitude and opposite in sign to the mismatch caused by the temperature excursion $T_0 - T_1$, that is, e_{ij}^T defined by Eq 27.

We have also computed the stresses and the Somigliana dislocations b_i for the case of spherical inhomogeneity. The material properties of this spherical

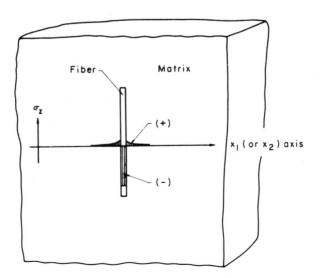

FIG. 5—*A schematic view of the residual stress σ_z in the matrix and fiber.*

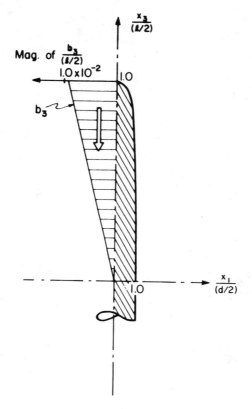

FIG. 6—*The distribution of the Somigliana dislocations, b_3 (the direction is indicated by white arrow) along the matrix-fiber interface (the corresponding data are given by Eq 28).*

inhomogeneity are taken as the same as those used by Lee et al [*13*] who have computed the stress field in and around the spherical inhomogeneity with misfit strain. The results based on the present model coincide exactly with those obtained by Lee et al [*13*].

Conclusions

The residual stresses induced in a short fiber metal matrix composite after a full temperature excursion, from room temperature to high temperature and back to the room temperature, are evaluated based on a simple model. In our model, Somigliana dislocations are used to simulate the mismatch along the matrix-fiber interface. It is found in this study that the Somigliana dislocations are equal in magnitude to the mismatch caused by $\Delta T \Delta \alpha$ (where Δ denotes the difference) but opposite in sign. Thus, our estimate of the residual stresses is the upper bound to the actual stresses.

Acknowledgments

This research was supported by U.S. Army Research Grant DAAG29-81-K-0090 and Honda R & D Company.

APPENDIX I

The Eshelby's tensor S_{ijkl} for an ellipsoidal fiber with major axis l and minor axis d are expressed by

$$S_{1111} = S_{2222} = \frac{3}{8(1-\nu)}\frac{\alpha^2}{(\alpha^2-1)} + \frac{1}{4(1-\nu)}\left\{1 - 2\nu - \frac{9}{4(\alpha^2-1)}\right\}g$$

$$S_{3333} = \frac{1}{2(1-\nu)}\left[1 - 2\nu + \frac{(3\alpha^2-1)}{(\alpha^2-1)} - \left\{1 - 2\nu + \frac{3\alpha^2}{(\alpha^2-1)}\right\}g\right]$$

$$S_{1122} = S_{2211} = \frac{1}{4(1-\nu)}\left\{\frac{\alpha^2}{2(\alpha^2-1)} - \left((1-2\nu) + \frac{3}{4(\alpha^2-1)}\right)g\right\}$$

$$S_{1133} = S_{2233} = -\frac{1}{2(1-\nu)}\frac{\alpha^2}{(\alpha^2-1)}$$

$$+ \frac{1}{4(1-\nu)}\left\{\frac{3\alpha^2}{(\alpha^2-1)} - (1-2\nu)\right\}g \quad (32)$$

$$S_{3311} = S_{3322} = -\frac{1}{2(1-\nu)}\left\{1 - 2\nu + \frac{1}{(\alpha^2-1)}\right\}$$

$$+ \frac{1}{2(1-\nu)}\left\{1 - 2\nu + \frac{3}{2(\alpha^2-1)}\right\}g$$

where

ν = Poisson's ratio of a matrix,
α = aspect ratio of a fiber ($= l/d$), and
g is given by

$$g = \frac{\alpha}{(\alpha^2-1)^{3/2}}\{\alpha(\alpha^2-1)^{1/2} - \cosh^{-1}\alpha\} \quad (33)$$

APPENDIX II

$$F_1 = 1 - \frac{1}{F}\left\{1 - \frac{1}{D}\left(\frac{\lambda^*+\mu^*}{\mu}C_{22} - \frac{\lambda^*}{\mu}C_{12}\right)\right\}$$

$$F_2 = 1 - \frac{1}{F}\left\{1 - \frac{1}{D}\left(\frac{\lambda^*+2\mu^*}{2}C_{11} - \frac{\mu^*}{2}C_{21}\right)\right\}$$

$$F_3 = -\frac{1}{FD}\left\{\frac{\lambda^*}{2\mu}C_{22} - \frac{\lambda^* + 2\mu^*}{2\mu}C_{12}\right\} \tag{34}$$

$$F_4 = -\frac{1}{FD}\left\{\frac{\lambda^*}{\mu}C_{11} - \frac{\lambda^* + 2\mu^*}{\mu}C_{21}\right\}$$

where

$$D = C_{11}C_{22} - C_{12}C_{21}$$

$$F = \left[1 - \frac{1}{D}\left\{\frac{\lambda^* + \mu^*}{\mu}C_{22} - \frac{\lambda^*}{\mu}C_{12}\right\}\right]$$

$$\left[1 - \frac{1}{D}\left\{\frac{\lambda^* + 2\mu^*}{2\mu}C_{11} - \frac{\lambda^*}{2\mu}C_{21}\right\}\right]$$

$$-\frac{1}{D^2}\left\{\frac{\lambda^*}{\mu}C_{11} - \frac{(\lambda^* + \mu^*)}{\mu}C_{21}\right\}\left\{\frac{\lambda^*}{2\mu}C_{22} - \frac{(\lambda^* + 2\mu^*)}{2\mu}C_{12}\right\} \tag{35}$$

In these equations, C_{11}, C_{22}, C_{12}, and C_{21} are given by Eq 17.
The explicit expressions for H_i used in Eq 20 are

$$H_1 = \frac{1}{2(1-\nu)}\left\{\frac{\alpha^2}{(\alpha^2-1)} - 1 - \nu\right\} - \frac{1}{4(1-\nu)}\left\{4\nu - \frac{3}{(\alpha^2-1)}\right\}g$$

$$H_2 = -\frac{1}{2(1-\nu)}\frac{\alpha^2}{(\alpha^2-1)} + \frac{1}{4(1-\nu)}\left\{\frac{3\alpha^2}{(\alpha^2-1)} - 1 - 2\nu\right\}g$$

$$H_3 = -\frac{1}{(1-\nu)}\frac{\alpha^2}{(\alpha^2-1)} + \frac{1}{(1-\nu)}\left\{1 - \nu + \frac{3}{2(\alpha^2-1)}\right\}g$$

$$H_4 = \frac{1}{(1-\nu)}\frac{\alpha^2}{(\alpha^2-1)} - \frac{1}{2(1-\nu)}\left\{1 + \frac{3\alpha^2}{(\alpha^2-1)}\right\}g$$

where g is defined in the Appendix I.

References

[1] Garmong, G., Metallurgical Transactions, Vol. 15, 1974, pp. 2183-2190.
[2] Takao, Y. and Taya, M., "Thermal Expansion Coefficients and Thermal Stresses in an Aligned Short Fiber Composite with Application to a Short Carbon Fiber/Aluminum," to appear in Journal of Applied Mechanics.
[3] Mori, T. and Tanaka, K., Acta Metallurgica, Vol. 21, 1973, pp. 571-574.
[4] Taya, M. and Mura, T., Journal of Applied Mechanics, Vol. 48, 1981, pp. 361-367.
[5] Eshelby, J. D., Proceedings, Royal Society of London, Vol. A241, 1957, pp. 376-396.
[6] Mura, T., Micromechanics of Defects in Solids, Martinus Nijhoff Publishers, The Hague, 1982, p. 33.

[7] Mura, T., *Micromechanics of Defects in Solids*, Martinus Nijhoff Publishers, The Hague, 1982, p. 39.

[8] Asaro, R. J., "Somigliana Dislocations and Internal Stresses; with Application to Second Phase Hardening," *International Journal of Engineering Science*, Vol. 13, 1975, pp. 271–286.

[9] Mura, T., *Micromechanics of Defects in Solids*, Martinus Nijhoff Publishers, The Hague, 1982, p. 66.

[10] Mura, T. and Cheng, P. C., *Journal of Applied Mechanics*, Vol. 44, 1977, pp. 591–594.

[11] Smith, R. E., *Journal of Applied Physics*, Vol. 43, 1972, pp. 2555–2561.

[12] Harris, S. J., and Marsden, A. L., in *Practical Metallic Composites*, The Institute of Metallurgists, Spring Meeting, Series 3, 1974, pp. B35–B42.

[13] Lee, J. K., Earmme, Y. Y., Aanonson, H. I., and Russell, K. C., *Metallurgical Transactions*, Vol. 11A, 1980, pp. 1837–1847.

*Hidehito Okumura,[1] Katsuhiko Watanabe,[1]
and Yoshiaki Yamada[2]*

Finite-Element Analyses of
Saint-Venant End Effects for
Composite Materials

REFERENCE: Okumura, H., Watanabe, K., and Yamada, Y., **"Finite-Element Analyses of Saint-Venant End Effects for Composite Materials,"** *Recent Advances in Composites in the United States and Japan, ASTM STP 864*, J. R. Vinson and M. Taya, Eds., American Society for Testing and Materials, Philadelphia, 1985, pp. 225-235.

ABSTRACT: Stress eigenstates at the loaded end in composite materials are discussed. It is shown that a characteristic matrix equation for end effects can be obtained by the usual finite-element formulation based on the principle of virtual work and, through several examples, that Saint-Venant end effects can be easily analyzed by the method developed. The rate of attenuation of nonuniform stress distribution in composite materials is slow in comparison with isotropic materials, and it is emphasized that the evaluation of this decay by the numerical method is important in relation to the estimation of the strength of composite materials.

KEY WORDS: composite material, Saint-Venant's principle, finite-element method, eigenvalue, mode, sandwich strip, laminate

There are various types of local stresses which must be considered to estimate the properties and the strength of composite materials, and interlaminar stresses at the free edges of a laminate and nonuniform stress distribution at the loaded end are typical. Many papers with respect to the former have been published [1-5], and Yamada and Okumura [6] have dealt with interlaminar stresses at the free edge of a hole in a laminate plate using the three-dimensional finite-element method, and emphasize that finite-element numerical analysis is useful in this region. On the other hand, Choi and Horgan [7-9] have discussed the latter analytically.

[1] Research associate and associate professor, respectively, Institute of Industrial Science, University of Tokyo, 22-1, Roppongi 7 chome, Minato-ku, Tokyo, Japan.
[2] Professor, Ikutoku Technical University, 1030 Shimoogino, Atsugi-shi, Kanagawa, Japan.

Generally, composite materials show a gradual decrement of nonuniformity in stress distribution at the ends owing to their inhomogeneities in comparison with homogeneous materials; therefore, analysis of end effects is important in relation to the estimation of the strength of composite materials. Recently these problems were investigated by Dong and Goetschel [10] using semi-analytical method with finite-element interpolations over the thickness.

This paper shows that the characteristic matrix equation for these problems can be obtained by finite-element formulations similar to the conventional finite-element displacement method based on the principle of virtual work. The matrix equation obtained can be solved by transformation into standard eigenvalue problems, and the method developed is effective for studying the end effects for composite materials through analyses of several examples.

Finite-Element Formulation

Consider a rectangular region V_e whose range is x_1 to x_2 and y_1 to y_2 in a semi-infinite elastic strip as shown in Fig. 1. For plane problems the virtual work principle to the region V_e is expressed as

$$\int_{x_1}^{x_2} \int_{y_1}^{y_2} (\sigma_x \delta\epsilon_x + \sigma_y \delta\epsilon_y + \tau_{xy} \delta\gamma_{xy}) t \, dx \, dy = \int_{y_1}^{y_2} (T_{Ax} \delta u_A + T_{Ay} \delta v_A) t \, dy$$

$$+ \int_{y_1}^{y_2} (T_{Bx} \delta u_B + T_{By} \delta v_B) t \, dy + \int_{x_1}^{x_2} (T_{cx} \delta u_c + T_{cy} \delta v_c) t \, dx \qquad (1)$$

$$+ \int_{x_1}^{x_2} (T_{Dx} \delta u_D + T_{Dy} \delta v_D) t \, dx$$

where

$$t = \text{thickness of the strip,}$$
$$T_A = (T_{Ax}, T_{Ay}) = \text{traction force vector acting on the surface } S_A \text{, and}$$
$$u_A = (u_A, v_A) = \text{displacement vector at the same point.}$$

Other traction force vectors and displacement vectors are defined in a similar way. The left-hand side of Eq 1 corresponds to the strain energy of region V_e, and the terms on the right-hand side are the virtual works of traction forces. The mechanical boundary conditions for each surface are given by

$$T_{Ax} = -\sigma_x|_{x=x_1}, \quad T_{Ay} = -\tau_{xy}|_{x=x_1}, \quad T_{cx} = -\tau_{xy}|_{y=y_1}, \quad T_{cy} = -\sigma_y|_{y=y_1}$$

$$\qquad (2)$$

$$T_{Bx} = \sigma_x|_{x=x_2}, \quad T_{By} = \tau_{xy}|_{x=x_2}, \quad T_{Dx} = \tau_{xy}|_{y=y_2}, \quad T_{Dy} = \sigma_y|_{y=y_2}$$

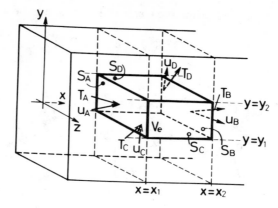

FIG. 1—*Semi-infinite elastic strip.*

The finite element used in the present analysis is a rectangular element with three nodes as shown in Fig. 2. The x-coordinate of this rectangular element is any value in the region $0 \leqq x < \infty$, and the y-coordinate is transformed to a natural coordinate with variable η by

$$y = \sum_{i=1}^{3} H_i y_i, \qquad dy = |J| d\eta \qquad (3)$$

where

$$H_1 = \frac{1}{2} \eta(-1 + \eta), \qquad H_2 = \frac{1}{2} \eta(1 + \eta), \qquad H_3 = (1 - \eta)(1 + \eta)$$

$$|J| = \frac{1}{2} (y_2 - y_1),$$

(4)

y_i is a nodal y-coordinate and $|J|$ is Jacobian.

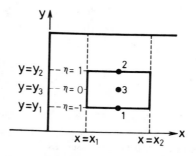

FIG. 2—*Three-node rectangular element.*

The displacement field inside the element is assumed to be as

$$\mathbf{u}(x,y) = e^{-\lambda x}\bar{\mathbf{u}}(y) = e^{-\lambda x}[H]\{\bar{u}\} \qquad (5)$$

where

$$\mathbf{u}^T(x,y) = \lfloor u(x,y), v(x,y) \rfloor, \qquad \bar{\mathbf{u}}^T(y) = \lfloor \bar{u}(y), \bar{v}(y) \rfloor \qquad (6)$$

$$[H] = \begin{bmatrix} H_1 & 0 & H_2 & 0 & H_3 & 0 \\ 0 & H_1 & 0 & H_2 & 0 & H_3 \end{bmatrix}, \qquad \{\bar{u}\}^T = \lfloor \bar{u}_1 \bar{v}_1 \bar{u}_2 \bar{v}_2 \bar{u}_3 \bar{v}_3 \rfloor$$

and

λ = parameter to show a measure of the attenuation and is the eigen-value to be obtained,

$[H]$ = a matrix composed of shape functions which are indicated using parametric coordinate η, and

$\{\bar{u}\}$ = a nodal displacement vector.

From Eq 5, the strain-displacement relations are given by

$$\{\epsilon\} = e^{-\lambda x}[B]\{\bar{u}\} \qquad (7)$$

where

$$\{\epsilon\}^T = \lfloor \epsilon_x \epsilon_y \gamma_{xy} \rfloor, \qquad [B]\{\bar{u}\} = \sum_{i=1}^{3} \mathbf{B}_i \bar{\mathbf{u}}_i = \sum_{i=1}^{3} (-\lambda \mathbf{B}_{ia} + \mathbf{B}_{ib})\bar{\mathbf{u}}_i$$

$$\mathbf{B}_{ia} = \begin{bmatrix} H_i & 0 \\ 0 & 0 \\ 0 & H_i \end{bmatrix}, \qquad \mathbf{B}_{ib} = \begin{bmatrix} 0 & 0 \\ 0 & \dfrac{1}{L_y}\dfrac{\partial H_i}{\partial \eta} \\ \dfrac{1}{L_y}\dfrac{\partial H_i}{\partial \eta} & 0 \end{bmatrix}, \qquad (8)$$

$$L_y = \frac{1}{2}(y_2 - y_1), \qquad \bar{\mathbf{u}}_i^T = \lfloor \bar{u}_i \bar{v}_i \rfloor$$

The so-called $[B]$ matrix has two parts: one is the first-order term in λ and the other is a constant term.

The stress-strain relationship is expressed as

$$\{\sigma\} = [D]\{\epsilon\} = [\mathbf{d}_1 \mathbf{d}_2 \mathbf{d}_3]^T\{\epsilon\} \qquad (9)$$

where d_1, d_2, d_3 are row vectors which correspond to stress components σ_x, σ_y, τ_{xy}, respectively.

Substituting Eqs 2 to 9 into Eq 1, we obtain the virtual work principle for an arbitrary element as

$$\{\delta\bar{u}\}^T\left(-\frac{t}{2\lambda}\right)(e^{-2\lambda x_2} - e^{-2\lambda x_1})(\lambda^2[a] + \lambda[b] + [c])\{\bar{u}\} = 0 \qquad (10)$$

where

$$[a] = [ka] - [ksa], \ [b] = -[kb] + [ksb] + [P_a], \ [c] = [kc] - [P_b]$$

$$[ka] = \int_{-1}^{1} [B_a]^T[D][B_a]|J|\,d\eta,$$

$$[kb] = \int_{-1}^{1} ([B_a]^T[D][B_b]$$

$$+ [B_b]^T[D][B_a])|J|\,d\eta$$

$$[kc] = \int_{-1}^{1} [B_b]^T[D][B_b]|J|\,d\eta$$

$$[ksa] = \int_{-1}^{1} 2[H]^T\begin{bmatrix} \mathbf{d}_1 \\ \mathbf{d}_3 \end{bmatrix}[B_a]|J|\,d\eta,$$

$$[ksb] = \int_{-1}^{1} 2[H]^T\begin{bmatrix} \mathbf{d}_1 \\ \mathbf{d}_3 \end{bmatrix}[B_b]|J|\,d\eta$$

$$[P_a] = [\tilde{P}_a]_{\eta=1} - [\tilde{P}_a]_{\eta=-1}, \qquad [P_b] = [\tilde{P}_b]_{\eta=1} - [\tilde{P}_b]_{\eta=-1}$$

$$[\tilde{P}_a] = [H]^T\begin{bmatrix} \mathbf{d}_3 \\ \mathbf{d}_2 \end{bmatrix}[B_a], \qquad [\tilde{P}_b] = [H]^T\begin{bmatrix} \mathbf{d}_3 \\ \mathbf{d}_2 \end{bmatrix}[B_b]$$

As the relation of Eq 10 holds to arbitrary nodal displacement vector $\{\delta\bar{u}\}$, the characteristic equation is given by

$$(\lambda^2[a] + \lambda[b] + [c])\{\bar{u}\} = 0 \qquad (12)$$

Following the ordinary manner in the finite-element method, individual element equations are assembled into the overall structural equation

$$\{\lambda^2[A] + \lambda[B] + [C]\}\{\bar{U}\} = 0 \qquad (13)$$

where

$$[A] = \Sigma[a], \qquad [B] = \Sigma[b], \qquad [C] = \Sigma[c], \qquad \{\bar{U}\} = \Sigma\{\bar{u}\} \quad (14)$$

The characteristic Eq 13, which is quadratic in λ, can be transformed to the standard eigenvalue problem as

$$[S]\begin{Bmatrix} \bar{V} \\ \bar{U} \end{Bmatrix} = \frac{1}{\lambda}\begin{Bmatrix} \bar{V} \\ \bar{U} \end{Bmatrix}, \qquad \{\bar{V}\} = \lambda\{\bar{U}\}, \qquad [S] = \begin{bmatrix} \mathbf{O} & \mathbf{I} \\ -\mathbf{C}^{-1}\mathbf{A} & -\mathbf{C}^{-1}\mathbf{B} \end{bmatrix}$$

$$(15)$$

Equation 15 is solved by the double QR method [12,13] in order to obtain the eigenvalue λ which includes the complex roots of the conjugate pair in general, because the matrix $[S]$ is unsymmetric.

Numerical Examples

For verification of the numerical formulation and computer program, several examples are considered.

First is the analysis of an homogeneous isotropic semi-infinite elastic strip. Material properties used in this case and the finite-element mesh division to the upper half-strip (symmetric mode is considered) are shown in Fig. 3. The displacement v_1 of node 1, of which the y-coordinate is zero, is constrained. In this study the nodal displacement $\{\bar{u}\}$ is independent of the x-coordinate.

The results are compared in Table 1 with analytical results [11] for the five lowest eigenvalues. The real part of the smallest eigenvalue which provides a measure of the attenuation for Saint-Venant's principle is about 2.1. It can be seen that the computed values coincide with the analytical values, although the number of elements is small.

The effect of anisotropy is discussed in the second example of an homoge-

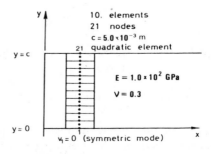

FIG. 3—*Mesh division.*

TABLE 1—*Eigenvalues for semi-infinite elastic strip (symmetric mode).*

Mode	Computed		Analytical [11]	
	$R_e(c\lambda)$	$I_m(c\lambda)$	$R_e(c\lambda)$	$I_m(c\lambda)$
1	2.1057	1.1252	2.1062	1.1254
2	5.3555	1.5514	5.3563	1.5516
3	8.5276	1.8041	8.5367	1.7755
4	11.9530	1.8374	11.6992	1.9294
5	14.5787	1.0904	14.8541	2.0469

neous anisotropic semi-infinite strip. Figure 4 shows material properties and the finite-element division taken in this case.

The lowest five (nonzero) eigenvalues are listed in Table 2 for symmetric deformations, and analytical values obtained by Ref *7* and computed values by Ref *10* are also employed. Four eigenvalues exist in the region where $R_e(c\lambda)$ is 2.1 or less. This indicates that nonuniform stress distribution affects a wide range far from the loaded edge.

Figures 5 and 6 show the displacement modes and the stress modes for an anisotropic semi-infinite strip. In the symmetric-mode case (Fig. 5), where the lowest eigenvalue $c\lambda_1$ is equal to 0.564, displacement u and stress σ_x, σ_y are symmetric with the x-coordinate, and displacement v and stress τ_{xy} are antisymmetric. At the upper surface, stress σ_y and τ_{xy} are zero, because the traction force of the top surface is free. In the antisymmetric-mode case (Fig. 6) for the same problem where the eigenvalue is 0.796, displacement u and stress σ_x, σ_y are antisymmetric, and displacement v and stress τ_{xy} are symmetric.

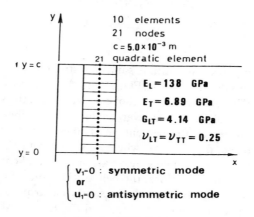

FIG. 4—*Mesh division.*

TABLE 2—*Eigenvalues for semi-infinite orthotropic strip (symmetric mode).*

Mode	Computed		Analytical [7] Computed [10]	
	$R_e(c\lambda)$	$I_m(c\lambda)$	$R_e(c\lambda)$	$I_m(c\lambda)$
1	0.5638	0	0.5643	0
			0.5643	0
2	1.1352	0	1.1345	0
			1.1345	0
3	1.7564	0	1.7510	0
			1.7510	0
4	2.0975	0.1778	2.1009	0.1847
			2.1009	0.1847
5	2.7157	0	2.7255	0
			2.7256	0

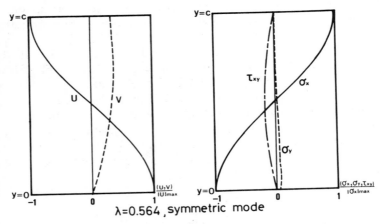

$\lambda=0.564$, symmetric mode

FIG. 5—*Displacement and stress mode for orthotropic strip.*

The third example, a semi-infinite sandwich strip composed of two dissimilar isotropic materials, is shown in Fig. 7, and the results are compared with theory in Ref 8. The smallest real part of eigenvalue λ in the symmetric case is shown as a function of E_1/E_2 for several values of V_f. Fiber-reinforced composite materials of which the percentage of fiber content is low (fiber: E_1, matrix: E_2) correspond to the case when V_f is small and E_1/E_2 is large and, especially in this case, the smallest real part of λ is small compared with that for isotropic materials. Therefore, the penetration zone of end effects is of a very wide range. On the other hand, the sandwich strip (face: E_2, core: E_1) corresponds to the case when V_f is large and E_1/E_2 is small, and the lowest real part of λH is 2.1 or less. Figure 8 shows displacement modes and stress

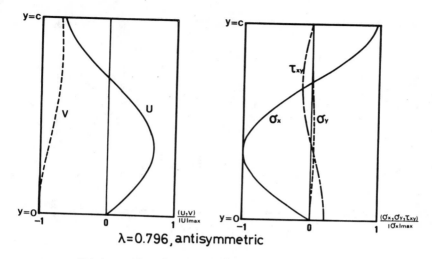

λ=0.796, antisymmetric

FIG. 6—*Displacement and stress mode for orthotropic strip.*

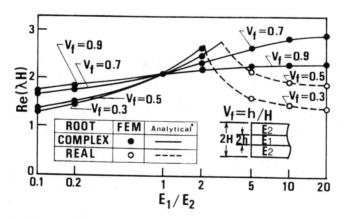

FIG. 7—*Smallest real part of eigenvalue λ (the asterisked word "Analytical" in the legend box refers to Ref 8).*

modes of laminate structures $[E_2/E_1]_s$. It is noticed that the value of stress τ_{xy} is maximum at the interface and the value of stress σ_y is maximum at midplane.

The last example is for the case of an eight-ply laminate strip $[E_2/E_1/E_2/E_1]_s$ as shown in Fig. 9. Material properties are $E_1/E_2 = 5$ and Poisson's ratio $\nu_1 = \nu_2 = 0.3$, and a finite-element division of 20 quadratic elements is used. It is noticed that the τ_{xy} stress distribution in the third layer becomes maximum, and the value of stress σ_y is maximum at the midplane.

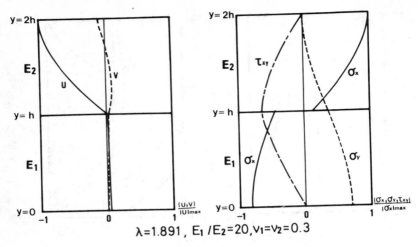

$$\lambda=1.891, \ E_1/E_2=20, \nu_1=\nu_2=0.3$$

FIG. 8—*Displacement and stress mode for sandwich strip (symmetric mode).*

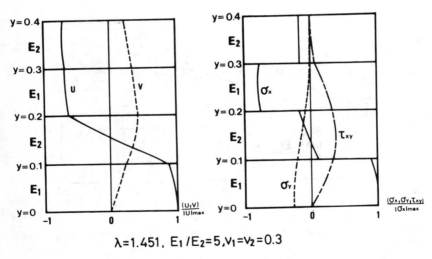

$$\lambda=1.451, \ E_1/E_2=5, \nu_1=\nu_2=0.3$$

FIG. 9—*Displacement and stress mode for laminate strip (symmetric mode).*

Conclusion

A method to analyze Saint-Venant end effects in composite materials following the procedure of the finite-element displacement method was developed, and its effectiveness was demonstrated through several examples. Generally, the rate of attenuation of nonuniform local stress distribution is small in composite materials; thus it is necessary to analyze end effects for the accurate estimation of composite material properties and strength. In the evaluation of actual fiber-reinforced laminate plate, a three-dimensional analysis should be carried out; the method developed can easily be extended to three dimensional problems.

References

[1] Puppo, A. H. and Evensen, H. A., *Journal of Composite Materials*, Vol. 4, 1970, pp. 204–220.

[2] Pipes, R. B. and Pagano, N. J., *Journal of Composite Materials*, Vol. 4, 1970, pp. 538–548.

[3] Pipes, R. B. and Daniel, I. M., *Journal of Composite Materials*, Vol. 5, 1971, pp. 255–259.

[4] Wang, S. S. and Choi, I., *Journal of Applied Mechanics*, Vol. 49, 1982, pp. 541–548.

[5] Wang, S. S. and Choi, I., *Journal of Applied Mechanics*, Vol. 49, 1982, pp. 549–560.

[6] Yamada, Y. and Okumura, H. in *Composite Materials: Mechanics, Mechanical Properties and Fabrication*, Japan Society for Composite Materials, Tokyo, 1981, pp. 55–64.

[7] Choi, I. and Horgan, C. O., *Journal of Applied Mechanics*, Vol. 44, 1977, pp. 424–430.

[8] Choi, I. and Horgan, C. O., *International Journal of Solids and Structures*, Vol. 14, 1978, pp. 187–195.

[9] Horgan, C. O., *Journal of Composite Materials*, Vol. 16, 1982, pp. 411–422.

[10] Dong, S. B. and Goetschel, D. B., *Journal of Applied Mechanics*, Vol. 49, 1982, pp. 129–135.

[11] Johnson, M. W. and Little, R. W., *Quarterly of Applied Mathematics*, Vol. 22, 1965, pp. 335–344.

[12] Francis, J. G. F., *Computer Journal*, Vol. 4, 1961, pp. 265–271.

[13] Francis, J. G. F., *Computer Journal*, Vol. 4, 1962, pp. 332–345.

Lawrence W. Rehfield,[1] *Erian A. Armanios,*[1] *and Qiao Changli*[1]

Analysis of Behavior of Fibrous Composite Compression Specimens

REFERENCE: Rehfield, L. W., Armanios, E. A., and Changli, Q., "**Analysis of Behavior of Fibrous Composite Compression Specimens,**" *Recent Advances in Composites in the United States and Japan, ASTM STP 864*, J. R. Vinson and M. Taya, Eds., American Society for Testing and Materials, Philadelphia, 1985, pp. 236–252.

ABSTRACT: Unidirectional laminated composite specimens reinforced with end tabs are analyzed. The stress field in the laminate-tab interface and within the laminate is predicted using a new analysis method. The method is simple and does not require extensive numerical computations. Results are obtained in closed form. Selected examples are modeled also by finite elements in order to verify that the simple analysis method yields good results. Analyses of previously tested specimens provide a plausible explanation of observed behavior. The present analysis method facilitates specimen design and data interpretation and, therefore, aids in overcoming many of the difficulties associated with the compression testing of composite laminates.

KEY WORDS: composite materials, compression testing, interlaminar stresses, finite-element analysis

Most fibrous composite compression tests have been performed on specimens of small unsupported gage length or specimens that are extrinsically stiffened or stabilized. The primary concern is to preclude overall specimen buckling during the test and, therefore, to determine intrinsic compressive strength and stiffness. The former testing methods employ elaborate fixtures which support and align the specimen over most of its length while leaving a small unsupported gage length portion. The most widely used fixtures are the Celanese (ASTM Test Method for Compressive Properties of Unidirectional or Crossply Fiber-Resin Composites D-3410-75) and the Illinois Institute of Technology Research Institute (IITRI) types [1]. Extrinsic stiffening has

[1] Professor, graduate research assistant, and visiting scholar, respectively, School of Aerospace Engineering, Georgia Institute of Technology, Atlanta, GA 30332.

taken two forms—continuous or intermittent specimen side support by fixtures and bonding laminates to a core to form a sandwich specimen which is tested in bending. Sandwich beam specimens are expensive to manufacture and generally provide the highest failure loads.

The practical consequences of the above approaches are rarely discussed. The small unsupported gage length in the first type of test raises questions about the uniformity of conditions achieved within the gage length. In the second type of test, the influence of the extrinsic stiffening on the outcome of the experiment remains unknown. Regardless of the method used, compressive strength properties are the most difficult material properties to measure. Experimental scatter is greater than with other tests, and differences in failure mode are often found. The difficulty is due to the requirements for buckling stabilization and for alignment to control load eccentricity and its resulting bending action.

A simple alternative approach is explored in this paper. It is to utilize a specimen with robust tabs bonded to the ends of the laminated composite coupon and to test it flat-ended in compression. The tabs eliminate brooming of the ends and stabilize against buckling. Alignment problems are controlled by machining the ends flat and parallel to within acceptable limits. Buckling resistance is increased by the end tabs in two ways: Clamped end conditions are achieved and the bending stiffness near the ends is increased. A similar approach was successfully used earlier [2] to evaluate the compression behavior of ribs of a continuous-filament advanced composite isogrid structure. A prerequisite for the use of this concept is an analysis method to predict the stress field inside the laminate. The laminate and the attached tabs must be modeled as a combined, composite structure.

The Analysis Problem

The presence of the tabs creates a stress diffusion/concentration problem that originates from tab-gage section junctures. There are three concerns: a failure at the tab-gage juncture due to localized overstress; disbonding of the tabs, which may be followed by overall laminate buckling and bending-related fracture; and a departure from stress uniformity in the central portion of the gage section.

In the present paper, a new analysis method is developed to predict laminate-tab interaction and the stress field within the laminate. The method is simple—expensive, complex numerical computations are avoided as results are obtained in closed form. It is based upon the ideas utilized in the theory developed by Rehfield and Murthy [3]. This theory assumes that the statically equivalent stresses obtained from the classical bending theory can be used to estimate transverse shear and normal strains. These strain estimates permit a suitable displacement field to be determined which further leads to improved stresses.

Outline of Analysis Methodology

In the present formulation, the tab and specimen will be considered as independent elements. Each element (tab or specimen) is treated as an homogeneous body in equilibrium. Interfacial stresses are assumed initially to be unknown. Enforcement of continuity conditions at interfaces and boundary conditions, in an overall sense, leads to a solution for these stresses. This procedure and the development of element equations are given in detail in the Appendix. In the present case, each element will be treated as an orthotropic sublaminate represented by its effective extensional modulus under a plane-strain condition. Also, the underlined terms in the equations will be neglected.

Examples

A quantitative demonstration of the present formulation is provided by the solution to the problem of the simple flat-ended compression specimen with tabs bonded to its ends as shown in Fig. 1. For other test methods, modeling of tab lateral pressure is important, but it depends heavily upon details of the specimen grips or fixtures used. The geometry and properties of the specimen are similar to those of Ref 4. A specimen of this geometry tested flat-ended is likely to buckle. The same dimensions and properties, however, have been used to investigate the stress and strain distributions using the present theory. Due to symmetry, only one quarter of the problem is analyzed. The global coordinates are x and z. The x-coordinate is measured from the specimen edge, and the z-coordinate is measured from the specimen midplane. The tab length is L and its thickness is h^1. The specimen half-length is $(L + a)$ and its half-thickness h^2.

It is convenient to refer to the tab and specimen in the region with three local coordinate systems, (x, z^k), as indicated in the generic element appearing in Fig. 2: $k = 1$ for the tab, $k = 2$ for the part of the specimen supported by the tab $(0 \leq x \leq L)$, and $k = 3$ for the gage section part $(L \leq x \leq L + a)$. Let $z^k = 0$ be the middle surface for each element (tab or specimen). Also let the superscript 1, 2 and 3 denote, respectively, the tab and both regions of the specimen and the variables associated with them.

The following conditions ensure symmetry of deformation

$$
\begin{array}{lll}
\text{At } x = L + a & \tilde{u}^3 = \bar{\phi}^3 = 0 & \\
\text{At } z = 0 & w^2(x, -h^2/2) = 0; & 0 \leq x \leq L \\
& w^3(x, -h^2/2) = 0; & L \leq x \leq L + a \qquad (1) \\
& t_1^2 = 0; & 0 \leq x \leq L \\
& t_1^3 = 0; & L \leq x \leq L + a
\end{array}
$$

The average axial displacement defined in Appendix Eq 28 is $\tilde{u}$ and the section rotation defined in Appendix Eq 29 is $\bar{\phi}$. The transverse displacement is

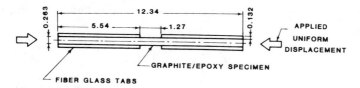

FIG. 1—*Specimen details and loading.*

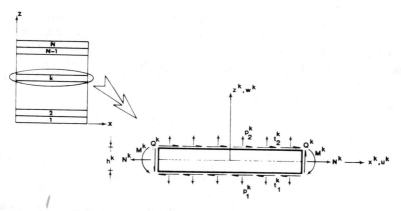

FIG. 2—*Notation and sign convention for kth element.*

w, and the interfacial shear stress on the specimen bottom surface is t_1^2 in $0 \leq x \leq L$ and t_1^3 in $L \leq x \leq L + a$.

The tab and specimen are assumed to be co-cured. The continuity requirements for the displacement components at the interface are

$$u^1(x, -h^1/2) = u^2(x, h^2/2); \qquad 0 \leq x \leq L$$

$$w^1(x, -h^1/2) = w^2(x, h^2/2); \qquad 0 \leq x \leq L$$

(2)

The reciprocity of interfacial stresses is given by

$$t_1^1 = t_2^2; \qquad 0 \leq x \leq L$$

$$q_1^1 = q_2^2; \qquad 0 \leq x \leq L$$

(3)

The interfacial shear and transverse normal stress on the tab bottom surface is t_1^1 and q_1^1, respectively. On the specimen upper surface, they are t_2^2 and q_2^2, respectively. Since there is no gripping of the tabs, the stress-free condition at the tab upper surface is ensured by setting the shear and transverse normal stresses to zero

$$t_2^1 = q_2^1 = 0; \qquad 0 \le x \le L \tag{4}$$

The stress-free condition on the upper surface of the specimen gage section is

$$t_2^3 = q_2^3 = 0; \qquad L \le x \le L + a \tag{5}$$

The equilibrium equations (9) for each element, the constitutive relations (30), the displacement distributions (31), the continuity requirements (2,3), and the stress-free surface conditions (4,5) make up the desired set of equations to be solved. These equations are supplemented by a set of boundary conditions. These boundary conditions reflect a uniform axial flat-ended shortening $\bar{u}$ at $x = 0$ and tab free edge at $x = L$

$$\text{At } x = 0, \qquad \tilde{u}^1 = \tilde{u}^2 = \bar{u}$$

$$\tilde{\phi}^1 = \tilde{\phi}^2 = 0 \tag{6}$$

$$\text{At } x = L, \qquad N^1 = M^1 = 0$$

Continuity of axial displacement and reciprocity of axial force and moment resultant at the face between the specimen-tab-supported region $(0 \le x \le L)$ and the gage section $(L \le x \le L + a)$ are also enforced. In this set of boundary conditions, axial and bending behavior have been considered. Transverse shear and transverse displacement boundary conditions cannot be satisfied independently as a consequence of the simplifications adopted in the governing equations and the physical considerations of the compression specimen problem.

In order to verify the accuracy of the present formulation, the problem has been modeled by finite elements. The finite element discretization is schematically shown in Fig. 3. The total number of elements used in the modeling is 2100. The element used is a constant-strain four-node rectangular element. The finite-element code used is the Engineering Analysis Language (EAL).

The geometry, dimensions, and material properties for the proposed compression specimen should be determined based upon requiring no buckling of the specimen and the following considerations:

1. preventing disbonding of tabs,
2. preventing failure at the tab-gage juncture, and
3. ensuring stress uniformity in the central portion of the gage section.

In order to check these a comparison of the stress distribution at the interface between tabs and specimen is shown in Fig. 4 for an AS/PR288 unidirec-

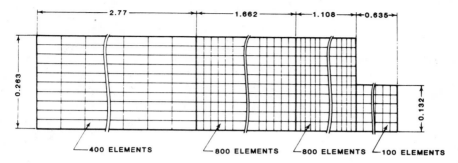

FIG. 3—*Finite-element discretization.*

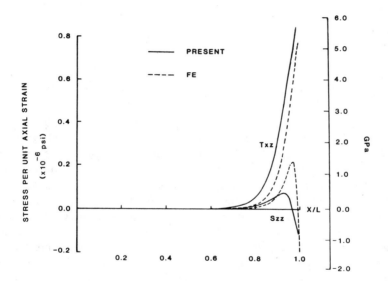

FIG. 4—*Interlaminar stresses at the tab-specimen interface: fiber glass tabs, AS/PR 288 graphite/epoxy specimen.*

tional graphite fiber/epoxy resin specimen and in Fig. 5 for T300/5208 unidirectional graphite/epoxy. The shear stress distribution obtained by the present formulation shows good agreement with the finite-element solution. Due to the tab free edge, the high shear stress value near the end may cause disbonding of tabs from the specimen. Tapering the tab ends may help reduce the peak shear stress value and avoid this problem. The agreement in the transverse normal stress is less satisfactory since it is the derivative of the shear stress.

The axial strain distribution at the surface of the gage section is shown in Fig. 6. This distribution may cause failure at the tab-specimen interface junc-

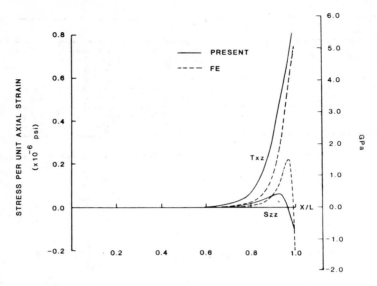

FIG. 5—*Interlaminar stresses at the tab-specimen interface: fiber glass tabs, T300/5208 graphite/epoxy specimen.*

ture. The tensile nature of the axial strain at the tab-specimen interface juncture is due to the nonuniform axial stress distribution across the thickness as a result of stiffness differences between the woven fiber glass tabs and the graphite/epoxy specimen. This distribution supplies a plausible reason for failures at this location and the nature of the fractured surfaces described in Ref *4*. The finite-element solution is based upon the use of simple constant-strain elements. As a result, the true nature of the stresses in the juncture region cannot be captured using such simple elements. To illustrate this fact, the problem has been modeled using 2200 and 2400 elements; the axial strain at the juncture region was −0.326 and −0.535, respectively. Moreover, to overcome computer core size limitation a region 0.635 cm from both sides of the tab-specimen juncture has been modeled with a refined mesh using 2700 elements; the strain at the juncture region was −0.845. This shows that the finite element used has difficulty in capturing the steep strain gradient at the juncture region since the absolute strain value increases as the element size progressively decreases.

Strain uniformity through the specimen depth in the central portion is checked by computing the extreme strain difference compared with the average strain value. The T300/5208 specimen shows 33.5% difference while in the AS/PR288 specimens the difference is 21.1%. A larger gage length will provide enough distance for the strain to decay, thereby ensuring more strain uniformity in the central portion. Also, tapering the tabs will enhance strain uniformity by reducing the strain value near the juncture region.

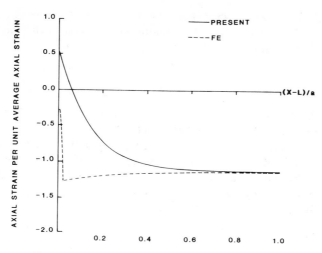

FIG. 6—*Axial strain distribution at gage section surface.*

Limited Experimental Verification

The essential features of the proposed approach were used in Ref 5 to determine the compressive behavior of a $[\pm 45, 0, 90]_2$ quasi-isotropic graphite/epoxy laminate composite specimen made of the Narmco T300/5208 material system. Typical stress-strain curves from specimen tests are shown in Fig. 9 of Ref 5. The strain values reported are based on the average values from two strain gages mounted back-to-back. The data exhibited consistent and smooth behavior, and no unanticipated events occurred during the test. This supports the proposed experimental approach.

Concluding Remarks

A simple analysis method has been developed and applied to compression specimens with end tabs that are tested flat-ended. The method has been validated using a finite-element analysis. The proposed analytical method is ideally suited for preliminary design studies, which require a large number of configurations to be evaluated quickly and economically. Specimen configurations should be determined based upon preventing overall buckling, disbonding of tabs, and failure at the tab-gage juncture, while ensuring stress uniformity in the central portion of the gage section. Disbonding of the tabs and failure at the tab-gage juncture were found to be possible modes of failure for the specimen studied. Moreover, the through-thickness strain distribution in the central portion of the specimen gage section was found to be highly nonuniform. In any case, the specimen and test arrangement must be treated as a structural system and analyzed thoroughly. The proposed simple method for doing this has been shown to describe the essential features of the behavior. A limited demonstration of the soundness of this approach is provided by the experimental results presented in Ref 5.

Acknowledgments

This work was supported by the U.S. Air Force Office of Scientific Research under Grants AFOSR 82-0080 and 83-0056. This support is gratefully acknowledged. The authors acknowledge the benefit of discussions with Dr. C.C. Chamis of the NASA Lewis Research Center. The finite-element results were obtained using the code Engineering Analysis Language (EAL) which was provided by Engineering Information Systems, Inc., Saratoga, CA; this assistance is also gratefully acknowledged.

APPENDIX

Development of the Analytical Model

Foundations of the Model

The development that follows is based upon the fundamental assumption that the statically equivalent stresses obtained from classical engineering theory can be used to estimate the transverse shear and normal strains ignored in the classical theory. All subsequent results follow from this premise. These strain estimates permit a suitable displacement field to be determined, which further leads to improved stresses. This approach has been thoroughly documented and evaluated for beams [3], plates [6], and composite laminates [7,8] by comparisons with exact solutions (when available), finite-element solutions, demonstrations of correspondence with other established theories, and order-of-magnitude error investigations [9,10].

In the present work, models appropriated to individual plies of a laminate or groups of plies ... sublaminates ... are developed using the same fundamental logic and assumptions. The development is restricted to static plane-stress states. The same results can be used, however, in a plane-strain situation by a proper transformation of the elastic constants. This restriction is not intrinsic to the present development. It allows the presentation of essential features in a simple way without confronting the additional complexities of fully three-dimensional behavior.

Statically Equivalent Stress Field

Consider a laminate made of N perfectly bonded layers or plies, each ply having a plane of material symmetry parallel to the plane of the laminate. Let k denote a particular ply that is singled out for study. The ply thickness is h^k. Let x^k and y^k represent the coordinates in the midplane of the ply. The interlaminar stresses σ_{zz} and σ_{xz} at the top of the ply are denoted by p_2 and t_2, respectively, while the corresponding stresses at the bottom of the ply are designated as p_1 and t_1. Notation and sign convention appear in Fig. 2.

The equilibrium equations that are valid within each ply are

$$(\sigma_{xx,x} + \sigma_{xz,z})^k = 0$$

$$(\sigma_{zx,x} + \sigma_{zz,z})^k = 0$$

(7)

The usual convention and notation are followed: σ_{xx} is the axial stress, σ_{zz} is the transverse normal stress, and σ_{xz} the transverse shear stress. Superscript k signifies that each variable within the parentheses is associated with the kth ply.

The above equations are supplemented by interlaminar conditions which ensure reciprocity of tractions and continuity of displacements at the interfaces between plies. At the interface between the kth and the $(k - 1)$th plies, they are

$$t_1^k = t_2^{k-1}$$

$$p_1^k = p_2^{k-1}$$

$$u^k\left(x^k, -\frac{h^k}{2}\right) = u^{k-1}\left(x^{k-1}, \frac{h^{k-1}}{2}\right)$$

$$w^k\left(x^k, -\frac{h^k}{2}\right) = w^{k-1}\left(x^{k-1}, \frac{h^{k-1}}{2}\right)$$

(8)

The axial component of the displacement is u^k and the lateral component w^k. In addition, traction boundary conditions are normally specified at the extreme upper and lower surfaces of the laminate. At the boundaries corresponding to constant values of x, the laminate sections, boundary conditions will be enforced in an overall sense as is commonly done in engineering theories of bending and stretching.

Overall equations of equilibrium in terms of force and moment resultants may be derived by appropriate integration of Eqs 7. They are

$$N_{,x}^k + t_2^k - t_1^k = 0$$

$$Q_{,x}^k + p_2^k - p_1^k = 0$$

(9)

$$M_{,x}^k - Q^k + \frac{h^k}{2}(t_2^k + t_1^k) = 0$$

The force, moment, and transverse shear stress resultants of the kth ply are N^k, M^k, and Q^k, respectively. These bear the classical definitions

$$(N^k, M^k, Q^k) = \int_{-h^k/2}^{h^k/2} (\sigma_{xx}, z\sigma_{xx}, \sigma_{xz})^k \, dz$$

(10)

Superscript k, which identifies the ply, will be dropped in the subsequent equations for convenience.

According to classical theory, the stresses that satisfy equilibrium Eqs 7 and interlaminar stress boundary conditions at the top and bottom surfaces of each ply are

$$\sigma_{xx} = \frac{N}{h} + \frac{12M}{h^3} z$$

(11)

$$\sigma_{xz} = \frac{z}{h}(t_2 - t_1) + \frac{3}{2h}\left(1 - \frac{4z^2}{h^2}\right)Q - \frac{1}{4}\left(1 - \frac{12z^2}{h^2}\right)(t_2 + t_1) \tag{12}$$

$$\sigma_{zz} = \frac{1}{2}(p_2 + p_1) + \frac{h}{8}\left(1 - \frac{4z^2}{h^2}\right)(t_2 - t_1)_{,x} + \frac{3z}{2h}\left(1 - \frac{4z^2}{3h^2}\right)(p_2 - p_1)$$

$$+ \frac{z}{4}\left(1 - \frac{4z^2}{h^2}\right)(t_2 + t_1)_{,x} \tag{13}$$

The following contractions which denote the sum and difference of interlaminar stresses are introduced for convenience

$$n = t_2 - t_1$$

$$m = \frac{h}{2}(t_2 + t_1)$$

$$\tag{14}$$

$$q = p_2 - p_1$$

$$r = \frac{1}{2}(p_2 + p_1)$$

where n can be regarded as an effective distributed axial force, m an effective distributed moment, q an effective lateral pressure, and r a mean flatwise tensile stress.

The stresses given in Eqs 11 through 13, although not exact, are statically equivalent to the applied surface loadings. They will be used subsequently to develop approximations for the displacement components.

Kinematics

The central assumption in the present development is that the statically equivalent stresses in Eqs 11 through 13 can be used to estimate the transverse normal strain and transverse shear strain. This is not a kinematic assumption, but an assumption regarding stresses.

The development will be carried out for orthotropic materials with principal material directions corresponding to axes of the ply. The appropriate form of Hooke's law for plane stress is

$$\epsilon_{xx} = S_{11}\sigma_{xx} + S_{13}\sigma_{zz} \tag{15}$$

$$\epsilon_{zz} = S_{13}\sigma_{xx} + \underline{S_{33}\sigma_{zz}} \tag{16}$$

$$\gamma_{xz} = S_{55}\sigma_{xz} \tag{17}$$

ϵ_{xx}, ϵ_{zz}, γ_{xz} are the axial strain, transverse normal strain, and transverse shear strain, respectively. The S-terms are flexibilities; in terms of engineering constants, they are

$$S_{11} = 1/E_{11}$$

$$S_{13} = -\nu_{13}/E_{11}$$

$$(18)$$

$$S_{33} = 1/E_{33}$$

$$S_{55} = 1/G_{13}$$

E_{11} and E_{33} are elastic moduli associated with the x- and z-directions. ν_{13} is Poisson's ratio and G_{13} is the transverse shear modulus. The underlined term in Eq 16 represents the influence of the transverse normal stress in estimating the transverse normal strain. The same convention will be followed throughout the development to identify this type of contribution. Its effect upon the derived equations will be discussed later.

The strain components are related to the displacement components by the following strain displacement relations

$$u_{,x} = \epsilon_{xx}$$

$$w_{,z} = \epsilon_{zz} \tag{19}$$

$$u_{,z} + w_{,x} = \gamma_{xz}$$

Equations 11, 13, 16, and 19 permit the transverse normal strain to be approximated as

$$w_{,z} \cong S_{13}\left(\frac{N}{h} + \frac{12M}{h^3}z\right)$$

$$+ S_{33}\left[r + \frac{h}{8}\left(1 - \frac{4z^2}{h^2}\right)n_{,x} + \frac{3z}{2h}\left(1 - \frac{4z^2}{3h^2}\right)q + \frac{z}{2h}\left(1 - \frac{4z^2}{h^2}\right)m_{,x}\right]$$

$$(20)$$

Integration of this equation results in the following expression for the lateral displacement component w

$$w = W(x) + S_{13}z\left(\frac{N}{h} + \frac{6M}{h^3}z\right)$$

$$+ \frac{1}{2}S_{33}z\left[2r + \frac{h}{4}\left(1 - \frac{4z^2}{3h^2}\right)n_{,x} + \frac{3z}{2h}\left(1 - \frac{2z^2}{3h^2}\right)q + \frac{z}{2h}\left(1 - \frac{2z^2}{h^2}\right)m_{,x}\right]$$

$$(21)$$

$W(x)$ is the lateral deflection of the ply axis $z = 0$.

The axial component of displacement u can be estimated as follows. Equations 17 and 19 permit $u_{,z}$ to be expressed as

$$u_{,z} = S_{55}\sigma_{xz} - w_{,x} \tag{22}$$

Substitution of Eqs 12 and 21 into Eq 22 and subsequent integration results in

$$u = U(x) - zW(x) + \frac{z^2}{2h} S_{55}Sn + \frac{3z}{2h}\left(1 - \frac{4z^2}{3h^2}S\right)S_{55}Q$$

$$- \frac{z}{2h}\left(1 - \frac{4z^2}{h^2}S\right)S_{55}m \tag{23}$$

$$- \frac{1}{4}S_{33}z^2\left[2r_{,x} + \frac{h}{4}\left(1 - \frac{2z^2}{3h^2}\right)n_{,xx} + \frac{z}{h}\left(1 - \frac{2z^2}{5h^2}\right)q_{,x} + \frac{z}{3h}\left(1 - \frac{6z^2}{5h^2}\right)m_{,xx}\right]$$

where

$$S = 1 + S_{13}/S_{55} \tag{24}$$

$U(x)$ is the axial deflection at the ply axis.

The static displacement field is completely described by Eqs 21, 23, and 24. This approximate displacement field was determined by estimating precisely the strains that are ignored in classical theory. U and W, the axis displacement components, emerge as natural kinematic variables. The kinematic compatibility equations of continuum theory are not satisfied exactly by this displacement field. The objective, however, is not rigor but a consistent, reliable engineering theory which features high accuracy together with simplicity. The approximations made are consistent with the other inherent approximations introduced [9, 10].

Refined Axial Stress

The axial stress in the ply is usually large and of the greatest importance. An accurate knowledge of it is essential in practical applications. A refined estimate that improves Eq 11 is central to the theoretical improvements that are sought.

Equations 13, 15, 19, and 23 can be utilized to produce a refined axial stress expression

$$\sigma_{xx} = \frac{1}{S_{11}}(u_{,x} - S_{13}\sigma_{zz})$$

$$= \frac{1}{S_{11}}\left\{U_{,x} - zW_{,xx} - S_{13}r - \frac{h}{8}S_{13}\left[1 - \frac{4z^2}{h^2}\left(2 + \frac{S_{55}}{S_{13}}\right)\right]n_{,x}\right. \tag{25}$$

$$- \frac{S_{13}z}{2h}\left[1 + \frac{S_{55}}{S_{13}} - \frac{4z^2}{h^2}\left(2 + \frac{S_{55}}{S_{13}}\right)\right]m_{,x} - \frac{3z}{2h}S_{13}\left[1 + \frac{S_{55}}{S_{13}} - \frac{4z^2}{3h^2}\left(2 + \frac{S_{55}}{S_{13}}\right)\right]q$$

$$\left. - \frac{1}{4}S_{33}z^2\left[2r_{,xx} + \frac{h}{4}\left(1 - \frac{2z^2}{3h^2}\right)n_{,xxx} + \frac{z}{h}\left(1 - \frac{2z^2}{5h^2}\right)q_{,xx} + \frac{z}{3h}\left(1 - \frac{6z^2}{5h^2}\right)m_{,xxx}\right]\right\}$$

Relationships for the axial force and moment resultants are obtained by using Eqs 10 and 25. The results are

$$N = \frac{h}{S_{11}}\left[U_{,x} - S_{13}r - \frac{h}{24}S_{13}\left(1 - \frac{S_{55}}{S_{13}}\right)n_{,x} - \frac{h^2}{16}S_{33}\left(\frac{2}{3}r_{,xx} + \frac{3h}{40}n_{,xxx}\right)\right]$$

$$M = \frac{h^3}{12S_{11}}\left[-W_{,xx} - \frac{3}{10h}S_{13}\left(3 + 4\frac{S_{55}}{S_{13}}\right)q + \frac{S_{13}}{10h}\left(1 - 2\frac{S_{55}}{S_{13}}\right)m_{,x}\right. \qquad (26)$$

$$\left. - \frac{h}{1120}S_{33}(39q_{,xx} + 11m_{,xxx})\right]$$

Equations (26) permit Equation (25) to be rewritten as

$$\sigma_{xx} = \frac{N}{h} + \frac{12M}{h^3}z - \alpha\left[\frac{h}{12}\left(1 - \frac{12z^2}{h^2}\right)n_{,x} + \frac{3z}{5h}\left(1 - \frac{20z^2}{3h^2}\right)(q + m_{,x})\right]$$

$$+ \frac{h^2}{16}\frac{S_{33}}{S_{11}}\left\{\frac{2}{3}\left(1 - \frac{12z^2}{h^2}\right)r_{,xx} + \frac{3h}{40}\left[1 - \frac{40z^2}{3h^2}\left(1 - \frac{2z^2}{3h^2}\right)\right]n_{,xxx}\right. \qquad (27)$$

$$\left. + \frac{39z}{70h}\left[1 - \frac{280}{39}\frac{z^2}{h^2}\left(1 - \frac{2z^2}{5h^2}\right)\right]q_{,xx} + \frac{11z}{70h}\left[1 - \frac{280}{33}\frac{z^2}{h^2}\left(1 - \frac{6z^2}{5h^2}\right)\right]m_{,xxx}\right\}$$

The parameter α is $(S_{55} + 2S_{13})/2S_{11}$; it is unity for an isotropic material.

It has been shown [9,10] that the transverse normal and shear stress distributions which are consistent with the refined axial stress distribution, Eq 27, are those given by the classical expression in Eqs 12 and 13.

Summary

The governing equation for the present model can be summarized now. They encompass four categories. Overall-type equations consist of the equilibrium Eqs 9 and the constitutive Eqs 26. In addition, two sets of equations provide the distributions of stresses and displacements throughout the ply. The first set for stresses consists of Eqs 27, 12, and 13. The second for displacements is composed of Eqs 21 and 23.

Considerable simplification is achieved if the underlined terms in the governing equations are neglected. This is equivalent to neglecting the influence of the transverse normal stress in estimating the transverse normal strain as indicated in Eq 16. The implications of this approximation and its accuracy are discussed in Ref 9.

In the place of axis-related kinetmatic variables, averaged variables can also be used. These are defined as

$$\tilde{u} = \int_{-h/2}^{h/2}\frac{u}{h}dz$$

$$\tilde{w} = \int_{-h/2}^{h/2}\frac{w}{h}dz \qquad (28)$$

In addition to the averaged kinematic variables in Eqs 28, it is convenient to introduce an intermediate kinematic variable that is related to ply section rotation. This is selected to be the mean rotation of the ply section

$$\tilde{\phi} = \int_{-h/2}^{h/2} \frac{u_{,z}}{h} dz = \frac{1}{h}\left[u(x,\, h/2) - u(x,\, -h/2) \right] \tag{29}$$

The above collection of equations requires the specification of boundary conditions at specified values of x. There are three boundary conditions per end. It is usual in an engineering theory to prescribe N or $\tilde{u}$, Q or $\tilde{w}$, and M or $\tilde{\phi}$. Ply-axis kinematic variables can also be used in specifying kinematic boundary conditions. The modeling of boundary restraint conditions in classical theory is straightforward since the displacement varies linearly through the entire thickness. The situation is more complex here. Experience with the use of the equations and specific study of the sensitivity of predictions to boundary restraint modeling are required. The corresponding problem for homogeneous structures is discussed in Ref 3 and 6.

For convenience, the constitutive relations, along with the stress and displacement distributions in terms of averaged kinematic variables and derivatives of resultant force and moment, are listed below.

Constitutive relations

$$N = \frac{h}{S_{11}}\left[\tilde{u}_{,x} - S_{13}r + \frac{h}{12}S_{13}N_{,xx} + \frac{7h^3}{2880}\alpha S_{13}N_{,xxxx} \right]$$

$$M = \frac{h^3}{12S_{11}}\left[-\tilde{w}_{,xx} + \frac{1}{h}(S_{13} + S_{55})Q_{,x} + \frac{2\alpha}{5h}S_{11}M_{,xx} - \frac{S_{33}h}{60}Q_{,xxx} \right. \tag{30}$$

$$\left. - \frac{S_{33}h}{10}\left(\frac{1}{21} - \frac{\alpha S_{13}}{8S_{33}} \right)M_{,xxxx} \right]$$

Displacement distribution

$$w = \tilde{w} + \frac{z}{h}S_{13}N - \frac{S_{13}}{2h}\left(1 - \frac{12z^2}{h^2} \right)M$$

$$+ \frac{S_{33}h}{2}\left[\frac{2z}{h}r - \frac{z}{4}\left(1 - \frac{4z^2}{3h^2} \right)N_{,xx} + \left(\frac{1}{12} - \frac{z^2}{h^2} \right)Q_{,x} + \left(\frac{7}{240} - \frac{z^2}{2h^2} + \frac{z^4}{h^4} \right)M_{,xx} \right]$$

$$u = \tilde{u} - z\tilde{w}_{,x} + \frac{z}{h}S_{55}Q + \frac{1}{2}(S_{55} + S_{13})\left[\frac{h}{12}\left(1 - \frac{12z^2}{h^2} \right)N_{,x} + \frac{z}{h}\left(1 - \frac{4z^2}{h^2} \right)M_{,x} \right]$$

$$+ \frac{h^2 S_{33}}{48}\left[2\left(1 - \frac{12z^2}{h^2} \right)r_{,x} - 3h\left(\frac{3}{40} - \frac{z^2}{h^2} + \frac{2z4}{3h^4} \right)N_{,xxx} - \frac{2z}{h}\left(1 - \frac{4z^2}{h^2} \right)Q_{,xx} \right.$$

$$\left. - \frac{z}{h}\left(\frac{7}{10} - \frac{4z^2}{h^2} + \frac{24}{5}\frac{z4}{h^4} \right)M_{,xxx} \right] \tag{31}$$

Stress distribution

$$\sigma_{xx} = \frac{N}{h} + \frac{12M}{h^3}z + \alpha\left[\frac{h}{12}\left(1 - \frac{12z^2}{h^2}\right)N_{,xx} + \frac{3z}{5h}\left(1 - \frac{20}{3}\frac{z^2}{h^2}\right)M_{,xx}\right.$$

$$+ \frac{S_{33}}{16S_{11}}h^2\left[\frac{2}{3}\left(1 - \frac{12z^2}{h^2}\right)r_{,xx} - \frac{3h}{40}\left(1 - \frac{40}{3}\frac{z^2}{h^2} + \frac{80}{9}\frac{z^4}{h^4}\right)N_{,xxxx}\right.$$

$$\left. - \frac{2z}{5h}\left(1 - \frac{20}{3}\frac{z^2}{h^2}\right)Q_{,xxx} - \frac{11z}{70h}\left(1 - \frac{280}{33}\frac{z^2}{h^2} + \frac{112}{11}\frac{z^4}{h^4}\right)M_{,xxxx}\right] \quad (32)$$

$$\sigma_{xz} = -\frac{z}{h}N_{,x} + \frac{Q}{h} + \frac{1}{2h}\left(1 - \frac{12z^2}{h^2}\right)M_{,x}$$

$$\sigma_{zz} = r - \frac{h}{8}\left(1 - \frac{4z^2}{h^2}\right)N_{,xx} - \frac{z}{h}Q_{,x} - \frac{z}{2h}\left(1 - \frac{4z^2}{h^2}\right)M_{,xx}$$

Method of Solution

The above equations are to be applied to individual plies of a laminate or to groups of plies ... sublaminates ... as needed to analyze a situation producing significant interlaminar stresses. The following are the solution steps in terms of averaged kinematic variables:

1. Divide the laminate into elements according to geometry and loading condition. The element length is selected such that within the element the geometry and loading are continuous as is commonly done in engineering theories of bending and stretching. Elements or sublaminates or both are characterized by their effective extensional or flexural moduli as appropriate.

2. The displacements, resultant force and moment, and interlaminar stresses in each element are governed by the equilibrium equations (Eqs 9), the constitutive relations (Eqs 26), and the displacement distributions (Eqs 31). Write these equations for each element in the analysis model.

3. Apply interlaminar continuity conditions (Eqs 8) and enforce tractions or displacement conditions at the extreme upper and lower surfaces of the laminate.

4. Solve the system of coupled ordinary differential equations for the elements variables.

5. Enforce the boundary conditions at constant values of x, the laminate sections, as well continuity requirements between element ends as is commonly done in engineering theories of bending and stretching in order to find the values of the arbitrary constants resulting from the solution in Step. 4.

6. Determine through-thickness stress and displacement distributions from the element variables using Eqs 32 and 31, respectively.

References

[1] Whitney, J. M., Daniel, I. M., and Pipes, R. B., *Experimental Mechanics of Fiber Reinforced Composite Materials*, Society for Experimental Stress Analysis Monograph No. 4, 1st ed., Brookfield Center, CT, 1982, pp. 175–185.

[2] Rehfield, L. W. and Reddy, A. D., "Design Information for Continuous Filament Advanced Composite Isogrid Structure," *Proceedings*, Fifth Department of Defense/National Aeronautics and Space Administration Conference on Fibrous Composites in Structural Design, New Orleans, 27–29 Jan. 1981; published in Report No. NADC-81096-60, Vol. II, Jan. 1981, pp. II-113 to II-143.

[3] Rehfield, L. W. and Murthy, P. L. N., "Toward a New Engineering Theory of Bending: Fundamentals," *AIAA Journal*, American Institute of Aeronautics and Astronautics, Vol. 20, No. 5, May 1982, pp. 693–699.

[4] Sinclair, J. H. and Chamis, C. C., "Compression Behavior of Unidirectional Fibrous Composites," *Compression Testing of Homogeneous Materials and Composites*, in *ASTM STP 808*, American Society for Testing and Materials, Philadelphia, 1983, pp. 155–174.

[5] Reddy, A. D., Rehfield, L. W., and Haag, R. S., "Effect of Large Delaminations on the Compressive Postbuckling Behavior of Laminated Composite Panels," *Proceedings*, Sixth Conference on Fibrous Composites in Structural Design, Report AMMRC MS 83-2, Army Materials and Mechanics Research Center, Nov. 1983, pp. VI-41 to VI-51.

[6] Rehfield, L. W. and Valisetty, R. R., "A Simple, Refined Theory for Bending and Stretching of Homogeneous Plates," *AIAA Journal*, American Institute of Aeronautics and Astronautics, Vol. 22, No. 1, Jan. 1984, pp. 90–95.

[7] Rehfield, L. W. and Valisetty, R. R., "A Comprehensive Theory for Planar Bending of Composite Laminates," *Computers and Structures*, Vol. 16, No. 1-4, 1983, pp. 441–447.

[8] Valisetty, R. R. and Rehfield, L. W., "A Theory for Stress Analysis of Composite Laminates," American Institute of Aeronautics and Astronautics, Paper 83-0833-CP, presented at the 24th AIAA/ASME/ASCE/AHS Structures, Structural Dynamics and Materials Conference, Lake Tahoe, NV, 2–4 May 1983.

[9] Valisetty, R. R., "Bending of Beams, Plates and Laminates: Refined Theories and Comparative Studies," Ph.D. thesis, Georgia Institute of Technology, Atlanta, GA, March 1983.

[10] Murthy, P. L. N., "A New Engineering Theory of Planar Bending and Applications," Ph.D. thesis, Georgia Institute of Technology, Atlanta, GA, Dec. 1981.

Mary E. Cunningham, [1] *Scott V. Schoultz,* [2] *and*
Joseph M. Toth, Jr. [2]

Effect of End-Tab Design on Tension Specimen Stress Concentrations

REFERENCE: Cunningham, M. E., Schoultz, S. V., and Toth, J. M., Jr., **"Effect of End-Tab Design on Tension Specimen Stress Concentrations,"** *Recent Advances in Composites in the United States and Japan, ASTM STP 864,* J. R. Vinson and M. Taya, Eds., American Society for Testing and Materials, Philadelphia, 1985, pp. 253–262.

ABSTRACT: An analytical and experimental study was performed to evaluate geometrical and material parameter effects on the stress concentrations of tension specimen end-tab design. The parameters were constrained to those falling within the options set forth in the ASTM Test for Tensile Properties of Fiber Resin Composites (D 3039-76). Geometrical parameters included tab angle, tab thickness, cutoff thickness, tab length, and adhesive thickness. Material parameters included tab material and adhesive material. The tension specimen and thickness were held constant.

KEY WORDS: graphite/epoxy, tension specimen, end tab, fiber glass

When tension tests of advanced composite materials are conducted to determine such mechanical properties as modulus of elasticity, Poisson's ratio, and strength, the frequent failure of the specimen in the tab region rather than the gage region results in inconsistent mechanical properties data. Tab region failures occur because of the stress concentration from the change in geometry at the tab/specimen interface and also from differences in material elastic constants where applicable.

The study described here began when graphite/epoxy tension specimens fabricated from the same material lot and same autoclaved panel showed widely varying strengths when tested within the requirements of ASTM Test for Tensile Properties of Fiber Resin Composites (D 3039-76). Study objectives were to

[1] Graduate student, Department of Mechanical Engineering, Massachusetts Institute of Technology, Cambridge, MA, 02139; formerly, associate engineer, Martin Marietta Corporation, Denver, CO 80201.

[2] Engineer and Manager-Composites Technology, respectively, Martin Marietta Corporation, Denver, CO 80201.

minimize the stress concentrations and thereby produce more consistent and realistic values of the material's strength.

Procedure

A five-step approach was used to determine the end-tab geometry and material that resulted in a minimum stress concentration. First, the stress distribution in the specimen attributable to various end-tab configurations was determined using finite-element analysis. Second, from the analytical results, the end tab having the lowest stress concentration effects was chosen. Third, specimens using this end-tab design were constructed. Fourth, tension tests of these specimens were conducted. Finally, the test results were compared with the analytical results and results from previous tests of advanced composite materials.

Analysis

Stress Distribution

The stress distribution in the specimen was analyzed using the MSC/NASTRAN finite-element program. Figures 1 and 2 show the specimen geometry and loading conditions. Because of the symmetry of the specimen, only one quarter of the specimen was modeled (Fig. 2) with two-dimensional plate elements having increasing element density in the interface region (Fig. 3). The required density was determined with the use of preliminary models in which the

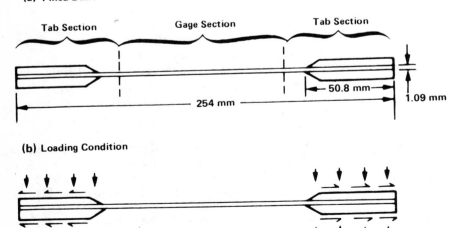

FIG. 1—*Tension specimen.*

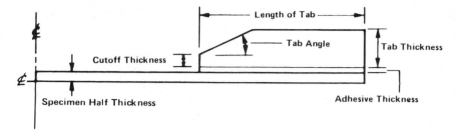

FIG. 2—*Geometry modeled.*

element density was increased until the peak stresses reached an asymptote. A plane-strain stiffness matrix was used in the modeling because of the large width/ thickness ratio (that is, 25.4 mm/1.10 mm).

Laminate

The specimen laminate analyzed was an eight-ply, 0° longitudinal laminate of graphite/epoxy. The fiber was HM-S produced by North America Court-aulds Inc., and the resin system was 934 produced by Fiberite Corporation. The end-tab materials evaluated were a (0/90) crossply laminate of Scotchply type 1003 fiber glass/epoxy produced by 3M Company and a 0°-longitudinal laminate of the HM-S/934 graphite/epoxy (Table 1).

Tab Characteristics

The end tab not only affected the longitudinal stress, but also induced inter-laminar normal and shear stresses in the specimen. A number of tab character-istics were analyzed, and their effect on these three stresses compared. The tab characteristics examined were thickness, angle, cutoff thickness, length, and stiffness. Adhesive stiffness and adhesive thickness were also examined. Fig-ures 4 through 6 show typical distributions of the longitudinal stress, interlami-nar normal stress, and interlaminar shear stress along the specimen.

Thickness—The thickness of the tab was found to have very little effect on the stress concentration. The results (Table 2) showed no difference using a 10° tab angle and little difference with the 45° tab angle. Because a 78% decrease in thickness caused only a 2% decrease in stress, thickness was not considered a significant factor.

Angle—The angle of the tab was found to be one of the two most significant factors affecting the stress concentrations. Increasing the tab angle increased all of the stresses (Table 3), but had the greatest effect on the interlaminar stresses. For example, increasing the angle from 10 to 25° increased the longi-tudinal stress by only 5%, but increased the interlaminar normal and shear

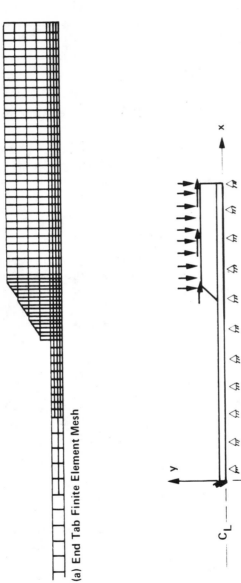

(a) End Tab Finite Element Mesh

(b) Applied Boundary Conditions

FIG. 3—*Specimen modeling.*

TABLE 1—*Laminate and tab material properties.*

	E_{11}, GPa	E_{33}, GPa	G_{13}, GPa	ν_{13}
HM-S/934	197.13	16.00	9.52	0.27
Scotchply, 0°	39.30	9.65	9.65	0.27
Scotchply, 90°	9.65	9.65	9.65	0.25

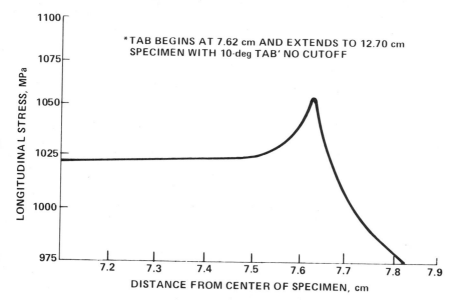

FIG. 4—*Typical longitudinal stress distribution.*

stresses by 85 and 70%, respectively. The smaller angle provided a more gradual change in geometry, therefore reducing the stress concentration. For the final tab analyses, the 10° angle was chosen to remain a constant.

Cutoff Thickness—The tab cutoff thickness was found to be the second significant factor (Table 4). By increasing the cutoff thickness from 0.00 to 0.254 mm, the longitudinal stress concentration increased 6%. This increase in cutoff thickness also increased the interlaminar normal and shear stress concentrations by 750 and 200%, respectively.

Length—The length of the tab did not affect the stresses. Again, the 10° angle tab was chosen for comparison. The length was varied from 38.1 to 50.8 to 63.5 mm (Table 5). The individual peak stresses obtained for these lengths were within 1% of those obtained for the other lengths. The 50.8 mm length was used throughout the remainder of the study.

Tab Stiffness—Changing the stiffness of the tab had a significant effect on

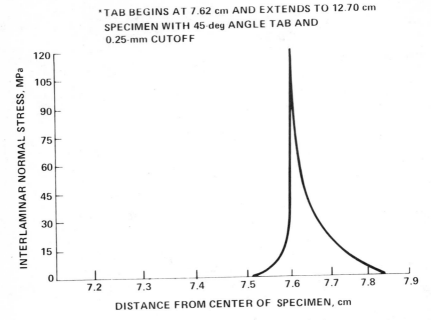

FIG. 5—*Typical interlaminar normal stress distribution.*

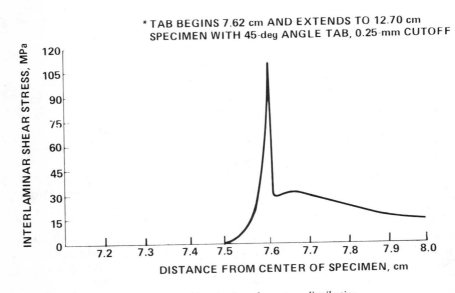

FIG. 6—*Typical interlaminar shear stress distribution.*

TABLE 2—*Effect of tab thickness.*

	Tab		Peak Stress		
Angle, °	Cutoff, mm	Thickness, mm	Longitude, GPa	Interlaminar Normal, MPa	Interlaminar Shear, MPa
45	0.25	0.23	1.23	110.80	115.08
45	0.25	0.41	1.26	125.21	117.84
10	no	0.13	1.06	4.21	12.34
10	no	0.18	1.06	4.21	12.34
10	no	0.23	1.06	4.21	12.34

TABLE 3—*Effect of tab angle.*

	Tab		Peak Stress		
Angle, °	Cutoff, mm	Thickness, mm	Longitude, GPa	Interlaminar Normal, MPa	Interlaminar Shear, MPa
10	0.25	2.29	1.12	35.65	39.72
25	0.25	2.29	1.18	65.99	67.57
45	0.25	2.29	1.23	110.80	105.08
90	4.07	4.07	1.30	176.93	156.45

TABLE 4—*Effect of tab cutoff.*

	Tab		Peak Stress		
Angle, °	Cutoff, mm	Thickness, mm	Longitude, GPa	Interlaminar Normal, MPa	Interlaminar Shear, MPa
10	0.00	2.29	1.06	4.21	12.34
10	0.25	2.29	1.12	35.65	39.72
10	0.51	2.29	1.16	68.67	65.09

TABLE 5—*Effect of tab length.*

	Tab		Peak Stress		
Angle, °	Cutoff, mm	Thickness, mm	Longitude, GPa	Interlaminar Normal, MPa	Interlaminar Shear, MPa
10	2.29	38.12	1.06	4.21	12.34
10	2.29	50.82	1.06	4.21	12.34
10	2.29	63.53	1.06	4.21	12.34

the peak stresses in the specimen. A 10° fiber glass tab and a graphite/epoxy tab were compared (Table 6). The graphite laminate stiffness was five times that of fiber glass, and this tab design produced stresses that were at least 34% higher at the graphite/fiber glass interface. Variations in the fiber glass layup were also compared. Because this made little difference (0.1%) in the stresses, the crossply fiber glass tab was retained.

Adhesive Stiffness—Two adhesive shear stiffnesses were examined, one being three times larger than the other. The stiffer adhesive produced peak longitudinal stresses only 1% higher, and interlaminar stresses 20% higher (Table 7). The adhesive currently used, Hysol Corporations EA 9309, was assumed to have a stiffness in the vicinity of 12.066 GPa. This is an estimate, however, because no firm data are available for the mechanical properties of this adhesive under confined conditions.

Adhesive Thickness—The preceding comparisons were based on a finite-element model that included no adhesive thickness but did account for compatibility of deformation across the end-tab/specimen boundary. (This was done to reduce the complexity of the finite-element modeling and the subsequent cost per computer run). An adhesive layer of 0.101 mm was subsequently added to the analytical model. The resulting stresses are shown in Table 7. The 0.102-mm layer had little effect on the longitudinal peak stress but increased

TABLE 6—*Effect of tab stiffness.*

Material	Laminate	Cutoff, mm	Longitude, GPa	Interlaminar Normal, MPa	Interlaminar Shear, MPa
Fiber glass	...[b]	0.25	1.12	35.65	39.72
Fiber glass	...[b]	0.00	1.06	4.21	12.34
Graphite/epoxy	[0]$_8$	0.25	1.50	149.62	179.96
Fiber glass	[0]$_9$	0.00	1.06	4.21	14.41

[a]10° angle.
[b](0/90/0/90/0/90/0/90/0).

TABLE 7—*Effect of adhesive thickness and stiffness.*

Thickness, mm	Stiffness, GPa	Longitude, GPa	Interlaminar Normal, MPa	Interlaminar Shear, MPa
0.076	12.07	1.06	14.96	14.41
0.102	12.07	1.06	9.17	15.31
No adhesive	No adhesive	1.06	4.21	12.34
0.102	36.20	1.07	20.62	18.41

[a]10° angle, no cutoff, cross ply fiber glass tab.

the peak interlaminar normal and shear stresses significantly. The thickness of the adhesive layer was changed from 0.102 to 0.076 mm with no resulting effect on peak longitudinal stress, but indicated an increase in interlaminar peak normal stress and a reduction in interlaminar peak shear stress.

Analysis Results

Results of the analysis indicated that with some compromise, the lowest stress concentrations could be obtained with an end-tab design that used a crossplied (0/90), unidirectional fiber glass laminate, with a 10° angle beginning at a "feathered" edge (that is, no cutoff) and an adhesive layer thickness of 0.076 mm.

Experimental

Tension specimens were fabricated from the same panel of graphite/epoxy material. The panel was fabricated with eight plies of HM-S/934 material. Two end-tab configurations were fabricated with a 10° angle having (1) no cutoff and (2) a 0.254 mm cutoff. The end-tab material was bonded to the panel material, and five specimens of each end-tab configuration were cut from the panel. Each specimen was then tested in accordance with the D 3039-76 procedure.

Upon testing, it was found that the cutoff distance significantly affected the experimental results. Tab failure occurred in 80% of the specimens tested with a 0.254 mm cutoff. This was much greater than the 20% tab failure found in the specimens with a 0.00 mm cutoff. The cutoff thickness appeared to affect the failure mode as well as the location of the failure. Specimens with a 0.00 mm cutoff failed in pure tension. There were no signs of bending or shear. The specimens with the 0.254 mm cutoff, however, did not exhibit the same "clean" fractures. Longitudinal cracking and delamination were evident in all of the specimens tested.

The test results (Table 8), along with data from a previous test of 12 speci-

TABLE 8—*Experimental results.*

	Nominal Tensile Strength, GPa		
Condition	10° Angle with No Cutoff	10° Angle with 0.254-mm Cutoff	45° Angle with 1.271-mm Cutoff
Values	1.08	1.33	
	1.10	1.02	(1.01 max)
	1.07	0.84	(0.72 min)
	1.08	0.97	
	1.09	0.94	
Average	1.09	1.02	0.87[a]
Coeffic. of variation	1.2%	18.1%	13%[a]

[a]12 specimens.

mens with 45° end tabs and a 1.27-mm cutoff, demonstrated the desirability of the 10° end tab with no cutoff. This configuration yielded the highest tensile strength of the material (1.09 GPa) as well as the most consistent data (that is, coefficient of variation of 1.2%).

Conclusions

The analytical procedure used has made it possible to evaluate pertinent composite material failure modes. The most efficient end-tab design lies within the scope of the recommendations of ASTM D 3039-76. The limited experimental results confirm the results of the analytical investigation that was directed toward obtaining more realistic and consistent strength values with an efficiently designed end-tab configuration.

Takashi Akasaka[1] *and Kazuo Asano*[1]

Stress and Deformation of the Sandwich Panel Having Curved Faceplates Under Pressure Loading

REFERENCE: Akasaka, T. and Asano, K., **"Stress Deformation of the Sandwich Panel Having Curved Faceplates Under Pressure Loading,"** *Recent Advances in Composites in the United States and Japan*, *ASTM STP 864*, J. R. Vinson and M. Taya, Eds., American Society for Testing and Materials, Philadelphia, 1985, pp. 263–277.

ABSTRACT: An analytical and experimental study is conducted on the stress and deformation of a pressurized square sandwich panel, simply supported by four corner hinges. This panel is constructed with a pair of curved faceplates of stainless steel and foam plastics core together with a reinforcing side frame. This is expected to be available for a lightweight roof panel of huge frame architecture like a gymnasium by virtue of its prominent ability in withstanding the pressure load due to the shell effect of curved faceplates in addition to the sandwich effect. In this analysis, we used the extremum principle of total potential under the subsidiary conditions of boundary and symmetry for displacement functions. Good agreement is obtained between the theoretical and the experimental results of strains and deflections.

KEY WORDS: sandwich panel, curved faceplate, side frame, Fourier series, strain energy, subsidiary conditions, shell effect

A sandwich panel constructed with curved faceplates of stainless steel and urethane foam plastics core has been recently developed for a lightweight roof structure to be available for huge steel frame architectures like a gymnasium. This roof panel is designed to be supported by four hinges protruding from the frame structure at the panel corners. Then, for reinforcing the roof panel, a side frame consisted of four members is furnished to the outer periphery of the core and sandwiched between the two faceplates.

The faceplates of being initially flat are plastically deformed into shallow shells by the inflation pressure and the heat generated in a foaming process of the plastic core. Those sandwich roof panels could be thus expected to have a

---

[1]Professor and assistant, respectively, Department of Precision Mechanical Engineering, Chuo University, Kasuga 1-13-27, Bunkyo-ku, Tokyo 112, Japan.

prominent ability in withstanding the pressure load by virtue of the "shell effect" of a convex faceplate together with the "sandwich effect" of a pair of faceplates.

Although a large number of research papers [1-2] were previously published on the strength and rigidities of the sandwich panel with flat faceplates, the stress analysis for that with curved faceplates and a side frame has not yet been conducted. In this paper, we presented an analysis on the stress and deformation of a pressurized square sandwich roof panel, simply supported by four corner hinges, by using the extremum principle of total potential. An experimental study was further carried out for those panel specimens with a 1 m² plane area, in order to confirm the theoretical results.

Analysis

Figure 1 shows a sandwich panel structure consisting of curved thin faceplates and foam plastics core together with a reinforcing side frame made of four channel members. Each corner of the side frame is stiffened by a pair of short diagonal members. We consider a case when this sandwich panel is simply supported by four corner hinges and subjected to the pressure load uniformly distributed within the area a^2 of the upper faceplate. The variation of the half thickness $h(x,y)$ of the square sandwich panel is approximated by

$$h(x,y) = h_0 + h_1 \cos \frac{\pi x}{1} \cos \frac{\pi y}{1} \tag{1}$$

Figure 2 shows the cross-section shape of a curved faceplate represented by $\bar{h}(x) = h_1 \cos(\pi x/1)$ comparing with the measured points for a specimen, indicating good agreement between them. Thus, Eq 1 can be used hereafter as an approximate expression for the thickness variation.

The contraction of the core in the thickness direction is ignored, and an antisymmetric deformation is assumed for inplane displacements of the upper and the lower faceplates with respect to the neutral plane of this panel in this analysis. The displacement functions of faceplates in the x, y, and z directions are denoted by $\pm u$, $\pm v$, and w, respectively, as depicted in Fig. 3. The extremum principle of total potential is utilized for analyzing stresses and deformations of the sandwich panel.

Strain Energy and Potential Energy

Various kinds of strain energy stored in the structural elements and the potential drop due to the external loading follow.

1. Bending and twisting strain energy of a faceplate U_{Bf} is expressed by

$$U_{Bf} = \frac{D_f}{2} \int_{-1/2}^{1/2} \int_{-1/2}^{1/2} [w_{,xx}^2 + w_{,yy}^2 + 2\nu w_{,xx} w_{,yy} + 2(1-\nu)w_{,xy}^2] dx dy$$

$$\tag{2}$$

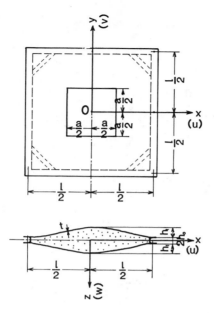

FIG. 1—*Square sandwich panel consisted of curved faceplates, foam plastics and side frame.*

where D_f denotes the bending rigidity of a faceplate

$$D_f = \frac{Et^3}{12(1 - \nu^2)} \tag{3}$$

expressed by its elastic modulus E and Poisson's ratio ν.

2. Membrane strain energy of a faceplate U_{Mf} is given by

$$U_{Mf} = \frac{C_f}{2} \int_{-1/2}^{1/2} \int_{-1/2}^{1/2} \Big[(u_{,x} - w_{,x}h_{,x})^2 + (v_{,y} - w_{,y}h_{,y})^2$$

$$+ 2\nu(u_{,x} - w_{,x}h_{,x})(v_{,y} - w_{,y}h_{,y})$$

$$+ \frac{1 - \nu}{2} (u_{,y} + v_{,x} - w_{,y}h_{,x} - w_{,x}h_{,y})^2 \Big] dxdy \tag{4}$$

where C_f is the extensional rigidity of a faceplate

$$C_f = \frac{Et}{1 - \nu^2} \tag{5}$$

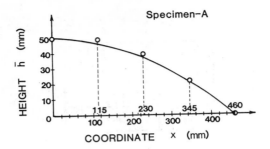

FIG. 2—*Cross-sectional curve of the faceplate.*

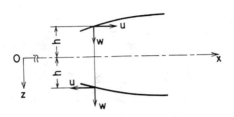

FIG. 3—*Displacements* u *and* w *of the faceplates.*

The components of strain involving $h_{,x}$ and $h_{,y}$ in the integrand of Eq 4 were obtained by linearizing the additional strains due to deflection w of a curved faceplate [3]. Those strain terms play a significant role in characterizing the deformation behavior of the sandwich panel noted as the "shell effect."

3. Transverse shear strain energy of the core U_c is written by

$$U_c = \frac{G_c}{2} \int_{-1/2}^{1/2} \int_{-1/2}^{1/2} 2h \left[\left(\frac{u}{h} - w_{,x} \right)^2 + \left(\frac{v}{h} - w_{,y} \right)^2 \right] dxdy \qquad (6)$$

where G_c denotes the shear modulus of the core.

4. Bending strain energy of a frame member U_{Bs} is represented by

$$U_{Bs} = \frac{D_s}{2} \int_{-1/2}^{1/2} \bar{w}_{,yy}^2 dy, \qquad \bar{w} = w(1/2, y) \qquad (7)$$

where D_s and $\bar{w}$ denote the bending rigidity and the vertical deflection of a frame member, respectively.

5. Torsional strain energy of a frame member U_{Ts} is expressed by

$$U_{Ts} = \frac{C_s}{2} \int_{-1/2}^{1/2} \left(\frac{\bar{u}_{,y}}{h_0} \right)^2 dy, \qquad \bar{u} = u(1/2, y) \qquad (8)$$

where C_s denotes St. Venant torsional rigidity of a frame member and $\bar{u}_{,y}/h_0$ implies its rate of twist.

6. Potential drop V due to the loading pressure p can be formulated as

$$V = \int_{-a/2}^{a/2} \int_{-a/2}^{a/2} pw \; dxdy \qquad (9)$$

Boundary Conditions

The following boundary conditions can be generally formulated, based on the symmetry and the antisymmetry regarding the deformation of each square faceplate with respect to x- and y-axes together with its diagonals.

1. Deflection w should be equal to zero at each panel corner being supported by the hinge. Then, we have

$$w(1/2, 1/2) = 0 \qquad (10)$$

2. A frame member might be assumed to be clamped at its both ends. Therefore

$$w_{,x}(1/2, 1/2) = 0 \qquad (11)$$

3. Displacements u and v of both faceplates have to vanish at each corner according to the clamped condition assumed for frame members given by Eq 11. Then

$$u(1/2, 1/2) = 0, \quad \text{or} \quad v(1/2, 1/2) = 0 \qquad (12)$$

4. The assumption that a pair of adjacent frame members are rigidly connected with each other in the x-y plane, yields

$$u_{,y}(1/2, 1/2) = 0, \quad \text{or} \quad v_{,x}(1/2, 1/2) = 0 \qquad (13)$$

5. The transverse shear deformation of each frame member is usually neglected in the elementary beam theory and then

$$u(x, 1/2)/h_0 = w_{,x}(x, 1/2) \qquad (14)$$

The system of those boundary conditions would presumably impose rather excessive restrictions to the deformation of the actual sandwich panel. Then, another system of boundary conditions only holding 1 and 5 is also taken into consideration in this analysis. The latter case is designated by Theory 1, while the former is denoted by Theory 2.

Displacement Functions

Considering the symmetry and antisymmetry conditions for displacement functions of u, v, and w, with respect to x- and y-axes together with diagonals, we obtain

$$u(x, y) = v(y, x), \qquad w(x, y) = w(y, x) \tag{15}$$

With referencing Eq 15, we can express u, v, and w by the following system of double Fourier Series.

$$u = \sum_m \sum_n a_{mn} \sin \frac{m\pi x}{1} \cos \frac{n\pi y}{1}$$

$$v = \sum_m \sum_n a_{mn} \sin \frac{m\pi y}{1} \cos \frac{n\pi x}{1} \tag{16}$$

$$w = \sum_m \sum_n c_{mn} \cos \frac{m\pi x}{1} \cos \frac{n\pi y}{1}$$

where

$$c_{mn} = c_{nm} \tag{17}$$

Then, we obtain the subsequent expressions for the total strain energy U and the potential drop V, by using Eq 16.

$$U = 2U_{Bf} + 2U_{Mf} + U_c + 4(U_{Bs} + U_{Ts})$$

$$= \sum_m \sum_n \sum_{m'} \sum_{n'} [c_{mn}c_{m'n'}\alpha_{mnm'n'} + a_{mn}c_{m'n'}\beta_{mnm'n'} + a_{mn}a_{m'n'}\gamma_{mnm'n'}]$$

$$V = \sum_m \sum_n c_{mn}\delta_{mn} \tag{18}$$

where $\alpha_{mnm'n'}$, $\beta_{mnm'n'}$, $\gamma_{mnm'n'}$, and δ_{mn} are constants shown in Appendix. It should be mentioned here that the notation Σ_m implies $\Sigma_{m=0}^{N}$ where N is the highest order of Fourier Series. The boundary conditions given by Eqs 10 to 14 are expressed as below by using Eq 16.

1. $f_1 = \sum_m \sum_n c_{mn} \cos \frac{m\pi}{2} \cos \frac{n\pi}{2} = 0$

2. $f_2 = \sum_m \sum_n c_{mn} m \sin \frac{m\pi}{2} \cos \frac{n\pi}{2} = 0$

3. $f_3 = \sum_m \sum_n a_{mn} \sin \dfrac{m\pi}{2} \cos \dfrac{n\pi}{2} = 0$ (19)

4. $f_4 = \sum_m \sum_n a_{mn} n \sin \dfrac{m\pi}{2} \sin \dfrac{n\pi}{2} = 0$

5. $g_m = \sum_n \left(a_{mn} + \dfrac{m\pi h_0}{1} c_{mn} \right) \cos \dfrac{n\pi}{2} = 0$

Equation 17 can be further written by

$$f_{mn} = c_{mn} - c_{nm} = 0, \qquad (m \neq n) \tag{20}$$

The modified total potential Π with subsidiary conditions is thus expressed as

$$\Pi = U - V + \lambda_1 f_1 + \lambda_2 f_2 + \lambda_3 f_3 + \lambda_4 f_4 + \sum_m \mu_m g_m + \sum_{m \neq n} \rho_{mn} f_{mn}$$

(21)

where λ_i $(i = 1, 2, 3, 4)$, μ_m, and ρ_{mn} are all Lagrange multipliers. The stationary conditions lead to a symmetric system of linear equations with respect to the unknowns of c_{mn}, a_{mn}, and Lagrange multipliers. Then, we have

$$\frac{\partial \Pi}{\partial c_{mn}} = \frac{\partial \Pi}{\partial a_{mn}} = \frac{\partial \Pi}{\partial \lambda_i} = \frac{\partial \Pi}{\partial \mu_m} = \frac{\partial \Pi}{\partial \rho_{mn}} = 0 \tag{22}$$

The first two conditions of Eq 22 yield

$$\sum_{m'} \sum_{n'} c_{m'n'} (\alpha_{mnm'n'} + \alpha_{m'n'mn}) + \sum_{m'} \sum_{n'} a_{m'n'} \beta_{mnm'n'} - \delta_{mn}$$

$$+ \lambda_1 \cos \frac{m\pi}{2} \cos \frac{n\pi}{2} + \lambda_2 m \sin \frac{m\pi}{2} \cos \frac{n\pi}{2} + \mu_m \frac{m\pi h_0}{1} \cos \frac{n\pi}{2}$$

$$+ \rho_{mn} - \rho_{nm} = 0 \tag{23}$$

$$\sum_{m'} \sum_{n'} c_{m'n'} \beta_{mnm'n'} + \sum_{m'} \sum_{n'} a_{m'n'} (\gamma_{mnm'n'} + \gamma_{m'n'mn})$$

$$+ \lambda_3 \sin \frac{m\pi}{2} \cos \frac{n\pi}{2} + \lambda_4 n \sin \frac{m\pi}{2} \sin \frac{n\pi}{2} + \mu_m \cos \frac{n\pi}{2} = 0 \tag{24}$$

The last three conditions of Eq 22 yield the same equations as Eqs 19 and 20 in which m and n were replaced by m' and n', respectively. The total number of unknowns M thus becomes

$$M = \frac{5}{2} N^2 + \frac{9}{2} N + 5, \tag{25}$$

since each number of c_{mn}, a_{mn}, λ_i, μ_m, and ρ_{mn} attains to $N + 1$, $N(N + 1)$, 4, N, and $N(N + 1)/2$, respectively.

Numerical Calculation

Numerical calculation was conducted for two examples which were designated by A and B, where the height h_1 was larger for A than B. Numerical values of A and B are the same as that of specimens which are named by the same notations, respectively. They are listed as follows

$E = 2.059 \times 10^5$ (N/mm^2), $\nu = 0.3$, $G_c = 2.501$ (N/mm^2)

$1 = 920$ (mm), $h_0 = 20$ (mm), $t = 0.3$ (mm), $a = 500$ (mm)

$pa^2 = 3.9226 \times 10^3$ (N), $h_1 = 50$ for A, 42.5 for B (mm)

$D_f = 5.092 \times 10^2$ (N $\cdot$ mm), $C_f = 6.789 \times 10^4$ (N/mm)

$\beta = h_1/h_0 = 2.5$ for A, 2.125 for B

$D_s = 2.003 \times 10^{10}$ (N $\cdot$ mm^2), $C_s = 6.761 \times 10^8$ (N $\cdot$ mm^2)

The highest order of Fourier Series N is confined to 3. Then, the total number of unknowns M becomes 41 for Theory 2 and 38 for Theory 1 according to Eq 25. Those equations were solved by Gauss-Jordan method by the aid of a microcomputer.

Deflection and strain distribution of the panel can be calculated from the displacement functions as follows.

1. Deflection at the panel center

$$w(0, 0) = \sum_m \sum_n c_{mn} \tag{26}$$

2. Strain in the x-direction at the center of the faceplate

$$\epsilon_x(0, 0) = u_{,x}(0, 0) = \frac{\pi}{1} \sum_m \sum_n ma_{mn} \tag{27}$$

3. Distribution of strain in the x-direction along the x-axis of faceplate

$$\epsilon_x(x, 0) = (u_{,x} - w_{,x}h_{,x})_{y=0}$$

$$= \sum_m \sum_n \left(\frac{m\pi}{1}\right) a_{mn} \cos \frac{m\pi x}{1} - h_1\left(\frac{\pi}{1}\right)^2 \sin \frac{\pi x}{1} \sum_m \sum_n c_{mn} m \sin \frac{m\pi x}{1} \quad (28)$$

4. Distribution of strain in the y-direction along the x-axis of faceplate

$$\epsilon_y(x, 0) = v_{,y}(x, 0) = \sum_m \sum_n a_{mn}\left(\frac{m\pi}{1}\right) \cos \frac{n\pi x}{1} \quad (29)$$

5. Distribution of bending strain in the longitudinal direction of frame member

$$\bar{\epsilon}(x) = -h_0 w_{,xx}(x, 1/2) = h_0\left(\frac{\pi}{1}\right)^2 \sum_m \sum_n c_{mn} m^2 \cos \frac{n\pi}{2} \cos \frac{m\pi x}{1} \quad (30)$$

6. Rotation angle of a frame member at its midsection

$$\phi = u(1/2, 0)/h_0 = \left(\sum_m \sum_n a_{mn} \sin \frac{m\pi}{2}\right)\bigg/h_0 \quad (31)$$

Experimental Results and Discussions

Experiments were conducted for two specimens of A and B of which structural data were already described. Four corners of a specimen were connected with the struts set up on the stand of the testing machine. The pressure load was applied to the central portion of the upper faceplate within the area of a^2 via a concrete block of which bottom was covered with a soft rubber sheet formed fit to the curved surface of the faceplate. Deflection and strains were measured by dialgages and strain gages, respectively, as shown in Fig. 4.

Figure 5 shows the relation of the load P to the maximum deflection $w(0,0)$ for specimen B, comparing the theoretical results depicted by full and broken lines with the experimental ones denoted by black points. It is seen here that a collapse begins to occur at the load of nearly 40 kN probably due to local buckling of the convex faceplate. This collapsing load is confirmed to be about ten times larger than the standard strength of 4 kN/m² for the roof panel structure. Good agreement is found between the theoretical and the experimental results within a range of linearity, where the theoretical curve of Theory 1 is shown slightly lower than that of Theory 2.

Figure 6 shows the relation of the strain $\epsilon_x(0,0)$ of specimen A to the load P. It is found that both theoretical curves obtained by Theories 1 and 2

FIG. 4—*Test specimen and experimental setup.*

roughly coincide with the mean curve of the experimental points, indicating that the curve of Theory 1 gives somewhat larger values of the strain than that of Theory 2.

Figures 7 and 8 illustrate the distribution of strains $\epsilon_x(x,0)$ and $\epsilon_y(x,0)$ of faceplates along the x-axis of specimen A, when $P = 4$ kN. Comparison between these theoretical and experimental results exhibits good agreement, with indication that there is no marked difference between Theories 1 and 2.

Figure 9 depicts the calculated distributions of bending strain $\bar{\epsilon}$ of a frame member along its longitudinal direction of specimen A under $P = 4$ kN, comparing with the corresponding experimental results. It is seen that Theory 1 is in better agreement with the measured points than Theory 2, which enables us to understand that Theory 1 having rather slackened boundary conditions is more adequate for this sandwich panel than Theory 2. It is further understood that the difference in boundary conditions between Theories 1 and 2 gives little effects on the deformation and the strain of curved faceplates but considerable effects on that of frame members.

Corresponding figures to Figs. 6, 7, 8, and 9 given for A specimen are also obtained for B specimen, which have the same trend as that of A. It should be noted further that frame members rotate in the direction of inside up and outside down during loading, for which the rotation angle ϕ becomes positive according to Eq 31. This unexpected phenomenon could be attributed to the flattening of the upper faceplate under the pressure loading, which extends the convex plate outward and causes rotation of frame members in such a direction described previously. Theoretical value of ϕ was found to be positive but nearly three times larger than the measured results for both specimens A and B. This difference in ϕ-values between the theoretical and the experi-

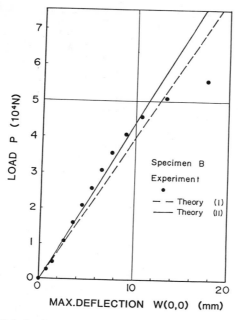

FIG. 5—*Relation between load P and panel center deflection* w *(0, 0).*

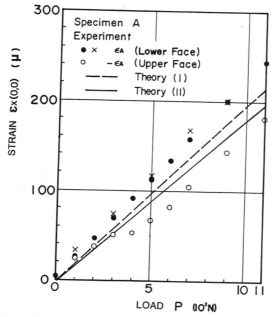

FIG. 6—*Relation between load P and strain* ϵ_x*(0, 0) of the faceplate.*

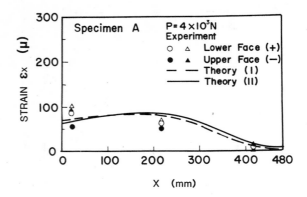

FIG. 7—*Distribution of strain ϵ_x along x-axis of the faceplate.*

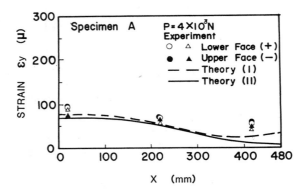

FIG. 8—*Distribution of strain ϵ_y along x-axis of the faceplate.*

mental might be thought mainly due to underestimation of the torsional rigidity C_s of a frame member by ignoring the reinforcing effect of a pair of diagonal members set at each frame corner. Considering the fact that the rotation angle ϕ of a frame member usually becomes negative for a sandwich panel with flat faceplates, it could be expected that there exists an optimum value of the maximum height of the curved faceplate h_1 to make ϕ equal to zero, in connection with the other dimensions of faceplate and frame member.

It is further found through additional calculations that the deflection of this sandwich panel slightly decreases with the increase of bending and twisting rigidities of a frame member denoted by D_s and C_s, respectively, while the absolute values of strain of curved faceplates and frame members decrease with the increase of D_s, but increase with the increase of C_s.

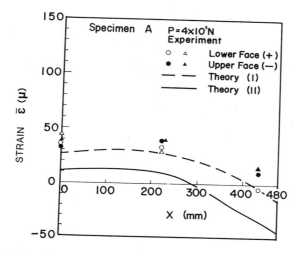

FIG. 9—*Distribution of bending strain $\bar{\epsilon}(x)$ of the frame member.*

Conclusion

A study on the deformation and strength was conducted for a square sandwich panel having curved faceplates subjected to pressure loading. Concluding remarks are summarized as follows.

1. A method of deformation analysis is presented for a sandwich panel with curved faceplates, which would be available for lightweight roof structure. In this analysis, the extremum principle of total potential including subsidiary conditions was used.

2. This sandwich panel is proved to have excellent structural properties of stiffness and strength against the pressure loading, which are attributed to the shell effect together with the sandwich effect of the curved faceplates.

3. Experimental results are verified better by Theory 1 than Theory 2, with stressing that the boundary conditions at each panel corner are more likely to be simply supported than clamped.

4. The shell effect is clearly confirmed by the fact that the frame member rotates in such a direction as inside up and outside down under the downward loading. Then, there remains a problem to obtain an optimum h_1-value to nullify the rotation angle of the frame member.

5. Good agreement is obtained between the theoretical and the experimental results; thus, this analytical method could be extensively applied for analyzing rectangular and trapezoidal sandwich roof panels with curved faceplates.

Acknowledgment

It should be mentioned here that the basic idea of this subject was developed by Mr. H. Kawai of Nikken Sekkei Ltd. and Mr. Y. Hirano of Sumitomo Corporation, of which patents are pending in several countries. The authors wish to express their hearty gratitude to the following companies for permission to publish this paper and for the cooperation in conducting the experimental works. Nikken Sekkei Ltd., Nisshin Spinning Co. Ltd., Sanko Metal Industrial Co., Ltd., and Sumitomo Corporation.

APPENDIX

Notations used in Eq 18 are given by

$$
\alpha_{mnm'n'} = (2D_f\pi^2/1^2)[\{(mm')^2 + \nu(mn')^2\}I_{mm'}I_{nn'} + (1-\nu)mnm'n'J_{mm'}J_{nn'}]
$$

$$
+ (2C_f\pi^2h_1^2/1^2)\{(mm')(M_{mm'}N_{nn'} + \nu P_{mn'}P_{m'n})
$$

$$
+ \frac{1-\nu}{2}(nn')(\bar{M}_{mm'}\bar{N}_{nn'} + P_{n'm}P_{nm'})\} + 2(G_ch_0)(mm')R_{mm'nn'}
$$

$$
+ (2D_s\pi^3/1^3)\cos\frac{m\pi}{2}\cos\frac{m'\pi}{2}(nn')^2I_{nn'} \tag{30}
$$

$$
\beta_{mnm'n'} = (4C_f\pi h_1/1)\{-(mm')(K_{mm'}L_{nn'} + \nu K_{nm'}L_{mn'})
$$

$$
+ \frac{1-\nu}{2}(nn')(K_{mm'}\bar{L}_{nn'} + K_{m'n}L_{mn'})\} + (4G_c1/\pi)m'I_{nn'}J_{mm'} \tag{31}
$$

$$
\gamma_{mnm'n'} = (2C_f)\{(mm')(I_{mm'}I_{nn'} + \nu I_{mn'}I_{nm'})
$$

$$
+ \frac{1-\nu}{2}(nn')(J_{mm'}J_{nn'} + J_{mn'}J_{nm'})\} + (2G_ch_0)(1/h_0\pi)^2Q_{mm'nn'}
$$

$$
+ (2C_s\pi/1h_0^2)(nn')\sin\frac{m\pi}{2}\sin\frac{m'\pi}{2}J_{nn'} \tag{32}
$$

$$
\delta_{mn} = p\left(\frac{1}{\pi}\right)^2 H_mH_n \tag{33}
$$

where

$$
I_{mm'} = \int_{-\pi/2}^{\pi/2}\cos(m\xi)\cos(m'\xi)d\xi,
$$

$$
J_{mm'} = \int_{-\pi/2}^{\pi/2}\sin(m\xi)\sin(m'\xi)d\xi,
$$

$$K_{mm'} = \int_{-\pi/2}^{\pi/2} \sin \xi \cos(m\xi)\sin(m'\xi)d\xi,$$

$$L_{mm'} = \int_{-\pi/2}^{\pi/2} \cos \xi \cos(m\xi)\cos(m'\xi)d\xi,$$

$$M_{mm'} = \int_{-\pi/2}^{\pi/2} \sin^2 \xi \sin(m\xi)\sin(m'\xi)d\xi,$$

$$N_{mm'} = \int_{-\pi/2}^{\pi/2} \cos^2 \xi \cos(m\xi)\cos(m'\xi)d\xi,$$

$$\bar{L}_{mm'} = \int_{-\pi/2}^{\pi/2} \cos \xi \sin(m\xi)\sin(m'\xi)d\xi,$$

$$\bar{M}_{mm'} = \int_{-\pi/2}^{\pi/2} \sin^2 \xi \cos(m\xi)\cos(m'\xi)d\xi,$$

$$\bar{N}_{mm'} = \int_{-\pi/2}^{\pi/2} \cos^2 \xi \sin(m\xi)\sin(m'\xi)d\xi,$$

$$P_{mm'} = \int_{-\pi/2}^{\pi/2} \sin \xi \cos \xi \sin(m\xi)\cos(m'\xi)d\xi,$$

$$Q_{mm'nn'} = \int_{-\pi/2}^{\pi/2} \int_{-\pi/2}^{\pi/2} \frac{\sin(m\xi)\sin(m'\xi)\cos(n\eta)\cos(n'\eta)}{1 + \beta \cos \xi \cos \eta} d\xi d\eta,$$

$$R_{mm'nn'} = \int_{-\pi/2}^{\pi/2} \int_{-\pi/2}^{\pi/2} (1 + \beta \cos \xi \cos \eta)\sin(m\xi)\sin(m'\xi)$$

$$\times \cos(n\eta)\cos(n'\eta)d\xi d\eta, \text{ and}$$

$$H_m = \int_{-\pi a/21}^{\pi a/21} \cos m\xi d\xi = \begin{cases} \dfrac{2}{m} \sin \dfrac{m\pi a}{21}, & (m \geqq 1). \\[2ex] \dfrac{\pi a}{1}, & (m = 0) \end{cases}$$

References

[1] Hertel., Leichtbau, Springer Verlag, 1960.
[2] Akasaka, T. *Theories and Applications for Light Weight Structures*, T. Hayashi, Ed., Japan Union of Scientists and Engineers Publication, Tokyo, 1966, Chapter 15, Vol. 2, pp. 1–87 (in Japanese).
[3] Marguerre, K., *Proceedings*, Fifth International Congress of Applied Mechanics, Cambridge, MA, 1938, pp. 93–101.

Dynamic Behavior

Jaak Soovere[1]

Dynamic Response of Flat Integrally Stiffened Graphite/Epoxy Panels Under Combined Acoustic and Shear Loads

REFERENCE: Soovere, J., **"Dynamic Response of Flat Integrally Stiffened Graphite/ Epoxy Panels Under Combined Acoustic and Shear Loads,"** *Recent Advances in Composites in the United States and Japan, ASTM STP 864*, J. R. Vinson and M. Taya, Eds., American Society for Testing and Materials, Philadelphia, 1985, pp. 281–296.

ABSTRACT: This project involved an experimental investigation into the effect of a high random acoustic loading, in combination with a static shear load near initial buckling, on the dynamic response of integrally stiffened graphite/epoxy panels. The program covered the design, fabrication, finite-element analysis, and testing of a *J*-stiffened monolithic panel and a *J*-stiffened minisandwich panel. The minisandwich panel had a syntactic core containing glass micro-balloons. The panels were designed to an initial shear buckling load of 17.51 kN/m (100 lb/in.). The predicted initial shear buckling load was, generally, in good agreement with the measured load. The damping, in all of the modes, increased significantly on approaching buckling. The root-mean-square strain, at the critical panel location, increased by as much as 30% in the vicinity of panel buckling, in spite of the increase in the damping.

KEY WORDS: testing, composite panels, integrally stiffened, shear buckling, dynamic response, damping, random acoustic loading, measured strain

Graphite/epoxy composite structures offer a potential for significant weight savings over current aluminum alloy structures. Minimum-weight structures are especially important for the vertical/short takeoff and landing (V/STOL) class of aircraft. Lightly loaded fuselage structure will be used on certain types of V/STOL aircraft. The design will involve the use of minimum-thickness advanced graphite/epoxy structures which will be required to operate well into the post-buckling region, with initial buckling just above 1 *g*.

---

[1]Research specialist senior, Lockheed-California Co., Burbank, CA 91520.

Due to the large power requirements of V/STOL aircraft, especially during vertical takeoff and landing, the fuselage and other lightly loaded structures will be subjected to compression and shear loading at 1 g, simultaneously with high thermal and acoustic environments. This combination of loads and environments is expected to affect the acoustic fatigue life of the fuselage panels.

Two recently completed programs [1,2] involved a study into the effect of elevated temperature on the acoustic fatigue life of integrally stiffened graphite/epoxy panels. This paper describes the result of an experimental investigation into the effect of a high random acoustic environment, in combination with static shear load, on the dynamic response of integrally stiffened graphite/epoxy panels near initial buckling.

Panel Design

A J-stiffened monolithic panel (Fig. 1) and a J-stiffened minisandwich panel (Fig. 2) were tested in this program. Both panels were fabricated from AS 3501-6 unidirectional graphite/epoxy tape. The monolithic panel had an eight-ply, 1-mm-thick (0.04 in.) skin with a quasi-isotropic layup and an 18.03-cm-wide (7.1 in.) center bay. The minisandwich panel had a six-ply, 1.52-mm-thick (0.06 in.) skin with a $(45/0/-45/SYNT)_s$ layup, an 0.762-mm-thick (0.03 in.) syntactic core (SYNT) containing glass microballoons [3,4], and a 30.48-cm-wide (12-in.) center bay. The subscript denotes a symmetric layup. The panel edges were reinforced with a 9.52-cm-wide (3.75 in.) doubler, bonded to the panel. The overall panel size was 68.5 by 68.5 cm (27 by 27 in.). The J-stiffener dimensions and layup are illustrated in Fig. 3.

FIG. 1—*J-stiffened monolithic panel.*

FIG. 2—*J-stiffened minisandwich panel.*

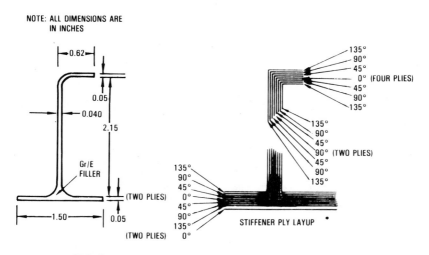

FIG. 3—*Stiffener dimensions and layup (1 in. = 25.4 mm).*

Analysis

The panels were designed to an initial buckling, a limit, and an ultimate shear load of 17.51, 61.29, and 91.94 kN/m (100, 350, and 525 lb/in.), respectively. An anisotropic finite-element program, developed by Lockheed-California Co., was used to predict the initial buckling load. The predicted buckling loads are listed in Table 1 and the corresponding mode shapes are illustrated in Figs. 4 and 5. The predicted and measured initial shear buck-

TABLE 1—*Initial anisotropic buckling strength.*

Type of Panel	Length, in.	Width, in.	Thickness, in.	Skin Layup	N_{XYCR} lb/in. Strap 5	N_{XYCR} lb/in. Test Avg	Program
J-stiffened monolithic	21.5	6.0	0.035	$(\pm 45/0/\overline{90})_s$	102[a]	98.5	NADC post-buckling
	21.0	7.1	0.040	$(45/0/-45/90)_s$	106[a]	100.2 to 103.1	this study
J-stiffened minisandwich	21.0	12.0	0.060	$(45/0/-45/synt)_s$	103[a]	135.5 to 150.2	this study

[a]Target value 11.3 N · m (100 lb/in.).
1 in. = 25.4 mm; 1 lb/in. = 175.13 N/m.

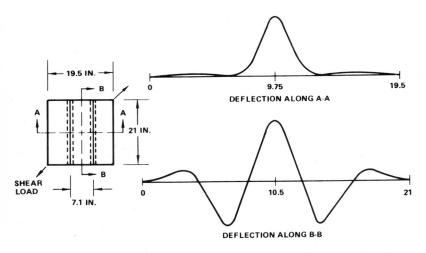

FIG. 4—J-*stiffened monolithic panel buckled mode shape (1 in. = 25.4 mm).*

ling loads, obtained in a previous post-buckling study [5] with twelve similar integrally *J*-stiffened four-bay panels, are included in Table 1 for comparison. The analysis method over-estimated the initial buckling load of the four-bay panels by approximately 3%.

Shear Buckling Tests

Test Setup

The test panels were mounted in turn into a 68.6-cm (27 in.) square shear frame, which featured pretensioned pivot fittings (Fig. 6) to react out the stiffener vibration loads. The shear frame was installed in a load frame, as illus-

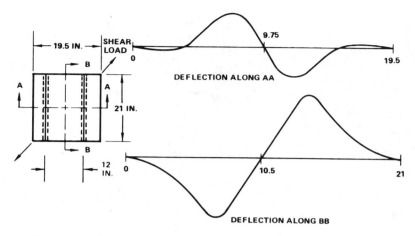

FIG. 5—*J-stiffened minisandwich panel buckled mode shape (1 in. = 25.4 mm).*

FIG. 6—*Reverse side of shear frame showing stiffener pivot attachments—J-stiffened mini-sandwich panel.*

trated in Fig. 7. The load frame was supported in a 1.8-m (6 ft) square test aperture of an acoustic progressive wave tunnel (PWT) by a steel frame assembly (Figs. 7 and 8). The load frame could be rotated down (Fig. 8) for inspection and for conducting the preliminary tests. The shear load was applied by means of a diagonally mounted hydraulic jack, with a 5400-kg (12 000 lb) load capacity, through pivoting beam fittings. The jack train contained universal joints and a load cell.

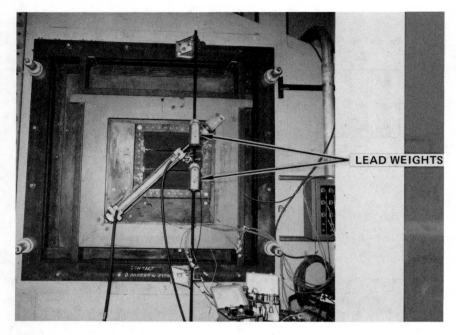

FIG. 7—*Load frame in acoustic progressive wave tunnel wall configured for shear load— J-stiffened monolithic panel.*

Shear Buckling Tests and Modal Studies

Initial static shear buckling tests were conducted to identify, by means of the Moiré fringe patterns, the buckled mode shapes (Figs. 9 and 10) and to select strain-gage locations. These tests were followed by modal studies using impedance head hammer taps to excite each of the panels, at the panel center, and at an antinode near the quarter panel length location, in the center bay. An average of five impacts per location was used and the response was analyzed with the Hewlett Packard 5451C Fourier Analyzer. This procedure was repeated for a range of shear loads through initial buckling. The static and dynamic responses of the panels were measured by a noncontacting displacement transducer (Fig. 11) located at an antinode of the critical shear buckling mode. The static strain was also recorded by a strain-gage rosette located, again, at an antinode in the center bay.

The monolithic panel center displacements, measured during the Moiré fringe pattern test and, a year later, during modal studies are illustrated in Fig. 12. The measured buckling loads differed only by 3%. The variation of the antinode strain with shear load is illustrated in Fig. 13. It was difficult to follow the monolithic panel critical mode frequency through buckling (Fig.

FIG. 8—*Load frame rotated down for inspection.*

14) because of the high order of the critical mode (Fig. 9), the increase in modal density on approaching buckling, and the increase in modal damping, even in the noncritical modes (Fig. 15).

The apparent drift in the displacement prior to buckling (see Fig. 12) was due to the motion of the shear frame. This conclusion was confirmed by dial-gage measurements (Fig. 16) on the minisandwich panel shear frame. Three dial gages would be required to minimize the divergence between the panel and shear frame displacement measurements since rotation of the shear frame was also observed. The second (2,1) mode was the critical mode (Fig. 10) in the minisandwich panel. In spite of the increase in modal density (Fig. 17) on approaching buckling, it was possible to follow the resonant frequency of this low-order critical mode through buckling. The critical buckling jack loads, obtained from the displacement and the resonant frequency plots, dif-

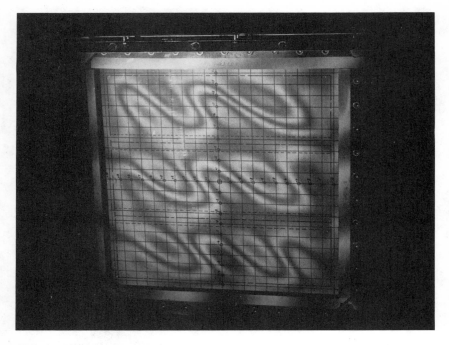

FIG. 9—*Moiré fringe pattern for J-stiffened monolithic panel at 1.8 times the critical buckling load.*

fered by 10%. The damping in all of the modes, and especially the critical mode (Fig. 18), increased significantly on approaching buckling.

The increase in the damping represents a problem for the hammer tap test method. As the damping increases, more energy needs to be imparted to the structure to obtain identifiable resonance peaks.

Correlation With Shear Buckling Analysis

The measured jack loads of 1530 and 1575 kg (3400 and 3500 lb) at buckling (Fig. 12) correspond to shear loads of 17.55 and 18.05 kN/m (100.2 and 103.1 lb/in.), respectively. These results are in good agreement with the predicted shear buckling load of 18.56 kN/m (106 lb/in.) for the monolithic panel. The accuracy is also consistent with the results (Table 1) obtained in a prior post-buckling program [5].

The measured shear buckling load for the minisandwich panel exceeded the predicted value by between 35 and 50%. The tool used in fabricating this panel affected the inner skin. The resulting increase in skin thickness is responsible for the higher measured initial shear buckling load.

The theoretically predicted buckled mode shapes (Figs. 4 and 5) were, gen-

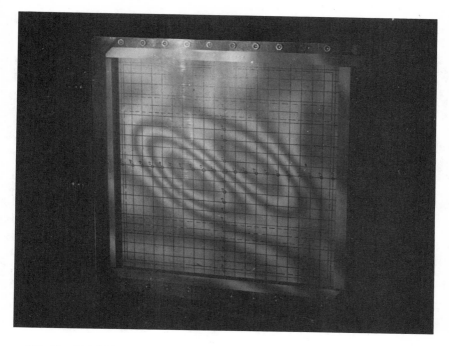

FIG. 10—*Moiré fringe pattern for J-stiffened minisandwich panel at 1.4 times the critical shear buckling load.*

erally, in good agreement with those indicated by the Moiré fringe patterns (Figs. 9 and 10).

Combined Shear and Random Acoustic Loading Tests

Test Procedure

The acoustic progressive wave tunnel was, initially, calibrated with the test aperture in the progressive wave tunnel blocked with a concrete plug. Typical measured random noise spectral density is illustrated in Fig. 19.

The setup for the combined loads tests is illustrated in Fig. 7. The panel response was measured by strain gages, and a noncontacting displacement transducer located at an antinode of the critical shear buckling mode. A low-frequency mounting was used for the displacement transducer in which the vibration levels were reduced by lead weights attached to the transducer support tube (Fig. 7).

Generally, two strain curves were developed as a function of sound pressure level, one at zero shear load and the other at a load level in the vicinity of

FIG. 11—*Typical noncontacting displacement transducer and dial gage setup for measuring panel and shear frame displacement.*

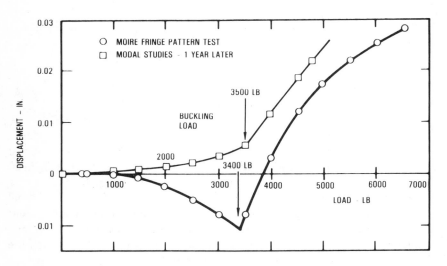

FIG. 12—*J-stiffened monolithic panel center displacement as a function of jack load (1 in. = 25.4 mm; 1 lb = 0.45 kg).*

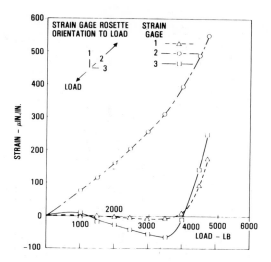

FIG. 13— *Variation of strain at the panel center antinode as function of jack load for* J-*stiffened monolithic panel* (*1 in.* = 25.4 mm; 1 lb = 0.45 kg).

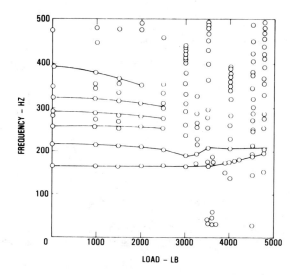

FIG. 14—*Measured resonant frequencies for* J-*stiffened monolithic panel* (*1 lb* = 0.45 kg).

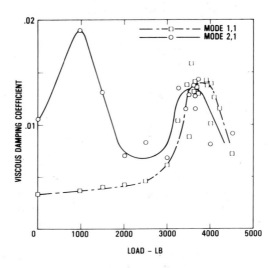

FIG. 15— *Variation of* J*-stiffened monolithic panel damping with jack load (1 lb = 0.45 kg).*

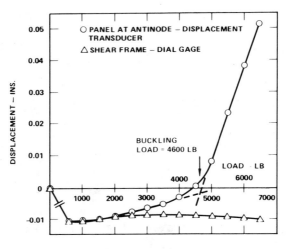

FIG. 16—J*-stiffened minisandwich panel and shear frame displacement as a function of jack load (1 in. = 25.4 mm; 1 lb = 0.45 kg).*

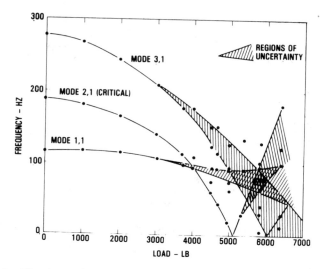

FIG. 17—J-*stiffened minisandwich panel frequency variation as a function of jack load* (1 lb = 0.45 kg).

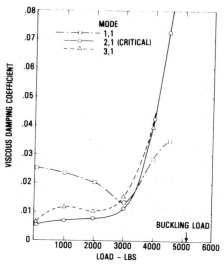

FIG. 18— *Variation of damping with jack load for* J-*stiffened minisandwich panel* (1 lb = 0.45 kg).

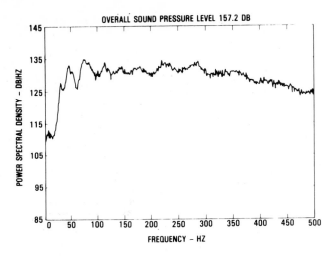

FIG. 19—*Typical random noise spectrum at panel test location in acoustic progressive wave tunnel.*

initial panel buckling. The zero load test was restricted to a strain level below 500 μmm/mm (500 μin./in.) to prevent premature strain-gage failure.

Test Results

The effect of the static shear buckling load on the panel dynamic response can be observed in Fig. 20 by comparing the two strain power spectral density plots obtained at the low overall acoustic excitation level of 140 dB. These plots represent the minisandwich panel response at zero and 2084-kg (4630 lb) jack load. As the overall sound pressure level was increased to 161 dB, the panel response at the 2084-kg (4630 lb) jack load became nonlinear (large amplitude). The general increase in frequency and the broadening of the spectral peaks are characteristic of nonlinear panel response [3,4]. The increase, at the most critical panel location, of the overall (root mean square) strain with overall sound pressure level is illustrated in Figs. 21 and 22 for the monolithic and minisandwich panels. The overall strain levels were increased, near initial buckling, by approximately 24% and 30% for the monolithic and minisandwich panels, respectively.

Conclusions

The results from the finite-element analysis are generally in good agreement with the test data. The noncontacting displacement transducer used in the tests is very effective in measuring the onset of buckling. The damping in the noncritical modes can increase by as much as a factor of four at initial

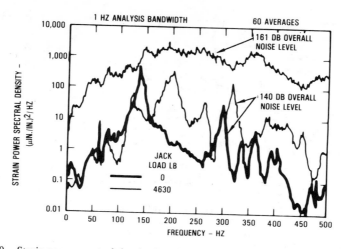

FIG. 20—*Strain power spectral density for minisandwich panel (1 in. = 25.4 mm; 1 lb = 0.45 kg).*

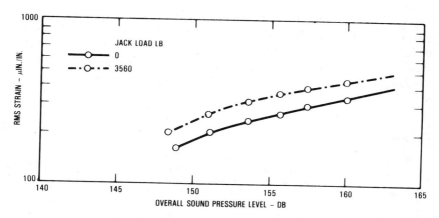

FIG. 21—*Variation of overall strain with overall sound pressure level—monolithic shear panel (1 in. = 25.4 mm; 1 lb = 0.45 kg).*

buckling. The increase in the panel damping is even greater in the critical buckling mode as it approaches buckling. The increase in the damping makes it very difficult to track the critical buckling mode frequency through buckling with the impedance head hammer tap test method.

The strain level in the acoustic fatigue critical location on the panel can increase by as much as 30% on approaching panel buckling in spite of the increase in panel damping. The critical location is usually at the edge of the longest stiffener flange, near the center of the panel. Nonlinear panel response is obtained at the higher noise levels.

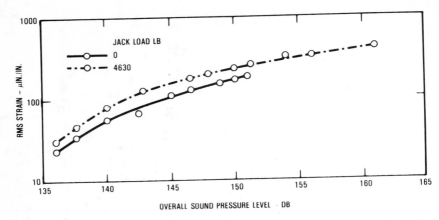

FIG. 22—*Variation of overall strain with overall sound pressure level—minisandwich shear panel (1 in. = 25.4 mm; 1 lb = 0.45 kg).*

An approximate margin of safety of 30% over the ambient temperature sonic fatigue strain levels should be applied to fuselage panels which are designed to buckle just above 1 *g*. Alternatively, the fuselage panels could be designed to buckle well above 1 *g*. Consequently, the combined acoustic and shear loads will affect the design of lightweight V/STOL fuselage structure. These loads, when combined with flight loads into the post-buckling region, could also affect the fuselage fatigue life.

References

[1] Soovere, J., "The Effect of Acoustic-Thermal Environments on Advanced Composite Fuselage Panels," Paper Presented at the 24th Structures, Structural Dynamics and Materials Conference, Lake Tahoe, Nevada, May 2–4, 1983.
[2] Jacobson, M. J., "Sonic Fatigue of Advanced Composite Panels in Thermal Environments," *Journal of Aircraft*, Vol. 20, No. 3, March 1983.
[3] Soovere, J., "Sonic Fatigue Testing of the NASA L-1011 Composite Aileron," *The Shock and Vibration Bulletin*, No. 4, Part 4, Sept. 1980.
[4] Soovere, J., "Sonic Fatigue Testing of an Advanced Composite Aileron," *Journal of Aircraft*, Vol. 19, No. 4, April 1982.
[5] Ostrom, R. B., "Post-Buckling Fatigue Behavior of Flat, Stiffened Graphite/Epoxy Panels Under Shear Loading," Naval Air Development Center Report No. NADC-78137-60, May 1981.

Hisaichi Ohnabe,[1] *Osamu Funatogawa,*[1] *and Mototsugu Itoh*[1]

Stability of Nonlinear Circumferential Waves in an Accelerated Spinning Rectilinearly Orthotropic Disk

REFERENCE: Ohnabe, H., Funatogawa, O., and Itoh, M., "**Stability of Nonlinear Circumferential Waves in an Accelerated Spinning Rectilinearly Orthotropic Disk**," *Recent Advances in Composites in the United States and Japan, ASTM STP 864,* J. R. Vinson and M. Taya, Eds., American Society for Testing and Materials, Philadelphia, 1985, pp. 297–308.

ABSTRACT: A system of two coupled nonlinear differential equations of the second and fourth order, respectively, governing motions of an accelerating membrane-like disk, made of a rectilinearly orthotropic material, is derived. A harmonic-type nonlinear transverse vibration and wave representing the gravest mode in the linear case is then studied. The membrane is assumed to be free, and its deflection is represented by a simple monomial. Free nonlinear vibrations represent the familiar decrease of period of vibrations with an increase of amplitude and show a dependence on the degree of anisotropy. By imposing temporal perturbations on the deflections, the equations for stability of the vibration and the forward and backward traveling waves are derived and discussed.

KEY WORDS: vibration, wave propagation, rotating disk, orthotropy, stability

Disks rotating at high speed form essential elements of many engineering structures, starting from rotors of steam and gas turbines and ending with solar sails, optical and radar reflectors, and precision gyroscopes in space vehicles. Since failures of such disks have been caused by flexural vibrations and circumferential waves, the motions induced by rotation have been investigated widely, at first in a linear isotropic disk [1-3]. They were later extended to include the nonlinear case [4-6]. Most of the earlier work concerned the isotropic materials, but with the development of composites it became necessary to reconsider the problem with regard to anisotropic structure.

[1] Chief research engineer, Advanced Engineering Group, research engineer, Advanced Engineering Group, and manager, Civil Aero-engine Development Department, respectively, Ishikawajima-Harima Heavy Industries Co., Ltd., Tokyo, Japan.

A study of this kind was first made in Ref 7 with regard to finite vibrations of a spinning orthotropic membrane. This work was extended to circumferential waves and their stability in a spinning cylindrically [8–10] and rectilinearly orthotropic disk [11].

On the other hand, with increasing application of high-speed and high-acceleration machinery, the effects due to angular acceleration have become an important design consideration. The influence of angular acceleration on the stresses induced in a rotating isotropic disk has been studied in [12]. The anisotropic characteristic of the material was taken into account in [13–14].

In the present study, we intend to extend these works to the problem in nonlinear vibrations and circumferential waves in an accelerated rectilinearly orthotropic disk. We first establish the system of the governing equations of the problem consisting of the equation of the compatibility of deformations and the equation of the linear momentum in the direction perpendicular to the plane of the disk. The equations are coupled nonlinear partial differential equations. The present investigation is confined to harmonic-type nonlinear vibrations and waves representing the gravest mode of vibrations in the linear isotropic material, that is, two nodal diameters and no nodal circles. Representing the deflection of the disk in the form of a simple monomial, a particular integral for the stress function is found from the compatibility equation. The procedure of Galerkin then provides a nonlinear differential equation for the time function whose solution is a Jacobian elliptic cosine function. Free nonlinear vibrations represent the familiar decrease of the period of vibration with an increase of amplitude and show a dependence on the degree of anisotropy. The decrease is less pronounced for higher velocities of spinning. By means of small temporal disturbances of transverse motion, the differential equations for stability of the vibration and the forward and backward traveling waves are derived and discussed.

Fundamental

Let us consider a flat circular disk of negligible flexural rigidity and radius a, having a uniform thickness h, and made of rectilinearly orthotropic material with a density ρ. The disk rotates around its axes of symmetry with an angular velocity Ω and an angular acceleration $\dot{\Omega}$. It is assumed that the transverse displacements of particles are finite and large as compared with the in-plane displacements, so that the inertia terms associated with the latter may be disregarded. On account of the negligible bending rigidity of the disk, its motions are controlled by the in-plane tensions. Let us locate the origin 0 of a cartesian rectangular coordinate system x, y, z at the center of the disk and direct the x and y axes of the system parallel to the principal elastic directions of the disk (Fig. 1). The equations of equilibrium with the body forces due to angular velocity and angular acceleration of disk with constant thickness are

$$\frac{\partial \sigma_{xx}}{\partial x} + \frac{\partial \tau_{xy}}{\partial y} + \rho \Omega^2 x + \rho \dot{\Omega} y = 0$$

$$\frac{\partial \tau_{xy}}{\partial x} + \frac{\partial \sigma_{yy}}{\partial y} + \rho \Omega^2 y - \rho \dot{\Omega} x = 0$$

(1)

The in-plane stress components $\sigma_{xx} \sigma_{yy}$, τ_{xy}, generated by the stretching of the disk, are the average values. Equations 1 are satisfied by introducing stress function F such that

$$\sigma_{xx} = \frac{\partial^2 F}{\partial y^2} - \frac{1}{2} \rho \Omega^2 (x^2 + y^2) - \frac{1}{2} \rho \dot{\Omega} xy - 2C \frac{xy}{(x^2 + y^2)^2}$$

$$\sigma_{yy} = \frac{\partial^2 F}{\partial x^2} - \frac{1}{2} \rho \Omega^2 (x^2 + y^2) + \frac{1}{2} \rho \dot{\Omega} xy + 2C \frac{xy}{(x^2 + y^2)^2}$$

(2)

$$\tau_{xy} = -\frac{\partial^2 F}{\partial x \partial y} + \frac{1}{4} \rho \Omega (x^2 - y^2) + C \frac{x^2 - y^2}{(x^2 + y^2)^2}$$

The associated strains e_{xx}, e_{yy} and e_{xy} have to satisfy the compatibility equation

$$\frac{\partial^2 e_{xx}}{\partial y^2} + \frac{\partial^2 e_{yy}}{\partial x^2} - 2 \frac{\partial^2 e_{xy}}{\partial x \partial y} = \left(\frac{\partial^2 w}{\partial x \partial y} \right)^2 - \frac{\partial^2 w}{\partial x^2} \frac{\partial^2 w}{\partial y^2}$$

(3)

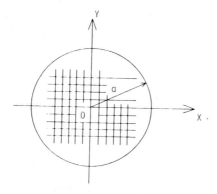

FIG. 1—*Dimensions and elastic directions.*

where $w(x, y; t)$ is the transverse deflection of the disk. Clearly, the generalized Hooke's law for the rectilinearly orthotropic material are (see Ref 15, p. 9)

$$e_{xx} = \frac{1}{E_1} (\sigma_{xx} - \nu_1 \sigma_{yy})$$

$$e_{yy} = \frac{1}{E_2} (\sigma_{yy} - \nu_2 \sigma_{xx}) \tag{4}$$

$$e_{xy} = \frac{1}{2G_{12}} \tau_{xy}$$

where

$E_1 \nu_2 = E_2 \nu_1$,
E_1 and $E_2 =$ Young's moduli in the x and y directions, respectively,
$G_{12} =$ shear modulus, and
$\nu_1 =$ for example, Poisson's ratio associated with tension in the y-direction.

According to Refs 13–14 and 16 we also confine our investigation to the case of the material satisfying the condition

$$\frac{1}{E_1} + \frac{1}{E_2} + \frac{2\nu_2}{E_2} - \frac{1}{G_{12}} = 0 \tag{5}$$

Substitution of Eqs 4 into Eq 3 with due regard to Eqs 2 and 5 gives the first fundamental equation of the problem

$$\frac{\partial^4 F}{\partial x^4} + P^2 \frac{\partial^4 F}{\partial x^2 \partial y^2} + k^2 \frac{\partial^4 F}{\partial y^4} = E_2 \left[\left(\frac{\partial^2 w}{\partial x \partial y} \right)^2 - \frac{\partial^2 w}{\partial x^2} \frac{\partial^2 w}{\partial y^2} \right]$$

$$- (2\nu_2 - k^2 - 1)\rho\Omega^2 \tag{6}$$

where

$$k^2 = \frac{E_2}{E_1}, \qquad P^2 = \frac{E_2}{G_{12}} - 2\nu_2 \tag{7}$$

The equation of transverse motion, that is, the second fundamental equation, is easily found [11] to be, with the acceleration terms

$$\frac{\partial^2 w}{\partial x^2}\frac{\partial^2 F}{\partial y^2} - \frac{\partial^2 w}{\partial x \partial y}\frac{\partial^2 F}{\partial x \partial y} + \frac{\partial^2 w}{\partial y^2}\frac{\partial^2 F}{\partial x^2}$$

$$- \frac{1}{2}\rho\Omega^2(x^2 + y^2)\left[\frac{\partial^2 w}{\partial x^2} + \frac{\partial^2 w}{\partial y^2}\right] - \rho\Omega^2\left[x\frac{\partial w}{\partial x} + y\frac{\partial w}{\partial y}\right]$$

$$- \frac{1}{2}\rho\dot{\Omega}xy\left[\frac{\partial^2 w}{\partial x^2} - \frac{\partial^2 w}{\partial y^2}\right] + \frac{1}{2}\rho\dot{\Omega}(x^2 - y^2)\frac{\partial^2 w}{\partial x \partial y} \tag{8}$$

$$- \rho\dot{\Omega}\left[y\frac{\partial w}{\partial x} - x\frac{\partial w}{\partial y}\right] - \rho\frac{\partial^2 w}{\partial t^2} = 0$$

This completes the formulation of the problem, which is now reduced to an integration of the system of two heavily nonlinear coupled equations, together with appropriate boundary conditions. Assuming the absence of external tractions, t_{nx} and t_{ny}, at the rim of the disk, $r = a$, we have

$$t_{nx} = \sigma_{xx}\frac{x}{a} + \tau_{xy}\frac{y}{a} = 0$$

$$\tag{9}$$

$$t_{ny} = \tau_{xy}\frac{x}{a} + \sigma_{yy}\frac{y}{a} = 0$$

corresponding to the edge of the disk free from external tractions.

For future purpose, it is convenient to transform Eq 8 from rectilinear to polar coordinates r, θ. The transformation easily leads to the equation

$$\frac{\partial^2 w}{\partial r^2}\left(\frac{1}{r}\frac{\partial F}{\partial r} + \frac{1}{r^2}\frac{\partial^2 F}{\partial \theta^2}\right) - 2\frac{\partial}{\partial r}\left(\frac{1}{r}\frac{\partial w}{\partial \theta}\right)\frac{\partial}{\partial r}\left(\frac{1}{r}\frac{\partial F}{\partial \theta}\right)$$

$$+ \left(\frac{1}{r}\frac{\partial w}{\partial r} + \frac{1}{r^2}\frac{\partial^2 w}{\partial \theta^2}\right)\frac{\partial^2 F}{\partial r^2}$$

$$- \frac{1}{2}\rho\Omega^2 r^2\left(\frac{\partial^2 w}{\partial r^2} + \frac{1}{r}\frac{\partial w}{\partial r} + \frac{1}{r^2}\frac{\partial^2 w}{\partial r^2}\right) - \rho\Omega^2 r\frac{\partial w}{\partial r} \tag{10}$$

$$+ \frac{1}{2}\rho\dot{\Omega}r^2\frac{\partial}{\partial r}\left(\frac{1}{r}\frac{\partial w}{\partial \theta}\right) + \rho\dot{\Omega}\frac{\partial w}{\partial \theta} - \rho\frac{\partial^2 w}{\partial t^2} = 0$$

We wish to examine a nonlinear vibration of an accelerating membrane disk having two nodal diameters and no nodal circles. For this we take

$$w(r, \theta; t) = Ar^2(\cos 2\theta)\tau(t) \tag{11}$$

where A is the amplitude and $\tau(t)$ a (so far unknown) function of time.
Substitution of Eq 11 into Eq 6 yields the right-hand side of the latter to be

$$4E_2A^2\tau^2(t) - (2\nu_2 - k^2 - 1)\rho\Omega^2 \tag{12}$$

Taking the general integral in the form

$$F(x, y; t) = C_1x^2 + C_2y^2 + C_3xy + A_1x^3 + A_2y^3 + A_3x^2y$$

$$+ A_4xy^2 + B_1x^4 + B_2y^4 + B_3x^2y^2 \tag{13}$$

it is not difficult to show that one has to set

$$A_i = 0, \quad i = 1, 2, 3, 4, \quad \text{and} \quad C_3 = 0. \tag{14}$$

Obeying the boundary conditions (Eq 9) and bearing in mind that $x^2 + y^2 = a^2$ on the rim of the disk, we easily arrive at the equations

$$B_1 = B_2 = \frac{1}{2} B_3 \tag{15}$$

$$C_1 = C_2 = -B_3a^2 + \frac{1}{4} \rho\Omega^2a^2 = C'$$

and

$$C = -\frac{1}{4} \rho\dot{\Omega}a^4 \tag{16}$$

Using Eqs 14 and 15, we obtain from equation (6)

$$B_3 = \frac{4A^2E_2\tau^2(t) + (1 + k^2 - 2\nu_2)\rho\Omega^2}{4(3 + 3k^2 + P^2)} \equiv B \tag{17}$$

Consequently, all of the constants in Eq 12 are obtained and the expression (12) for the stress function simplifies to

$$F = C'r^2 + \frac{1}{2} Br^4 \tag{18}$$

where the coefficients B and C' are given by Eqs 17 and 16.
In view of this, the stress components are calculated from Eq 2 to be

$$\sigma_x = 2C' + 2B(x^2 + 3y^2) - \frac{1}{2}\rho\Omega^2(x^2 + y^2) + \frac{1}{2}\rho\dot{\Omega}a^4 \frac{xy}{(x^2 + y^2)^2}$$

$$\sigma_y = 2C' + 2B(3x^2 + y^2) - \frac{1}{2}\rho\Omega^2(x^2 + y^2) - \frac{1}{2}\rho\dot{\Omega}a^4 \frac{xy}{(x^2 + y^2)^2} \qquad (19)$$

$$\tau_{xy} = -4Bxy + \frac{1}{4}\rho\dot{\Omega}(x^2 - y^2) - \frac{1}{4}\rho\dot{\Omega}a^4 \frac{x^2 - y^2}{(x^2 + y^2)^2}$$

and in the case of no angular acceleration these coincide with Eq 2.18 in Ref *11*.

Governing Time Equation

We now turn to the only remaining field equation of motion, Eq 10. To this end we proceed in the following way. Upon substitution of Eqs 11 and 18 into 10, we apply the procedure of Galerkin. One multiplies both members of the latter equation by the spatial part $r^2 \cos 2\theta$ of Eq 11 and integrates the results over the domain of the plate. A trivial calculation yields the following nonlinear differential equation of the second order for the time function $\tau(t)$:

$$\frac{d^2\tau}{dt^2} + \alpha\tau + \beta\tau^3 = 0 \qquad (20)$$

where

$$\alpha = \frac{2\Omega^2\{4(1 + k^2) + P^2 - 2\nu_2\}}{3 + 3k^2 + P^2}$$

$$\beta = \frac{8A^2E_2}{(3 + 3k^2 + P^2)\rho} \qquad (21)$$

Let us suppose that $\alpha, \beta > 0$. Then the solution of the differential equation (Eq 20) is the Jacobian elliptic

$$\tau(t) = \text{cn}(\omega^*t, k^*) \qquad (22)$$

provided the following normalized initial conditions for the time function are satisfied:

$$\tau(0) = 1, \qquad \left[\frac{d\tau(t)}{dt}\right]_{t=0} = 0 \qquad (23)$$

and the notation

$$\omega^* = (\alpha + \beta)^{1/2}, \qquad k^* = \frac{\beta}{2(\alpha + \beta)} \tag{24}$$

and cn is the cosine type of Jacobian elliptic functions. The period of the function cn$(\omega^* t, k^*)$ is $T^* = 4K/\omega^*$, where $K(k^*) = F(k^*, \Pi/2)$ is a complete elliptic integral of the first kind. Since the modulus k^* is dependent on the amplitude A, it follows that the periodic time of the nonlinear vibrations depends on the amplitude, a fact which is well known.

Free Linear Vibrations

Let us now reject the nonlinear term in the Eq 20. Then the equation is reduced to that governing free linear vibrations with the relative circular frequency

$$\frac{\omega}{\Omega} = \left[2 \frac{4(1 + k^2) + P^2 - 2\nu_2}{3 + 3k^2 + P^2} \right]^{1/2} \tag{25}$$

where $\omega = \sqrt{\alpha}$. In the case of isotropic material with two nodal diameters and no nodal circle, this result (Eq 25) is in full accord with the known results derived for the linear case (compare Eqs 25 of Ref *17*).

Free Nonlinear Vibrations

We now turn to the investigation of free nonlinear vibrations of the membrane-like disk.

Since the corresponding period of the linear vibrations is $T = 2\Pi/\omega = 2\Pi/\sqrt{\alpha}$, the ratio of the two periods is given by the equation

$$\frac{T^*}{T} = \frac{2K}{\Pi \left(1 + \dfrac{\beta}{\alpha} \right)^{1/2}} \tag{26}$$

In the present case we have

$$\frac{T^*}{T} = 2K \bigg/ \left\{ \Pi \left[1 + \frac{4A^2 E_2}{\rho \Omega^2 \{4(1 + k^2) + P^2 - 2\nu_2\}} \right]^{1/2} \right\} \tag{27}$$

In the isotropic case, the above ratio is carried into

$$\frac{T^*}{T} = \frac{2K}{\Pi} \; 1 \Big/ \Big\{\Big[1 + \frac{2EA^2}{\rho\Omega^2(5 - \nu)}\Big]^{1/2}\Big\} \qquad (28)$$

which is in agreement with the result obtained by Nowinski [7].

To illustrate the procedure, the orthotropy as well as the isotropic case is chosen as listed in Table 1. For definiteness, we also use the values of ρ, a, and h in Table 1.

Figure 2 displays the dependence of the relative period T^*/T on the relative amplitude A/h for isotropic and orthotropic material. It is noted that the same phenomenon occurs in both the isotropic and orthotropic cases as the speed of the spin of the disk is increased.

Stability of Vibration

The stability of the vibration given by Eq 11 can be studied in the following way. We impose on the deflection $w(r, \theta; t)$ of the disk a small temporal

TABLE 1—*Two types of anisotropy and disk dimensions.*

	E_1, Gpa	E_2	G_{12}	ν_1	ν_2
Isotropy	200	200	83.3	0.2	0.2
Orthotropy	200	25	21.1	0.24	0.03

NOTE: $\rho = 7.9 \times 10^3$ (Ns^2/m^4); $a = 1$ (m), $h = 0.005$ (m).

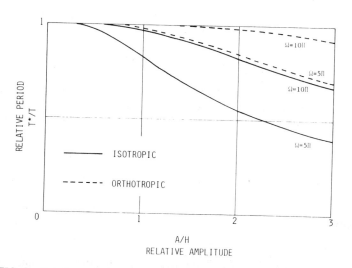

FIG. 2—*Relative period versus relative amplitude for various angular speeds.*

perturbation, $\delta(t)$, of the time function $\tau(t)$. The deflection is now given by the form

$$w(r, \theta; t) = Ar^2 \cos 2\theta [\tau(t) + \delta(t)] \tag{29}$$

The corresponding stress function is altered only by the substitution of $[\tau(t) + \delta(t)]$ for $\tau(t)$. We find that $\delta(t)$ is governed by

$$\ddot{\delta}(t) + [\alpha + 3\beta\tau^2(t)]\delta(t) = 0 \tag{30}$$

where α, β are expressed in Eq 21.

Matters relating to the stability of solutions to Eq 30 are well known.

Stability of Transverse Waves

We now confine our investigation to the mode of propagation of circumferential waves that exhibits two nodal diameters and no nodal circles. This mode is of special interest inasmuch as it is the gravest mode in the linear isotropic case; it is represented in the expanded form by

$$w(r, \theta; t) = Ar^2 [\cos 2\theta \cos 2ct \mp \sin 2\theta \sin 2ct] \tag{31}$$

where the upper (lower) sign refers to the backward (forward) traveling wave. Following Ref 5, we impose on the deflection $w(r, \theta; t)$ of the disk two slight temporal perturbations, $\delta_1(t)$ and $\delta_2(t)$, of the time functions $\cos 2ct$ and $\sin 2ct$. Then we have

$$w(r, \theta; t) = Ar^2 \{ \cos 2\theta [\cos 2ct + \delta_1(t)]$$

$$+ \sin 2\theta [\mp \sin 2ct + \delta_2(t)] \} \tag{32}$$

Following a line of argument very similar to that used for the derivation of Eqs 15-17, we substitute Eq 32 into Eq 6 and find the perturbed stress function for a disk with its edge free of tractions that $\tau^2(t)$ in Eq 17 is replaced by

$$\{ [\cos 2ct + \delta_1(t)]^2 + [\mp \sin 2ct + \delta_2(t)]^2 \}$$

We obtain, after linearization with respect to powers and products of δ_1 and δ_2

$$B = \frac{4E_2A^2[1 + 2\cos \tau \cdot \delta_1 \mp 2 \sin \tau \cdot \delta_2] + (1 + k^2 - 2\nu_2)\rho\Omega^2}{4(3 + 3k^2 + P^2)} \tag{33}$$

It now remains to insert Eq 32 and the perturbed stress function F into the remaining governing equation (Eq 10) and apply the orthogonalization procedure. Consequently, disregarding powers of δ_1 and δ_2 higher than the first one, we finally arrive at

$$\dot{\eta}^2\delta_1'' + \ddot{\eta}\delta_1' + \delta_1\{4\xi(1 + \cos 2\eta) + 4\xi + 2\zeta\Omega^2\}$$

$$= \pm(4\xi \sin 2\eta \pm 3\dot{\Omega})\delta_2 + (-4\xi - 2\zeta\Omega^2 + \dot{\eta})\cos \eta + (\ddot{\eta} \mp 3\dot{\Omega})\sin \eta$$

$$\tag{34}$$

$$\dot{\eta}^2\delta_2'' + \ddot{\eta}\delta_2' + \delta_2\{4\xi(1 - \cos 2\eta) + 4\xi + 2\zeta\Omega^2\}$$

$$= \pm(4\xi \sin 2\eta \mp 3\dot{\Omega})\delta_1 + (-4\xi - 2\zeta\Omega^2 + \dot{\eta})\sin \eta + (\pm\ddot{\eta} - 3\dot{\Omega})\cos \eta$$

where

$$\xi = \frac{2E_2A^2}{(3 + 3k^2 + P^2)\rho}$$

$$\tag{35}$$

$$\zeta = \frac{4(1 + k^2) + P^2 - 2\nu_2}{3 + 3k^2 + P^2}$$

In the preceding equations, $\eta = 2ct$, and the prime denotes differentiation with respect to η. We say that the wave motions are *stable* with respect to the given perturbations if the solutions of the above "variational" equations satisfied by the perturbations $\delta_1(t)$ and $\delta_2(t)$ are bounded functions of time. On the contrary, we say that the motion is *unstable* if the solutions of the equations are unbounded functions of time. It is impossible to solve the above equations in a closed form in the present study. We leave them for future study. But in the case of constant speed, they become

$$\delta_1'' + \delta_1[d + \gamma(1 + \cos 2\eta)] = \pm\gamma \sin 2\eta \cdot \delta_2 + (1 - d)\cos \eta$$

$$\tag{36}$$

$$\delta_2'' + \delta_2[d + \gamma(1 - \cos 2\eta)] = \pm\gamma \sin 2\eta \cdot \delta_1 \mp (1 - d)\sin \eta$$

where

$$d = \gamma + \epsilon$$

$$\gamma = \frac{2E_2A^2}{(3 + 3k^2 + P^2)c^2\rho}$$

$$\tag{37}$$

$$\epsilon = \frac{\Omega^2}{2c^2} \frac{4(1 + k^2) + P^2 - 2\nu_2}{3 + 3k^2 + P^2}$$

They are discussed in detail by Nowinski [11].

Conclusions

As far as we consider the entire disk and we assume that the deflection function $w(r, \theta; t)$ is $Ar^2 \cos 2\theta \cdot \tau(t)$ for the gravest mode of vibrations, that is, two nodal diameters and no nodal circles, the acceleration effect for the transverse vibration in a spinning rectilinearly orthotropic disk is not found. But it is necessary to confirm it by experimental study.

The solvable equation is derived for stability of transverse vibration. The complicated variational equations for stability of wave propagation are derived, but the solution is left for future study.

Acknowledgment

The authors would like to thank Dr. J. L. Nowinski, H. Fletcher Brown, Professor Emeritus, University of Delaware, for the many helpful comments and suggestions.

References

[1] Lamb, H. and Southwell, R. V. in *Proceedings*, Royal Philosophical Society of London, Series A, Vol. 99, 1921, pp. 272-280.
[2] Campbell, W., *Transactions*, American Society of Mechanical Engineers, Vol. 46, 1924, pp. 31-160.
[3] Stodola, A., *Steam and Gas Turbines*, McGraw-Hill, New York, 1927.
[4] Nowinski, J. L., *Journal of Applied Mechanics*, Vol. 31, March 1964, pp. 72-78.
[5] Bulkeley, P. Z., *Journal of Applied Mechanics*, Vol. 40, 1973, pp. 133-136.
[6] Nowinski, J. L., Technical Report No. 232, University of Delaware, Newark, DE, March 1980.
[7] Nowinski, J. L. and Woodall, S. R., *Journal of Acoustical Society of America*, Vol. 36, No. 11, Nov. 1964, pp. 2113-2118.
[8] Nowinski, J. L., Technical Report No. 231, University of Delaware, Newark, DE, May 1980.
[9] Nowinski, J. L., *Journal of Applied Mechanics*, Vol. 49, Sept. 1982, pp. 570-572.
[10] Ohnabe, H., Funatogawa, O., and Nowinski, J. L., "*Progress in Science and Engineering of Composites*," in *Proceedings*, Fourth International Conference on Composite Materials (ICCM-IV), T. Hayashi, K. Kawata, and S. Umekawa, Eds., Tokyo, 1982, pp. 423-430.
[11] Nowinski, J. L., Technical Report No. 245, University of Delaware, Newark, DE, April 1982 (to appear in the *International Journal of Mechanical Science* (in Britain)).
[12] Tang, S., *International Journal of Science*, Vol. 12, 1970, pp. 205-207.
[13] Reddy, T. Y., Lakshminarayana, H. V., and Srinath, H., *Journal of Applied Mechanics*, Sept. 1974, pp. 817-819.
[14] Genta, G., Gola, M., and Gugliotta, A., *Journal of Applied Mechanics*, Vol. 49, Sept. 1982, pp. 658-661.
[15] Lekhnitskii, S. G., *Anisotropic Plates* (in Russian), GITTL, Moscow, 1957; English Translation No. 50, American Iron and Steel Institute, New York, 1965.
[16] Genta, G. and Gola, M., *Journal of Applied Mechanics*, Vol. 48, Sept. 1981, pp. 559-562.
[17] Prescott, J., *Applied Elasticity*, Dover, New York, 1961.

Vikram K. Kinra[1]

Dispersive Wave Propagation in Random Particulate Composites[2]

REFERENCE: Kinra, V. K., **"Dispersive Wave Propagation in Random Particulate Composites,"** *Recent Advances in Composites in the United States and Japan, ASTM STP 864*, J. R. Vinson and M. Taya, Eds., American Society for Testing and Materials, Philadelphia, 1985, pp. 309–325.

ABSTRACT: Propagation of ultrasonic waves has been studied in random particulate composites at low, intermediate, and high frequencies. The principal objective of this experimental work is to demonstrate the existence of two separate branches of wave propagation: a low-frequency acoustical branch where the inclusions move essentially in phase with the matrix and a high-frequency optical branch where inclusions move essentially out of phase with the matrix. The two are separated by a well-defined cutoff frequency which corresponds to the rigid-body-translation resonance of the relatively heavy inclusions. The measured values of the cutoff frequency were found to be in *excellent* agreement with an elementary model of Moon and Mow. To the best of our knowledge this is the first experimental observation of the optical branch and of the cutoff frequency.

KEY WORDS: composite materials, wave propagation, ultrasonics, dispersion, attenuation, acoustical branch, optical branch, cutoff frequency

Nomenclature

C_1, C_2 Longitudinal and shear wavespeed in the matrix

$\langle \ \rangle, (\ )'$ Composite and inclusion property

α Attenuation of longitudinal waves

$\tilde{C}$ Volume fraction of spherical inclusions

a Radius of spherical inclusions

k_1, λ_1 Longitudinal wavenumber and wavelength in the matrix: $k_1 = 2\pi/\lambda_1$

Ω, ξ Normalized frequency and wavenumber: $\Omega = k_1 a; \xi = <k_1>a$

ρ Density

[1]Associate professor, Department of Aerospace Engineering and Mechanics and Materials Center, Texas A&M University, College Station, TX 77843.
[2]This paper is dedicated to my intellectual mentor, Mrs. Ved Swaroop Narang.

ν Poisson's ratio
n Frequency
Ω_c Cutoff frequency

Recently, there appears to have been a renewed interest in the calculation of effective properties of particulate composites [1-10]; for an extensive bibliography the reader is referred to a book by Mura [1] and recent review articles by Christensen [2], McCoy [3], and Cleary et al [4]. Relatively little has been done in the way of experiments [11-15]. Much of the theoretical work is concerned with the *static* properties. It is only recently that (necessarily approximate) theories which predict frequency-dependent moduli have been constructed [5,6,16,17]. In the absence of any exact solution to this rather complex problem, the deviation of the approximate models from the true value is difficult to estimate, hence the need for accurate experimental results.

The purpose of this paper is to report some *experimental* results concerning propagation of ultrasonic waves in *random* particulate composites. We will demonstrate the existence of classical acoustical branch, an optical branch, and a cutoff frequency which separates the two. *To the best of our knowledge this is the first experimental observation of the optical branch, and of the cutoff frequency.*

Three acoustically disparate composite systems have been studied: glass spheres/epoxy; lead spheres/epoxy; and steel spheres/Plexiglass, (polymethylmethacrylate or PMMA). At very low frequencies the measured longitudinal wavespeed was found to be in excellent agreement with the classical bound calculations, for example, those of Hashin and Shtrikman [18]. The measured cutoff frequency, Ω_c, agrees remarkably well with the predictions of Moon and Mow [19]. Across Ω_c the wavespeed takes a large positive jump. At the higher frequencies, the wave propagation occurs along the optical branch. For very high Ω the wavespeed appears to approach a frequency-independent optical limit.

Experimental Procedures

A detailed description of the experimental procedures used in this investigation has been given elsewhere [11,14]; therefore, only a brief description is included in the following.

Two types of through-transmission ultrasonic apparatuses were used: (1) water immersion [11], and (2) direct contact [14]. The heart of this system is a pair of accurately matched (in the sense of their frequency response), broad-band, X-cut, piezoelectric transducers. The *phase velocity* was measured as follows. The time taken by a *reference peak* near the center of a toneburst (typically 20 cycles long) to trasverse the known thickness of a specimen was measured with an instrument accuracy of five nanoseconds (ns). Two time measurements, t_1

and t_2, for specimens of thickness W_1 and W_2, were taken; then $\langle C_1 \rangle = (W_2 - W_1)/(t_2 - t_1)$. This procedure minimizes the error due to the boundary-layer effect [12]. The attenuation was measured similarly. Let A_1 and A_2 be the amplitudes received through W_1 and W_2, respectively; then $\langle \alpha \rangle = (W_2 - W_1)^{-1} \ell n (A_1/A_2)$.

To ensure that we were measuring the phase velocity and not the group velocity, the following precautions were taken. (1) The reference peak was selected near the middle of the toneburst, that is, away from the transients at the two ends. (2) From the time domain measurements it was ensured that a phase difference of 2π occurred between the successive peaks of both the incident and the transmitted waveforms in the vicinity of the reference peak. (3) Measurements were carried out only if, starting from the leading edge of the toneburst, we could establish an unequivocal, one-to-one correspondence between the peaks of the incident and the transmitted waves. The velocity measurements are accurate to 1%; see Ref 14 for a detailed and systematic error analysis.

Ultrasonic Spectroscopy—Redundant measurements were obtained by working in the frequency domain. Here, instead of a toneburst the transmitter is excited with a pulse of about 50-ns duration; this results in a broad-band pulse. Both the incident and the received signals are fed into a microprocessor, a Fast Fourier Transform (FFT) is computed, a suitable deconvolution is carried out, and both the velocity and the attenuation data are obtained as functions of frequency.

The properties of the constituents are listed in Table 1.

Results and Discussion

Since the concept of the cutoff frequency is central to the subject of this paper, it will be discussed in the first subsection. In the second subsection will be presented the results concerning acoustical and optical branches. To guard against fortuitous results three disparate composites were tested: glass/epoxy; lead/epoxy; and steel/PMMA.

TABLE 1—*Acoustic properties of the constituents.*

Material	C_1, mm/μs	C_2, mm/μs	ρ, g/cm^3	ν,	α, nepers/mm
Epon 828Z epoxy	2.64	1.20	1.202	0.372	0.043 @ 1 MHz
Lead	2.21	0.86	11.3	0.411	0.026 @ 2 MHz
Tra-cast 3012 epoxy	2.54	1.16	1.180	0.370	0.045 @ 1 MHz
Glass	5.28	3.24	2.492	0.200	negligible
PMMA	2.63	1.32	1.16	0.334	0.02 @ 1 MHz
Steel	5.94	3.22	7.80	0.292	negligible

Cutoff Frequency

Following Moon and Mow [*19*], consider a single rigid sphere in an unbounded elastic medium. Suppose the sphere is displaced from its equilibrium position (without rotation) and let go. Heuristically, it will undergo damped natural vibrations like a mass on a viscoelastic spring ("zeropole" resonance). The damping is due to the radiation of waves and is known as radiation damping. Now consider a random distribution of spheres. Consider an incident time-harmonic longitudinal wave. When the excitation frequency coincides with the cutoff frequency, Ω_c, the scattering cross section goes through a maximum or the through-transmitted amplitude goes through a minimum. Typical results of an experiment designed around this simple heuristic argument are shown in Fig. 1. EPON 828Z epoxy specimens containing lead spheres (1.32 mm diameter) were tested. The volume fraction of inclusions, $\tilde{C}$, was varied from a "dilute suspension" to a "concentrated mixture": $\tilde{C} = 5.4\%$, 15.9%, 26.1%, 34.0%, and 52%. For each $\tilde{C}$ two specimens of different thickness were tested. In Fig. 1, we see a sharp minimum in the received amplitude; the location of the mini-

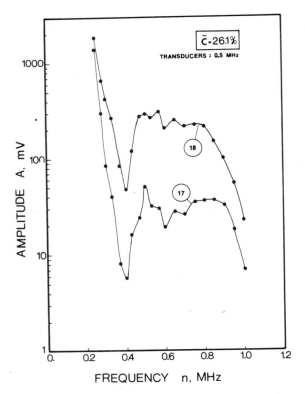

FIG. 1—*Amplitude of the signal received through the thickness of a specimen exhibits a very sharp minimum corresponding to the cutoff frequency. Lead spheres/epoxy.*

mum is the same for both specimens. These measurements of the cutoff frequency, n_c, were found to be reproducible to 0.01 MHz.

Moon and Mow [19] have given an approximate analytical solution of this
problem under the following assumptions: (1) The inclusions are distributed in
a random homogeneous manner; (2) the inclusions are heavy, $\rho'/\rho \rightarrow \infty$ (hence
the choice of lead inclusions in our experiments); (3) the inclusions are rigid;
(4) a dilute suspension, that is, $\tilde{C} \ll 1$; and (5) the frequency is small, that is, Ω
$\ll 1$. They gave expressions for the cutoff frequency Ω_c. After a bit of algebraic
manipulation their expressions reduce to the following simple formula

$$\Omega_c^2 = \frac{9(1 - 2\nu)}{(5 - 6\nu)}\left[\frac{\rho}{\rho'} + \frac{\tilde{C}}{1 - \tilde{C}}\right] \tag{1}$$

Here, ν and ρ are, respectively, the Poisson's ratio and the density of the matrix,
and the prime refers to the inclusion material. Our experiments are compared
with the calculation of Ref 19 in Fig. 2. In view of a number of assumptions of
the theory not satisfied in our experiments, the comparison is considered surprisingly good. For two cases, namely, $\tilde{C} = 15.9\%$ and 52%, two different experimental values of Ω_c are reported, and the reasons are as follows. For $\tilde{C} =$
15.9%, two neighboring minima were observed in both specimens. For $\tilde{C} =$
52%, the location of the (single) minimum was slightly different for the two
specimens.

Evidently, the model of Moon and Mow is based on some rather restrictive
assumptions listed in the preceding paragraph; the last three assumptions are
not satisfied in our experiments. We now raise an appropriate question: Why
does the model still predict Ω_c so well? Before proceeding to offer some plausible explanations, we point out situations where the model *completely fails* to
predict the experimental results. In addition to Ω_c, we also calculated the dispersion and the attenuation curves for all composites studied. In all three cases,
at all concentrations (with one minor exception) the comparison was found to
be extremely poor. The exception was the case of a dilute suspension (5%) of
lead spheres in an epoxy matrix. Even here, the model predicts only the dispersion curve with some accuracy. Thus, the Moon and Mow model can, at best,
be considered a *qualitative* model. Its value lies in the fact that it correctly predicts the phenomena of cutoff frequency and acoustical and optical branches in
a *random* particulate composite, and in that it provides a framework in which
the results of the present experimental investigation can be interpreted.

As was seen above, the model accurately predicts the cutoff frequency (Fig. 2).
However, this is true only for the case of heavy inclusions (lead and steel); for the
relatively light inclusions (glass/epoxy) even the Ω_c predictions of the model
were found to be in significant disagreement with the experimental observations.

For lead the Poisson's ratio, $\nu = 0.411$, that is, the ratio of the shear modulus
to bulk modulus, is ~ 0.19. Thus lead may be viewed as a "fluid." The model
of Moon and Mow, on the other hand, treats perfectly rigid inclusions. This

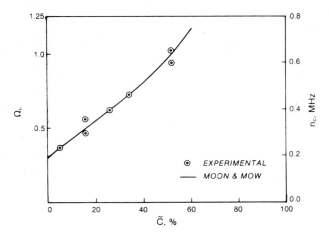

FIG. 2—*Cutoff frequency increases with* $\tilde{C}$. *Comparison of theory and experiment is considered excellent. Lead spheres/epoxy.*

raises another interesting question: For the case of a lead/epoxy composite, in spite of this discrepancy why does the Moon and Mow model predict Ω_c so well?

We now offer some plausible explanations. The total scattering cross section of an inclusion may be viewed as consisting of two distinct physical effects: the scattering due to elastic mismatch between the inclusion and the matrix, and the scattering due to the density mismatch. We conjecture that when the density ratio ρ'/ρ is large, as is the case for lead/epoxy composites ($\rho'/\rho = 9.4$), the density-mismatch effects dominate the elastic-mismatch effects. If this is true, then it should not matter whether the lead inclusions are theoretically treated as rigid, elastic, or fluid. In this context, we mention the work of Datta [26], who treated the problem of high-frequency wave propagation in an elastic solid containing a random distribution of spheres of an elastic *fluid*. He assumed $k_1'a$, $k_1a, k_2a \gg 1$. Even though he treats *fluid* inclusions, the comparison for $\langle C_1 \rangle$ between his theory and our experiments with lead/epoxy composites was found to be quite good [26].

The cutoff frequency of a composite, Ω_c, is roughly the same as the resonance frequency of a single sphere in an unbounded matrix, Ω_0, slightly modified by the concentration $\tilde{C}$. The starting point of the theory of Moon and Mow is the equation of motion of a rigid inclusion and this equation is exact. This may explain why the model predicts Ω_c so well. Furthermore, at one point in the derivation, all terms involving the density of the matrix are dropped under the assumption that the effective inertia of the matrix is negligibly small. This may explain why, for the relatively light inclusions in glass/epoxy composites, even Ω_c is not accurately predicted.

Acoustical and Optical Branches of Wave Propagation

In this sub-section we show that the model of Moon and Mow predicts, albeit qualitatively, all the salient features of the experiments; these features are summarized next.

At low frequencies, $\Omega \ll \Omega_c$, the particles move essentially in phase with the matrix and the wave propagation occurs along the familiar (or classical) acoustical branch. At high frequencies, $\Omega \gg \Omega_c$, the particles move essentially out of phase with the matrix and the wave propagation occurs along the optical branch.[3] The two branches are separated by Ω_c. (For the unrealistic case of a perfectly elastic or undamped system the phase difference along the acoustical and the optical branch is precisely 0 and π, respectively. For a real system with damping, there is a gradual transition from 0 to π across Ω_c.)

In the following we report our observation of the acoustical and optical branches of wave propagation in three disparate systems: (1) glass spheres/epoxy, (2) lead spheres/epoxy, and (3) steel spheres/PMMA.

Glass Spheres/Epoxy—Extensive results concerning this composite were presented in Ref *12*; here typical results for $\tilde{C} = 15\%$ are presented in Fig. 3. The acoustical branch is $\Omega < 2.5$, the optical branch is $\Omega > 3.75$, and we have elected to call the range $2.5 < \Omega < 3.75$ "the forbidden zone" because, here, we could not measure $\langle C_1 \rangle$ due to severe harmonic distortion (the limits are, of course, experimental).

In Fig. 3a, the point at $\Omega = 0$ is the static lower bound due to Hashin and Shtrikman [18]; the agreement of the theory [18] with our low-frequency data is quite good. As Ω increases, the wave propagation remains weakly dispersive up until $\Omega = 2.5$. Across the forbidden zone the phase velocity is seen to take a substantial positive jump. At high frequencies it appears that $\langle C_1 \rangle$ is approaching a high-frequency or optical limit. We now introduce a dimensionless wavenumber $\xi = \langle k_1 \rangle a$; thus, $\langle C_1 \rangle / C_1 = \Omega / \xi$. A plot of Ω versus ξ is shown in Fig. 3b. The data exhibit the following properties of the acoustical and optical branches. (1) The slope of the optical branch is significantly higher (about 30%) than that of the acoustical branch. (2) The velocity approaches a frequency-independent limit at high frequencies along the optical branch, and at low frequencies along the acoustical branch.

As noted above, the value of Ω_c as predicted by the Moon and Mow model was found to be in substantial disagreement with our "forbidden zone." This, how-

---

[3]There appears to be a fair amount of confusion concerning the use of the terms acoustical and optical. These appellations come about originally in the context of ionic crystals such as sodium chloride. Solution to the wave equation in such materials yields two branches. In the first, sodium and chlorine ions move in phase with each other; in the second, they move 180 deg out of phase with each other. The early physicist thought of the elastic (or acoustic) wave as a means of exciting the first (in phase) branch, hence the designation "acoustical," and of an electromagnetic (or optical) wave as a means of exciting the second (out of phase) branch, hence the designation "optical." For more details the reader is referred to the solid-state physics literature [22,23,25].

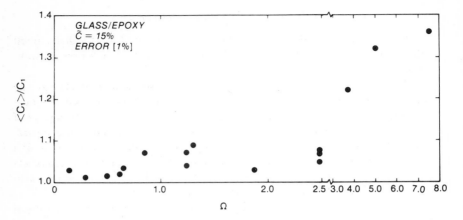

FIG. 3a—For $\Omega < 2.5$ the wave propagation is weakly dispersive; this is the slower, acoustical branch. For $2.5 < \Omega < 3.75$, velocity could not be measured; this is the forbidden zone. For $\Omega > 3.75$, velocity is very much higher; this is the faster, optical branch.

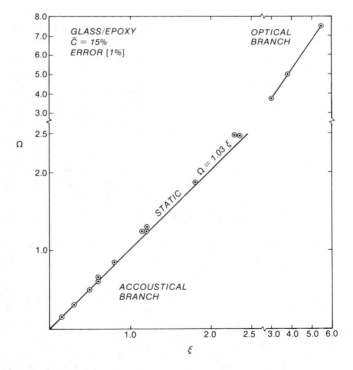

FIG. 3b—Acoustical and optical branches of wave propagation in the $\Omega - \xi$ plane. The straight line marked STATIC is the wavespeed calculated from the Hashin-Shtrikman bounds [18].

ever, is not at all disturbing because the theory assumes extremely heavy inclusions, that is, $\rho'/\rho \to \infty$, whereas for glass/epoxy ρ'/ρ is only about two.

We note that in addition to the zeropole (rigid-body-translation) resonance of a rigid inclusion studied in Ref 19, there are other resonances that will get excited in an elastic inclusion, for example, a monopole resonance or a dipole resonance. These have been studied by Überall and his co-workers (see, for example, Ref 20). These resonances are excited for $\Omega > 0.5$. Inclusion of these resonance effects into a theory for the aggregate behavior is critically important in order for the theory to be effective up to high frequencies. Gaunaurd and Überall [21] recently included the lowest of these resonances, namely, the monopole resonance (isochoric expansion and contraction, breathing in and out). Unfortunately, the effect on $\langle C_1 \rangle$ was found to be very small ($\sim 5\%$) compared with the experimental observations ($\sim 35\%$), indicating that the monopole effects in elastic inclusions are rather small. Based on our laboratory experience we offer an intuitive *conjecture* that the inclusion of the zeropole resonance (rigid-body-translation) will significantly improve the theory. At the present time there is no theory which quantitatively predicts the acoustical and optical branches of wave propagation.

Attention is now turned to the attenuation; results for the case of $\tilde{C} = 25.5\%$ are included in Fig. 4 over a broad range of frequencies: $0.7 \leq \Omega \leq 11$. The matrix attenuation is negligibly small compared with the scattering attenuation (recall that α is a logarithmic measure of the amplitude decay). The attenuation curve is characterized by some well-defined peaks. This is very much like the plot of normalized scattering cross section versus Ω; see, for example, Appendix H of Ref 24. The rapid changes in the velocity are accompanied by rapid changes in attenuation as expected on the basis of Kramers-Kronig relations derived by Weaver and Pao [26].

Lead Spheres/Epoxy—As mentioned above, one of the key assumptions of the Moon and Mow model is $\rho'/\rho \gg 1$, hence the choice of lead inclusions where ρ'/ρ is about ten; see Table 1. The dispersion data are shown in Figs. 5 for 5% concentration and in Figs. 6 for 15% concentration. The smallest lead balls available to us were of radius $a = 0.66$ mm; hence the paucity of data for small Ω (the smallest frequency tested was 0.17 MHz with a 0.25 MHz center-frequency homemade crystals). The wavespeed calculated from the Hashin-Shtrikman bound [18] is indicated by an arrow marked *STATIC* in Figs. $5a$ and $6a$. Evidently 0.17 is not a sufficiently small Ω; this is a rather surprising observation because for both glass/epoxy and steel/PMMA composites at similar values of the frequency, the comparison was noted to be excellent. Apparently, as Ω increases from zero, the velocity first decreases. It then takes a substantial positive jump across Ω_c (which is also indicated). For 5% concentration, the jump is accompanied by a small overshoot; for 15% concentration the overshoot is absent. (These phenomena are reminiscent of underdamped and overdamped second-order systems). For large Ω, $\langle C_1 \rangle / C_1$ appears to have reached a frequency-independent optical limit. Interestingly, the optical limit was

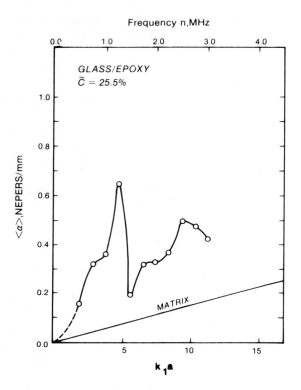

FIG. 4—*Attenuation versus frequency curve exhibits peaks which are most likely due to the excitation of the elastic resonances of the inclusions.*

found to be about 1.0 for all five $\tilde{C}$ tested, namely, 5%, 15%, 25%, 35%, and 52% (nominal). In other words, at high frequencies, the speed of sound is insensitive to the volume fraction of lead spheres in the epoxy. In the frequency-wavenumber plots, the broken lines are the hypothetical lines of extension for the unrealistic case of zero damping [9]. In reality the damping is quite large due to two sources: the visoelastic nature of the matrix, and attenuation due to scattering. This damping is the reason for the merger of the two branches as observed in our experiments. The straight lines marked *STATIC* are the predictions of Hashin and Shtrikman [18]. Unlike the case of glass/epoxy, here we *were* able to collect data connecting the acoustical and the optical branches; see Fig. 5b.

Note that Ω_c is deduced from the amplitude data. The transition from the acoustical to the optical branch is a phase-related phenomenon. Therefore, it is reassuring to note that the transition occurs approximately at Ω_c. This implies that the independent measurements of amplitude and phase corroborate each other.

Steel Spheres/PMMA—With reference to Table 1, PMMA is significantly less attenuative than epoxy. As will be seen in the sequel, reducing the matrix attenuation unmasks higher resonances (besides the zeropole resonance). The choice of steel inclusion was governed by the following considerations. For glass/epoxy $E'/E \gg 1$ whereas $\rho'/\rho > 1$; for lead/epoxy just the reverse is true, that is, $E'/E > 1$ whereas $\rho'/\rho \gg 1$; for steel spheres both the inclusion stiffness and the density are large, that is, $E'/E \gg 1$ and $\rho'/\rho \gg 1$. Thus the three systems represent extreme cases. Here, E is a measure of the elastic stiffness, for example, the Young's modulus.

The dispersion results are shown in Figs. 7. At low Ω, there is excellent agreement between the experiments and the lower (that is, the appropriate) bound of Hashin and Shtrikman [18]. In Fig. 7a this value of $\langle C_1 \rangle / C_1 (= 0.83)$ is marked by an arrow labelled *STATIC*, in Fig. 7b this is the slope of the *STATIC* curve, $\Omega = 0.83\xi$. The wave propagation is weakly dispersive up to about $\Omega = 0.6$ when the velocity takes a substantial jump; $\Omega < 0.6$ is the acoustical branch. The value of $\Omega_c (= 0.57)$ calculated by the model of Moon and Mow [19] is also indicated; note that the increase in $\langle C_1 \rangle$ begins to occur around Ω_c. Recall that in the case of lead inclusion the velocity became essentially constant after the jump. To the contrary, for the present case of steel inclusions, the wave propagation continues to be dispersive along the optical branch, that is, after the first

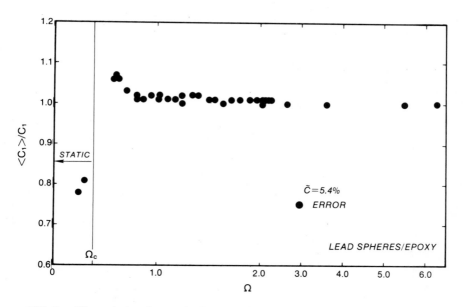

FIG. 5a—*Wave velocity takes a substantial jump across the cutoff frequency even though the concentration of inclusions is rather small. Surprisingly, $\langle C_1 \rangle$ becomes frequency-independent for $\Omega > 1$. Note change of scale at $\Omega = 2$. Lead spheres/epoxy. The straight line marked STATIC is the wavespeed of calculated from the Hashin-Shtrikman bounds [18].*

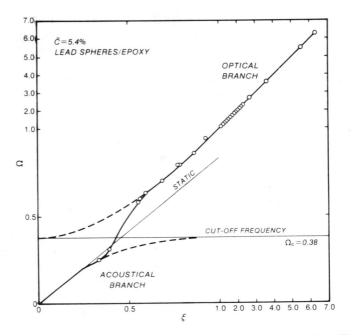

FIG. 5*b*—*Acoustical and optical branches of wave propagation in the* $\Omega - \xi$ *plane. The broken lines are the hypothetical lines of extension for the unrealistic case of an ideal elastic system with no damping. The damping causes the two branches to merge. Note change of scale at* $\Omega = 1$. *Lead spheres/epoxy. The straight line marked STATIC is the wavespeed calculated from the Hashin-Shtrikman bounds [18].*

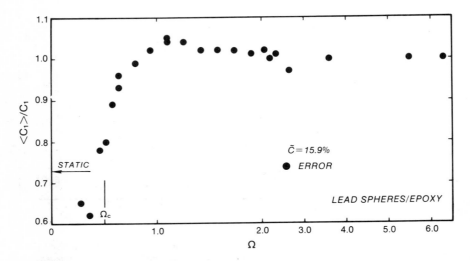

FIG. 6*a*—*Wave velocity takes a substantial jump across the cutoff frequency. Lead spheres/ epoxy. The straight line marked STATIC is the wavespeed calculated from the Hashin-Shtrikman bounds [18].*

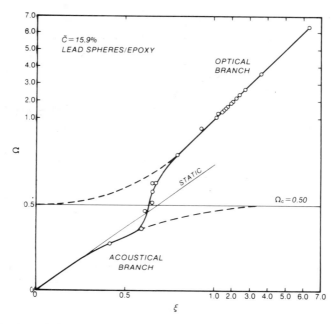

FIG. 6b—*Acoustical and optical branches of wave propagation in the $\Omega - \xi$ plane. Lead spheres/epoxy. The straight line marked STATIC is the wavespeed calculated from the Hashin-Shtrikman bounds [18].*

jump. In particular, there are two additional positive jumps at $\Omega = 1.2$ and $\Omega = 1.5$. A plausible explanation is now given. Flax and Überall [20] have studied the excitation of *all* resonances of a single elastic sphere in an elastic matrix. They showed that these resonances begin to get excited for $\Omega > 0.5$ (their numerical values are material-specific). We conjecture that the dispersion along the optical branch may be due to the excitation of these resonances. That the phenomenon was absent for the case of lead inclusions is rather surprising and may possibly be due to the following two important differences: (1) Lead is heavy, hence one may expect the zeropole (rigid-body-translation) resonance effects to be the dominant one; and (2) the fact that PMMA is much less attenuative than epoxy may make the higher resonances (and their influence on the overall wave propagation behavior) relatively more pronounced.

Returning to Figs. 7, at very high Ω (≈ 3.0), $\langle C_1 \rangle / C_1$ goes to 1.0, which is just like the case of (heavy) lead inclusions, but unlike the case of (light) glass inclusions.

With reference to Fig. 7b we make an interesting observation. Near $\Omega = 1.2$ note that $d\Omega/d\xi < 0$; that is, even though the *phase velocity is positive* and undergoes a small (innocuous) jump, the *group velocity is negative*. Put another way, two different frequencies produce the same wavenumber ξ.

The subject is now changed to attenuation, see Fig. 8. The volume fraction of

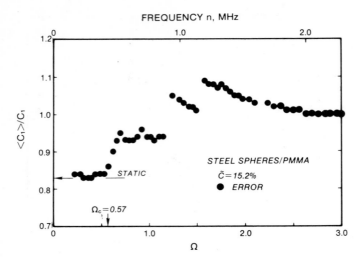

FIG. 7a—*Wave velocity takes a substantial jump across* Ω_c. *In addition, there is fairly strong dispersion for* $\Omega > \Omega_c$. *As* $\Omega \to 3.0$, $\langle C_1 \rangle / C_1 \to 1.0$. *The straight line marked STATIC is the wavespeed calculated from the Hashin-Shtrikman bounds [18].*

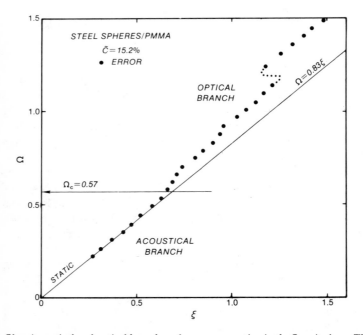

FIG. 7b—*Acoustical and optical branches of wave propagation in the* $\Omega - \xi$ *plane. The group velocity is negative near* $\Omega = 1.2$. *The straight line marked STATIC is the wavespeed calculated from the Hashin-Shtrikman bounds [18].*

inclusions $\tilde{C} = 15.2\%$, as in Figs. 7. The matrix attenuation is negligibly small compared with the scattering attenuation. Therefore, this choice of constituents closely models the theoretically treated case of elastic inclusions in an elastic matrix. Data collected by two *independent* techniques, namely, toneburst (or steady time-harmonic motion) and spectroscopy (or transient response) may be seen to corroborate each other within the 1% error of measurement. The most interesting feature of this data is that they are characterized by a series of peaks. In keeping with the calculations of scattering cross section of an elastic sphere, these peaks occur for $\Omega > 0.5$ [24]. The rapid fluctuations in attenuation are accompanied by changes in velocity; this is consistent with what one might expect on the basis of Kramers-Kronig relations [26]. The first peak, at $\Omega = 0.84$, is most probably due to zeropole (rigid-body-translation) resonance. The subsequent peaks are probably due to the excitation of the higher resonances, for example, monopole, dipole, and so forth [20–24].

Conclusions

Ultrasonic wave propagation up to k_1a (or Ω) = 10 has been studied experimentally in three disparate systems: glass/epoxy, lead/epoxy, and steel/PMMA. At low frequencies the speed of wave propagation is predicted ade-

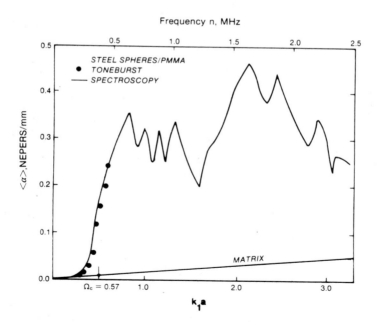

FIG. 8—*Attenuation versus frequency. The peaks are most probably due to the excitation of the particle resonances. The theoretically calculated cutoff frequency, Ω_c, is indicated by an arrow. The attenuation of the neat matrix is indicated by the straight line marked MATRIX.*

quately by the classical static bounds. *At high frequencies, however, a faster branch of wave propagation is excited which, so far as we know, has been experimentally observed for the first time.* Following the literature in solid-state physics, the slower branch has been labeled the acoustical branch, and the faster branch has been labeled the optical branch. Along the acoustical and optical branch, respectively, the inclusions move in phase and out of phase with the excitation. The two branches are separated by a cutoff frequency; this is the damped natural frequency of the inclusions in their rigid-body-translation mode. The measured values of the cutoff frequency were found to be in good agreement with the predictions of an elementary model by Moon and Mow across the entire range of volume fraction of inclusions. At the cutoff frequency the attenuation goes through a maximum. *So far as we know this is also the first experimental observation of the cutoff frequency.*

These phenomena were observed in all three systems tested.

Acknowledgment

We gratefully acknowledge the financial support of the National Science Foundation under Grants ENG 78-10168 and ENG 78-10869 (Program Director, Dr. Clifford J. Astill). Thanks are due to Dee Meier for her immense patience in carefully preparing the manuscript.

References

[1] Mura, T., *Micromechanics of Defects in Solids*, Martinus Nijhoff, The Hague-Boston, 1982.

[2] Christensen, R. M., "Mechanical Properties of Composite Materials" in *Mechanics of Composite Materials, Recent Advances, Proceedings*, 1st International Union of Theoretical and Applied Mechanics Symposium, Z. Hashin and C. T. Herakovich, Eds., Pergamon Press, New York, 1983.

[3] McCoy, J. J., "Microscopic Response of Continua with Random Microstructures" in *Mechanics Today*, Vol. 6, S. Nemat-Nasser, Ed., Pergamon Press, New York, 1981, pp. 1–40.

[4] Cleary, M. P., Chen, I.-N, and Lee, S.-M, "Self-Consistent Techniques of Heterogeneous Media," *Journal of the Engineering Mechanics Division* of the American Society of Civil Engineers, Vol. 106, 1980, p. 861.

[5] Sayers, C. M. and Smith, R. L., "The Propagation of Ultrasound in Porous Media," Harwell Report AERE-R-10359, Atomic Energy Research Establishment, Harwell, Oxfordshire, U.K., Nov. 1981.

[6] Beltzer, A. I., Bert, C. W., and Striz, A. G., "On Wave Propagation in Random Particulate Composites," *International Journal of Solids and Structures* (to appear).

[7] Talbot, D. R. S. and Willis, J. R., "Variational Estimates for Dispersion and Attenuation of Waves in Random Composites—I. General Theory, and II. Isotropic Composites," *International Journal of Solids and Structures*, Vol. 18, No. 8, 1982, pp. 673–698.

[8] Gubernatis, J. E. and Domany, E., "Effects of Microstructure on the Speed and Attenuation of Elastic Waves" in *Proceedings of the Review of Progress in Quantitative NDE*, D. Thomson, Ed., Plenum Press, New York, 1983 (to appear).

[9] Nemat-Nasser, S., Iwakuma, T., and Hejazi, M., "On Composites with Periodic Structures," *Mechanics of Materials*, Vol. 1, 1982, pp. 239–267.

[10] Nomura, S., "Statistical Aspects of Heterogeneous Materials" in *Proceedings*, Fourth International Conference on Composite Materials, Vol. II, T. Hayashi et al, Eds., Japan Society of Composite Materials, 1982, pp. 1083–1089.

[*11*] Kinra, V. K., Petraitis, M. S., and Datta, S. K., "Ultransonic Wave Propagation in a Random Particulate Composite," *International Journal of Solids and Structures*, Vol. 16, 1980, pp. 301–312.

[*12*] Kinra, V. K. and Anand, A., "Wave Propagation in a Random Particulate Composite at Long and Short Wavelengths," *International Journal of Solids and Structures*, Vol. 18, 1982, pp. 367–380.

[*13*] Kinra, V. K., Ker, E. L., and Datta, S. K., "Influence of Particle Resonance on Wave Propagation in a Random Particulate Composite," *Mechanics Research Communications*, Vol. 9, 1982, pp. 109–114.

[*14*] Kinra, V. K. and Ker, E. L., "Effective Elastic Moduli of a Thin-Walled Glass Microsphere/PMMA Composite," *Journal of Composite Materials*, Vol. 16, 1982, pp. 117–138.

[*15*] Kinra, V. K. and Ker., E. L., "An Experimental Investigation of Pass Bands and Stop Bands in Two Periodic Particulate Composites," *International Journal of Solids and Structures*, Vol. 19, No. 5, 1983, pp. 393–410.

[*16*] Sayers, C. M., "On the Propagation of Ultrasound in Highly Concentrated Mixtures and Suspensions," *Journal of Physics D: Applied Physics*, Vol. 13, 1980, pp. 179–184.

[*17*] Fu, L. S. and Shue, Y. C., "Ultrasonic Wave Propagation in Two-Phase Media: Spherical Inclusions," Unpublished NASA Contractor Report, June 1983.

[*18*] Hashin, Z. and Shtrikman, S., "A Variational Approach to the Theory of the Elastic Behavior of Multiphase Materials," *Journal of the Mechanics and Physics of Solids*, Vol. 11, 1963, pp. 127–140.

[*19*] Moon, F. C. and Mow, C. C., "Wave Propagation in a Composite Material Containing Dispersed Rigid Spherical Inclusions," Rand Corp. Report, RM-6139-PR, Rand, Santa Monica, CA, 1970.

[*20*] Flax, L. and Überall, H., "Resonant Scattering of Elastic Waves From Spherical Solid Inclusions," *Journal of the Acoustical Society of America*, Vol. 67, 1980, pp. 1432–1442.

[*21*] Gaunaurd, G. and Überall, H., "Dependence on Ultrasonic Resonances of the Effective Properties of Composites" in *Proceedings*, Ninth U.S. National Congress on Applied Mechanics, Cornell University, Ithaca, NY, June 1982, p. 484.

[*22*] Kittel, C., *Introduction to Solid State Physics*, Wiley, New York, 1968, p. 148.

[*23*] Maradudin, A. A., Montroll, E. W., Weiss, G. H., and Ipatova, I. P., "Theory of Lattice Dynamics in the Harmonic Approximation," 2nd ed., Supplement 3 of *Solid State Physics. Advances in Research and Applications*, H. Ehrenreich, F. Seitz, and D. Turnbull, Eds., Academic Press, New York, 1971, p. 21.

[*24*] Truell, R., Elbaum, C., and Chick, B. B., *Ultrasonic Methods in Solid State Physics*, Academic Press, New York, 1969.

[*25*] Brillouin, L., *Wave Propagation in Periodic Structures*, Dover, New York, 1953, p. 15.

[*26*] Weaver, R. L. and Pao, Y.-H., *Journal of Mathematical Physics*, Vol. 22, 1981, p. 1909.

[*27*] Datta, S. K., "Wave Propagation in the Presence of a Random Distribution of Inclusions" in *Continuum Models of Discrete Systems*, University of Waterloo Press, Waterloo, Ont., Canada, 1980.

Christos C. Chamis[1] and John H. Sinclair[1]

Impact Resistance of Fiber Composites: Energy-Absorbing Mechanisms and Environmental Effects

REFERENCE: Chamis, C. C. and Sinclair, J. H., "**Impact Resistance of Fiber Composites: Energy-Absorbing Mechanisms and Environmental Effects,**" *Recent Advances in Composites in the United States and Japan, ASTM STP 864*, J. R. Vinson and M. Taya, Eds., American Society for Testing and Materials, Philadelphia, 1985, pp. 326–345.

ABSTRACT: Energy-absorbing mechanisms are identified and evaluated using several approaches. The energy-absorbing mechanisms considered are those in unidirectional composite beams subjected to impact. The approaches used include mechanistic models, statistical models, transient finite-element analysis, and simple beam theory. Predicted results are correlated with experimental data from Charpy impact tests. The environmental effects on impact resistance are also evaluated. Working definitions for energy-absorbing and energy-releasing mechanisms are proposed and a dynamic fracture progression is outlined. Possible generalizations to angleplied laminates are described.

KEY WORDS: composite mechanics, predictive models, mechanistic, statistical, finite element, transient analysis, dominant mechanisms, energy absorption, energy release, correlation, fracture initiation, fracture progression, generalization, definitions

Nomenclature

C Unknown coefficients in statistical model; subscripts 1 to 5 correspond to five terms in equation

d_f Fiber diameter

E Modulus

E_P Undegraded modulus

G Shear modulus

h Beam thickness

IED Impact energy density (in. $\cdot$ lb/in.^{3} or ft $\cdot$ lb/ft^3)

[1]National Aeronautics and Space Administration, Lewis Research Center, Cleveland, OH 44135.

k Volume ratio
ℓ Length
N Number of individual layers
$N_{\ell D}$ Number of delaminated layers
P Property subscripts define reference
S Strength, subscripts define material type and direction and sense
T Temperature subscripts define reference
V_{DI} Local damage volume
W Work done, beam width
X, Y, Z Structural axis coordinates
$1, 2, 3$ Material axis coordinates
δ Beam deflection
σ Stress; subscripts denote type, direction and sense

Superscripts

a Average through-the-thickness
s Surface ply properties

Subscripts

C Compression
c Core property, composite property
F Flexural
FC Composite flexure
FN Individual beam layer flexure
f Fiber property
fd Fibers pulled out
gd Glass transition, dry
gw Glass transition, wet
H Hygrothermally degraded property
ℓ Ply property
N Individual layer beam
S Shear
SC Composite shear
T Tension
0 Reference property
$1, 2, 3$ Material axis coordinate directions

Conversion Factors of Units used in Text

1 ft · lb   = 1.35 Joules (J)
1 in.   = 2.54 cm
1 in.^{2} = 6.5 cm^2

$$1 \text{ in.}^3 \quad = 16.4 \text{ cm}^2$$
$$1 \text{ in.} \cdot \text{lb} \quad = 0.112$$
$$1 \text{ ksi} \quad = 6.9 \text{ MPa}$$
$$1 \text{ lb} \quad = 4.45 \text{ N}$$
$$1 \text{ lb/in.}^3 \quad = 27.7 \text{ gm/cm}^3$$
$$1 \text{ psi} \quad = 6.9 \text{ kPa}$$

Fiber composites have several attractive design attributes for use in structures and structural components. In addition to the well-known high strength and stiffness-to-density ratios, fiber composites exhibit high tensile fatigue resistance, notch insensitivity, ease of fabrication, and low scrap rate. Furthermore, fiber composites provide the unique opportunity to simultaneously optimize structural configuration, material makeup, fabrication process, and structural integrity. On the other hand, fiber composites have some disadvantages which include complex constitutive relationships, coupled structural responses, relatively high material cost, moisture degradation, and low impact resistance. Mechanisms contributing to low impact resistance are receiving considerable research attention [1–4].

There is still no general consensus in the composites community as to what constitutes energy-absorption mechanisms relative to impact. The lack of consensus arises, in part, from the difficulty of identifying and quantifying the multitude of mechanisms in which composites fail and through which composites absorb energy during impact. This multitude of failure or energy-absorption mechanisms includes micromechanics mechanisms, such as fiber and matrix fractures, intra- and interlaminar shear delaminations, fiber pull-out, and local indentation. The difficulty in achieving a consensus is further compounded by a school of thought which advocates that the energy release rates of fracture surfaces associated with fracture modes I, II, and III are related to some energy-absorption mechanisms for composite impact resistance.

The objective of this paper is to evaluate the significance of various mechanisms contributing to impact resistance using mechanistic models, statistical models, transient finite-element analysis, and simple beam analysis associated with the Charpy impact tests and data. The hygrothermal effects on composite impact resistance are also evaluated.

Mechanistic Model

The mechanistic model for the assumed energy-absorption mechanisms was derived using elementary beam theory in conjunction with the composite mechanics procedures described in Refs 1 and 5. The beam schematic used, the boundary conditions, and the resulting equation for impact energy density (IED) are summarized in Fig. 1. The coefficients representing the energy-

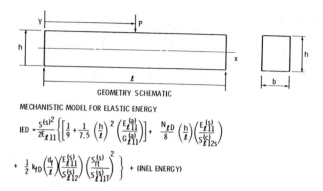

FIG. 1—*Simple beam schematic and mechanistic model.* (Note: all dimensions *in inches*).

absorption mechanisms included in the IED equation (assuming a unidirectional composite with fibers aligned parallel to the x-axis) are as follows:

1. Area under the stress strain curve, $S^{(s)2}_{\ell 11T}/2E_{\ell 11}$, where S^s_ℓ represents the tensile strength on the surface ply, E_ℓ the corresponding modulus, numerical subscripts 11 the direction, and subscript T the tension.

2. Flexural bending, 1/9.

3. Interlaminar shear, $(1/7.5) \times (h/\ell)^2 \times (E^{(a)}_{\ell 11}/G^{(a)}_{\ell 11})$, where the superscript a denotes integrated averages for E and G and where $G_{\ell 11}$ is the interlaminar shear modulus. (The interlaminar shear modulus equals the intralaminar shear modulus for undirectional composites.)

4. Delamination, $(N_{\ell D}/8) \times (h/\ell) \times (E^{(s)}_{\ell 11}/S^{(c)}_{\ell 12 S})$, where $N_{\ell D}$ denotes the number of delaminated layers and $S^{(c)}_{\ell 12 S}$ denotes the interlaminar shear strength of the respective delaminated layer at the core of the beam.

5. Fiber pullout, $(1/2) (k_{fD}) \times (d_f/\ell) \times (E^{(s)}_{\ell 11}/S^{(s)}_{\ell 12 S}) \times (S^{(s)}_{fT}/S^{(s)}_{\ell 11T})^2$, where k_{fD} denotes the fiber pullout volume ratio, d_f the fiber diameter, superscript s the surface plies or plies failing by longitudinal tension, and S_{fT} the respective fiber tensile strength.

6. Inelastic energy (INEL ENERGY) associated with transverse splitting and local indentation is roughly proportional to $V_{DI}S^2_{\ell 22T}/E_{\ell 22}$ where V_{DI} is the local damage volume, $S_{\ell 22T}$ the transverse tensile strength, and $E_{\ell 22}$ the undirectional transverse modulus. The V_{DI} and proportionality constant can be determined using an available contact law [6,7]. This term may also be expressed as $2\alpha h S^2_{\ell 22C}E^{(s)}_{\ell 11}/\ell E_{\ell 22}S^{(s)2}_{11T}$ where α is the ratio of the indentation area δw to the square of the beam thickness h^2 or $\alpha = \delta w/h^2$.

The energy-absorbing mechanisms included in the mechanistic model are those mechanisms which are assumed herein to absorb impact energy prior to any impact load drop (from a current level) that would be observed in the impact load versus time trace. This is a reasonable assumption and is consistent with energy absorbed prior to initial or subsequent damage. Any additional energy absorbed subsequent to initial damage would be indicated by corresponding impact load rise versus time. On the other hand, constant or decreasing impact load versus time indicates release of energy already stored in the composite beam rather than energy absorbed. Each term in the mechanistic model can be further expressed in terms of constituent material properties using composite micromechanics [1,5,8].

It is instructive to note that each term in the mechanistic model, including the inelastic energy (item 6) but excluding the flexural, includes beam length ℓ in the denominator. This indicates that for relatively long contact-time impact events the flexural term will present the dominant energy absorbing mechanism since ℓ is large compared with the other dimensions. Long contact-time impact events are those in which the shear waves (x,y-plane, Fig. 1) traverse the beam span several times prior to the separation of the projectile from the target. On the other hand, for relatively short contact-time events, the impact resistance is highly localized and the inelastic energy term represents the dominant energy-absorbing mechanism.

Predicted results from the mechanistic model were correlated with Charpy impact data [9]. The geometry of the Charpy test specimen is depicted schematically in Fig. 2. The constituent material parameters and properties data, used for the correlation, are summarized in Table 1. These data are for five different composite systems ranging from ultra-high-modulus graphite fiber (T75/EP) to S-glass/epoxy (S-GL/EP). The correlation results for two composite systems are summarized in Table 2 for each energy-absorbing mechanism term in the mechanistic model excluding the inelastic term. This term was excluded from the summary for two reasons: (1) unavailability of experimental data (mainly due to difficulty in measuring such data) on each local indentation, and (2) insignificant contribution to the Charpy impact energy compared with the contributions of other terms. For example, if a local indentation one ply deep by one ply wide across the width of the specimen is

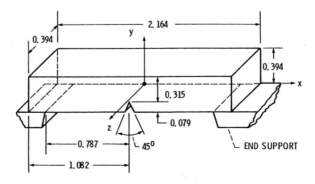

FIG. 2—*Charpy test specimen schematic. (Dimensions in inches; 1 in. = 2.54 cm).*

TABLE 1—*Constituent material and composite properties.*

Property and Units	Composite					
	Mod-II/EP	Mod-I/EP	T75/EP	B/EP	S-GL/EP	KEV/EP
d_f—Fiber diameter, in.	0.0003	0.0003	0.0003	0.004	0.00036	0.00045
S_{ft}—Fiber tensile strength, ksi	350	250	345	460	670	400
FVR—Fiber volume ratio	0.55	0.53	0.59	0.55	0.59	0.63
$E_{\ell 11}$—Composite longitudinal modulus, 10^6 psi	18.9	31.3	39.2	31.5	7.2	13.2
$G_{\ell 12}$—Composite shear modulus, 10^6 psi	0.66	0.64	0.48	0.96	0.83	0.27
$S_{\ell 11T}$—Composite longitudinal strength, ksi	161	129	154	222	222	170
$S_{\ell 12C}$—Composite SBS strength, ksi	10.3	8.4	7.8	8.5	14	6.9
CIE—Charpy impact energy, ft-lb	6.5	2.4	3.0	8.5	35	13.0

NOTE: Composite properties were generated by Pratt and Whitney Aircraft under NASA Contract NAS 3-15568.

Conversion factors:
1 in. = 2.54 cm
1 psi = 0.69 N/cm^2 = 6.9 kPa
1 ksi = 6.9 MPa
1 ft · lb = 1.35 Joules (m · N)

assumed, the energy absorbed due to this indentation is about 0.000112 J (0.001 in. · lb) for HM/EP composites. This is only 1% of that due to flex [0.012432 J (0.111 in. · lb)], which is the smallest contribution in Table 2.

The following significant points are observed from the results in Table 2:

1. The projected combined contribution of all energy-absorbing mechanisms is too high compared with measured values for both composite systems (about 240 times higher for HM/EP and 30 times higher for S-GL/EP).

2. The contributions due to flex and shear mechanisms correlate almost exactly with the measured value for the HM/EP composite.

3. The flex and shear mechanism contributions are about one-third the measured value for the S-GL/EP composite.

4. The interply delamination mechanism contribution is excessively high, even for one delaminated layer, for both composite systems compared with the measured values.

5. The fiber-pullout mechanism contribution is about 40% of the flex-plus-shear contribution for the HM/EP composite and about 120% of the corresponding contribution for the S-GL/EP composite.

Based on the correlation of the mechanistic model results with Charpy impact data, the following conclusions are made:

1. Flex and shear deformations appear to be energy-absorbing mechanisms.

TABLE 2—*Mechanistic model-experimental data correlation.*

$$\left.\begin{array}{l}\text{Impact}\\\text{energy}\\\text{density}\\\text{(IE/V)}\end{array}\right\} = \underbrace{\frac{S_{\ell11T}^{(s)2}}{2E_{\ell11}^{(s)}}}_{\text{AUC}}\left\{\underbrace{\frac{1}{9}}_{\text{FLEX}} + \underbrace{\frac{1}{7.5}\left(\frac{h}{\ell}\right)^2\left(\frac{E_{\ell11}^{(a)}}{G_{\ell11}^{(a)}}\right)\right]}_{\text{SHEAR}} + \underbrace{\frac{N_{\ell D}}{8}\left(\frac{h}{\ell}\right)\frac{E_{\ell11}^{(s)}}{S_{\ell12s}^{(c)}}}_{\text{DEL}}\right.$$

$$+ \underbrace{\frac{1}{2}k_{fD}\left(\frac{d_f}{\ell}\right)\left(\frac{E_{\ell11}^{(s)}}{S_{\ell12s}^{(s)}}\right)\left(\frac{S_{ft}^{(s)}}{S_{\ell11T}^{(s)}}\right)^2}_{\text{FPO}}\left.\right\}$$

Term	Units	Composite	
		HM/EP[a]	S-GL/EP
Measured impact energy	ft-lb	2.4	35
Volume, V	in.^{3}	0.220	0.220
Area under curve, AUC	in.^{2}	266	3420
Flex	...	0.111	0.111
Shear	...	0.396	0.072
Delamination, DEL (I-layer)	...	116.6	16.1
Fiber pullout, FPO	...	0.207	0.211
Computed impact energy (all terms)	ft-lb	572	1034
Computed impact energy (flex and shear)	ft-lb	2.47	11.47

[a] Mod-I/EP, Table 1. Conversion factors: 1 ft · lb = 1.35 Joules; 1 in. = 2.54 cm; 1 in.^{2} = 6.5 cm^2; 1 in.^{2} = 16.4 cm^3.

2. Interply delamination and fiber pullout do not appear to be energy-absorbing mechanisms.

3. Advanced composites usually fail in a brittle manner under impact once the outer plies have failed in tension or compression.

4. S-GL/EP composites most likely fail by flexure of progressively delaminated layers (individual beams). This conclusion is supported additionally by the following example. If it is assumed that all the fibers fail in tension above the notch in the Charpy specimens and that a parabolic distribution of stress exists along the specimen span from the notch to the supports (proportional to an approximate specimen displacement shape), the predicted impact energy is about 49.2885 J (36.51 ft. · lb) compared with the measured value of 47.25 J (35 ft · lb). This is *too* close to be purely coincidental. Furthermore, this conclusion is consistent with the high-speed movie observations [2].

5. The area under the unaxial tension/compression stress-strain curve appears to be a reliable quantity for assessing composite energy absorbed under impact, which is consistent with findings from previous investigations [1,2].

Statistical Model

The energy-absorption mechanisms were also evaluated using statistical methods in conjunction with multiple linear regression. The statistical model used consists of five terms, which are the same as in the mechanistic model (Fig. 1), but each term has an unknown coefficient. The contribution of the inelastic energy due to local indentation is included in this model.

The experimental data used to evaluate the unknown coefficients are summarized in Table 3 [9]. These data include maximum and minimum values for advanced composites, S-glass composites, and two intraply hybrids. At least two specimens were used for each strength or modulus property, and five specimens were used in cases where considerable scatter was observed. Most of the data show a difference of about 10% between maximum and minimum values, indicating relatively small scatter from specimen to specimen. Noticeable exceptions are the data for tensile strength of the Modmor II at 47% fiber volume and for Charpy impact of the Modmor II at 65% fiber volume. Though these data may appear insufficient for statistical analysis, their use is considered justifiable for three reasons:

1. The data scatter is relatively small, indicating composite uniformity and consistent properties.

2. The statistical model used is derivable from the physics of the problem using composite mechanics.

3. The statistical analysis is pursued primarily to establish the relative significance of each mechanism and thereby to discriminate among energy-absorbing and other mechanisms in a formal, unbiased manner.

The specific statistical analysis used [10] is in the form of a computer code and is available at the Lewis Computer Center. Inputs to this code are the appropriate data from Table 3 (except the data for the two intraply hybrids) and the statistical model. The outputs include values of the unknown coefficients, the coefficient of correlation, the level of confidence, and the significant (dominant) term retained by the multiple regression. It is noted that the level of confidence is *not* considered an important parameter in this statistical analysis since the emphasis is on the relative significance of the contribution of each energy-absorbing-mechanism term and *not* on absolute values.

The form of the statistical model, the dominant term retained by the multiple regression, and the correlation coefficient are shown in Fig. 3. Only the flex (elastic bending) term coefficient C_1 was retained with a correlation value of 0.994, which is considered very good. This C_1-value was used to predict the Charpy impact energy for all the composites used in the statistical analysis. It was also used to predict the Charpy impact energy for the two intraply hybrids which were not used in the analysis. The comparisons with measured data are summarized in Table 4. The comparisons are excellent.

The most significant conclusion from the statistical analysis is that flexure

TABLE 3—Experimental data for statistical model.

Fiber	Mod-II/E	Mod-II/E	Mod-II/E	Mod-I/E	T-75/E	B/E	S-GL/E	KEV-49-E	S-GL/E// T-75/E	KEV-49/E// T-75/E
Avg. composite density, lb/in.³	0.054	0.056	0.053	0.058	0.058	0.075	0.071	0.049	0.063	0.052
Avg. void content	1.0	1.4	1.2	1.7	0.3	1.1	0.8	1.9	1.0	
Avg. fiber volume %	55	65	47	53	59	55	59	63	31 (S-Glass) / 22 (T-75-S)	39 (PRD-49-1) / 17 (T-75-S)
Longitudinal tensile strength, 10³ psi	161[b]	149	130[a]	129	154	222	222	170	107	104
	146[a]	143	97[a]	146	142	205	216	154	107	101
Longitudinal tensile modulus, 10⁶ psi	18.9[b]	20.1	18.2[a]	31.3	39.2	31.5	7.2	13.2	19.5	18.0
	19.6[a]	19.7	15.2[a]	32.8	38.2	32.9	7.3	13.7	19.6	17.8
Transverse tensile strength, 10³ psi	5.8	6.3	7.4	5.0	4.5	9.3	14	3.7	5.6	2.2
	6.0	5.6	7.3	5.1	3.8	7.2	12	2.4	6.1	2.8
Transverse tensile modulus, 10⁶ psi	1.46	1.52	1.42	1.01	0.77	4.20	3.21	0.91	2.10	0.86
	1.34	1.43	1.27	0.97	0.82	3.80	3.28	0.86	2.05	0.83
Beam shear strength (L/D=4:1), 10³ psi	10.3	11.5	12.1	8.4	7.8	15.3	14.0	6.9	3.3	4.5
	9.4	12.0	11.8	8.2	7.8	14.7	12.6	6.8	3.2	4.5
Charpy impact strength, ft-lb	6.0	4.0	2.4	2.4	3.0	8.5	35.0	13.0	12.9	7.8
	7.0	2.9	2.3	2.7	3.0	8.5	35.0	13.0	12.1	8.6
Torsion rod shear strength, 10³ psi	10.0	10.1	11.3	7.2	7.3	12.8	16.7	5.0	3.6	3.3
	10.8	13.3	11.3	6.8	7.0	13.2	17.1	4.0	3.6	3.7
Shear modulus, 10⁶ psi	0.660	0.762	0.620	0.640	0.481	0.956	0.830	0.267	0.686	0.349
	0.680	0.835	0.697	0.640	0.451	1.03	0.900	0.261	0.762	0.395

[a]Average of two tests from one panel.
[b]Average of five tests from one panel.
Conversion factors:
$\quad$ 1 lb/in.³ = 27.7 gm/cm³
$\quad$ 1 psi = 6.9 kPa
$\quad$ 1 ft·lb = 1.35 Joules

COMPOSITE MECHANICS MODEL

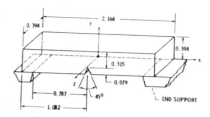

$$\begin{Bmatrix} \text{CHARPY} \\ \text{IMPACT} \end{Bmatrix} = C_1 \frac{S_{\ell 11T}^2}{E_{\ell 11}} + C_2 \frac{S_{\ell 11T}^2}{G_{\ell 12}} + C_3 \frac{S_{\ell 11T}^2}{S_{\ell 12s}} + C_4 \frac{S_{\ell 11T}^2}{k_f^2 S_{\ell 12s}} + C_5 \frac{S_{\ell 22T}^2}{E_{\ell 22}}$$

ELASTIC BENDING	ELASTIC SHEAR	INTERPLY DELAMINATION	FIBER PULL-OUT	TRANSV. SPLIT + LOCAL INDENT

ABOVE MODEL WITH REGRESSION ANALYSIS AND WITH EXPERIMENTAL DATA RANGING FROM HM/EP TO S-GL/EP YIELDED

CHARPY IMPACT$| = 5.3 \times 10^{-3} S_{\ell 11T}^2 / E_{\ell 11}$ (ft-lb)
WITH A CORRELATION COEFFICIENT .994

CONVERSION FACTOR: 1 ft-lb = 1.35 Joules

FIG. 3—*Simple beam schematic statistical model and correlation.*

is the dominant energy-absorbing mechanism in Charpy impact. The contributions of the remaining mechanisms in the statistical model are relatively insignificant and therefore are excluded from being energy-absorbing mechanisms. Additionally, the statistical value of 0.0053 for C_1 is about five times the corresponding coefficient of 0.00103 for the mechanistic model (Table 2) in terms of foot · pound. This indicates that flexural fracture of progressively delaminated layers occurs, which is consistent with the conclusion for S-glass/epoxy composite fracture progression mentioned previously. However, it is not consistent with the previous conclusions for the HM/EP (Table 2). Apparently, the interlaminar shear contribution for the HM/EP composite, and similar composites, appears to be (1) relatively significant if the impact-induced fracture is primarily brittle, or (2) insignificant if the impact-induced fracture is flexural in a progressively delaminated layer. This latter case represents independent beam behavior of each delaminated layer as will be described later.

Transient Finite-Element Analysis

Transient finite-element analysis of the Charpy impact specimen was performed to assess the possibility of flexural failure of progressively delaminated layers. NASTRAN was used for the finite-element analysis. The re-

TABLE 4—*Correlation of statistical model with experimental data.*

Composite	UDC Longitudinal Properties		Charpy Impact, ft · lb	
	Strength, ksi	Modulus, 10^6 psi	Measured	Predicted
Mod-II/EP	153	19.2	6.5	6.5
Mod-I/EP	129	31.3	2.4	2.8
T75/EP	148	38.7	3.0	3.0
B/EP	222	31.5	8.5	8.2
S-GL/EP	219	7.2	35.0	35.0
KEV/EP	170	13.2	13.0	11.6
Mod-II/POLY	183	21.3	6.5	8.4
S-GL/E, T75/E, S-GL/EP	[a]219	[b]19.5	12.5	13.0
KEV/E, T75/E, KEV/EP	[a]170	[b]17.9	8.2	8.6

[a]Composite shell strength.
[b]Hybrid composite modulus.
Conversion factors:
  1 ksi= 6.9 MPa
  1 psi = 6.9 kPa
  1 ft · lb = 1.35 Joules

quired input was obtained from the experimental data (Tables 1 and 3). The interlaminar shear-stress variation in the specimen due to a static load was also determined as a part of the transient analysis. The transient analysis was performed using the direct time integration capability of NASTRAN. The transient analysis was performed by assuming that the specimen remains intact during the impact event. This assumption is reasonable since the interest was on determining the interlaminar shear stress magnitude prior to initial damage. Transient analysis including progressive damage is under current study and will be reported in the future.

The finite-element model and the interlaminar shear-stress contour plot for KEV/EP are summarized in Fig. 4. The static load of 8054.5 N (1810 lb) for the interlaminar shear-stress contour plot was selected to produce tensile stress fracture at the notch tip [11]. The interlaminar shear strength for KEV/EP is about 48.3 MPa (7 ksi). This strength is exceeded by the interlaminar shear stresses in most of the specimen (Fig. 4), indicating a high probability of progressive delamination. The other composite systems exhibit similar behavior [11].

Transient finite-element analysis results for interlaminar shear stress for the S-GL/EP composite are summarized in Fig. 5. The load-time trace [9] used for the impact force in the transient analysis and the static interlaminar shear strength are shown in the figure. The transient interlaminar shear stress exceeds the corresponding static strength near the impact and notch tip regions at early times of the impact event. This will cause delaminations in these regions and corresponding load decreases in the load-time trace. Additional delaminations will occur toward the center of the specimen as the impact load

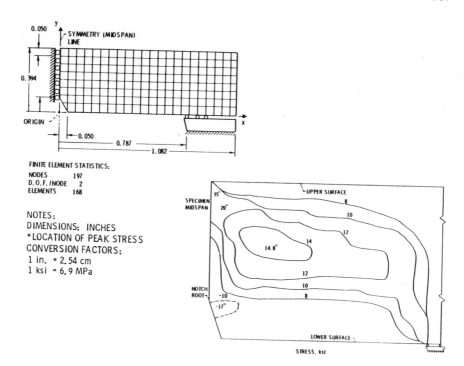

FINITE ELEMENT STATISTICS:
NODES   197
D.O.F. /NODE   2
ELEMENTS   168

NOTES:
DIMENSIONS: INCHES
*LOCATION OF PEAK STRESS
CONVERSION FACTORS:
1 in. = 2.54 cm
1 ksi = 6.9 MPa

FIG. 4—*Finite-element model and interlaminar shear stresses.*

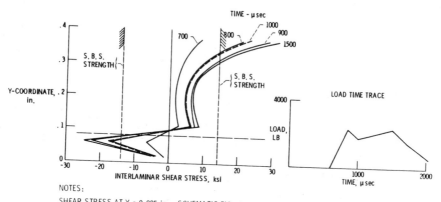

NOTES:
SHEAR STRESS AT X = 0.025 in., SCHEMATIC FIG. 4
LOAD TIME TRACE: APPROXIMATION TO S-G/E, FIG. 8
CONVERSION FACTOR: 1 ksi = 6.9 MPa; 1 in. = 2.54 cm

FIG. 5—*Transient finite-element analysis results.*

increases. The delaminated portions will more than likely behave as individual beams transmitting normal stress but not shear stress at the delaminated surfaces. The individual beams will continue to absorb energy until they fail in flexure. It is interesting to note that the significant initial load drop in the load-time trace (Fig. 5) is probably associated with the rapid displacement increase as the specimen delaminates into individual beams.

The important conclusion from the transient analysis results and discussion is that the transient (dynamic) interlaminar shear stresses can cause progressive delamination early in the impact event, leading to subsequent individual layer beam action. This conclusion is consistent with conclusions made previously from the results of the mechanistic and statistical models.

Individual-Layer-Beam Behavior and Fracture Progression

The individual-layer-beam behavior was evaluated in order to obtain a better assessment of the fracture progression during impact. The individual-layer-beam behavior was evaluated using finite-element and simple beam theory analyses. The finite-element analysis was used on a four-layer beam with progressively degraded interply layer modulus (E). The simple beam theory analysis was applied to a four-individual-layer beam.

The results of the finite-element analysis are summarized in Fig. 6. The bending deflection of the beam increases nonlinearly as the interlaminar layer modulus E is progressively degraded. The equation describing the curve is shown in the figure. For the individual-layer-behavior limiting case ($E/E_{EP} = 0$) the bending deflection of the beam with the four individual layers is 16 times that of the undegraded composite beam. The curve in Fig. 6 is very significant since it shows substantial increases in the bending deflection as the

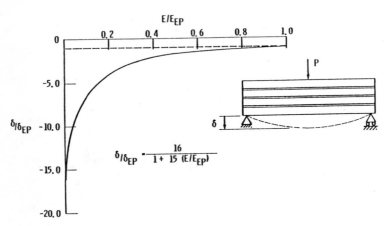

FIG. 6—*Bending deflections of a beam with progressively degraded interlaminar layer modulus.*

interlaminar layer modulus degrades to 0.2 of its undegraded composite value. This will definitely increase the energy absorption considerably but will degrade structural integrity as assessed by deflection limits, buckling, and resonant frequencies.

The results of the simple beam theory analysis are summarized in Fig. 7. The results are generalized to the behavior of a beam with N individual layers as compared to a composite beam. The following are observed for the same load:

1. The bending deflection and the work done (energy absorbed) increase as the square of the number of individual (delaminated) layers.

2. The flexural stress in each individual layer increases as the number of individual beams increases.

3. The interlaminar, or short beam shear (SBS), stress is independent of the number of individual layers. This is very significant since progressive doubling of the individual-layer beams can occur by delamination without increase in the load.

4. Progressive individual-layer-beam doubling will occur so long as the interlaminar (SBS) strength satisfies the inequality shown in the figure.

The finite-element transient analysis results and the individual-layer beam analysis results can now be used to postulate dynamic fracture progression:

1. Dynamic delamination occurs in one plane initially.

2. The dynamic flexural stress in each individual layer will increase.

3. Progressive doubling of delaminated layers will continue so long as the inequality in Fig. 7 is satisfied.

4. The bending deflection and the external work done will increase progressively.

5. The impact load will decrease during delamination because of displacement "jumps" as additional individual layers are formed.

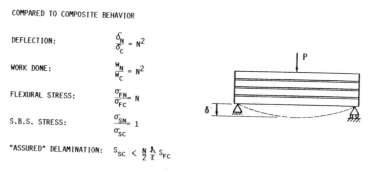

COMPARED TO COMPOSITE BEHAVIOR

DEFLECTION: $\dfrac{\delta_N}{\delta_C} = N^2$

WORK DONE: $\dfrac{W_N}{W_C} = N^2$

FLEXURAL STRESS: $\dfrac{\sigma_{FN}}{\sigma_{FC}} = N$

S.B.S. STRESS: $\dfrac{\sigma_{SN}}{\sigma_{SC}} = 1$

"ASSURED" DELAMINATION: $S_{SC} < \dfrac{N}{2}\dfrac{\lambda}{\ell} S_{FC}$

FIG. 7—*Effects of simulated individual layer behavior on bending.*

6. Intralayer flexural fracture will occur when the flexural stress equals or exceeds the corresponding strength.

7. Specimen fracture will follow either by progressive intralayer flexural fracture or by combinations of progressive intralayer delamination and flexural fracture.

8. The impact load will progressively decrease due to progressive deflection "jumps" if these jumps are relatively large (S-G/E) or remain level if these jumps are relatively small (HMS/E).

The impact load-time traces in Fig. 8 appear to corroborate, in part, the postulated fracture progression which is consistent with that hypothesized in Ref *11*. Needless to say, additional experimental data using instrumented impact and high-speed movies will be required to verify the postulated fracture progression.

Environmental Effects

The environmental effects on impact resistance can be evaluated using the hygrothermomechanical theory [8]. This is done by determining the environmental effects in the terms representing (1) the dominant energy mechanisms in the mechanistic model at the micromechanics level and (2) the interlaminar shear and flexural strengths which influence the fracture progression.

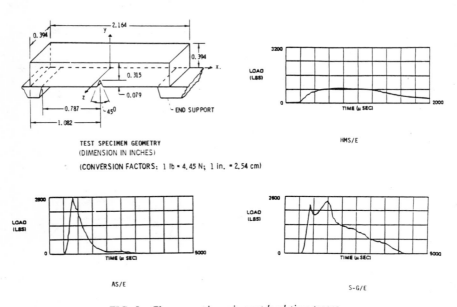

FIG. 8—*Charpy specimen impact load-time traces.*

Results of some general trends may be summarized as follows:

1. The energy absorption prior to delamination will decrease when the hygrothermal environment degrades the interlaminar shear strength relative to a reference value and vice versa.

2. The impact load, prior to delamination, will be less when the interlaminar shear strength degrades and vice versa.

3. The total work done prior to fracture will increase when the hygrothermal environment (i) degrades the interlaminar shear strength at a faster rate than the flexural strength and (ii) has a relatively small effect on the tensile strength.

Specific applications need to be examined on an individual case basis. An illustrative example that may be used to assess the environmental effects on impact resistance is outlined in the Appendix.

It is worth noting that extreme environments, which substantially degrade (1) the compression and shear moduli and (2) the transverse and longitudinal compression strengths, will change the impact resistance dramatically. These types of environments will probably change the relative significance of the contributions of the energy-absorbing mechanisms, especially that due to local indentation. The degradation of impact resistance at extreme hostile environments is academic, however. These hostile environments will also substantially degrade the structural integrity of the composite to make it impractical for structural applications.

Generalizations and Proposed Definitions

The collective results, attendant conclusions, and illustrative example for the impact resistance of unidirectional composites described herein should be applicable in assessing, or estimating, the impact behavior of angleplied laminates as well. The significant supporting reasons are: (1) the consistency of the theoretical results obtained from the three different methods (mechanistic model, statistical model, and transient finite-element analysis); (2) the excellent correlation of predicted results with measured data (mechanistic model—energy absorbed prior to onset of damage, statistical model—total work done, transient analysis—delamination influence on impact load); (3) the inherent features of the Charpy impact specimen [interlaminar shear stress dominance, stress concentration at notch tip, highly localized impact (impact load point above notch tip), and delamination initiated fracture]; (4) fracture progression (progressive delamination followed by flexural failure of the delaminated layers); and (5) the localized nature of impact in angle plied laminates which is similar to Charpy impact near field behavior in terms of stress, energy-absorbing mechanisms, failure initiation, and fracture progression.

Considerable skepticism exists in the composites community (including the authors) as to the significance of Charpy impact to angleplied laminate impact. The main reason for this skepticism is the difficulty associated in relating the impact energy absorbed in the angleplied laminate to that in the Charpy specimen. As discussed herein, the local behavior needs to be assessed at several levels during the impact event. In this way fundamental composite material characteristics, which lead to impact resistance (defined by energy absorption prior to delamination and subsequent fracture progression), are identified and quantified. As expected, fundamental composite material characteristics are not altered by the type of impact.

Considering the collective results and discussions of this investigation, the following working definitions are proposed for energy-absorption mechanisms and energy-release mechanisms:

• Energy-absorption mechanisms are those composite micromechanics mechanisms through which composites continue to absorb impact energy while the impact load rises from a current value without any or additional damage, as observed in an impact load versus time trace (Fig. 5).

• Energy-release mechanisms are those composite micromechanics mechanisms through which composites release impact energy. This release in energy is associated with sustained or progressive damage and corresponds to impact load leveling or decrease as observed in an impact load versus time trace (Fig. 5).

Both of these definitions emanate from and are consistent with the notion that sustained and progressive damage reduces the structural integrity of a composite as measured by deflection and frequency response limits and buckling resistance.

Fracture surfaces related to fracture modes I, II, and III are thought to constitute or represent energy-absorption mechanisms during impact. However, released energy associated with these fracture surfaces is an integrated effect of the energy-absorbing micromechanics mechanisms just defined. Mode I, II, and III types of mechanisms are gross energy release mechanisms and, as such, do not represent fundamental material characteristics.

Summary of Results and Conclusions

The significant results and conclusions of an investigation on the impact resistance of unidirectional fiber composites, including energy-absorbing mechanisms and environmental effects, are now summarized:

1. Flexure and interlaminar shear deformations are the dominant energy-absorbing mechanisms during impact; the area under the linear stress-strain diagram represents the most significant contribution.

2. Fiber pullout, interply delamination, and local indentation are energy-release mechanisms.

3. Interply delamination occurs early in the impact event.

4. Interply delaminations increase the deflection, external work, and flexural stress in the delaminated layers, but they decrease buckling resistance and vibration frequencies.

5. The external work in delaminated composites increases approximately as the square of the numbers of delaminated individual layers (beams) increases compared with the undelaminated composite.

6. Dynamic fracture during impact is a progression of interlaminar delaminations and intralayer fracture of the delaminated layers.

7. The environmental effects on impact resistance can be assessed using hygrothermomechanical theory.

8. The mechanical behavior of fiber composites under impact can be assessed using mechanistic models, statistical models, finite-element transient analysis, and available experimental data.

APPENDIX

Illustrative Example of Environmental Effects on Impact Resistance

The hygrothermal effects on impact resistance may be assessed using the following approach. First it is necessary to estimate the hygrothermal degradation effects on unidirectional composite properties which are resin dominated. This approach is briefly described below, including a numerical example.

The hygrothermal degradation is estimated using [3]

$$\frac{P_H}{P_0} = \left[\frac{T_{gw} - T}{T_{gd} - T_0} \right]^{1/2} \tag{1}$$

where

P_H = hygrothermally degraded property,
P_0 = a reference property,
T_{gw} = glass transition temperature of the wet resin,
T_{gd} = glass transition temperature of the dry resin,
T = temperature at which P_H is needed, and
T_0 = reference temperature corresponding to P_0.

The dominant energy absorbing mechanism ($S_{\ell 11T}^2/E_{\ell 11}$) is fiber dominated. However, the interlaminar shear ($S_{\ell 12S}$) and flexural ($S_{\ell 11F}$) strengths, which control delaminations and fracture progression, are resin dominated. The flexural strength is resin dominated through longitudinal compression as given by [12]

$$S_{\ell 11F} = \frac{3S_{\ell 11C}}{1 + (S_{\ell 11C}/S_{\ell 11T})} \tag{2}$$

where

$S_{\ell11F}$ = longituding flexural strength,
$S_{\ell11C}$ = longitudinal compressive strength, and
$S_{\ell11T}$ = longitudinal tensile strength.

As a numerical example, consider the hygrothermal effects on the dynamic fracture progression of a Charpy specimen made from an S-glass/epoxy composite with the following properties:

$$T_{gw} = 177°C \ (350°F)$$
$$T_{gd} = 216°C \ (420°F)$$
$$T = 93°C \ (200°F)$$
$$T_0 = 21°C \ (70°F)$$

$$S_{\ell11To} = 1531.8 \ \text{MPa} \ (222 \ \text{ksi})$$
$$S_{\ell11Co} = 1242 \ \text{MPa} \ (180 \ \text{ksi})$$
$$S_{\ell12So} = 96.6 \ \text{MPa} \ (14 \ \text{ksi})$$

Using these numerical values in Eqs 1 and 2 we obtain the following:

Degraded property ratio, P_H/P_0 0.655
Degraded interlaminate shear strength, $S_{\ell11SH}$, ksi   9.2
Degraded flexural strength, $S_{\ell11FH}$, ksi   231
Degraded longitudinal compressive strength, $S_{\ell11CH}$, ksi   118
Undegraded flexural strength, $S_{\ell11FO}$, ksi   298

The interlaminar shear strength was degraded. Therefore, the impact load to initial damage will be reduced. The degraded ratio of the interlaminar shear and flexural strengths relative to reference values are, respectively:

$$S_{\ell12SH}/S_{\ell12SO} = 0.655$$
$$S_{\ell11FH}/S_{\ell11FO} = 0.775$$

The interlaminar shear strength degrades at a faster rate than the flexural strength since the longitudinal tensile strength is fiber dominated and thus not affected. Therefore, the work done to fracture will increase relative to the reference value. This example illustrates that it is not intuitively apparent what effects the environment has on peak load or total work to fracture. This example also illustrates the advantage of having formal procedures, even if approximate, to assess complex composite behavior such as impact.

References

[1] Chamis, C. C., Hanson, M. P., and Serafini, T. T. in *Composite Materials: Testing and Design (Second Conference), ASTM STP 497*, American Society for Testing and Materials, Philadelphia, 1972, pp. 324–349.

[2] Broutman, L. J. and Rotem, A. in *Foreign Object Impact Damage to Composites, ASTM STP 568*, American Society for Testing and Materials, Philadelphia, 1975, pp. 114–133.

[3] Yeung, P. and Broutman, L. J., *Polymer Engineering and Science*, Vol. 18, No. 2, Mid-Feb. 1978, pp. 62–72.

[4] Sun, C. T. and Wang, T., "Dynamic Responses of a Graphite/Epoxy Laminated Beam to Impact of Elastic Spheres," NASA CR 165461, National Aeronautics and Space Administration, Washington, D.C., Sept. 1982.

[5] Chamis, C. C., Hanson, M. P., and Serafini, T. T., "Designing for Impact Resistance with Unidirectional Fiber Composites," NASA TN D-6463, National Aeronautics and Space Administration, Washington, D.C., Aug. 1971.

[6] Chamis, C. C. and Smith, G. T., "Environmental and High-Strain Rate Effects on Composite for Engine Applications," NASA TM-82882, National Aeronautics and Space Administration, Washington, D.C., May 1982.

[7] Sun, C. T. and Yang, S. H., "Contact Law and Impact Responses of Laminated Composites," NASA CR-159884, National Aeronautics and Space Administration, Washington, D.C., Feb. 1980.

[8] Chamis, C. C., Lark, R. F., and Sinclair, J. H. in *Advanced Composite Materials-Environmental Effects, ASTM STP 658*, American Society for Testing and Materials, Philadelphia, 1978, pp. 160–192.

[9] Friedrich, L. A. and Preston, J. L., Jr., "Impact Resistance of Fiber Composite Blades Used in Aircraft Turbine Engines," NASA CR-134502, National Aeronautics and Space Administration, Washington, D.C., May 1973.

[10] Sidik, S. M., "An Improved Multiple Linear Regression and Data Analysis Computer Program Package," NASA TN D-6770, National Aeronautics and Space Administration, Washington, D.C., April 1972.

[11] Chamis, C. C., "Failure Mechanics of Fiber Composite Notched Charpy Specimens," NASA TM X-73462, National Aeronautics and Space Administration, Washington, D.C., Sept. 1976.

[12] Sinclair, J. H. and Chamis, C. C., "Compression Behavior of Unidirectional Fibrous Composites," NASA TM-82833, National Aeronautics and Space Administration, Washington, D.C., March 1982.

Fumio Yamauchi[1] *and Shigeo Emoto*[1]

Ferrite-Resin Composite Material for Vibration Damping

REFERENCE: Yamauchi, F. and Emoto, S., **"Ferrite-Resin Composite Material for Vibration Damping,"** *Recent Advances in Composites in the United States and Japan, ASTM STP 864,* J. R. Vinson and M. Taya, Eds., American Society for Testing and Materials, Philadelphia, 1985, pp. 346–354.

ABSTRACT: Newly developed ferrite-resin composite materials have high rigidity ($E = 1.8 \times 10^9$ N/m^2) and high mechanical damping constant (in a logarithmic decrement, $\Delta = 1.1$). The composite materials were used effectively for loudspeaker enclosure humming suppression. It is also convenient to use the composite materials in a paste state. They were applied to a hopper and a machine base for vibration control.

KEY WORDS: ferrite-resin composite materials, mechanical properties, high mechanical damping, resistance to environment, humming suppression, vibration control

Noise and vibration pollution is a serious problem in modern urban life since the population is housed in a concentrated narrow space near cities and towns, where traffic and industry noise vibrations are always being generated. In some factories such as those manufacturing integrated circuit (IC) and large-scale integration (LSI) devices, small vibrations on the part of the machines are also serious, since they cause low device yields in the production line. To solve these noise-vibration problems, new damping material development is necessary as well as systematic engineering technology. Newly developed ferrite-resin composite materials are effective for vibration control even at low frequencies, because they have heavy densities, high rigidity, and high mechanical damping constants. This paper presents material properties and their applications.

Manufacturing Process and Measuring Method

A lot of ferrite powder, mostly composed of magnetite (Fe_3O_4), is produced subsidiarily in iron and steel factories, titanium mining factories, waste water

[1]Resources and Environment Research Labs., NEC Corp., Miyazaki, Miyamae-ku, Kawasaki-City, Kanagawa 213, Japan.

treatment systems, and elsewhere [1]. As the powder products usually contain water, they have to be dried to within 0.5 weight % water content. The water contained in the powder often hinders the composites from hardening [2]. Ferrite-resin composite materials are made from dried powder mixed with resin.

Logarithmic decrements (Δ) and Young's modulus (E) were measured from decay curves for damped mechanical oscillation and resonance frequencies for the fundamental bending vibration for a specimen hung at the modal points. They were calculated as follows

$$\Delta = 1/(ft) \times \ln (a_0 \cdot a_n)$$

where

f = mechanical vibration frequency, Hz

a_0, a_n = maximum acceleration amplitude for 0th and nth vibrations, respectively, and

t = time interval between maximum amplitude for 0th and nth vibrations, s

and Δ is connected to the loss factor Q^{-1} with $\Delta = 2 \pi \times Q^{-1}$:

$$E = 0.95 \times f_0^2 \times l^2 \times P/h^2$$

where

f_0 = mechanical resonance frequency, Hz,

l = length, m,

h = thickness, m, and

P = specimen density, kg/m^3.

Specimens about 0.3 m long, 0.05 m wide, and 0.01 m thick were prepared and used in the experiments. The resonance frequencies of the specimens were about 50 to 1000 Hz.

Material Properties

Logarithmic decrements (Δ) and Young's moduli (E) for the newly developed materials are summarized in Fig. 1 with those for other materials. Metal materials have high rigidity but low damping factors shown with logarithmic decrements, while rubber materials have low rigidity and high damping factors. In ferrite-resin composite materials, on the other hand, rigidity and damping factors are both high.

The typical ferrite epoxy-resin composite in the new materials has a very high logarithmic decrement (0.9 to 1.1), Young's modulus (1.8×10^9 N/m^2), and density (1.9 to 2.1×10^3 kg/m^3). The epoxy-resin material is composed

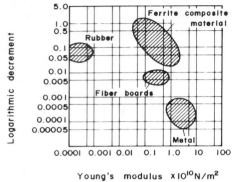

FIG. 1—*Logarithmic decrements and Young's modulus values for ferrite-resin composite and common materials.*

of a hardening accelerator and a principal ingredient containing 60 weight % ferrite powder. The material can be pasted on iron or aluminum plates after mixing the principal ingredient and the hardening accelerator.

Damping effects for the composite are shown in Fig. 2, when the material was coated 16 mm thick on iron plates (300 by 300 by 16 mm). The inertia (acceleration/force) levels for the bare plates are 40 to 50 dB at resonance frequencies. On the other hand, the levels for plates coated with the material are 10 to 20 dB and improved by above 30 dB.

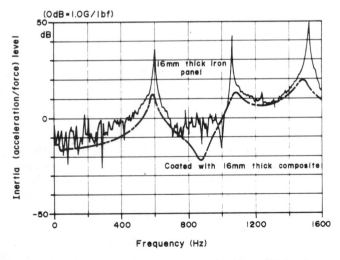

FIG. 2—*Typical vibration damping effects for iron panel (300 by 300 by 16 mm) coated with epoxy-resin composite.*

Thickness effects on damping are shown in Fig. 3. Damping in the aluminum plates is greater than in iron plates. The effects depend theoretically upon the ratio of Young's modulus. The effects are maximum at the ratio of 1 and become lower when the ratio becomes larger or smaller than 1. It is preferable that Young's modulus for the damping material be close to that for vibration plates.

Usually, damping materials have to be used for a wide range of frequencies in which vibrations are induced. Frequency dependences on damping are shown in Fig. 4. Damping coefficients ($\zeta = Q^{-1}/2$) are constant in a frequency range lower than 6 kHz.

As damping materials are often used in adverse weather surroundings, they have to be weather-proof. Compressive strengths and static Young's modulus values are shown in Fig. 5 in weatherometer tests. In the tests, the specimens were exposed alternately for 12 min exposure to ultraviolet rays after 120 min sprinkling with water. The strengths and Young's modulus value measurements were started after one week of curing at room temperature. After 400 h, in the weatherometer tests, the strengths and modulus values were saturated at 3.2×10^7 and 3×10^8 N/m^2, respectively.

Since the vibration suppression pads are used to mount engines and generators, their ability to withstand soaking in gasoline, diesel fuel, oil, and seawater must be checked. Weight changes in specimens soaked in gasoline, oil, or salt water were measured. The changes are within 1 weight % in 150 days. Adhesion to iron was also measured in 5 weight % sodium chloride solution. The strength values were the weakest at 2.0×10^6 N/m^2 after a 10-day soak-

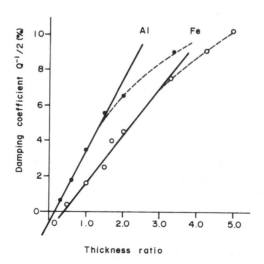

FIG. 3—*Thickness effects on vibration damping for aluminum and iron panels coated with ferrite epoxy-resin composite.*

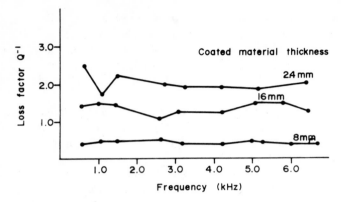

FIG. 4—*Frequency dependence of damping coefficients for 16-mm-thick iron panels coated with epoxy-resin composite.*

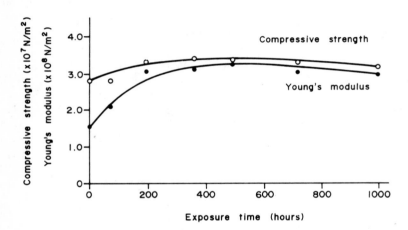

FIG. 5—*Compressive strengths and static Young's modulus values for ferrite epoxy-resin composites in weatherometer tests. Specimens were exposed alternatively to 12-min ultraviolet rays after 120-min sprinkling with water.*

ing. From these results, it is concluded that the ferrite epoxy-resin composite can be used in adverse weather surroundings, in gasoline, oil, or saltwater exposure conditions.

Application

The epoxy-resin ferrite composite is not only effective for vibration control when the composite material is used on vibration suppression pads but also effective when it is pasted on the plates or pads.

Loudspeaker Enclosure

The composite material is expected to be applied to loudspeaker enclosures in audio instruments. An enclosure has to hold loudspeakers themself tightly and to exclude out-of-phase acoustic waves radiated from the back of the speakers. Otherwise, the enclosure vibrations will generate hummings near the resonance frequencies and sounds differing from the original sounds. The enclosure made of the composite material reduced the unwanted hummings and generated the original sounds at high fidelity. Figure 6 shows inertia values for the speaker enclosure made of the composite material and plywood, respectively. The inertia values for the composite material are smoother and lower than those for the plywood.

Hopper for Coal Powder

Recently, the authors used the composite material for vibration control in a hopper system in an urban area, which was generating low-frequency ground vibrations and acoustic waves at 20 Hz and its overtone frequencies. Vibration acceleration amplitudes decreased to $1/2$ to $1/3$ at the surface after the outer surface was plastered with 5-mm-thick composite. The plastered hopper is shown in Fig. 7. The acoustic noise level also decreased to about 5 dB at the borderline facing a residential area.

Complex Structure for Enclosure

A complex structure panel for the enclosure was coated with about 5-mm-thick composite, where the panel was composed of a 1.5-mm-thick stainless

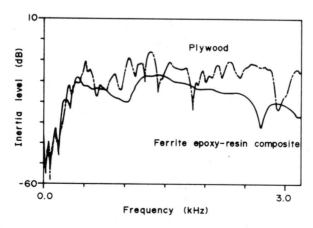

FIG. 6—*Inertia spectra for speaker enclosure made from ferrite epoxy-resin composite and plywood.*

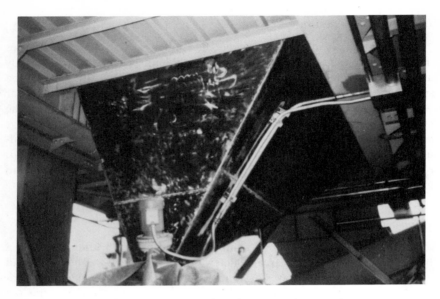

FIG. 7—*Hopper coated with ferrite epoxy-resin composite.*

steel surface panel supported by a 5-mm-thick punched plate and a 13-mm-radius steel truss, as shown in Fig. 8. Typical inertia values are shown in Fig. 9. Coated panel inertia values are about 10 dB lower than the uncoated panel. Their frequency dependences are smooth, without the spike responses that are generated by the stainless steel film vibrations.

Summary

Ferrite-resin composite materials have high rigidity and great vibration damping. They can reduce mechanical vibrations effectively. A typical ferrite

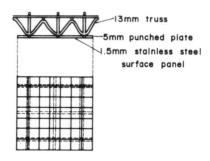

FIG. 8—*Complex structure panel.*

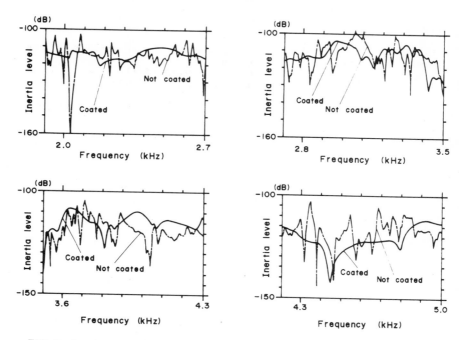

FIG. 9—*Inertia spectra for complex structure panel coated or uncoated with ferrite epoxy-resin composite.*

epoxy-resin composite can be coated on iron or aluminum panels. For example, bending vibrations for 16-mm-thick iron panels reduced to below about 30 dB when coated 16 mm thick with the composite material. These damping effects are constant in a wide frequency range. Furthermore, the composite is resistant to gasoline, oil, salt water, and ultraviolet rays.

The composite material was applied effectively for vibration control in a hopper, a loudspeaker enclosure, an engine base pad, and so on. The ferrite-resin composites are not so expensive when appropriate resin is mixed with the ferrite powder. Therefore, the composite can be used in civil engineering for road or structure vibration control.

Acknowledgment

The authors wish to thank Dr. Toshiro Tsuji, Mr. Izuru Sugano, and Mr. Naotaka Sakakibara, NEC Corp., for their encouragement in this research, and are grateful to individuals in NEC Corp. and MPI Epoxy Corp. for their enormous contribution to this paper, particularly to Mr. Minoru Yoshida and Mr. Tadao Iwata.

References

[1] Okuda, T., Sugano, I., and Tsuji, T., "Removal of Heavy Metals from Wastewater by Ferrite Co-Precipitation," *Filtration and Separation*, Sept./Oct. 1975, pp. 472–478.

[2] Yamauchi, F., Nakano, S., and Sugano, I., "Magnetic Marker Using Ferrite-Byproduct and its Application" in *Proceedings*, International Conference on Ferrites, Japan, Sept./Oct. 1980, pp. 894–897.

[3] Ungar, E. E., "Loss Factors of Viscoelastically Damped Beam Structures," *Journal of the Acoustical Society of America*, Vol. 34, No. 8, Aug. 1962, pp. 1082–1089.

*C. Allen Ross,[1] Lawrence E. Malvern,[1] Robert L.
Sierakowski,[1] and Nobuo Takeda[2]*

Finite-Element Analysis of Interlaminar Shear Stress Due to Local Impact

REFERENCE: Ross, C. A., Malvern, L. E., Sierakowski, R. L., and Takeda, N.,
"Finite-Element Analysis of Interlaminar Shear Stress Due to Local Impact," *Recent Advances in Composites in the United States and Japan*, *ASTM STP 864*, J. R. Vinson and M. Taya, Eds., American Society for Testing and Materials, Philadelphia, 1985, pp. 355–367.

ABSTRACT: Interlaminar delamination may occur as a result of a nonperforating low-velocity impact of a foreign object against a continuous-fiber composite laminate. Experimental evidence shows that for a crossplied laminate consisting of three lamina and subjected to a central localized impact, one delamination region will extend farther from the impact point than the second region.

To establish a quantitative value for the interlaminar shear stress of a crossplied fibrous composite plate subjected to a central impact, a finite-element computer program has been utilized. Specifically, a three-dimensional dynamic-elastic-orthotropic-material finite-element program has been used in which the interlaminar shear stress can be calculated directly. This program was used to evaluate the interlaminar shear stresses for a $[0_5/90_5/0_5]$ glass/epoxy plate with built-in edges, and subjected to a load simulating a central localized impact of 45.7 m/s by a cylindrical steel projectile.

Results of the calculations show a region of large shear stress (exceeding the estimated shear strength of the matrix material) extending from the impact point toward the boundary in the 0-deg direction while a region of large shear stress in the transverse 90-deg direction was found to be much smaller. Also, only minor differences in calculated shear stresses were evident in the interlaminar plane immediately below the impact point when compared with the interlaminar plane farthest from the impact point. The high-shear-stress regions agreed qualitatively with the areas observed in experiments.

KEY WORDS: composite materials, finite elements, analysis, delamination, continuous filament, crossplied composite

The central impact of nonperforating right circular cylinders against semi-transparent crossplied glass/epoxy plates produces delaminations in in-

[1]Professor, University of Florida, Gainesville, FL 32611.
[2]Research fellow, University of Tokyo, Tokyo, Japan.

terlaminar planes that are visible when the plates are backlighted. A typical pattern is shown in the photograph of Fig. 1. This photograph shows a five-lamina plate with four interlaminar planes and four delaminated areas of different size. The experimental results of these tests are omitted here and the reader is referred to Refs *1–3* for a detailed discussion. The present study is concerned with an effort to determine analytically the interlaminar shear stresses of a $[0_5/90_5/0_5]$ glass/epoxy 15.24-cm square plate having an average thickness of 0.64 cm when subjected to a central impact of 45.7 m/s by cylindrical steel projectiles 0.97 cm diameter and 2.54 cm long. The following sections describe this analysis.

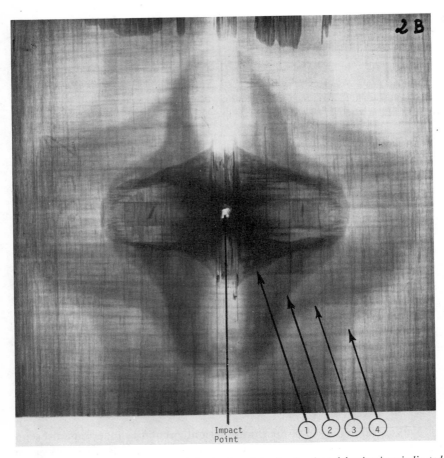

FIG. 1—*Photograph of backlighted five-lamina plate showing four delaminations indicated by circled numbers.*

Computer Model

The analysis was performed using the finite-element computer program SAPIV [4]. This program incorporates a three-dimensional elastic-orthotropic-material model with dynamic analysis capability that allows the calculation of transverse shear stresses directly. Due to the symmetry of loading, only one quarter of the plate was analyzed. The lines of symmetry and conditions for ensuring symmetry of the solution are shown in Fig. 2. The plate was analyzed using three-dimensional eight-node elements shown schematically in Fig. 3. The fiber is in the x-direction for the top and bottom laminas and in the y-direction for the middle lamina. The cell size near the lines of symmetry was selected as 0.635 by 0.635 by 0.159 cm. At other points the size was taken as

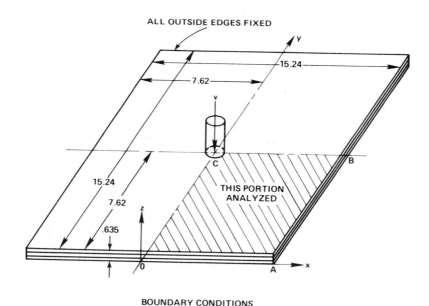

ALL OUTSIDE EDGES FIXED

15.24

7.62

15.24

7.62

.635

THIS PORTION ANALYZED

BOUNDARY CONDITIONS

Line or Point	Condition
0A	All rotations and displacements are zero
AB	All rotations and displacements are zero
0C	Rotations about 0C and all displacements in x direction are zero
BC	Rotations about BC and all displacements in y direction are zero
C	All rotations and displacements, except displacement in z direction, are zero

FIG. 2—*Loaded plate showing lines of symmetry (all dimensions in centimetres).*

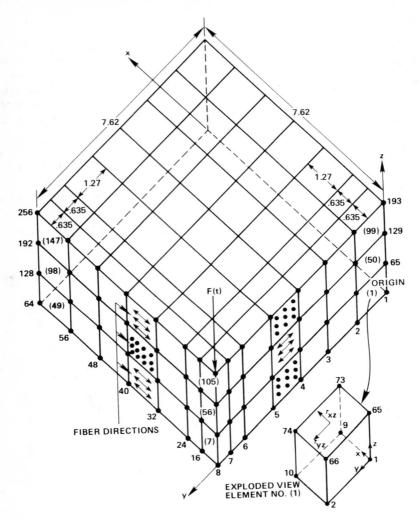

FIG. 3—*Plate schematic showing Nodes 1, 2 etc. and Elements (1), (2) etc. Drawing not to scale. Node 8 is bottom center point. Load is applied at Node 200, top center point (all dimensions in centimetres).*

1.27 by 1.27 by 0.159 cm. Nodes were assigned along the lines of symmetry, implying that the center of the closest elements to the lines of symmetry was 0.318 cm from the lines of symmetry.

The stress output for the program was identified with the center of each element and at the center of each face of the element. The interlaminar shear stresses shown later in the "Results" section were then obtained by taking the average between the evaluated lower-surface stress values of the top element and the upper-surface stress values of the middle element, and correspondingly the average between the lower surface of the middle element and the upper surface of the bottom element. Thus, interlaminar shear stresses were not calculated on the lines of symmetry of the plate. The plotted results were always at least 0.318 cm from the line of symmetry.

The solution thus represents a displacement type with three translational degrees of freedom at each node. To ensure no rotations occur about lines along and normal to the fixed sides of symmetry, zero-rotation boundary elements were used at these places.

Material Properties

The constitutive relations of the material used for the plate are written in terms of the global axes x, y, z as shown in Fig. 3. These equations are given in the SAPIV manual [4] as

$$\epsilon_x = (1/E_x)\sigma_x - (\nu_{xy}/E_y)\sigma_y - (\nu_{xz}/E_z)\sigma_z$$

$$\epsilon_y = -(\nu_{yx}/E_x)\sigma_x + (1/E_y)\sigma_y - (\nu_{yz}/E_z)\sigma_z$$

$$\epsilon_z = -(\nu_{zx}/E_x)\sigma_x - (\nu_{zy}/E_y)\sigma_y + (1/E_z)\sigma_z \qquad (1)$$

$$\gamma_{zy} = \tau_{zy}/G_{zy}, \qquad \gamma_{yz} = \tau_{yz}/G_{yz}, \qquad \gamma_{zx} = \tau_{zy}/G_{zy}$$

where

ϵ = unit extension, dimensionless,
γ = shear strain, dimensionless,
σ = normal stress, MPa,
τ = shear stress, MPa,
E = Young's modulus, MPa,
ν = Poisson's ratio, dimensionless, and
G = shear modulus, MPa.

The top and bottom laminas with the fibers aligned in the x-direction (0 deg) have the elastic properties of Material 1. In the middle lamina the fibers are aligned along the y-direction (90 deg) and have the elastic properties given as Material 2. The elastic properties of these two materials are given in Table 1.

Loading Function

The loading-time function was determined from the mass of the projectile and the contact time of the projectile and the plate. The mass of the projectile was 14.07 g and the contact time for impact of $v_0 = 45.7$ m/s was 0.25 ms. A linearly varying force-time relationship was assumed as shown in Fig. 4. The

TABLE 1—*Material properties.*

Elastic Property	Material 1	Material 2
E_x, GPa	40.00	8.28
E_y, GPa	8.28	40.00
E_z, GPa	8.28	8.28
ν_{xy}	0.05	0.24
ν_{xz}	0.05	0.30
ν_{yz}	0.30	0.05
G_{xy}, GPa	4.14	4.14
G_{yz}, GPa	3.03	4.14
G_{xz}, GPa	4.14	3.03

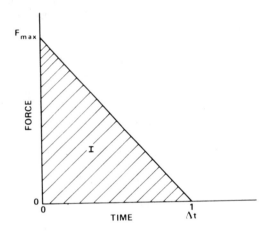

FIG. 4—*Assumed force-time relationship.*

impulse magnitude I is the area under the force-time curve and may be expressed as

$$I = \frac{F_{max}(\Delta t)}{2} = |\Delta(mv)| = mv_0 \tag{2}$$

where $|\Delta(mv)|$ is the magnitude of the change in momentum of the projectile and Δt is the contact time of the projectile on the plate. Thus

$$F_{max} = 2\frac{(mv_0)}{\Delta t}$$

and for the 45.7-m/s impact of a 14.07-g projectile

$$F_{max} = \frac{2(45.7)(0.0141)}{(0.00025)} \cong 5148 \text{ N}$$

gives an impulse approximately equal to that imparted by the projectile. One fourth of this is assumed to be transmitted to each quadrant of the centrally impacted plate. Thus the loading at Point C (Fig. 2) for the quarter-plate is initially 1287 N and decays linearly to zero in 0.25 ms.

Results and Discussion

Based upon the preceding discussion, the SAPIV program was run for the case of a 45.7-m/s impact. Interlaminar shear stresses for both the top and bottom interlaminar planes were determined and are shown in Figs. 5–9. The top interlaminar plane is the one closest to the impact side and the bottom interlaminar plane is the one farthest from the impact side. The signs of the shear stresses are shown as positive in the direction on the top of Element 1 of Fig. 3.

The lower-right portion of each figure shows the locations of the stations, A through G, for which interlaminar shear stress histories are shown in the plotted curves. Ordinate scales are MPa. It is seen in Figs. 5 and 6 for an elastic solution that the shear stress τ_{xz} attains or exceeds the estimated interlaminar shear strength of 17.24 MPa at almost every station along the line parallel to the x-axis (fiber direction of the top and bottom laminas) in both interlaminar planes. The shear stress τ_{yz} is much smaller along this line, which is near the symmetry plane where $\tau_{yz} = 0$, except at Station A near the impact point. Station A is at the same distance from the two symmetry planes, and the two shear stress components at Station A are approximately equal in magnitude there, as would be expected. From Figs. 7 and 8 the calculated shear stress along the line at 90 deg to the top and bottom fiber direction exceeds the interlaminar shear strength only locally near the impact point. Experimental studies of the impact simulated by this calculation [1–3] show that in the bottom interlaminar plane the delamination extends about halfway to the boundary in the 0-deg direction parallel to the fibers of the bottom lamina, while the delamination in the 90-deg direction extends approximately half as far as that in the 0-deg direction. The difference between experiment and calculation is even larger in the top lamina, where the experiment actually shows delamination extending at least as far in the 90-deg direction as in the 0-deg direction. The top interlaminar plane has less delamination than the bottom one in the experiment, while the elastic solution of Figs. 5–8 indicates approximately the same interlaminar shear stresses in the two interlaminar planes. The different delamination behaviors in the two planes are attributed to the sequential delamination mechanism initiated by a generator strip on the top lamina as described in Refs 5–7. When actual delamination cracks begin to propagate, the elastic stress field can be changed considerably.

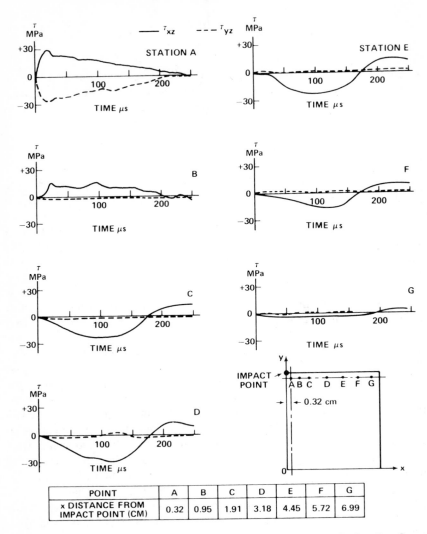

FIG. 5—*Shear τ_{xz}, τ_{yz}, top interlaminar plane versus time. Stresses for Stations A to G parallel to x-axis.*

The spatial stress distribution parallel to the x-axis for the shear stress τ_{xz} of the top interlaminar plane is shown in Fig. 9 for various times. This curve (Fig. 9) was generated as a crossplot of the data of Fig. 5. The plots of Fig. 9 show that the shear stress τ_{xz} changes sign at about $x = 1.27$ cm. It is remarkable that this distance is essentially the same in all the plots. At 110 μs the negative shear stress magnitude exceeds 17.24 MPa almost all the way to the boundary. These results are all based on elastic response and take no account

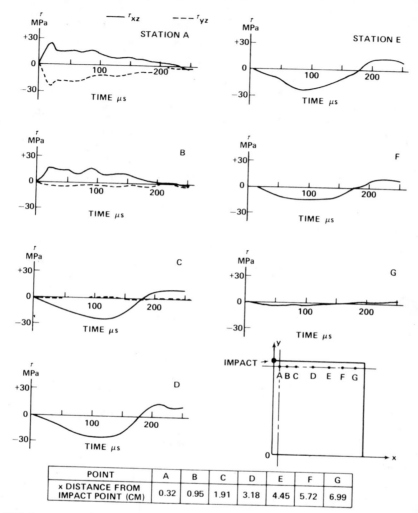

FIG. 6—*Shear stress* τ_{xz}, τ_{yz}, *bottom interlaminar plane versus time. Stresses for Stations A to G parallel to x-axis.*

POINT	A	B	C	D	E	F	G
x DISTANCE FROM IMPACT POINT (CM)	0.32	0.95	1.91	3.18	4.45	5.72	6.99

of the delamination that would occur at these stress levels and that would modify the actual shear stress distributions as the delamination crack propagates. Again, these results, while quantitative in nature, are based upon elastic-material models that may prove useful in extending the analytical methodology to provide insight into ultimately incorporating failure models into the evaluative procedures. For example, the computer code NONSAP [8] has a two-dimensional nonlinear material model that may be modified to pro-

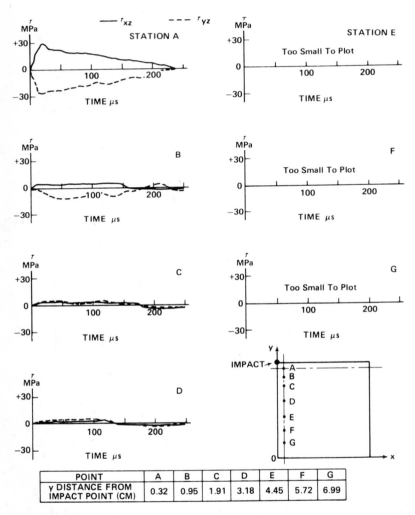

FIG. 7—Shear stress τ_{xz}, τ_{yz}, top interlaminar plane versus time. Stresses for Stations A to G parallel to y-axis.

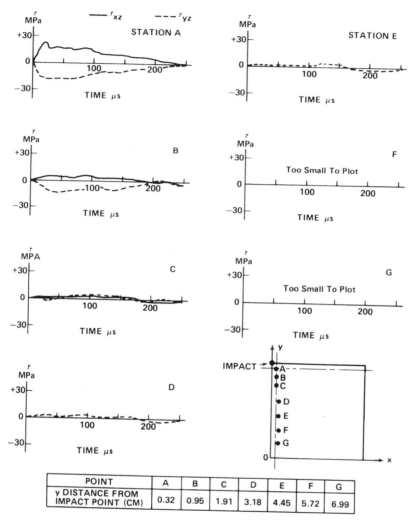

FIG. 8—*Shear stress τ_{xz}, τ_{yz}, bottom interlaminar plane versus time. Stresses for Stations A to G parallel to y-axis.*

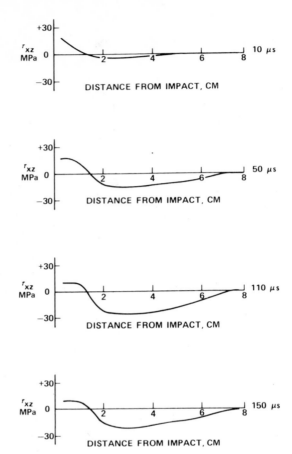

FIG. 9—*Shear stress* τ_{xz} *versus* x. *Distance from impact point for given times shown on right-hand side.*

vide tensile and shear stiffness reduction using a maximum-stress criterion. Also, NONSAP-C [9], developed just recently for concrete, has a nonlinear three-dimensional orthotropic-material model that allows cracking and subsequent stiffness reduction.

Acknowledgment

The authors acknowledge the U.S. Army Research Office, Durham, NC, for support of this research under Grant No. DAAG29-79-G-0007.

References

[1] Sierakowski, R. L., Ross, C. A., Malvern, L. E., and Cristescu, N., "Studies on the Penetration Mechanics of Composite Plates," Final Report, Grant No. DAAG29-76-G-0085, U.S. Army Research Office, Durham, NC, Dec. 1976.

[2] Sierakowski, R. L., Ross, C. A., Malvern, L. E., and Cristescu, N., "Studies on the Fracture Mechanisms in Partially Penetrated Filament Reinforced Laminated Plates," Final Report, Grant DAAG29-79-G-0007, U.S. Army Research Office, Durham, NC, Dec. 1981.

[3] Takeda, N., "Experimental Studies on the Delamination Mechanisms in Impacted Fiber-Reinforced Composites Plates," Ph.D. Dissertation, University of Florida, Gainesville, FL, 1980.

[4] Bathe, K. J., Wilson, E. L., and Peterson, F. E., "SAPIV, A Structural Analysis Program of Static and Dynamic Response of Linear Systems," EERC 73-11 Earthquake Engineering Research Center, University of California, Berkeley, CA, April 1974.

[5] Cristescu, N., Malvern, L. E., and Sierakowski, R. L., "Failure Mechanisms in Composite Plates Impacted by Blunt-Ended Penetrators" in *Foreign Object Damage to Composites*, *ASTM STP 568*, American Society for Testing and Materials, Philadelphia, 1975, pp. 159–172.

[6] Malvern, L. E., Sierakowski, R. L. Ross, C. A., and Cristescu, N., "Impact Failure Mechanisms in Fiber-Reinforced Composite Plates" in *High Velocity Deformation of Solids*, K. Kawata and J. Shiori, Eds., Springer-Verlag, Berlin, 1978, pp. 120–130 (*Proceedings* of IUTAM Symposium, Tokyo, 1977).

[7] Takeda, N., Sierakowski, R. L., and Malvern, L. E., Studies of Impacted Glass Fiber Reinforced Composite Laminates," *SAMPE Quarterly*, Vol. 12, No. 2, Jan. 1981, pp. 9–17.

[8] Bathe, K. J., Wilson, E. L., and Iding, R. H., "NONSAP: A Structural Analysis Program for Static and Dynamic Response of Nonlinear Systems," Report No. UCSESM 74-3, College of Engineering, University of California, Berkeley, CA, Feb. 1974.

[9] Smith, P. D. and Anderson, C. A., "NONSAP-C: A Nonlinear Stress Analysis Program for Concrete Containments Under Static, Dynamic, and Long-Term Loadings." NUREG/CR-0416, LA-7496-MS, Los Alamos Scientific Laboratory, Los Alamos, NM, Oct. 1978; upgraded March 1981.

Dale W. Wilson[1] and Jack R. Vinson[1]

Viscoelastic Buckling Analysis of Laminated Composite Columns

REFERENCE: Wilson, D. W. and Vinson, J. R., **"Viscoelastic Buckling Analysis of Laminated Composite Columns,"** *Recent Advances in Composites in the United States and Japan, ASTM STP 864*, J. R. Vinson and M. Taya, Eds., American Society for Testing and Materials, Philadelphia, 1985, pp. 368-383.

ABSTRACT: A linear viscoelastic buckling analysis for laminated composite columns was developed using the quasi-elastic approach. The analysis includes transverse shear deformation (TSD), transverse normal deformation (TND), and bending-extensional coupling. The Rayleigh-Ritz method of solution was employed to solve the governing equation derived from the Theorem of Minimum Potential Energy. Viscoelastic column buckling behavior was investigated for four laminate configurations, $[0]_{24}$, $[0/\pm45/0]_{3s}$, $[0/\pm45/90]_{3s}$, $[\pm45]_{6s}$, and for column length-to-thickness ratios (ℓ/t) ranging between 35.8 and 150.0. The results for graphite/epoxy show that viscoelastic effects are significant, reducing the critical buckling load by 10% to 20%, depending upon laminate configuration. For geometries where TSD is important ($\ell/t < 50.0$), the effects of viscoelasticity are magnified by the matrix-controlled shear stiffness contribution.

KEY WORDS: composite, stability, viscoelasticity, column buckling

The analytic treatment of stability in laminated composite structures is well developed under the assumptions of elastic behavior. However, many composite laminates exhibit significant viscoelastic behavior [1-4] which leads to nonconservative predictions of stability performance by elastic analysis. The effects of viscoelasticity on the stability of plates was addressed by Sims [5] using a linear viscoelastic formulation employing the quasi-elastic approximation coupled with a solution based on classical laminated plate theory.

More recently, investigators have shown that transverse shear deformation (TSD), bending-extensional coupling, and transverse normal deformations (TND) significantly affect laminated plate bending, buckling and vibration behavior [6-9]. These findings, coupled with the realization that shear properties of composite systems are the most strongly viscoelastic properties,

[1]Associate scientist, Center for Composite Materials, and H. Fletcher Brown professor, Department of Mechanical and Aerospace Engineering, University of Delaware, Newark, DE 19711.

prompted the authors to investigate the viscoelastic buckling response of laminated plates using a solution which includes TSD, TND, and bending-extensional coupling [10]. The results of the investigation showed that for geometries where TSD effects are significant and for states of biaxial loading the viscoelastic effects on stability are magnified. Under these conditions, classical theory can nonconservatively predict critical buckling loads by 20% or greater.

The present paper investigates viscoelastic column buckling behavior of laminated composites. The present analysis treats general instability of the laminate, not ply buckling. A viscoelastic column buckling analysis is formulated using the simplifying assumption of linear viscoelasticity and employing the quasi-elastic approximation. The governing equation is developed using the Theorem of Minimum Potential Energy and solved using the Rayleigh-Ritz method. Results from case studies using the material properties for a typical graphite/epoxy system are reported for columns having simply supported or clamped boundary conditions. In addition, the effects of column length-to-thickness aspect ratio (ℓ/t) and laminate configuration on critical buckling load are discussed.

Viscoelastic Considerations

The viscoelastic properties of continuous fiber reinforced plastic composite systems are orthotropic. While the response of the material ranges from nearly elastic in the fiber direction to nonlinear viscoelastic in shear, linear viscoelastic response is assumed in the current investigation. Research reported in the literature [1,5] indicates that this assumption is reasonable for glass/epoxy and graphite/epoxy composite laminates.

The constitutive relation for a linear viscoelastic anisotropic material is

$$\epsilon_{ij}(t) = \int_{-\infty}^{t} S_{ijk\ell}(t - t') \frac{d\sigma_{k\ell}(t)}{dt'} \, dt' \tag{1}$$

Formulation of the viscoelastic governing equation for column buckling is based on these relations. Sims [5] showed that for cases where the applied load is nearly timewise constant, the correspondence principle results reduce approximately to those obtained by direct substitution of time-varying properties into the elastic formulation of the problem. The error incurred by making this "quasi-elastic" approximation is less than 10% for a typical graphite/epoxy composite, and results in considerable simplification by negating the need to perform inverse Laplace transformations of the "associated elastic problem."[2] This error was deemed satisfactory in light of the analytical sim-

---

[2]The associated elastic problem is the term applied to the elastic formulation of a viscoelastic problem in the Laplace transform (LT) domain allowed by the correspondence principle.

plifications achieved, since the study is a comparative one investigating the significance of including transverse shear deformation (TSD) effects on the viscoelastic buckling behavior of columns.

Viscoelastic Property Determination

Since the quasi-elastic method of solution has been employed, the viscoelastic effects are expressed in the buckling analysis solely by the creep compliance or relaxation moduli. A survey of the literature quickly revealed that a complete set of viscoelastic data for graphite/epoxy (or any other system) is not available. Viscoelastic data were available for epoxy resin, and therefore a micromechanical materials modeling approach was used to determine the necessary properties.

The Tsai-Halpin equations have been used assuming that orthotropic elastic fibers are embedded in a viscoelastic, isotropic matrix. The properties assumed for graphite fibers are given in Table 1 along with all other important material property data. The following power-law formulation is assumed for the creep compliance [5]

$$S_m(t) = b' t^\gamma \tag{2}$$

where values for b' and γ are given in Table 1. The Young's modulus is a more convenient form so this expression is inverted using the correspondence principle to give

$$E_m(t) = \frac{1}{\Gamma(1 + \gamma) \, \Gamma(1 - \gamma) S_m(t)} \tag{3}$$

TABLE 1—*Material properties assumed for a graphite/epoxy composite (similar to ASI/3501 material).*

	SI	Engineering
Fiber modulus, longitudinal (E_{fl})	165.5 GPa	24.0×10^6 psi
Fiber modulus, transverse (E_{ft})	13.8 GPa	2.0×10^6 psi
Fiber Poisson's ratio (ν_{LT})	0.3	0.3
Fiber shear modulus (G_{LT})	27.5 GPa	4.0×10^6 psi
Matrix modulus, initial (E_{mo})	4.6 GPa	0.67×10^6 psi
Matrix Poisson's ratio (ν_{mo})	0.3	0.3
Constants for power-law compliance formulation		
$\quad S_{11} = b' t^\gamma \quad (b')$	42.2×10^{-9} Pa^{-1}	0.291×10^{-5} psi^{-1}
$\quad\quad\quad\quad\quad (a)$		0.20
Fiber volume fraction (V_f)		0.62

which is the exact inversion of Eq 2 accomplished using LT theory, where Γ is the gamma function. This was done to minimize error in $E_m(t)$ which could subsequently propagate into the relaxation moduli of the components.

Two other matrix properties are necessary to completely characterize its viscoelastic response: $\nu_m(t)$ and $G_m(t)$, the Poisson's ratio and shear modulus, respectively. The Poisson's ratio is found by assuming the bulk modulus to be a constant and determined from

$$K_m = E_m(0)/3[1 - 2\nu_m(0)] \tag{4}$$

Knowing K_m, the Poisson's ratio is found to be

$$\nu_m(t) = 1/2(1 - E_m(t)/3K_m) \tag{5}$$

and the shear modulus is found to be

$$G_m(t) = E_m(t)/2[1 + \nu_m(t)] \tag{6}$$

Using the Tsai-Halpin relations [11], the effective viscoelastic properties of the composite $E_1(t)$, $E_2(t)$, $E_3(t)$, $\nu_{12}(t)$, $\nu_{13}(t)$, $\nu_{23}(t)$, $G_{12}(t)$, $G_{13}(t)$, $G_{23}(t)$ are determined. Again the property information is given in Table 1. Effectively, the longitudinal properties (E_1, ν_{12}, ν_{13}) conform to the rule of mixtures

$$P_1(t) = P_f V_f + P_m(t)[1 - V_f] \tag{7}$$

The transverse and shear properties (E_2, E_3, G_{12}, G_{23}, etc.) are approximated by

$$P(t) = P_m(t)\left(\frac{P_f(1 - \xi_i V_f) + \xi_i P_m(t)(1 - V_f)}{P_f(1 - V_f) + P_m(t)(\xi_i + V_f)}\right) \tag{8}$$

where ξ_i is an empirical parameter related to the packing geometry, and P_f and P_m refer to the fiber[3] and matrix properties (ξ_i is 1 for the packing geometry assumed).

These basic viscoelastic lamina properties are substituted for their elastic equivalents in laminated plate theory under the quasi-elastic assumption. Laminated plate theory is then used exactly as in elastic analyses.

---

[3]Note that P_f is timewise constant in accordance with the earlier assumption of elastic behavior.

Derivation of the Buckling Equation

The buckling analysis formulated herein is for a column of the geometry described in Fig. 1. Derivation of the governing equation comes through a reduction of small-deflection plate theory [11,12] to one-dimensional beam theory. This means that

$$\frac{\partial(\ )}{\partial y} = V = F_y = N_{xy} = N_y = M_{xy} = M_y = 0 \qquad (9)$$

The strain-displacement relations used are

$$\epsilon_x(t) = \frac{\partial u^0(t)}{\partial x}$$

$$\gamma_{xz}(t) = \frac{\partial u(t)}{\partial z} + \frac{\partial w(t)}{\partial x} \qquad (10)$$

$$\epsilon_z(t) = \frac{\partial w(t)}{\partial z}$$

and the displacement fields in Eq 2 are defined

$$u(x, y, z, t) = u^0(x, y, t) + zF_x(x, y, t)$$
$$w(x, y, z, t) = w^0(x, y, t) + f_1F_z(z, t) \qquad (11)$$

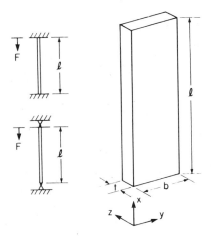

FIG. 1—*Column geometry description.*

where f_1 is a tracing constant employed to include TND effects ($f_1 = 1$), or in its neglect ($f_1 = 0$).

Using contracted tensor notation, the constitutive relation including thermal effects for the kth viscoelastic lamina (assuming the quasi-elastic approximation) in a laminated plate, transformed to the plate coordinate system, is

$$\sigma_i^k(t) = \bar{Q}_{ij}^k(t) \cdot \epsilon_j^k(t) + \alpha_i \Delta T, \qquad i, j = 1, 2, \ldots, 6 \qquad (12)$$

where the contracted notation is defined as usual, $\sigma_{ii} = \sigma_i$ for $i = 1, 2, 3$, $\sigma_{23} = \sigma_4$, $\sigma_{13} = \sigma_5$, and $\sigma_{12} = \sigma_6$.

The theorem of minimum potential energy was used to derive the governing equation of buckling. The validity of using minimum potential energy in viscoelastic analysis has been shown by Christensen [13] and an exact viscoelastic formulation given. The present analysis uses the correspondence principle, which is slightly different from the exact transform version and is stated

$$V(t) = \int_R W(t) \, dR - \int_S T_i(t) \, u_i(t) \, dS - \int_R F_i u_i(t) \, dR \qquad (13)$$

For a column, the body forces are assumed negligible and for buckling, only the single in-plane force resultant $N_x(t)$ is considered. Expanding Eq 13, the total potential energy is expressed

$$V = \frac{1}{2} \sum_{k=1}^{n} \int \int \{\sigma_x \epsilon_x + \sigma_z \epsilon_z + \sigma_{xz} \epsilon_{xz}\}^k \, dA$$

$$- \int \int_A N_x \left[\frac{\partial u^0}{\partial x} + \frac{1}{2} \left(\frac{\partial u^0}{\partial x} \right)^2 \right] dA \qquad (14)$$

Following the usual procedure of substituting the constitutive equations (12) and the strain displacement relations (10) into Eq 14, the total potential energy was found in terms of viscoelastic stiffnesses, beam displacements, and applied load. This equation with all terms not containing $F_z(z, t)$ integrated through the plate thickness is

$$V(t) = \frac{b}{2} \int_0^L \left\{ A_{11}(t) \left(\frac{\partial u(t)}{\partial x} \right)^2 + A_{55}(t) \right.$$

$$\times \left[F_x(t)^2 + 2F_x(t) \frac{\partial w(t)}{\partial x} + \left(\frac{\partial w(t)}{\partial x} \right)^2 \right]$$

$$+ 2B_{11}(t) \left(\frac{\partial u(t)}{\partial x} \frac{\partial F_x(t)}{\partial x} \right) + D_{11}(t) \left(\frac{\partial F_x(t)}{\partial x} \right)^2$$

$$-2\left[\frac{\partial u(t)}{\partial x} + \frac{1}{2}\left(\frac{\partial w(t)}{\partial x}\right)^2\right](N_x^t) - 2\left(\frac{\partial F_x(t)}{\partial x}\right)(M_x^t)\Bigg\} dx$$

$$+ f_1\frac{1}{2}\sum_{k=1}^{h}\int_0^L\int_{h_k}^{h_{k+1}}\Bigg\{\left[2\left(\frac{\partial u^0(t)}{\partial x}\;\bar{Q}_{13}^k(t)\;\frac{\partial F_z(t)}{\partial z}\right)\right]$$

$$+ 2\left(\frac{\partial F_x(t)}{\partial x}\right)\bar{Q}_{13}^k(t)z\;\frac{\partial F_z(t)}{\partial z} + \bar{Q}_{33}^k(t)\left(\frac{\partial F_z}{\partial z}\right)^2$$

$$-2\bar{Q}_{3j}^k\alpha_j^k(t)\;\theta\;(z,\;t)\;\frac{\partial F_z(t)}{\partial z} + \bar{Q}_{ij}^k\alpha_i\theta\;(z,\;t)\Bigg\}\;dzdx$$

$$- \int_0^L N_x\left[\frac{\partial u^0(t)}{\partial x} + \frac{1}{2}\left(\frac{\partial w^0(t)}{\partial x}\right)^2\right]dx \qquad (15)$$

where

A_{ij} = extensional stiffnesses,
D_{ij} = flexural stiffnesses, and
B_{ij} = bending-extensional coupling stiffnesses

defined by

$$A_{ij} = \sum_{k=1}^{n}\bar{Q}_{ij}^k(t)(h_{k+1} - h_k)$$

$$B_{ij} = \sum_{k=1}^{n}\bar{Q}_{ij}^k(t)(h_{k+1}^2 - h_k^2)$$

$$D_{ij} = \sum_{k=1}^{n}\bar{Q}_{ij}^k(t)(h_{k+1}^3 - h_k^3)$$

and the thermal stress and moment resultants are given by

$$N_i^t = \sum_{k=1}^{n}\int_{h_k}^{h_{k+1}}\bar{Q}_{ij}^k(t)\;\alpha_j^k(t)\;\theta\;(z,\;t)\;dz$$

$$\qquad (16)$$

$$M_i^t = \sum_{k=1}^{n}\int_{h_k}^{h_{k+1}}\bar{Q}_{ij}^k(t)\;\alpha_j^k(t)\;\theta\;(z,\;t)\;dz$$

Buckling occurs when the total potential energy is a minimum at a stationary point (the first variation is zero), and the second variation exists and is positive definite. Taking the variation of the total potential energy results in a set of Euler-Lagrange equations and associated natural boundary conditions. Mathematically this is stated

$$\delta V(t) = \int_0^L [(\ ) \delta u^0(t) + (\ ) \delta w^0(t) + (\ ) \delta F_x(t)]\ dx$$

$$+ \sum_{k=1}^n \int_0^L \int_{h_k}^{h_{k+1}} (\ ) \delta F_x(t)\ dz dx$$

$$+ \text{natural boundary conditions}\quad (17)$$

where $(\ )$ represent the cogent equations. In order to satisfy the conditions that $\delta V(t) = 0$, each bracketed equation must be equal to zero.

The Euler-Lagrange equations were used to determine approximate expressions for $F_x(x, y, t)$ and $F_z(z, t)$ as a function of the midplane deflection, w^0. This was accomplished for $F_x(x, y, t)$ using the following Euler-Lagrange equations

$$A_{11}(t) \frac{d^2 u^0(t)}{dx^2} + B_{11}(t) \frac{d^2 F_x(t)}{dx^2} = 0$$

$$A_{55}(t) \frac{dF_x(t)}{dx} + A_{55}(t) \frac{d^2 w^0(t)}{dx^2} - \bar{N}_x(t) \frac{d^2 w^0(t)}{dx^2} = 0$$

$$(18)$$

$$-A_{55}(t)F_x(t) - A_{55}(t) \frac{dw^0(t)}{dx} + B_{11}(t) \frac{d^2 u^0(t)}{dx^2}$$

$$+ D_{11}(t) \frac{d^2 F_x(t)}{dx^2} = 0$$

and solving for $F_x(x, t)$

$$F_x(x, t) = -\frac{dw^0(t)}{dx} - \left(\frac{D_{11}(t)A_{11}(t) - B_{11}^2(t)}{A_{11}(t)} \right)$$

$$\times \left(\frac{A_{55}(t) - N_x(t)}{A_{55}^2(t)} \right) \frac{d^3 w^0(t)}{dx^3}\quad (19)$$

An approximation for the function $\partial F_z(t)/\partial z$ was found by integrating the Euler-Lagrange equation associated with ∂F_z with respect to z. The Euler-Lagrange equation is

$$f_1\left[\bar{Q}_{13}^k \frac{\partial F_x(t)}{\partial x} + \bar{Q}_{33}^k \frac{\partial^2 F_z(t)}{\partial z^2}\right] = 0 \qquad (20)$$

Integrating this, the result is

$$\frac{\partial F_z(t)}{\partial z} = \frac{f_1}{\bar{Q}_{33}^k}\left[-\bar{Q}_{13}^k \frac{\partial F_x(t)}{\partial x}\right] + k \qquad (21)$$

Substituting into the natural boundary condition, the constant k is found to be

$$k = \frac{1}{\bar{Q}_{33}^k}\left[-\bar{Q}_{13}^k \frac{\partial u^0(t)}{\partial x}\right] \qquad (22)$$

With the mathematical form of $F_z(z, t)$ defined, it can be substituted into the total potential energy, Eq 15, and the integration of terms containing $F_z(z, t)$ across the column thickness performed.

The Rayleigh-Ritz method is used to solve the governing equation. For the simply supported column, the assumed displacement functions are chosen to satisfy the following boundary conditions

$$w^0 = 0, \qquad M_n = 0, \qquad u_n^0 = 0$$

where n and t refer to the normal and tangential direction with respect to the boundary. The displacement functions for a pinned or simply supported column are selected to satisfy the geometric boundary conditions, and are of the following form

$$u^0(x, t) = A(t) \sin \frac{\pi x}{L}$$

$$\qquad (23)$$

$$w^0(x, t) = C(t) \sin \frac{\pi x}{L}$$

For clamped boundary conditions, the following displacement functions were assumed

$$u^0(x, t) = A(t) \sin \frac{\pi x}{L}$$

$$w^0(x, t) = C(t)\left[1 - \cos \frac{2\pi x}{L}\right]$$

Substitution of these assumed displacement functions back into the governing equation and taking the variation with respect to the unknown displacement amplitudes $A(t)$ and $C(t)$ sets forth the eigenvalue problem

$$\begin{bmatrix} E_{11} & E_{12} \\ E_{21} & E_{22} \end{bmatrix} \begin{Bmatrix} A \\ C \end{Bmatrix} = N_x \begin{bmatrix} 0 & 0 \\ 0 & F_{22} \end{bmatrix} \begin{Bmatrix} A \\ C \end{Bmatrix} + N_x^2 \begin{bmatrix} 0 & 0 \\ 0 & G_{22} \end{bmatrix} \begin{Bmatrix} A \\ C \end{Bmatrix} \qquad (24)$$

Each E_{ij}, F_{ij}, and G_{ij} term above is an algebraic sum of the products of laminate stiffnesses and mode shape parameters of the form π/L. They are the coefficients of the displacement amplitudes resulting from taking the first variation of the total potential energy with respect to each of the unknown displacement amplitudes.

Notice that for the Rayleigh-Ritz method to give approximate solutions to the governing equation, the assumed form of the displacement functions need only satisfy the geometric boundary conditions. For the cases where $D_{16}(t)$ and $D_{26}(t)$ do not vanish, the natural boundary conditions are not satisfied. In this case, convergence of the approximate solution is slow.

Using a Gauss-Jordan reduction, Eq 24 can be simplified to give

$$[E(t)]\{C(t)\} = N_x(t)[F(t)]\{C(t)\} + N_x^2(t)[G(t)]\{C(t)\} \qquad (25)$$

which is a function of the single unknown displacement amplitude, C. Here the terms $E(t)$, $F(t)$, and $G(t)$ are the sums of the known E_{ij}, F_{ij}, and G_{ij} terms resulting from the above algebraic manipulation.

The critical buckling load is determined by finding the minimum eigenvalue from the solution of Eq 25. Since the eigenvalue problem contains quadratic terms, the solution must be found by an iterative approach. This can be accomplished by first solving the uncoupled problem

$$[E(t)]\{C(t)\} = N_x(t)[F(t)]\{C(t)\} \qquad (26)$$

the solution of which is used as a first approximation in the iterative solution of the whole problem. Numerical methods have been employed in the actual solution of the eigenvalue problem.

Results and Discussion

Two sets of boundary conditions were investigated, pinned and clamped. For each of these fixity conditions, laminate configuration and column length-to-thickness aspect ratio (ℓ/t) were varied and the effects of viscoelastic buckling response recorded. Critical buckling load results are presented in nondimensional form, the critical buckling load at time (t) divided by the zero time critical buckling load.

Figures 2 and 3 show the column buckling load as a function of time for four laminate configurations with pinned and clamped fixity, respectively. The results were generated using room temperature properties and a column ℓ/t geometry of 75.8. As expected, the stiffest column $[0]_{24}$, exhibited the highest critical buckling load at zero time, N_x^0, and was least viscoelastic. For all of the laminates the first four hundred hours were the most critical, followed by a very slow degradation of buckling strength. These results were obtained including TSD in the analysis; however, with ℓ/t of 75.8, only small differences exist between these curves and those obtained using classical theory. The major differences between the pinned and clamped columns are: (a) approximately a four-fold increase in critical buckling load for a given geometry, and (b) the stiffer laminates show slightly greater viscoelastic sensitivity.

The effect of beam aspect ratio on viscoelastic buckling behavior is shown in Figs. 4 and 5 for the $[0/\pm45/90]_{3s}$ column with pinned and clamped boundary conditions, respectively. Results are presented at three times, 200 h, 1300 h, and 2000 h, employing the analysis which includes TSD effects. At small aspect ratios, the viscoelastic effects are more pronounced, both for the pinned column and the clamped column. The clamped column, however, is more sensitive to the viscoelastic effects than the pinned column. This is because the TSD is more significant for the clamped fixity condition

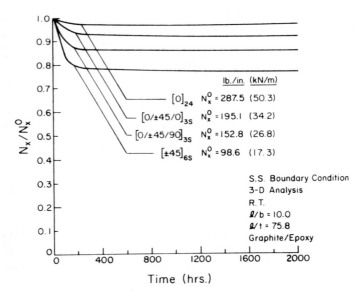

FIG. 2—*Nondimensionalized column buckling response as a function of time for four laminate configurations (pinned fixity condition).*

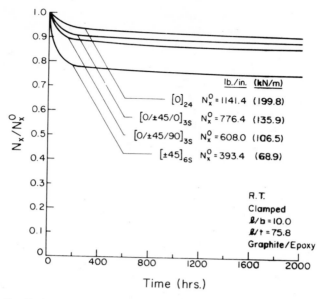

FIG. 3—*Nondimensionalized column buckling response as a function of time for four laminate configurations (clamped fixity condition).*

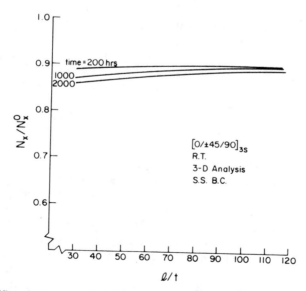

FIG. 4—*Viscoelastic column buckling as a function of length-to-thickness ratio for a quasi-isotropic laminate simply supported at each end.*

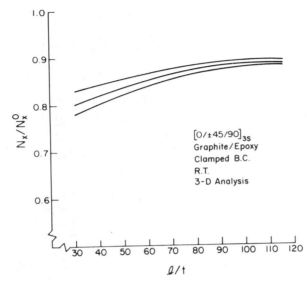

FIG. 5—*Viscoelastic column buckling of a quasi-isotropic laminate under clamped fixity conditions as a function of length-to-thickness aspect ratio (ℓ/t).*

and the TSD response is governed by $G_{xz}(t)$, which is the strongest viscoelastic property.

The effects of elevated temperatures on the viscoelastic buckling response were also evaluated. Elevating the temperature accelerated the rate process responsible for the viscoelastic behavior of the material, resulting in a larger viscoelastic response for a given interval of time. Figures 6 and 7 show critical buckling load versus ℓ/t for a $[0/\pm45/90]_{3s}$ column subjected to a single surface temperature exposure of 375 K. The viscoelastic effects are qualitatively similar to the previous case, only more pronounced. This is especially evident in Fig. 7, with the clamped boundary conditions, where for $\ell/t = 35.25$ the critical buckling load is reduced to 64% of its zero time value after 2000 h.

In Figs. 8 and 9, a comparison is made between viscoelastic buckling response predicted by classical analysis and analysis including transverse shear deformation, bending extensional coupling, and transverse normal deformation. The laminate configuration used is $[0/\pm45/90]_{3s}$, the ℓ/t is 38.5, and the boundary conditions are pinned (Fig. 8) and clamped (Fig. 9). It is seen that at 400 h there is a 1.0% difference in the results between the two analyses for the pinned fixity condition and a 4.0% difference for the clamped end constraint. This progresses to 2.0% and 6.0%, respectively, after 2000 h. This represents a significant difference, which is nonconservative if classical analysis is used indiscriminately.

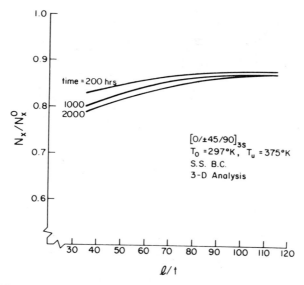

FIG. 6—*Effect of elevated temperature on the viscoelastic buckling behavior of a pinned column as a function of ℓ/t.*

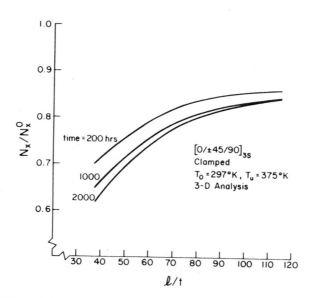

FIG. 7—*Effect of elevated temperature on the viscoelastic buckling behavior of a clamped column as a function of ℓ/t.*

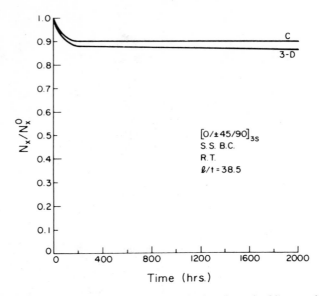

FIG. 8—*Comparison of classical analysis viscoelastic column buckling results with results from an analysis including TSD and TND effects (pinned fixity).*

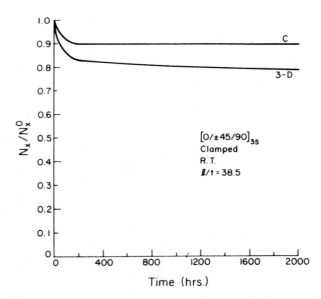

FIG. 9—*Comparison of classical analysis viscoelastic column buckling results with results from an analysis including TSD and TND effects (clamped fixity).*

Conclusions

The viscoelastic buckling analysis of columns has been investigated for two fixity conditions, pinned and clamped. Results show that the primary reduction in critical buckling load occurs during the first 400 h; between 400 and 2000 h, only a small additional decrease occurs. This result suggests that a knockdown factor could be applied to account for viscoelastic effects on stability in design of columns.

Classical analysis underpredicts the reduction in buckling load caused by viscoelastic effects, especially for ℓ/t ratios less than 50. When $\ell/t < 50$, the effect of transverse shear deformation is significant and TSD is controlled by the strongly viscoelastic shear modulus. For similar reasons, viscoelastic effects for clamped columns are greater than for pinned columns. In light of these findings, an analysis should be used which incorporates TSD and bending extensional coupling.

References

[1] Schapery, R. A., "Viscoelastic Behavior and Analysis of Composite Materials," *Composite Materials*, Vol. 2, G. P. Sendeckyj, Ed., Academic Press, New York, 1974, pp. 85–168.
[2] Schapery, R. A., "Stress Analysis of Viscoelastic Composite Materials," *Journal of Composite Materials*, Vol. 1, 1967, p. 228.
[3] Halpin, J. C. and Pagano, N. J., "Observations on Linear Anisotropic Viscoelasticity," *Journal of Composite Materials*, Vol. 2, No. 1, 1968, p. 68.
[4] Hashin, Zvi, "Viscoelastic Fiber Reinforced Materials," *AIAA Journal*, Vol. 4, No. 8, 1966, pp. 1411–1417.
[5] Sims, D. F., "Viscoelastic Creep and Relaxation Behavior of Laminated Composite Plates," Ph.D. Dissertation, Southern Methodist University, Dallas, TX, May 1972.
[6] Wu, C. I. and Vinson, J. R., "Influence of Large Amplitudes, Transverse Shear Deformation and Rotatory Inertia on Lateral Vibrations of Transversely Isotropic Plates," *Journal of Applied Mechanics*, June 1969, pp. 254–260.
[7] Whitney, J. M., "The Effects of Boundary Conditions on the Response of Laminated Composites," *Journal of Composite Materials*, Vol. 4, April 1970.
[8] Whitney, J. M., "The Effect of Transverse Shear Deformation on the Bending of Laminated Plates," *Journal of Composite Materials*, Vol. 3, July 1969.
[9] Flaggs, D. L. and Vinson, J. R., "Hygrothermal Effects on the Buckling of Laminated Composite Plates," *Fiber Science and Technology*, Vol. 11, 1978.
[10] Wilson, D. W. and Vinson, J. R., "Viscoelastic Analysis of Laminated Plate Buckling Including Transverse Shear and Normal Deformations," *American Institute of Aeronautics and Astronautics Journal*, Vol. 22, No. 7, July 1983, p. 982.
[11] Jones, Robert M., *Mechanics of Composite Materials*, McGraw-Hill, New York, 1975, p. 114.
[12] Wilson, D. W., "Viscoelastic Buckling Behavior of Laminated Plates," M.MAE Thesis, University of Delaware, Newark, DL, Aug. 1982.
[13] Christensen, R., *Theory of Viscoelasticity: An Introduction*, Academic Press, New York, 1971, pp. 165–172.

Design

Mitsunori Miki[1]

Design of Laminated Fibrous Composite Plates with Required Flexural Stiffness

REFERENCE: Miki, M., **"Design of Laminated Fibrous Composite Plates with Required Flexural Stiffness,"** *Recent Advances in Composites in the United States and Japan, ASTM STP 864*, J. R. Vinson and M. Taya, Eds., American Society for Testing and Materials, Philadelphia, 1985, pp. 387–400.

ABSTRACT: A new material design method of laminated fibrous composite plates with required flexural stiffness is proposed. The selection of kinds of fibers and matrices and the determination of fiber orientation angles can be done directly from the imposed conditions on flexural stiffness of laminated plates.

The laminated composites intended for this study are multiple balanced angle-ply symmetric laminates with one kind of unidirectional lamina.

A newly constructed diagram for flexural lamination parameters which are functions of the fiber orientation angles, and which appeared in Tsai's expression for evaluating the flexural stiffness of laminates, made it possible to determine the design specifications based on flexural effective engineering constants of laminates. A certain stacking sequence of laminates is indicated as a point in the diagram on which contours of flexural stiffness can be drawn. Using these contour lines, the design specification is easily determined as a point on this diagram. Thus, the point can be called a design point.

It became possible to specify the fiber orientation angle of each ply and the stacking sequence using the flexural lamination parameters indicated as the point on the diagram.

KEY WORDS: composite materials, laminated plates, material design, flexural stiffness, stacking sequence

In recent years, composite materials sometimes have been referred to as "tailored materials." This means that the physical properties of composite materials such as stiffness, strength, and the coefficient of thermal expansion can be varied by changing the kinds and the orientation of reinforcements. However, the material design methods for designing the composites with the required physical properties are few and no approaches to the design of fibrous laminated composites with required stiffness have been proposed.

[1] Professor, Materials System Research Laboratory, Kanazawa Institute of Technology, 7-1 Ogigaoka Nonoichi, Kanazawa 921, Japan.

Physical properties such as elasticity and strength are given as inherent material constants for conventional materials. For composites, however, these properties can be given as design specifications. When we take full advantage of this point, structures with higher functions can be designed and design optimization can be done.

Housner and Stein [1] calculated the optimum fiber direction of graphite/epoxy sandwich panels under axial compression, assuming that the angle of all the plies is the same. Hirano [2] calculated the optimum fiber direction of all the plies of laminated plate with orthotropic layers under uniaxial and biaxial compression. Their works showed the merit of composite materials, but their approaches are not efficient in designing composite materials under many constraints for strength and stiffness.

Tsai and Hahn [3] developed a new evaluation method of mechanical properties of laminated fibrous composites. Since stiffness and strength of laminated plates can be evaluated easily by using their expressions, we can choose the optimal constructions from many proposed ones.

Chamis [4] proposed a simple evaluation method of stiffness and strength by using some charts. This method can also be used to choose the optimal constructions.

When these evaluation procedures are used iteratively, the best construction can be found from a number of trials. However, the material design method using these evaluation procedures is essentially a method of trial and error. And it is very difficult to propose the trial construction of laminates for certain applications. To avoid this difficulty, the author has proposed a new approach [5] for designing laminated composite plates with required in-plane stiffness.

In this paper, a new material design method of laminated fibrous composite plates with required flexural stiffness is proposed. The selection of kinds of lamina and the determination of fiber orientation angles can be done directly from the imposed conditions on flexural stiffness of laminated plates.

Flexural Stiffness of Laminated Composites

The composite laminates intended for this study are multiple balanced angle-ply symmetric laminates with $2N_h$ plies of unidirectional lamina of a kind. The stacking sequence of the laminates is represented by

$$[(+\theta_n/-\theta_n)_{N_n}/(+\theta_{n-1}/-\theta_{n-1})_{N_{n-1}}/\cdots/(+\theta_2/-\theta_2)_{N_2}/(+\theta_1/-\theta_1)_{N_1}]_s \tag{1}$$

$$N_h = \sum_{k=1}^{n} N_k \tag{2}$$

This type of laminate is free from couplings between extension and bending, and between shear and extension, and has very little coupling between bending and twisting. That is, the coupling stiffness $B_{ij} = 0$ and the elements D_{16} and D_{26} of bending stiffness are nearly equal to zero when the number of plies is large. This condition is considered to hold in this paper.

The flexural effective engineering constants, that is, E_1^f (effective Young's modulus along the 1-axis), E_2^f (effective Young's modulus along the 2-axis), E_6^f (effective twisting modulus along the 1,2-axes), and ν_{21}^f (coupling coefficient between bending along 1- and 2-axes), are considered as flexural stiffness of the laminates. The subscript 1 and 2 represent the major and minor principal material axes of the laminates.

The relation between the moment M_i and the curvature k_i of a laminated plate is represented as

$$\mathbf{M} = \mathbf{Dk}, \qquad \mathbf{M} = \begin{bmatrix} M_1 \\ M_2 \\ M_6 \end{bmatrix}, \qquad \mathbf{k} = \begin{bmatrix} k_1 \\ k_2 \\ k_6 \end{bmatrix}, \qquad \mathbf{D} = \begin{bmatrix} D_{11} & D_{12} & 0 \\ & D_{22} & 0 \\ \text{Sym.} & & D_{66} \end{bmatrix}$$

(3)

The elements of the moment are shown in Fig. 1.

Tsai and Hahn [3] gave the expression for the normalized flexural stiffness $D_{ij}^* (=12D_{ij}/h^3)$ as

$$\begin{bmatrix} D_{11}^* \\ D_{22}^* \\ D_{12}^* \\ D_{66}^* \end{bmatrix} = \begin{bmatrix} U_1 & W_1^* & W_2^* \\ U_1 & -W_1^* & W_2^* \\ U_4 & 0 & -W_2^* \\ U_5 & 0 & -W_2^* \end{bmatrix} \begin{bmatrix} 1 \\ U_2 \\ U_3 \end{bmatrix}$$

(4)

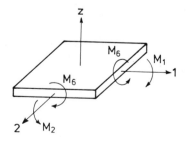

FIG. 1—Element of the moment.

TABLE 1—*Material constants of typical composites [3].*

Material	Engineering constants, E GPa				Linear Combinations of On-Axis Modulus, GPa				
	E_x	E_y	E_s	ν_x	U_1	U_2	U_3	U_4	U_5
Graphite/epoxy (T300/5208)	181	10.3	7.17	0.28	76.37	85.73	19.71	22.61	26.88
Aramid/epoxy (Kevlar49/epoxy)	76	5.5	2.3	0.34	32.44	35.55	8.65	10.54	10.95
Glass/epoxy (Scotchply 1002)	38.6	8.27	4.14	0.26	20.47	15.40	3.33	5.53	7.47

NOTE: E_x = longitudinal Young's modulus along fiber direction; E_y = transversal Young's modulus; E_s = shear modulus along fiber direction; ν_x = Poisson's ratio along fiber direction.

where U_m $(m = 1, \ldots, 5)$ is the linear combination of on-axis modulus of a lamina and can be considered as a material constant, and $W*$'s are the factors represented as

$$W_1^* = \frac{24}{h^3} \int_0^{h/2} \cos 2\theta \, z^2 dz, \qquad W_2^* = \frac{24}{h^3} \int_0^{h/2} \cos 4\theta \, z^2 dz \qquad (5)$$

where

h = thickness of laminated plate,
θ = fiber orientation angle, and
z = direction of normal to middle surface of laminated plates.

The flexural effective engineering constants are obtained from D_{ij}^* as follows

$$E_1^f = 1/d_{11}^*, \qquad E_2^f = 1/d_{22}^*, \qquad E_6^f = 1/d_{66}^*$$

$$\nu_{21}^f = -d_{21}^*/d_{11}^*, \qquad [d^*] = [D^*]^{-1} \qquad (6)$$

Table 1 shows the values of U_m of typical composites.

Flexural Lamination Paramater Diagram

Flexural Lamination Parameters

In this paper, W_1^* and W_2^* in Eq 5 are referred to as flexural lamination parameters. Figure 2 shows the section of the laminates represented by Eq 1. In this figure z_k $(k = 1, \ldots, n)$ represents the distance from the middle surface to the upper surface of the kth ply group in which the fiber orientation is $\pm\theta_k$. Then W_1^* can be rewritten as

$$W_1^* = \frac{24}{h^3} \int_0^{h/2} \cos 2\theta \, z^2 dz$$

$$= \frac{24}{h^3} \left(\int_{z_0}^{z_1} \cos 2\theta_1 z^2 dz + \cdots + \int_{z_{k-1}}^{z_k} \cos 2\theta_k z^2 dz \right.$$

$$\left. + \cdots + \int_{z_{n-1}}^{z_n} \cos 2\theta_n z^2 dz \right)$$

$$= \frac{8}{h^3} (\cos 2\theta_1 (z_1^3 - z_0^3) + \cdots + \cos 2\theta_k (z_k^3 - z_{k-1}^3)$$

$$+ \cdots + \cos 2\theta_n (z_n^3 - z_{n-1}^3))$$

$$= \sum_{k=1}^{n} \left(\left(\frac{2}{h} z_k \right)^3 - \left(\frac{2}{h} z_{k-1} \right)^3 \right) \cos 2\theta_k \qquad (7)$$

where $z_0 = 0$ and $z_n = h/2$.

When a factor is defined as

$$\xi_k = (2z_k/h)^3 - (2z_{k-1}/h)^3 \qquad (8)$$

Eq 7 is represented as

$$W_1^* = \sum_{k=1}^{n} \xi_k \cos 2\theta_k \qquad (9.1)$$

Similarly

$$W_2^* = \sum_{k=1}^{n} \xi_k \cos 4\theta_k \qquad (9.2)$$

The summation of ξ_k equals unity

$$\sum_{k=1}^{n} \xi_k = \sum_{k=1}^{n} (2z_k/h)^3 - (2z_{k-1}/h)^3 = \frac{8}{h^3} \sum_{k=1}^{n} (z_k^3 - z_{k-1}^3) = 1 \quad (10)$$

Allowable Region of Flexural Lamination Parameters

The design of composite laminates with required flexural stiffness begins with the determination of the design specifications for the flexural effective engineering constants. The author proposes a flexural lamination parameter diagram to investigate the possible design specifications. The diagrams of the

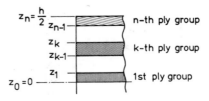

FIG. 2—*Sectional view of laminates.*

flexural lamination parameters W_1^* (as abscissa) and W_2^* (as ordinate) are considered. The lamination parameters are determined by Eq 9 when the stacking sequence of a laminate is given, and then the point on the flexural lamination parameter diagram is determined. This point is referred to as the lamination point.

In order to investigate the possibility on flexural stiffness of composite laminates, the allowable region of flexural lamination parameters should be clarified.

From Eq 9, the following condition holds

$$-1 \leqq W_r^* \leqq 1 \qquad (r = 1,2) \tag{11}$$

When the number of the kind of fiber orientation angle is one, the stacking sequence is represented as

$$[(+\theta/-\theta)_{N_h}]_s \tag{12}$$

Then Eq 9 is reduced to

$$W_1^* = \cos 2\theta, \qquad W_2^* = \cos 4\theta \tag{13}$$

In this case, the following relation holds from a trigonometric identity

$$W_2^* = 2W_1^{*2} - 1 \tag{14}$$

Consequently, the allowable region of lamination parameter is represented by the curve ABC in Fig. 3.

When the number of kinds of fiber orientation angle is two, the stacking sequence is represented as

$$[(+\theta_2/-\theta_2)_{N_2}/(+\theta_1/-\theta_1)_{N_1}]_s \tag{15}$$

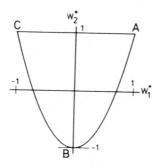

FIG. 3—*Allowable region (curve ABC) of the flexural lamination parameters. The stacking sequence is $[(+\theta/-\theta)_{N_h}]_s$.*

The flexural lamination parameters are given as

$$W_1^* = \xi_1 \cos 2\theta_1 + \xi_2 \cos 2\theta_2$$

$$W_2^* = \xi_1 \cos 4\theta_1 + \xi_2 \cos 4\theta_2$$

(16)

The necessary and sufficient conditions to obtain the orientation angles θ_1 and θ_2 are

$$-1 \leqq \cos 2\theta_r \leqq 1 \qquad (r = 1,2)$$

(17)

From Eqs 16 and 17, the following inequality can be obtained

$$W_2^* \geqq 2W_1^{*2} - 1$$

(18)

Then the allowable region of the flexural lamination parameters is shown as the hatched region in Fig. 4. The region can be obtained with the assumption that ξ in Eq 16 can vary continuously from zero to unity; that is, the total number of plies is supposed to be sufficiently large.

When the number of kinds of orientation angles is n, the evaluation $W_2^* - (2W_1^{*2} - 1)$ yields the allowable region of lamination parameters

$$W_2^* - (2W_1^{*2} - 1) = \sum_{k=1}^{n} \xi_k \cos 4\theta_k - \left[2\left(\sum_{k=1}^{n} \xi_k \cos 2\theta_k \right)^2 - 1 \right]$$

$$= 2 \sum_{k=1}^{n} \xi_k \left(\cos 2\theta_k - \sum_{k=1}^{n} \xi_k \cos 2\theta_k \right)^2 \geqq 0 \quad (19)$$

From Eq 19, the term $W_2^* - (2W_1^{*2} - 1)$ is nonnegative, and is zero only in the case that the orientation angle θ_k is the same. Therefore, the allowable

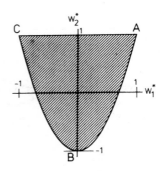

FIG. 4—*Allowable region (hatched region) of flexural lamination parameters. The number of the kinds of fiber orientation is two.*

region of the flexural lamination parameters is the same region of the laminates with two kinds of orientation angles.

It is found from the points mentioned above that two kinds of orientation
angles are necessary and sufficient for designing the laminates with required
flexural stiffness.

Thus, the lamination point is determined on the flexural lamination parameter diagram when the stacking sequence is given, and the stacking sequence can be determined when the lamination point is given from the design
specifications. In the latter case, the information on flexural stiffness is
needed on the allowable region of the flexural lamination parameters for the
determination of the lamination point.

Contours of Flexural Effective Engineering Constants
on the Flexural Lamination Parameters

When the flexural lamination parameters W_1^* and W_2^* are determined, the
flexural effective engineering constants are evaluated by using Eqs 4 and 6.
The contours of effective Young's modulus E_1^f can be drawn from the relations. The expression for the E_1^f-contour is obtained as

$$E_1^f = \frac{D_{11}^* D_{22}^* - D_{12}^{*2}}{D_{22}^*}$$

$$= \frac{(U_1 + U_2 W_1^* + U_3 W_2^*)(U_1 - U_2 W_1^* + U_3 W_2^*) - (U_4 - U_3 W_2^*)^2}{U_1 - U_2 W_1^* + U_3 W_2^*}$$

Then

$$W_2^* = \frac{U_2^2 W_1^{*2} - U_2 E_1^f W_1^* + U_1 E_1^f + U_4^2 - U_1^2}{U_3(2U_1 + 2U_4 - E_1^f)} \qquad (E_1^f\text{-contour}) \quad (20.1)$$

Similarly, E_2^f-contour, E_6^f-contour, and ν_{21}^f-contour are obtained as

$$W_2^* = \frac{U_2^2 W_1^{*2} + U_2 E_2^f W_1^* + U_1 E_2^f + U_4^2 - U_1^2}{U_3(2U_1 + 2U_4 - E_2^f)} \qquad (E_2^f\text{-contour})$$

$$(20.2)$$

$$W_2^* = \frac{U_5 - E_6^f}{U_3} \qquad (E_6^f\text{-contour}) \qquad (20.3)$$

$$W_2^* = \frac{U_2 \nu_{21}^f W_1^* + U_4 - U_1 \nu_{21}^f}{U_3(1 + \nu_{21}^f)} \qquad (\nu_{21}^f\text{-contour}) \qquad (20.4)$$

Figure 5 shows the contours of flexural effective engineering constants on the flexural lamination parameter diagram of typical composites. The design specifications for flexural stiffness can be easily determined by using the diagrams. It should be noted that there is E^f_6-scale instead of E^f_6-contour because the E^f_6-contour becomes parallel to the abscissa.

Determination of the Stacking Sequence of Composite Laminates

Suppose that the following conditions are given for flexural effective engineering constants

$$E^f_1 \geqq 40, \qquad E^f_2 \geqq 20, \qquad E^f_6 \geqq 6 \text{ (GPa)}, \qquad \nu^f_{21} \leqq 0.3 \qquad (21)$$

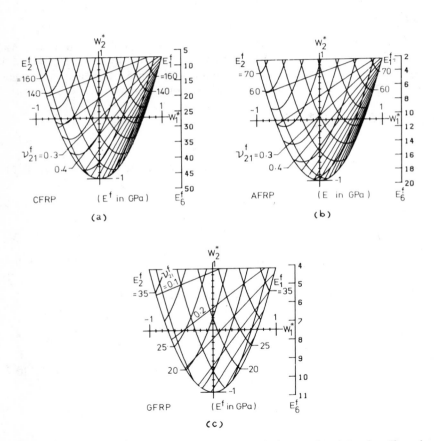

FIG. 5—*Flexural lamination parameter diagrams of typical composites: (a) carbon fiber reinforced plastic (CFRP). (b) aramid fiber reinforced plastic (AFRP). (c) glass fiber reinforced plastic (GFRP).*

Since a region which satisfied the above conditions is obtained in each lamination parameter diagram of T300/5208 and Kevlar/epoxy, these materials can be used.

Figure 6 shows the feasible design regions where the conditions in Eq 19 are satisfied. In such a case where the feasible design region can be obtained for more than two kinds of materials, the determination of the material can be done in considering the other factors—for example, the cost of the material.

Suppose that Kevlar/epoxy is used. Figure 7 shows the detail of the feasible design region on the flexural lamination parameter diagram of Kevlar/epoxy. One design point in the feasible design region which gives the values of W_1^* and W_2^* is determined when another condition for effective engineering constants is imposed. The following condition is supposed to be imposed here

$$\text{Maximization of } E_6^f \qquad\qquad (22)$$

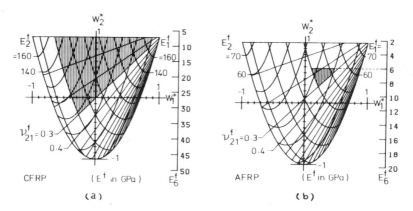

FIG. 6—*Feasible design region on the flexural lamination parameter diagrams: (a) CFRP. (b) AFRP.*

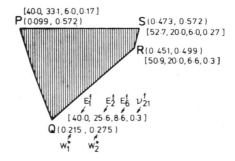

FIG. 7—*Detail of the feasible design region of AFRP.*

Expression 22 yields the design point as point Q in Fig. 7. The values of the flexural lamination parameters W_1^* and W_2^* are obtained easily by using the W^*-scales of the diagram, or from the calculation for the intersecting point of contour lines exactly. In this case, the values of W_1^* and W_2^* for the point Q are obtained as

$$W_1^* = 0.215, \qquad W_2^* = 0.275 \qquad (23)$$

The stacking sequence of the laminate can be determined from these values.

Since the allowable region of the flexural lamination parameters of the laminates with two kinds of fiber orientation angles is the same as that of the laminates with n kinds of fiber orientation angles, two kinds of orientation angles are necessary and sufficient for designing composite laminates with required flexural stiffness.

When the kinds of orientation angles are two, the flexural lamination parameters are evaluated by Eq 16. The design variables are ξ_1, ξ_2, θ_1, and θ_2 in the equation, but the number of independent design variables is three because ξ_1 and ξ_2 are not independent of each other. Therefore, the values of ξ_1 or ξ_2 can be given arbitrarily when Eq 16 is solved for θ_1 and θ_2 under the condition of Eq 23. This is very important because the value of ξ becomes discrete when the number of plies is finite.

The orientation angles θ_1 and θ_2 are obtained from Eq 16, as

$$\theta_1 = \frac{1}{2}\cos^{-1}T_1, \qquad \theta_2 = \frac{1}{2}\cos^{-1}T_2$$

$$T_1 = \{2\xi_1 W_1^* \pm \sqrt{[4\xi_1^2 W_1^{*2} - 2\xi_1(2W_1^{*2} - \xi_2 W_2^* - \xi_2)]}\}/2\xi_1 \qquad (24)$$

$$T_2 = \{2\xi_2 W_1^* \mp \sqrt{[4\xi_2^2 W_1^{*2} - 2\xi_2(2W_1^{*2} - \xi_1 W_2^* - \xi_1)]}\}/2\xi_2$$

The factors ξ_1 and ξ_2 are obtained from Eq 8 as

$$\xi_1 = (2z_1/h)^3, \qquad \xi_2 = 1 - \xi_1 \qquad (25)$$

Suppose that the total number of plies $2N_h$ equals 20. In this case, the possible value of $2z_1/h$ is a multiple of 0.1.

Table 2 gives the values of design variables which satisfy the condition of Eq 23. Then, the stacking sequence is represented as

$$[(+19.80/-19.80)_3/(+74.06/-74.06)_7]_s \qquad (26.1)$$

or

$$[(+62.47/-62.47)_2/(+7.52/-7.52)_8]_s \qquad (26.2)$$

TABLE 2—*Design variables.*

$2z_1/h$	ξ_1	ξ_2	θ_1, deg	θ_2, deg
0.7	0.657	0.343	74.06	19.80
0.8	0.512	0.488	7.52	62.47

TABLE 3—*Flexural properties of the laminate.*

Flexural Stiffness, GPa	Flexural Compliance, GPa^{-1}	Flexural Effective Engineering Constants, GPa
$D^*_{11} = 42.46$	$d^*_{11} = 0.02502$	$E^f_1 = 40.0$
$D^*_{22} = 27.18$	$d^*_{22} = 0.03911$	$E^f_2 = 25.57$
$D^*_{12} = 8.16$	$d^*_{12} = -0.00747$	$E^f_6 = 8.55$
$D^*_{66} = 8.57$	$d^*_{66} = 0.117$	$\nu^f_{21} = 0.30$

Table 3 gives the details of flexural stiffness of the laminate of which the stacking sequence is represented by Eq 26.1. It can be seen from Table 3 that the design specifications given in Eq 21 are satisfied.

Conclusion

The design procedure of composite laminates with required flexural stiffness is summarized as follows:

Step 1: Draw the flexural lamination parameter diagram of available materials.

Step 2: Make a design specification in relation to flexural stiffness.

Step 3: Find the feasible design region on the flexural lamination parameter diagram. When no design region is found on the diagram of a material, change the material. In case no design region can be found for all available materials, the design specification should be relaxed. In this step the ply material is determined.

Step 4: Determine one point in the feasible design region (referred as a design point) with another condition such as the maximization of E^f_6.

Step 5: Give the number of plies and the number of the kinds of orientation angles. Two kinds of orientation angles are enough.

Step 6: Obtain the fiber orientation angle of each ply from the flexural lamination parameters of the design point.

References

[1] Housner, J. M. and Stein, M., NASA Report TN D-7996, National Aeronautics and Space Administration, Washington, DC, Oct. 1975.

[2] Hirano, Y., *AIAA Journal*, Vol. 17, No. 9, Sept. 1979, p. 1017.
[3] Tsai, S. W. and Hahn, H. T., *Introduction to Composite Materials*, Technomic, Westport, CT, 1980, p. 167.
[4] Chamis, C. C., *Design Engineering*, May 1980, p. 39.
[5] Miki, M., *Progress in Science and Engineering of Composites*, T. Hayashi et al, Eds., Japan Society for Composite Materials, Tokyo, 1982, pp. 1725–1731.

Mikio Morita,[1] *Kumi Ochiai,*[1] *Hisato Kamohara,*[1] *Itsuo Arima,*[1] *Tatsuyoshi Aisaka,*[1] *and Hiromichi Horie*[1]

Short Fiber Reinforced Magnetic Powder Cores

REFERENCE: Morita, M., Ochiai, K., Kamohara, H., Arima, I., Aisaka, T., and Horie, H., **"Short Fiber Reinforced Magnetic Powder Cores,"** *Recent Advances in Composites in the United States and Japan, ASTM STP 864*, J. R. Vinson and M. Taya, Eds., American Society for Testing and Materials, Philadelphia, 1985, pp. 401–409.

ABSTRACT: Short fiber reinforced magnetic powder cores were investigated. The powder cores were made of iron powders and insulating resin by a compression molding process. Aramide fiber (Kevlar) chopped strand reinforced cores show very tough characteristics compared with those made without the fiber, and also show that they have suitable magnetic properties for magnetic wedges.

KEY WORDS: composite materials, magnetic cores, iron powder, chopped strands, aramide fiber, epoxy resin, permeability, toughness, magnetic wedges

Compressed iron powder cores are made of iron, iron based magnetic alloys, or permalloy powder and insulating resin. The magnetic properties of the cores are varied by chemical compositions and particle sizes of magnetic alloy powder and the kinds of insulating and binding resins. Also, they are influenced by the core densities, that is, by the volume fraction of alloy powder in the core and the compacting pressures. In other words, the magnetic properties of these cores can be designed and manufactured to meet user requirements.

So-called dust cores were one of the old uses for iron powder cores. One of the authors has developed another kind of core with large saturation induction and low iron losses for high-speed motor cores [1,2]. Also, the other kind of powder core with high permeability at higher frequencies can be used for some filter cores.

This paper introduces a newly developed tough magnetic core for an induc-

[1]Toshiba Research and Development Center, Toshiba Corp., 1, Komukai Toshiba-cho, Saiwai-ku, Kawasaki 210, Japan.

tion motor's magnetic wedges (Fig. 1). Magnetic wedges are used to decrease slot ripple loss, and also to fix the stator coils in the slots [3].

Properties demanded for the magnetic wedges are to have a required maximum permeability value ranging from 10 to 40 μH/m, and to have mechanical toughness. The permeability can be easily adjusted by changing the iron powder volume fraction. The amount of binder resin needed to insulate iron powder is only several percent. However, too high a volume fraction of iron powder produced high permeability, so it was necessary to add some amount of filler or resin to adjust the permeability. The authors have used short fibers for this filler, to make the cores tough as well as to adjust the permeability.

Experimental Procedure

Magnetic cores are prepared by a compression molding process. Materials used are listed in Table 1. Iron powder used is atomized pure iron powder

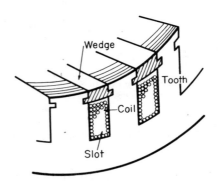

FIG. 1—*Magnetic wedges in an induction motor.*

TABLE 1—*Constitution of specimens.*

Constituents	Materials	Remarks
Iron powder	atomized power	105 ~ 74 μm (150 ~ 200 mesh) 50 ~ 90 volume %
Resin powder	B-stage epoxy resin (Epiform®)	10 ~ 40 volume %
Fibers	Kevlar® chopped strands	3, 6, 9, 12 mm 2.5 ~ 15 volume %
	glass fiber chopped strands	3 mm 10 volume %
Filler	alumina powder	15 μm in diameter 0 ~ 20 volume %

whose particle size lies between 74 and 105 μm in diameter (150 to 200 mesh). B-stage epoxy resin powder is used for insulator and binder. Polyaramide fiber (Kevlar) chopped strands and glass fiber chopped strands are used for reinforcements.

The specimens were made up from the mixtures of these materials. After they were compacted at a pressure of 500 MPa, they were heat-treated for 2400 s at 423 K to harden the resin.

Magnetic induction and permeability were measured by a conventional BH-tracer. Iron losses at 50 to 400 Hz were examined by a Watt meter. These measurements were made on circular specimens. Specimen dimensions were 30 mm in inner diameter, 40 mm in outer diameter and 5 mm in thickness. See Table 2.

Mechanical properties were obtained from 10 by 55 by 3.5-mm pillar specimens. Flexural strength σ_B was measured by the three-point bending test (span length: 48 mm), and calculated using

$$\sigma_B = \frac{3}{2} \cdot \frac{P}{W} \cdot \frac{\ell}{t^2} \tag{1}$$

where P means maximum flexural load and w, t, and ℓ are width, thickness, and span length of the specimen, respectively.

One of the definitions of toughness is the energy to propagate a crack in the specimen. This is assumed to be proportional to the area under the load-deflection curve. But the cores with fiber didn't crush in the three-point bending test, so we adopted toughness as shown by

$$\text{Toughness} = \sigma_B \times D_H \tag{2}$$

where D_H indicates the width of the chart (deflection) at a half flexural strength, as indicated by the load-deflection curve in Fig. 2.

Toughness will be changed with resin content, too. So, relative toughness is

TABLE 2—*Test specimens.*

Tests	Properties	Dimensions of Specimens, mm
Magnetic test	permeability induction iron loss	30ϕ × 40ϕ × 5
Flexural test (3-point bending)	strength toughness	10 × 55 × 3.5

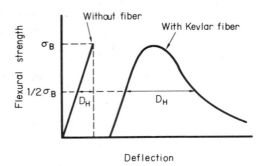

FIG. 2—*Stress-deflection curve for a magnetic core.*

adopted to compare the toughness of the cores with each other. Relative toughness is defined as

$$\text{Relative toughness} = \frac{T_f}{T_0} \tag{3}$$

Where T_f and T_0 indicate toughness for the fiber reinforced core and for the core without fiber, respectively.

Results and Discussion

Magnetic Properties

Figure 3 shows the relationship between maximum permeability and resin content in the cores. To produce a suitable permeability, iron powder fraction lies below 60% in volume. The maximum permeability was not affected too much by adding fibers, as shown in Table 3. Table 4 shows iron loss of the cores. Iron loss is sufficiently small, and shows little difference with and without the reinforcements.

Mechanical Properties

The influence of the added chopped strands on the mechanical properties was examined. Figure 4 shows flexural strengths and relative toughnesses for the powder cores. Though the flexural strengths are not much increased by adding fibers, the relative toughnesses are fairly improved. Glass fiber reinforced cores show only twice more toughness than the cores without fiber. The relative toughness of the core reinforced with Kevlar chopped strands reaches more than eight times greater than that without the reinforcements.

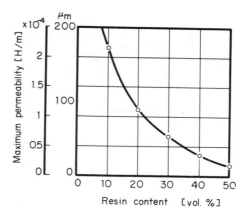

FIG. 3—*Relationship between* μ_m *and resin content. Iron powder: atomized powder 150 to 200 mesh; resin: Epiform.*

TABLE 3—*Maximum permeability of cores* ($\mu H/m$).

Iron powder, volume %	Without Fiber	10 Volume % Kevlar				10 Volume % Glass
		3 mm	6 mm	9 mm	12 mm	3 mm
90	217	...	...	...	...	...
80	112	...	...	...	...	...
70	68.0	57.4	...	...	...	55.6
60	36.3	29.5	31.8	34.3	33.8	27.5
50	24	19.1	...	...	...	16.6

TABLE 4—*Iron loss of cores* (W/kg).

Induction, T	Without Fiber	10 Volume % Kevlar				10 Volume % glass
		3 mm	6 mm	9 mm	12 mm	3 mm
0.1	0.56	0.56	0.60	0.58	0.59	0.62
0.2	1.76	1.72	1.79	1.77	1.85	1.91

Frequency: 50 Hz.

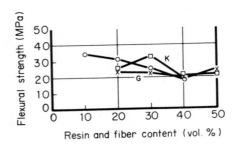

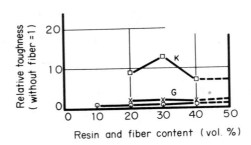

FIG. 4a—*Flexural strength and relative toughness for powder cores with 10 volume % of chopped strands.*

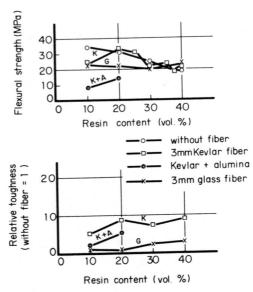

FIG. 4b—*Relationship between mechanical properties and resin content.*

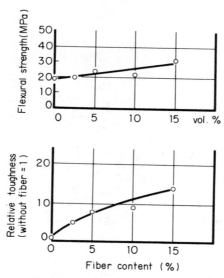

FIG. 5—*Flexural strength and relative toughness for powder cores with 3-mm-length Kevlar chopped strands. Iron powder: 60 volume %; resin + fiber: 40 volume %; compacting pressure: 500 MPa.*

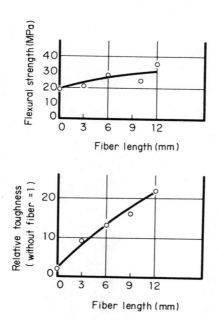

FIG. 6—*Relationship between flexural strength and fiber lengths and between relative toughness and fiber lengths. Iron powder: 60 volume %; resin: 30 volume %; Kevlar chopped strands: 10 volume %.*

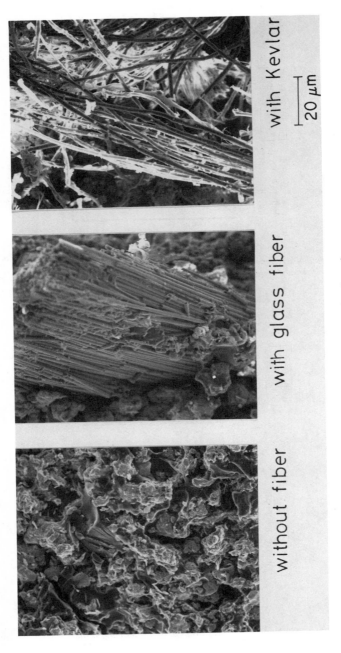

FIG. 7—SEM fractographs of powder cores.

Figure 5 shows one of the typical examples of the influence of Kevlar fiber content on flexural strength and relative toughness. Even by mixing only 2.5 volume % of chopped strands, relative toughness increases five times more than that without fibers. Figure 6 shows the influence of the fiber length on relative toughness. Relative toughness increases as fiber length increases. However, some limitations in fiber length will occur, because fibers that are too long cannot be mixed uniformly in the magnetic wedges. Also, limitation comes from the strip-like shape of wedges as shown in Fig. 1.

There seems to be two reasons why the relative toughness of Kevlar chopped strand reinforced magnetic cores has improved. The first reason is that the strength of Kevlar fiber is not as sensitive to a surface flaw as that of glass fiber, so it withstands more compressive pressure during the compacting process than when glass fiber is used. This is clearly shown in Fig. 4. The second reason is that the toughness depends on the pullout force for the fiber, and this force is proportional to the fiber length. Cross-sectional observation of fractured surfaces by scanning electron microscopy (SEM) shows that the pullout fiber lengths for Kevlar are longer than those for glass fiber (Fig. 7).

Conclusion

Iron powder cores were investigated. Cores were made of mixture of 50 and 60 volume percent of iron powder, 30 to 50 volume percent of epoxy resin powder and 10 volume percent of Kevlar chopped strands. They have a maximum permeability of 10 to 40 μH/m, and also have required toughness for magnetic wedges

Acknowledgment

Authors wish to thank Dr. J. Nagai, director of the Toshiba Research and Development Center, for permission to publish this paper.

References

[1] Fukui, K., Morita, M., Watanabe, I., and Ooshima, T., "Iron Powder Cores for Electic Machines and Apparatus" in *Proceedings*, 2nd Conference on Applied Magnetics, Japan, 1970, p. 15.
[2] Fukui, K., Watanabe, I., and Morita, M., "Compressed Iron Powder Cores for Electric Motors," *Transactions*, Institute of Electrical and Electronics Engineers, 1972, p. 682.
[3] *Energy Consumption and Saving in Electrical Apparatus*, S. Ushiku, Ed., Denkishoin, Tokyo, 1983, p. 9 (in Japanese).

Tsuneo Kawashima,[1] *Masataka Yamamoto,*[1] *Toshihiko Yamawaki,*[1] *and Yoshinori Yoshimura*[1]

Developmental Researches on the Lightweight Structure for Future Satellite in the National Development Agency of Japan

REFERENCE: Kawashima, T., Yamamoto, M., Yamawaki, T., and Yoshimura, Y., "**Developmental Researches on the Lightweight Structure for Future Satellite in the National Development Agency of Japan,**" *Recent Advances in Composites in the United States and Japan, ASTM STP 864*, J. R. Vinson and M. Taya, Eds., American Society for Testing and Materials, Philadelphia, 1985, pp. 410–427.

ABSTRACT: In recent years we are making some developmental studies on lightweight spacecraft structure and several lightweight components in National Space Development Agency of Japan (NASDA). Presented here are the test results of the structural components which were made of graphite fiber reinforced composites, carbon fiber reinforced composites, and aluminum alloy plate and honeycomb core, etc.

We describe the three items as well as the manufacturing process. First, graphite-epoxy tube trusses were manufactured and tested for use as struts and main frames of the satellite structure. Second, lightweight honeycomb panels were designed and fabricated to provide spacecraft structural components which are used as some equipment panels or stiffened panels. Third, lightweight solar panel was developed and tested for 3-axis stabilized spacecraft with sunoriented solar arrays which are expected to meet the power requirement of more than 1.8 kW. The panel has successfully been tested under acoustic noise environment, thermal cycling, and sinusoidal vibration, and thus qualified for space applicability.

KEY WORDS: satellite, lightweight structure, development, subsystems, strut tube, honeycomb panel, solar panel, CFRP, launch, vehicle, spacecraft

On the first step in 1975, we launched the Engineering Test Satellite-I (ETS-I) by N-rocket Type I launch vehicle, after that we are developing several kinds of satellites such as geostationary meteorological satellite, broad-

[1] Senior structural engineer, assistant senior engineer, engineer, and engineer, respectively, National Space Development Agency of Japan, 2-4-1, Hamamatsu-cho, Minato-ku, Tokyo 105, Japan.

casting satellite, communication satellite for commercial use, ETS-II-ETS-V, marine observation satellite, and so on to get any technical attainment.

In these developmental works, it is necessary to materialize lightweight bus equipments (spacecraft, solar array, buttery cell, and so on) for the sake of carrying many mission equipments under the limitation of payload weight. To meet the mechanical and thermal requirements in launching and orbiting environments to the satellite, we choose characteristic materials such as carbon fiber reinforced composites (CFRP), Kevlar fiber reinforced composites (KFRP), and any light metals (for example, aluminum, titanium, beryllium) which have high specific stiffness (stiffness/density), specific strength (fracture strength/density), and low-thermal expansion properties.

Figure 1 illustrates developmental schedule in respect to lightweight bus equipments. It contains design and fabrication of graphite fiber reinforced composites (GFRP) tubular truss, honeycomb panel, and solar array. As shown in Fig. 2, the preliminary study on the structural model of ETS-V started in 1980. The result should be reflected in the development of bus equipment of ETS-V.

Tubular Strut

The tubular struts are one of the main structural components which are made of graphite/epoxy in the full utilization of the preferable mechanical properties of the composite material. Several compression tests were carried out on the tubular components made of high modulus graphite/epoxy laminae, together with tension tests of the dumbell type specimens cut out of the tube. Table 1 shows the properties of the tube.

The test results [1] in both cases were shown to enable suitable choice of graphite layers for the lightweight satellite structure. Table 2 shows experimental results of tension and compression tests.

Lightweight Honeycomb Sandwich Panel

Honeycomb core sandwich plate has the excellent properties in lightweight, high rigidity, shape stability, and assembling easiness. Therefore, it performs an important role as structural components of the spacecraft. Panel is used for carrying mission equipments, substrates of solar array, antenna reflector, boxes of electronic equipment, and supporting frames of thermal louver.

Requirements for Design

The spacecraft and its components should be designed to satisfy the mechanical and thermal environments for launching and orbiting. Taking into account the launch condition by the H-1 rocket, which is under development,

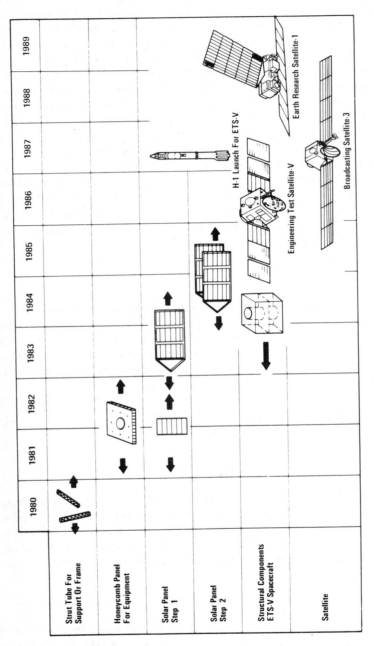

FIG. 1—*Lightweight spacecraft and components development program.*

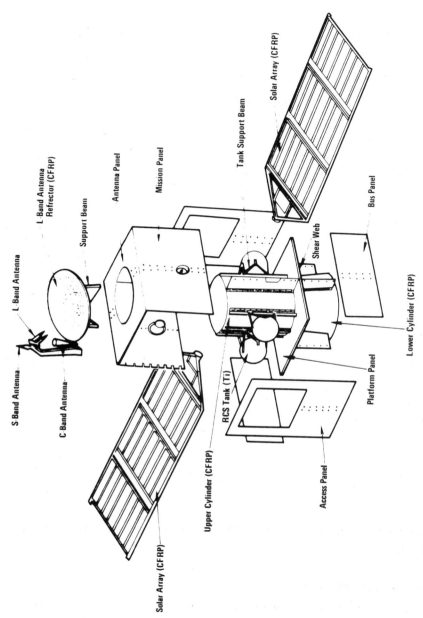

FIG. 2—*ETS-V structures and solar array subsystems, component installation.*

TABLE 1—*Materials and properties.*

Material	Graphite/Epoxy	GFRP
Specification	M 1547	M 1548
Type	unidirectional tape, 0.13 mm/ply	plain woven cloth, 0.02 mm/ply
Prepreg property	high performance epoxy resin resin content = 40%	high performance epoxy resin resin content = 33%
Laminate property	$F_{tu} = 1128$ N/mm^2 $E_t = 176500$ N/mm^2	$F_{tu} = 343$ N/mm^2 $E_t = 19\,600$ N/mm^2

TABLE 2—*Experimental results of tube test.*

	Specimen	Fracture Stress, N mm^{-2}	Young Modulus, $E_L \times 10^4$ N mm^{-2}
Tension	dumbell type width: 13 mm length: 400 mm	539 568 519	11.0 11.6 11.7
Compression	tube type diameter: 59.4 mm length: 800 mm	509 508	11.6 12.2

TABLE 3—*Design requirements for honeycomb panel.*

Shape	1000 × 1000 mm^2 through hole on the center of panel (diameter = 150 mm)
Minimum frequency	50 Hz
Equipment weight	100 kg
Loading	20 g, x-, y-, z-direction, respectively (g: gravity acceleration)
Number of fixed point	16
Supporting condition	determine as result of calculations and coupon tests

the design requirements for the panel were expediently specified as shown in Table 3.

Design and Fabrication

Main parts of honeycomb core sandwich panel are skin plates, honeycomb core, adhesive, fillers, stiffeners, and fasteners. Aluminum alloys are used as skin plates, core, stiffeners, and fasteners, and adhesive is an epoxy film of cell edge type from a lightweight viewpoint.

Rigidity of the panel was calculated by finite-element simulation analysis as shown in Fig. 3. Simulation model had 93 nodes, 108 elements, and eigenvalues and stress distribution were calculated by inputting material properties, boundary conditions, and load. Table 4 shows the result of calculations including dimension, weight, and so on. The predicted panel weight (3.87 kg) coincided with fabricated values (3.91 and 3.89 kg, respectively).

On the fabrication process, the formation conditions comply with the properties of adhesive. Because of honeycomb core foil being so thin (0.018 mm),

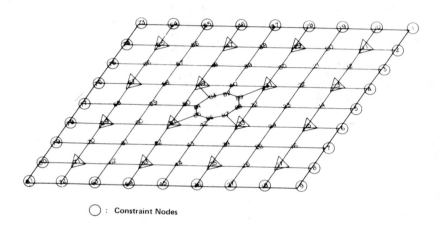

◯ : Constraint Nodes

△ : Load Nodes

FIG. 3—*Finite element simulation model of honeycomb panel.*

TABLE 4—*Specification of honeycomb panel after design.*

Constituent Parts	Weight, kg	Materials, Dimension
Skin plates	1.68	7075-T6 aluminum thickness = 0.3 mm
Honeycomb core	0.65	3/8-5056-0.0007P aluminum 3/8-5056-0.001P aluminum height = 42.2 mm
Mechanical inserts	0.71	7075-T6 aluminum 32 pieces
Integration inserts	0.36	7075-T6 aluminum 16 pieces
Adhesive	0.24	HEXABOND-III, FM24, FM123
Total weight of panel	calculated value = 3.87 fabricated panel weight = 3.89, 3.91	

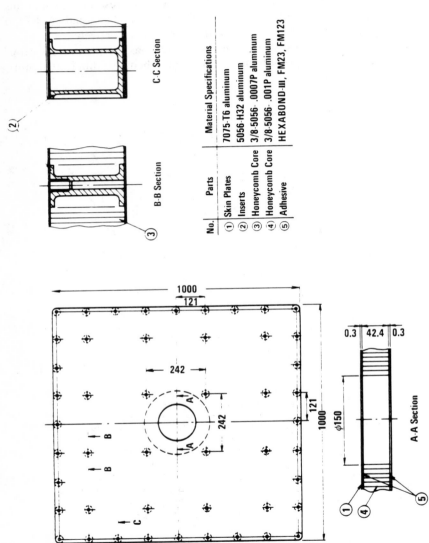

No.	Parts	Material Specifications
1	Skin Plates	7075-T6 aluminum
2	Inserts	5056-H32 aluminum
3	Honeycomb Core	3/8-5056-.0007P aluminum
4	Honeycomb Core	3/8-5056-.001P aluminum
5	Adhesive	HEXABOND-III, FM23, FM123

FIG. 4a—Configuration and constitutive materials of honeycomb panel (scale in millimetres).

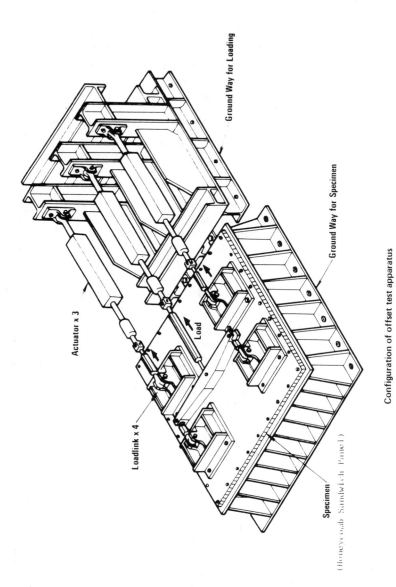

Configuration of offset test apparatus

FIG. 4b—*Configuration of offset test apparatus.*

the bonding conditions were like that; pressure, less than 0.078 N/mm^2, heating for about 90 min at 122 to 135°C.

Rigidity and Strength Testing

After several strength tests of each constituent material (coupon testing), directing towards the critical design or nearly zero margin of safety, two panels of 1000 by 1000 by 42 mm size were manufactured, and two types of static load tests were conducted on them to certify the specific requirements shown in Table 3.

Figure 4a shows the configuration and the constitutive materials of the panel specimen. The outline of the test rig is given in Fig. 4b. Although the figure shows only the setup for the offset load test, the test for the upward panel bending was also performed by the vertical action of the same actuators. Then all edges of each panel were rigidly supported with nine screw bolts, respectively, and the loads were applied to the inner equally spaced at 16 points. In the bending and offset load test, displacements and strains of the specimen were measured by dial gages and wire strain gages, respectively. Measuring points were decided and counted up about 40 corresponding to the result of simulation. Figure 5 shows modes of the strain along the periphery of central hole by bending load. In this test, the local fracture appeared at the loading point close to hole of 2.7 × 10^4 N (correspond to 140% of design

TABLE 5—*Comparison of numerical result with experimental displacement under bending and offset loading.*

Load	Measuring Points[a]	Displacement, mm		Difference, %
		Numerical	Experiment	
Bending	88	3.47	3.44	0.8
	84	3.47	3.41	1.7
	66	1.03	1.04	1.0
	67	1.38	1.77	22.0
	49	2.98	3.09	3.5
	59	2.42	2.49	2.0
	61	1.80	1.89	4.8
	21	1.80	1.86	3.2
Offset	88	−0.12	−0.17	29.4
	84	0.12	0.09	13.3
	47	0.26	0.24	8.3
	65	−0.16	−0.22	27.3
	49	0.26	0.30	13.3
	67	−0.16	−0.16	0
	92	0.29	0.34	14.7
	91	0.29	0.30	3.3

[a]Number correspond to it in Fig. 3. Nos. 91 and 92 are position of tool for loading.

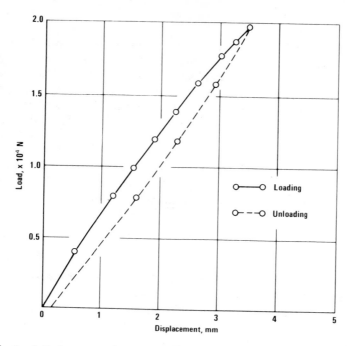

FIG. 5—*Load-displacement relation of bending load test. measured place is loading point (design load = 19 600 N. residual displacement = 0.2 mm).*

load), after that the fracture expanded to all face of the panel at 2.9×10^4 N. Shear mode fracture of the honeycomb core was observed in this test.

Comparison of simulation analysis to experimental result is shown in Table 5. Manufactured panels had 50 to 52 Hz minimum natural frequency. It satisfied the specific requirement as shown in Table 3.

Lightweight Solar Panel (LSP)

In view of the weight efficiency, lightweight solar panel (LSP) is considered as one of the most useful component to generate electric power for 3-axis stabilized spacecraft. So far panel substrate was constituted with aluminum or CFRP sandwich plate which was called rigid type and the power ratio was kept below 20 watts/kg, whereas the semirigid LSP easily attains more than 30 watts/kg, at the power requirement of over 800 W. The new type of solar panels will be applied to a NASDA Satellite launched within 4 years (1986) on this account.

Design and Fabrication

LSP has been designed to satisfy the requirement shown in Table 6. Only one panel was designed and fabricated as a trial in 1982 to 1983, but 5 panels will be manufactured at the final stage shown in Fig. 6. Figure 7 shows the configuration and size of the panel as well as its components. The panel is a lattice work with three kinds of rigid beams and many membranes stretched inside of each section. Many solar cells are bonded on a side of the membrane which is termed here the flexible blanket. The beams 1, 2, and 3 are constructed as skeleton, and they were made by the CFRP filament winding method. Figure 8 shows the cross section of beams 1, 2, and 3, Table 7 describes the cross-sectional properties of beams. The weight of the panel trially made was 3.93 kg, and the weight of the silicon solar cells of total 3500 chips including adhesive was about 3.82 kg.

Mechanisms

During launch each panel is held to the side panel of spacecraft by bolts. Subsequently panels deploy by means of pretensioned spiral springs which are attached to each panel hinges. Blanket which carries the solar cells is suspended by the frames [2,3]. Figure 9 shows breakdown views of hinge and suspended mechanism of the panel.

Test of LSP

The first stage of the LSP developmental test program was to investigate the stability of the carbon fiber structure under the space environment on the material base, and, subsequently, components and a full size panel. The following tests were performed especially with one panel to satisfy mechanical and thermal conditions in the phase of launching and orbiting of the satellite. For status reference and performance of the array prior to and after panel testing, electrical measurements and visual inspections were performed.

Sinusoidal Wave Vibration Test—The sinusoidal mode vibration test was made to measure resonance frequency of the panel, after that the qualification level test was performed to confirm the strength of critical items and for verifying the correctness of simulation model. Figure 10 shows the relation of acceleration to frequency for the modal survey test when the panel is vibrated in the normal direction to panel plane (Y-direction). X- and Z-axis were in the longitudinal and transverse directions of the panel, respectively. Performed test levels for X-, Y-, and Z-axis are given in Table 8.

Acoustic Noise Test—The acoustic noise test is considered to be important as well as the sinusoidal vibration test to verify the integrity of solar cell panels at the launch environment. Thus, a large area blanket is exposed to high energy acoustic noise in the vehicle during launching (H-1 project). Table 9

TABLE 6—*Specific requirements of lightweight solar panel.*

Acceleration load	lift off	body axis direction	7.2 g[a]
		perpendicular to body axis	6.0 g
	MECO/pogo	body axis direction	22.0 g
		perpendicular to body axis	3.0 g
Temperature range	on orbit	$-100 \sim 90°C$	
Minimum frequency	folding	body axis direction	35 Hz
		perpendicular to body axis	15 Hz
	deploying		0.2 Hz
Generated power	on orbit	1800 (beginning of life)	1350 W (end of life)

[a]g = gravity acceleration.

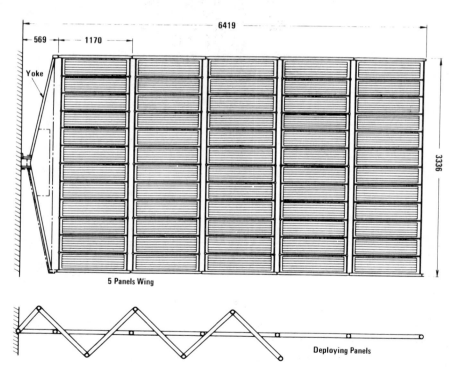

FIG. 6—*Panel wing and deploying configurations (scale in millimetres).*

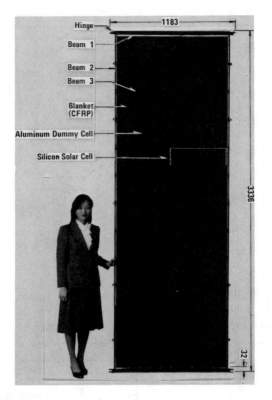

FIG. 7—*Setup lightweight solar panel (scale in millimetres).*

TABLE 7—*Mechanical properties of composite beams.*[a]

Type	Section, mm^2	$E_L \times 10^4$, N mm^{-2}	$G_T \times 10^2$, N mm^{-2}	$I_{xx} \times 10^3$, mm^4	$I_{yy} \times 10^3$, mm^4	$J \times 10^3$, mm^4
Beam 1	234	16.6	74.5	31.1	23.2	18.3
Beam 2	60	12.5	60.0	0.86	4.1	1.4
Beam 3	76	12.9	109.8	0.37	1.8	1.0

[a]E_L = longitudinal Young's modulus.
G_T = transverse shear modulus.
I_{xx}, I_{yy} = secondary moment of inertia of x-, y-axis, respectively.
J = polar secondary moment of inertia.

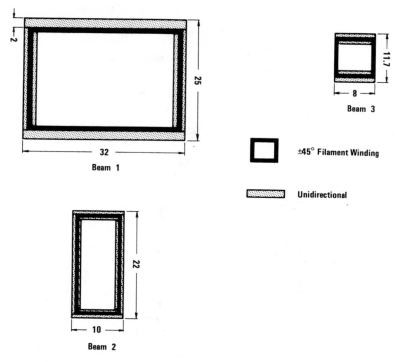

FIG. 8—*Cross section of beam 1, 2, and 3 (scale in millimetres).*

demonstrates the input sound level of acoustic spectrum; Fig. 11 shows the test result.

Thermal Cycle Test—Two types of tests are made to simulate the orbital environment. Thermal shock test was repeated up to 200 cycles, using small specimens sized 190 by 150 mm, between the temperature of −150 and 110°C. On the other hand, the full size panel partially covered with live solar cells was subjected in the evacuated container to 34 thermal cycles in the temperature range between 90 and −100°C at the cyclic period of 4 h. The specimen survived the environment without degradation.

Visual inspection and electrical measurement of the specimen including panel and solar array were made, and all test phases were passed without damages or degradation.

Conclusion

In carrying out the developmental study to construct the lightweight spacecraft, we are making an effort to develop the ETS-V which is to be launched in 1987. The developmental technique described herewith with respect to sub-

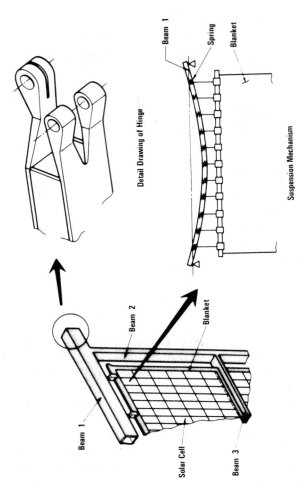

Detail Drawing of Hinge

Suspension Mechanism

Beam 1

Spring

Blanket

Beam 2

Blanket

Beam 1

Solar Cell

Beam 3

FIG. 9—*Detail drawing of a panel.*

TABLE 8—*Sinusoidal vibration test levels.*

	X-axis		Y-axis		Z-axis	
	Frequency, Hz	Displacement or Acceleration	Frequency, Hz	Displacement or Acceleration	Frequency, Hz	Acceleration
Modal survey	5 ~ 100	0.5 g[a]	5 ~ 100 15 ~ 40	0.5 g 0.8 g	5 ~ 100	0.5 g
Flight acceptance level	5 ~ 16 16 ~ 35 38 ~ 100	12.7 mm DA[b] 6.7 g 1.0 g	5 ~ 12.5 12.5 ~ 16 18 ~ 22 24 ~ 30 32.5 ~ 100	12.7 mm DA 4 g 0.7 g 4 g 0.7 g	NA[c]	NA
Qualification level	5 ~ 20 20 ~ 35 38 ~ 100	12.7 mm DA 10 g 1.5 g	5 ~ 15 15 ~ 16 18 ~ 22 24 ~ 30 32.5 ~ 100	12.7 mm DA 6 g 1 g 4 g 1 g	NA	NA

[a] g = gravity acceleration.
[b] DA = displacement amplitude.
[c] NA = no application.

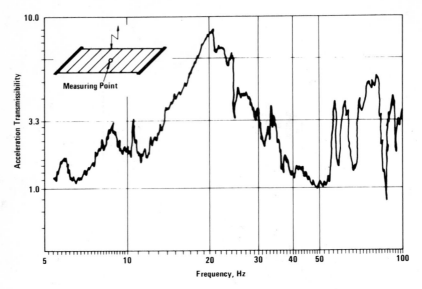

FIG. 10—*Y-direction modal survey test, input acceleration is 0.5 g.*

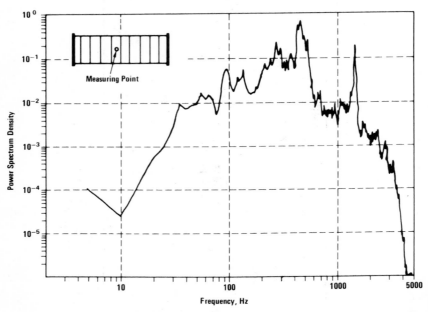

FIG. 11—*Acoustic noise test of one panel (full scale* $= g^2/Hz = 1$ *g: acceleration).*

TABLE 9—*Acoustic noise test levels.*

Octave Band Center Frequency, Hz	Sound Pressure Level, dB (Ref 0.0002 dyne/cm^2)		Difference, dB	Tolerance, dB	Time, S
	Specific	Measured			
31.5	126.0	123.3	−2.7	4.0 ~ −6.0	30
63	129.0	129.0	0	4.0 ~ −2.0	30
125	133.0	133.2	0.2	4.0 ~ −2.0	30
250	138.0	138.1	0.1	4.0 ~ −2.0	30
500	142.0	142.0	0	4.0 ~ −2.0	30
1000	136.0	136.0	0	4.0 ~ −2.0	30
2000	133.0	133.1	0.1	4.0 ~ −2.0	30
4000	129.0	129.9	0.9	4.0 ~ −6.0	30
8000	126.0	124.5	−1.5	4.0 ~ −6.0	30
Overall	145.0	144.9	−0.1	3.0 ~ −1.0	30

systems or components will be applied to the design of ETS-V which is a geostational satellite with the weight of 550 kg in orbit.

For the future developmental schedule in NASDA shown in Fig. 1, we intend to design the ERS-1 satellite which is operated on an orbit of relatively lower altitude with the weight of 1400 kg, and there is also a program to develop the ETS-VI satellite with the weight of 1000 to 2000 kg on orbit.

Finally the authors are thankful for the cooperation of Dr. B. Tomita and Mr. S. Morimoto in carrying out these projects.

References

[1] Kawashima, T., Sakatani, Y., and Yamamoto, T., *Proceedings*, Japan-U.S. Conference, Tokyo, K. Kawata and T. Akasaka, Eds., Japan Society for Composite Materials, 1981, pp. 453–460.
[2] Schneider, K., *Proceedings*, European Symposium on Photovoltaic Generators in Space, Nov. 1978, European Space Agency, SP-140, pp. 231–242.
[3] Koelle, D. E. and Bassewitz, H. V., AIAA/CASI 6th Communications Satellite Systems Conference, April 1976, AIAA Paper No. 76-286, American Institute of Aeronautics and Astronautics.

Frank K. Ko[1] and Christopher M. Pastore[1]

Structure and Properties of an Integrated 3-D Fabric for Structural Composites

REFERENCE: Ko, F. K. and Pastore, C. M., **"Structure and Properties of an Integrated 3-D Fabric for Structural Composites,"** *Recent Advances in Composites in the United States and Japan, ASTM STP 864*, J. R. Vinson and M. Taya, Eds., American Society for Testing and Materials, Philadelphia, 1985, pp. 428–439.

ABSTRACT: Based on experimental observations, a theoretical framework has been established to quantify the geometry of a three-dimensional braided fabric. Predictions of the strength of the fabric and composite were made using a geometric model.

KEY WORDS: three-dimensional fabric, composite, microstructure, complex shape, fully integrated (fabric), orientation distribution, tensile strength, Magnaweave

The need for innovative design and selection of reinforcement material/structure was never greater for advanced composites as a result of the current demand for significant improvement of inter/intralaminar shear strength, impact toughness, capability to form complex shapes, and the reduction of fabrication cost.

Recent progress in fabric formation technology has made available additional reinforcement concepts which may enable us to meet the requirements of advanced composites. For thin composite structures, two-dimensional (2-D) biaxial, triaxial woven fabric, or knitted fabrics as introduced by Ko et al [1–4] may be quite adequate to meet the requirements. For thick composite structures, that is, of more than two yarn thicknesses, 3-D integrated structures can be seriously considered. Although considerable work was done on 3-D structures during the sixties [5], most of these 3-D fabrics are orthogonal structures with limited shapeability and processibility. There is a revival of interest in 3-D fabrics in recent years due to the introduction of a few semi- and fully integrated 3-D fabrics [6–8].

[1]Director and research mathematician, respectively, Fibrous Materials Research Center, Drexel University, Philadelphia, PA 19104.

In this paper, the structure and properties of 3-D fabrics formed by an irregular braiding process is presented. Basically this 3-D braiding system is formed by orthogonal interlacing of yarns in a rectangular or circular configuration followed by a compacting or combing motion along the yarn axis. Accordingly, a 3-D fully integrated structure is formed with yarns intimately interlaced together to assume various net shape structures. One way of forming such a fabric system is referred to as the Magnaweave process, which has been described by Florentine [8]. Some of the mechanical properties and general features of this structure have also been presented by Ko [9,10]. Specifically, the macro- and microstructural geometry of a 3-D braided fabric is analyzed in this study. As a first approximation, a geometric model was employed to predict the tensile strength of the fabric and the composite. An experimental fabric was prepared and tested to verify this model.

Geometry

One of the most interesting features of the integrated 3-D fabric prepared by the irregular braiding process is its structural geometry—both in the macroscopic level (shapes) and in the microscopic level (microstructure or material orientation). The structural geometry of this 3-D fabric is a result of loom (machine) configuration and the manipulative sequence of loom components. To facilitate discussion, the two basic loom configurations and components are shown in Figs. 1a and 1b. The position of each yarn (bobbin) is identified with a vertical or "column" (V) and horizontal or "track" (U) coordinate as shown in the rectangular or circular loom arrangement. Accordingly, the movement and hence the geometric arrangement of the yarn segments can be monitored as the tracks and columns of the loom are manipulated.

Structural Shapes

One of the obstacles which prevents laminated and other conventional composites from being used more extensively for structural applications is their lack of ability to assume complex structural shapes in an integrated manner. Furthermore, since 3-D braids do not have separate layers of filaments or fabrics, composites reinforced by integrated structures should potentially have a higher level of through-the-thickness strength and damage tolerance. In our laboratory we have demonstrated that a wide range of structural shapes can be prepared in an integral manner with the irregular braiding process. A summary of the shapes which have been developed successfully in our laboratory is shown in Fig. 2.

Microstructure

In addition to the matrix and fiber material, the microstructure or the spatial orientation of the material plays a significant role in determining the pro-

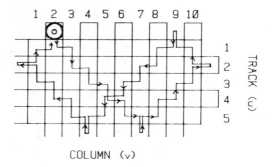

FIG. 1a—Rectangular loom.

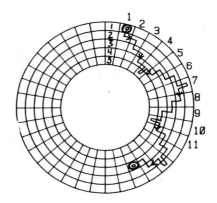

FIG. 1b—Circular loom.

cessibility and the performance of a composite. To analyze the microstructure, computer codes have been developed in our laboratory to graphically display the path of yarns in the 3-D fabric. Figures 3a and 3b illustrate the integrated nature and the spatial orientation of one of the yarns in a computer simulation of a T-beam structure. To assist design, formation, and analysis, the spatial orientation of the 3-D fabric is quantified and related to fabric construction and processing variables.

Based on our computer simulation, the microstructure of the 3-D fabric can be represented by the repeating unit of a given yarn as illustrated in Fig. 4. The orientation of each yarn segment is defined by the angle with respect to the Z-axis, θ, and the angle with respect to the Y-axis, α, on the XY-plane of the unit cell. The height of the unit cell is defined by h, the width dimension p, and the thickness dimension q. The height of a unit cell h can also be thought of as the inverse of the number of surface yarn interlacings

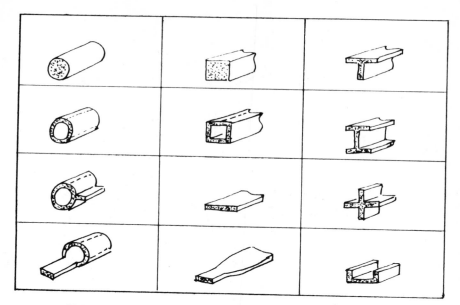

FIG. 2—*3-D fabric structural shapes fabricated successfully at FMRC.*

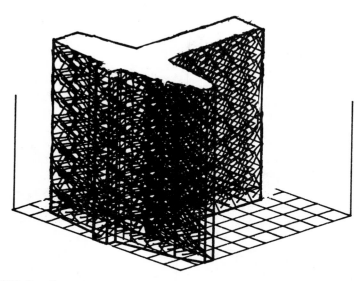

FIG. 3a—*Computer simulation showing the integrated nature of a 3-D fabric.*

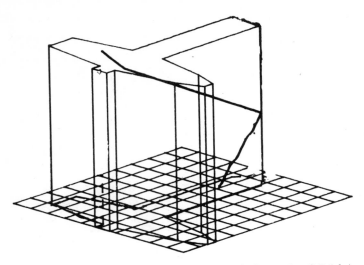

FIG. 3b—*Computer simulation of the path of a single yarn in a 3-D fabric.*

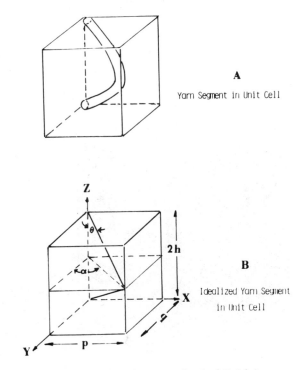

A

Yarn Segment in Unit Cell

B

Idealized Yarn Segment
in Unit Cell

FIG. 4—*Unit cell structure for the 3-D fabric.*

per unit length. The fabric constructed in this manner is composed of yarns or groups of yarns—yarn bundles. To facilitate computation, the height, width, and thickness dimension of the unit cell geometry can be normalized by the diameter of yarn bundle, d, to obtain dimensionless quantities w, u, and v, respectively. w is a function of the compacting action during the fabric formation, while u and v are the yarn displacement values in terms of the number of yarn diameters resulting from track and column motion, respectively.

The orientation of the yarns in the 3-D fabric depends on the fabric construction, fabric shape, and the dimension of the loom. For example, in a simple rectangular structure, there are three possible orientations as shown below for the yarn segment with respect to the Z-axis of the unit cell:

$$\theta_1 = \tan^{-1} (\sqrt{u^2 + v^2}/w) \tag{1}$$

$$\theta_2 = \tan^{-1} (u/w) \tag{2}$$

$$\theta_3 = \tan^{-1} (v/w) \tag{3}$$

By considering the loom dimension related to the number of tracks, m, and the number of columns, n, and the fabric formation parameters u, v, and w, one can determine the number of yarn n_i which assumes a certain orientation θ_i. With the orientation distribution of the material established, one can easily develop a geometric model for strength prediction as described in this paper or extend to a more general mathematical model analogous to the laminate theory as that proposed by Halpin et al [11].

Further examination of the distribution of material orientation indicated that the material orientation can be simply represented by an average angle $\bar{\theta}$, which is defined as

$$\bar{\theta} = \sum_{i=1}^{N} (\theta_i)/N \tag{4}$$

where n is the total number of yarns in the cross section of the fabric.

With the yarn orientation angle defined, the relationship between microstructure and fabric construction and processing parameters can now be examined.

As shown in the contour diagrams in Figs. 5a and 5b, plotting yarn orientation angle θ as a function of loom dimensions for various unit cell widths, w, the average yarn orientation tends to increase as the number of columns or the number of yarn bundles in the thickness direction increases. It should be noted that this trend is true only for constructions with less than ten columns or ten yarn diameters thick. Above ten columns, the average yarn orientation is insensitive to loom dimension. The effect of the density of yarn interlacing expressed in terms of normalized unit cell width, w, on yarn orientation is

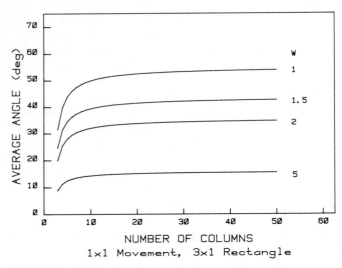

FIG. 5a—*Relationship of yarn orientation and processing parameters.*

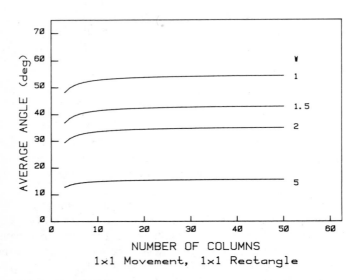

FIG. 5b—*Relationship of yarn orientation and processing parameters.*

significant. As w increases, the average yarn orientation angle decreases. As shown in Fig. 5, the average yarn orientation appears to be inversely proportional to w; for example, an increase of w-value from 1 to 5 brings about a decrease of the average angle from 55 to 15 deg. This means that the combing or compacting process plays a critical role in determining the material orientation of the braided fabric structure.

Prediction of Strength

With the orientation of the yarn in the 3-D fabric defined, we can now proceed to analyze the stress-strain behavior of the fabric before matrix impregnation (fabric preform) as well as the composite. For this study, the failure stress of the fabric and composite is presented with reference to fabric construction and processing variables.

Strength of the Fabric Preform

For a given yarn which has breaking load P_{yi} and oriented to angle θ_i with respect to the Z-axis or the direction of load on the fabric, a simple relationship can be established between fabric breaking load P_f and yarn breaking load P_y, as

$$P_f = \sum_{i=1}^{N} P_{yi} \cos \theta_i \qquad (5)$$

where n is the total number of yarns in the fabric cross section.

N can be related to the loom and processing parameters such as m, n, u, and v. It should be noted that P_{yi} can represent different types of materials. If the average orientation θ and only one type of yarn is used, the equation can be rewritten, as

$$P_f = NP_y \cos \theta \qquad (6)$$

or in terms of stress

$$\sigma_f = \sigma_y \cos^2 \theta \qquad (7)$$

Strength of the Composite

Knowing the strength of the fabric preform, the failure strength of the fabric reinforced composite can be estimated for a given matrix material using the rule of mixtures

$$\sigma_c = V_f \sigma_f + V_m \sigma_m \qquad (8)$$

or

$$\sigma_c = V_f \sigma_y \cos^2 \bar{\theta} + (1 - V_f)\sigma_m \qquad (9)$$

where

σ_c = strength of composite,
σ_y = strength of yarn,
σ_m = strength of matrix,
θ = average yarn orientation,
V_f = volume fraction of fabric, and
V_m = volume fraction of matrix.

Experimental Observations

In order to study the tensile stress-strain behavior of the 3-D fabric and its composite, several experimental 3-D fabrics were prepared using Celion 12K yarns. In order to maintain the continuity of the yarns, the fabrics were formed to exactly the test specimen dimension (3.2 by 25.4 mm). Fiberglass end tabs were used for both the fabric and composite test specimens. Tension testing of the specimens was carried out at 2 mm/min using a gage length of 152 mm under standard textile laboratory conditions (21°C, 65% relative humidity). A 1 × 1 fabric construction was used with a total of 42 yarns in the fabric. It can be seen, as expected, in Fig. 6 that the preform and the composite have lower strength than the yarn but a higher breaking elongation. Even though the modulus of the composite was low compared with the yarn, its modulus is reasonably greater than that of the fabric preform. The strength and elongation of the fabric preform were slightly greater than the composite. It should be noted that the measured elongations were based on the jaw separation of the testing machine rather than based on strain gage; consequently, the elongation measurement should be treated with caution. Nevertheless, the relative performance for the yarn, fabric and composite did indeed follow a reasonable trend. A summary of the properties of the various components of the composite is given in Table 1.

A closer examination of the stress-strain behavior of the 3-D fabric composites, as illustrated in Fig. 7, indicates at least two distinct regions of performance. Below 1% elongation, maximum contribution is obtained from both the resin and the fabric. Above 1% elongation the contribution from the matrix to resist deformation diminishes. The strength of the composite was 710 MPa at approximately 5% elongation.

A comparison of experimental results and theoretical predictions can be made for the fabric and the composite using Eqs 6 and 10. For $N = 42$, $P_y = 165$ kg, and $\bar{\theta} = 14$ deg, the predicted fabric strength was 792 MPa. As shown in Fig. 8, the predicted fabric strength is in close agreement (within 5% error) with experimental observation.

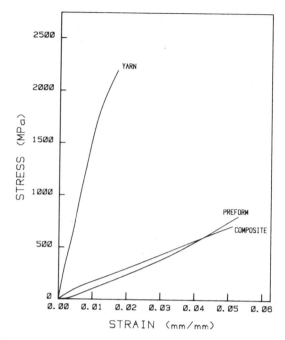

FIG. 6—*Stress-strain curve of 12K Celion graphite yarn, preform, and composite.*

Using the rule of mixture as shown in Eq 10, the calculated strength of the composite was 587 MPa using the values $\sigma_f = 810$ MPa, $\sigma_m = 13.8$ MPa, and $V_f = 0.72$.

A comparison of experimental and theoretical composite strength is also presented in Fig. 8. The prediction curves illustrate the familiar decreasing trend of the strength/material orientation relationship for both the 3-D fabric and the composite.

Conclusions

An introduction of a fully integrated 3-D fabric structure prepared by an irregular braiding process has been made.

The microstructure of the 3-D fabric preform has been examined and quantified to lay a foundation for future computer-aided design, manufacturing, and mechanistic analysis of this structure. Prediction of the tensile strength of the fabric preform and the composite were made based on a maximum-stress failure criterion. Experimental observations made on the tensile properties of the fabric and the composite indicated good agreement (within 5% to 25% error) with the predicted results.

TABLE 1—*Summary of properties of experimental materials.*

Specimen	Density, g/cm³	Strength, ksi/MPa	Modulus, Msi/GPa
Yarn (Celion 12K)	1.77	515/3552	37/255
Fabric	0.97	115/790	1.5/10
Composite	1.02	103/710	3.3/23
Resin (Epon 828/V40)	1.14	2/14	0.04/0.28

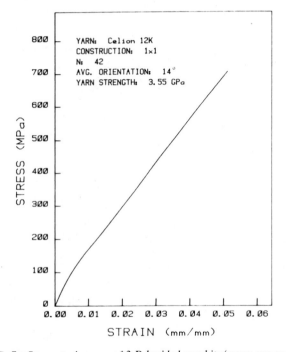

FIG. 7—*Stress-strain curve of 3-D braided graphite/epoxy composite.*

With its unique combination of integrated structure and an ability to assume complex structural shapes, the 3-D braided fabric is an interesting candidate preform for structural composites.

Acknowledgments

The authors wish to thank Ms. Ping Fang for her careful preparation and testing of the experimental fabrics and composites.

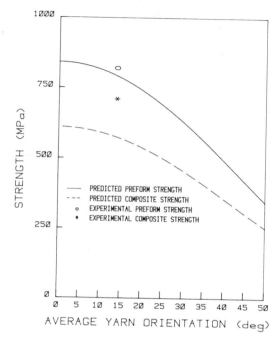

FIG. 8—*Prediction of preform and composite strength.*

References

[1] Ko, F. et al, "Development of Multi-Bar Weft Insertion Warp Knit Glass for Industrial Applications," *Journal of Engineering for Industry*, Vol. 102, 1980, p. 333.
[2] Ko, F., "Engineering Properties of Triaxial Fabrics," presented at the Textile Technology Forum, 68th Annual Canvas Products Association International Convention (Industrial Fabrics Association International), 1980.
[3] Scardino, F. and Ko, F., "Triaxial Woven Fabrics," *Textile Research Journal*, Vol. 51, No. 2, 1981, p. 80.
[4] Ko, F., Krauland, K., and Scardino, F., "Weft Insertion Warp Knit for Hybrid Composites" in *Proceedings*, ICCM-IV (Fourth International Conference on Composite Materials), Tokyo, 1982.
[5] Adsit, N. R., Carnahan, K. R., and Green, J. E., "Mechanical Behavior of Three-Dimensional Composite Ablative Materials" in *Composite Materials: Testing and Design (2nd Conference), ASTM STP 497*, American Society for Testing and Materials, Philadelphia, 1971.
[6] Fukuta, K., Onooka, R., Aoki, E., and Nagatsuka, Y., "Application of Latticed-Structural Composite Materials with Three-Dimensional Fabrics to Artificial Bones," *Bulletin of Research Institute for Polymers and Textiles*, No. 131, 1982-2, p. 151.
[7] U.S. Patent 4,346,741, U.S. Patent Office, Washington, DC, Aug. 31, 1982.
[8] U.S. Patent 4,312,261, U.S. Patent Office, Washington, DC, Jan. 26, 1982.
[9] Ko, F., "Three-Dimensional Fabrics for Composites" in *Proceedings*, ICCM-IV (Fourth International Conference on Composite Materials), Tokyo, 1982.
[10] Ko, F., "Tensile Evaluation of the Mechanical Properties of Magnaweave Composites," presented at the Annual Meeting of the American Helicopter Society, Philadelphia, 1983.
[11] Halpin, J. C., Jerina, K., and Whitney, J. M., "The Laminate Analogy for 2- and 3-Dimensional Composite Materials," *Journal of Composite Materials*, Vol. 5, Jan. 1971, p. 36.

Fabrication Methods

Toshoku Cho[1] *and Akimitsu Okura*[1]

Manufacture of Carbon-Carbon Composites by Using Fine Coke and Its Properties

REFERENCE: Cho, T. and Okura, A., **"Manufacture of Carbon-Carbon Composites by Using Fine Coke and Its Properties,"** *Recent Advances in Composites in the United States and Japan, ASTM STP 864*, J. R. Vinson and M. Taya, Eds., American Society for Testing and Materials, Philadelphia, 1985, pp. 443–455.

ABSTRACT: A fabrication of a carbon-carbon (C/C) composite, in which fine pulverized coke is used as carbon material of matrix, was carried out using three types of carbon fibers, that is, short fibers, long fibers, and cloth. The bending strength of the composite is highest with long fibers 3 to 5 times as high as of ordinary carbon material, then cloth and short fibers. A still stronger C/C composite could be obtainable with long fibers by improving dispersion of these fibers in the matrix. Accordingly, the C/C composite obtained through this study imply a success in development of such composite materials by means of a simple process.

KEY WORDS: carbon-carbon composite, pitch coke, coal-tar pitch, carbon fiber, hot press, high temperature treatment, apparent density, bending test, reflex polarization microscope, fracture behavior, brittle fracture, microtexture, crack

One of the most attractive applications of carbon fibers is for carbon fiber reinforced carbon composites (C/C composites). This carbon material as the matrix has been provided from thermosetting resins or the deposited carbon obtained through the chemical vapor desposition (CVD) process [1,2] while attempts for use of coal-tar pitch have been made for the same purpose [3–5]. Thermosetting resins or coal-tar pitch creates a problem: the dissolved gas generated under the process of carbonation may form as blowholes in the resultant composite material. Accordingly a cycle of impregnation and carbonation can not provide the composite material with a sufficient strength, but several cycles are usually repeated. The current methods to manufacture C/C composites thus require a complicated series of processes combined with ad-

[1]Technical official and assistant professor, respectively, Institute of Industrial Science, University of Tokyo 22-1, Roppongi 7-chome, Minato-ku, Tokyo 106

vanced techniques, so that they can not be judged as favorable even with respect to the productivity and costs.

Thus in this study with the aim of finding any way to improve these points, a simple process for manufacturing C/C composites is sought by using fine pulverized coke as matrix for C/C composites. That is to say, the fine pulverized coke is kneaded with a binder pitch, which is kneaded or layered with carbon fibers, and are directly turned to a C/C composite by densifying with a hot press and heat treatment so as to obtain a higher bending strength. The composites so obtained are examined for their apparent density, bending strength, and the bonding condition at the interface between carbon fibers and matrix. The relations between them, and further the influence of blending ratio of pitch, volume fraction of carbon fibers, and the treatment temperature upon the strength, are examined.

Experimental Method

Raw Material

Pitch coke and petroleum coke are used to form the matrix. They are both finely pulverized at three stages by a vibratory ball mill. The size of fine coke is obtained in the form of a specific surface area diameter (D_M) through measurement of the specific surface area (S) by Blaine's method using the formula as follows [6]

$$D_M = \frac{6}{\rho \times S}$$

where ρ represents real density of the coke.

The sizes of fine coke thus obtained are 3.50, 5.82, and 9.35 μm for pitch coke and 3.52, 5.46, and 8.91 μm for petroleum coke. Two types of binder pitches are used, coal-tar pitch for the pitch coke and naphtha-tar pitch for the petroleum coke. The properties of the cokes and binder pitches are given in Table 1.

The carbon fiber is high-tension polyacrylonitrile (PAN) type fiber in the form of short fibers (5 mm), long fibers (a bundle of 12 000 filaments), and cloth.

Method of Manufacture

Figure 1 shows the outline of manufacturing processes of a C/C composite. Manufacturing process for the matrix was conducted to determine the most appropriate type of the coke and its optimum size.

Baked blocks were fabricated in all cases by making first cold blocks at

TABLE 1—*Properties of cokes and binder pitches.*

Coke

	Real Density, g/cm^3	Fixed Carbon, %	Volatile Matter, %	Ash, %	S, %
Pitch coke	2.002	99.15	0.56	0.29	0.19
Petroleum coke	...	98.23	0.35	0.35	1.01

Binder Pitch

	Density, g/cm^3	Softening Point, °C	Fixed Carbon, %	Ash, %	QI,^{a} %	BI,^{b} %
Coal tar pitch	1.2	30	31	0.02	0.02	...
Naphtha tar pitch	...	...	10 ~ 20	...	0.03 ~ 0.5	0.03 ~ 1.5

^{a}QI = quinoline insoluble.
^{b}BI = benzene insoluble.

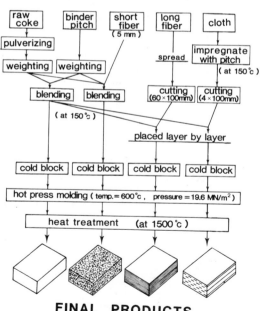

FINAL PRODUCTS
(size=100 × 60 × about 5 mm)

FIG. 1—*Manufacturing process of C/C composites.*

room temperature, molding them by hot press, and subjecting them to a high-temperature treatment. The final products have the dimension of 60 mm wide, 100 mm long, and 5 to 6 mm thick. Cold blocks for the matrix are made by mixing the coke powder and binder pitch with ratios up to 1 : 0.45 in weight (hereinafter expressed simply as "blending ratio of pitch"), kneading them thoroughly, and molding them into a case of 60 × 100 mm under heating at 150°C. When short fibers are used, the raw material of the matrix and the short fibers are kneaded simultaneously and molded at 150°C in the same manner as previously described. When long fibers are used, on the other hand, the bundle of fibers is spread 60 mm in width and cut 100 mm in length to form layers, which are placed layer by layer with the matrix admixture alternately and formed into a cold block in the same manner. When cloth is used, the cloth impregnated in the pitch is cut into 4 × 100 mm to form these strips, which are put in the matrix at a constant spacing to form a cold block as shown in Fig. 1.

The hot pressing conditions of these cold blocks were a maximum temperature and a pressure of 600°C and 19.6 MN/m^2, respectively, where the heating cycles thereto were a little different according to the blending ratio of the pitch. The high-temperature treatment is so performed burying the baked blocks in the coke powder and holding the atmosphere of nitrogen (N_2) gas at 1500°C for 30 min. The heating rate was 80°C/min and 400°C/h up to 800 and 1500°C, respectively.

Measurement of Bending Strength and Observation

Each specimen cut out of the sample was dimensionally finished accurately with an emery paper, its weight and dimensions were measured, and then its apparent density (ρ_A) was calculated. The bending strength (σ_{max}) was determined by the three-point bending test method using a specimen approximately 5 mm thick and 8 mm wide. The span was set equal to 35 mm. The testing machine was a Shimadzu Autograph IS-2000, in which the cross-head speed was set at 0.5 mm/min.

Three specimens of those samples which had undergone bending test were subjected to observation. They were fixed with resins, and, after surface grinding and polishing, observations were made on the specimens with a reflex polarization microscope at a magnification of 200 and 500.

Results and Discussions

Determination of Carbon Material for Matrix

In order to determine the optimum type and particle size of the coke to be used as matrix, first baked blocks of matrix only were prepared and its bending strength was investigated. The results obtained are shown in Fig. 2, which

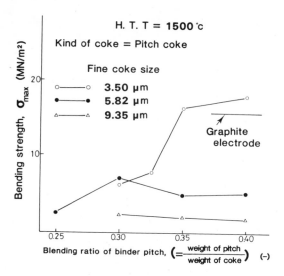

FIG. 2—*Relation between blending ratio of binder pitch and bending strength of matrix for each fine coke size.*

indicates that smaller size of coke particles give higher σ_{max}-value, maximizing with a pitch ratio of 0.4 if the smallest size (3.5 μm) is used.

The same tendency was observed with petroleum coke, but its σ_{max}-values were a little lower than those with pitch coke. Therefore, pitch coke with 3.5-μm particles was employed as the matrix for C/C composites in this experiment.

Properties of Composite Material

Figure 3 shows the relation between bending strength (σ_{max}) and the volume fraction of short fibers (hereinafter called solely V_f) for two kinds of temperature using heat treatment. As seen clearly from this figure, the σ_{max} becomes the maximum for $V_f = 3$ to 10%, then drops remarkably at V_f larger than 10%. It is indicated also that this composite material loses its heat treatment strength with a higher temperature. A comparison of the strength at $V_f = 2.65\%$ with that of a graphite electrode rod indicates that the σ_{max} is a little higher than the graphite electrode rod. As far as this comparison indicates, it can be judged that no good effect is expected from short fibers as a reinforcement for a matrix.

Figure 4 shows the relation between the bending strength σ_{max} and apparent density (ρ_A) of the composite material prepared under the condition that the V_f of the cut cloth is set at 2.65%, the blending ratio of pitch (the one with carbon fibers signifies the ratio of added pitch to the total amount of fibers

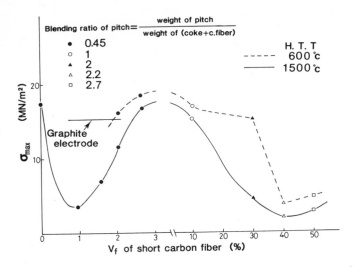

FIG. 3—*Relation between bending strength (σ_{max}) and volume fraction of short carbon fiber in C/C composites.*

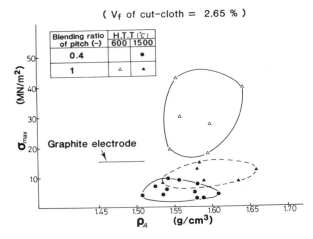

FIG. 4—*Relation between apparent density (ρ_A) and bending strength (σ_{max}) of C/C composites, using cut cloths.*

and coke powder) is 0.4 and 1, and the treatment temperature is 600 and 1500°C. It can be seen from Fig. 4 that higher blending ratio of pitch and lower treatment temperature provides higher strength and that the maximum value of σ_{max} with the blending ratio of pitch 1 and treatment temperature 600°C is three times as high as that of a graphite electrode rod.

Figure 5 shows the relation between the V_f and the bending strength σ_{max}

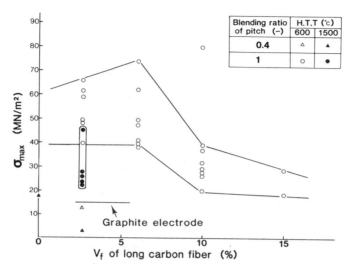

FIG. 5—*Relation between volume fraction of long carbon fiber and bending strength in C/C composites.*

for various temperature levels of the heat treatment, when the blending ratio of pitch is changed at two stages and long fibers are placed in layers. It follows from Fig. 5 that higher blending ratio of pitch and lower treatment temperature provides higher strength and that the σ_{max} with the pitch ratio 1 and treatment temperature 600°C yields notably high values for $V_f = 2.65$ to 6%, but the strength drops at $V_f \geq 10\%$. The σ_{max} at $V_f = 6\%$ becomes 40 to 75 MN/m², being 3 to 5 times stronger than graphite electrode rod approximately.

Figure 6 shows the relation between σ_{max} and apparent density (ρ_A) of C/C composite materials with different forms of carbon fibers, which indicate that there is a positive correlation between σ_{max} and ρ_A when short fibers are used and a similar correlation is seen for the case of the matrix only. When long fibers or cut cloth is used, however, no significant relationship is noticed between these two factors, and it can be supposed that composite materials have different properties from the one with short fibers or the matrix. According to this figure, a comparison of the bending strength of three types of composite materials for the range of ρ_A investigated, 1.55 to 1.65 g/cm³, indicates that the bending strength is highest with long fiber, next cut cloth, and short fibers, the matrix only being the lowest.

Figure 7 shows the residual carbon from the pitch, in the matrixes of three types of C/C composite materials with raised blending ratio of pitch and in the matrix with less blending ratio of pitch. The data indicate that the residual carbon is within the range of 2 to 8% regardless of the blending ratio of

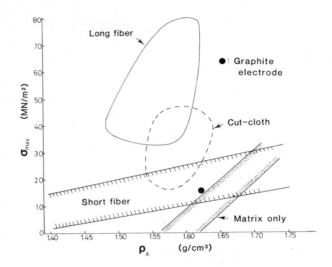

FIG. 6—*Relation between apparent density (ρ_A) and bending strength (σ_{max}) of various C/C composites.*

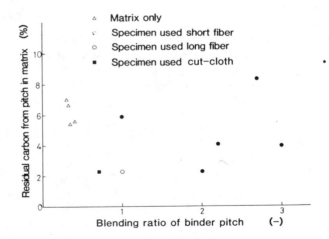

FIG. 7—*Relation between blending ratio of binder pitch and residual carbon from pitch in matrix of specimens after hot-pressed at 600°C.*

pitch. The reason why the ratio of residual carbon is low despite raised blending ratio of pitch is because the pitch has oozed out of the die at the time of hot press, and the fact that the specimen (mark ○ in the figure with $V_f =$ 2.65%) yields a remarkable increased strength when the blending ratio of pitch is increased, and, in the case of long fibers, is not due to the influence of the residual carbon in the pitch but to other factors.

Fracture Behavior of the Composite Material

Figure 8 shows typical three-point bending load-deflection curves of C/C composites with various types of carbon fiber. According to these curves, the fracture behavior depends completely upon the form of the carbon fibers. The matrix without carbon fiber shows brittle fracture in the same manner as ordinary carbon materials as shown by *D* in the figure. All composite materials with carbon fibers, however, do not rupture into two pieces and provide a load-deflection curve with complicated configuration. The load-deflection curves for composite material with long fibers (*A* in the figure) and short fibers (*C* in the figure) consist of two stages, the first being monotonic increase (going up hill), the second being gradual decrease (going down hill). When cut cloth is used, the maximum load is attained after minor striation of the load to be followed by a stepped fall as shown by *B* in the figure, and this rise and fall is repeated for several times.

Result of Observation with Polarization Microscope

Figure 9*a* shows the polarized light microphotographs of a specimen made by 3.5-μm pitch coke, with blending ratio of pitch and heat treatment temperature 0.4 and 1500°C, respectively. As known clearly from this photo, the microtexture is composed of the leaflet and fibrous textures originating from the raw coke and binder pitch, and structure of matrix signifies that the unit particles of the fine coke are tightly bonded by the binder pitch with the form

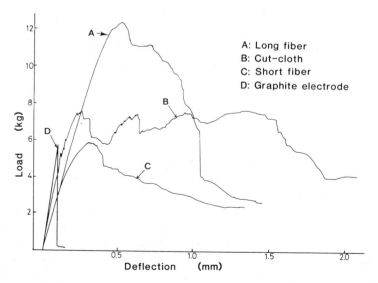

FIG. 8—*Load-deflection curves of various C/C composites.*

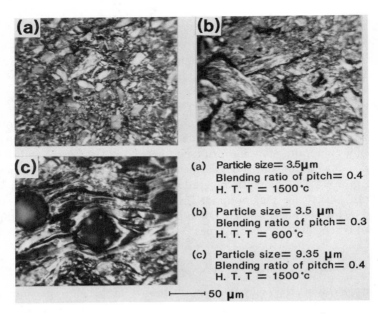

(a)   Particle size= 3.5μm
      Blending ratio of pitch= 0.4
      H. T. T = 1500°c

(b)   Particle size= 3.5 μm
      Blending ratio of pitch= 0.3
      H. T. T = 600°c

(c)   Particle size= 9.35 μm
      Blending ratio of pitch= 0.4
      H. T. T = 1500°c

|——————| 50 μm

FIG. 9—*Polarized light microphotographs of matrix cross section.*

of each particle kept as it was. This bonding condition, however, degrades as the blending ratio of pitch or treatment temperature is lowered, and in the former some places are perceived where unit particles in the matrix are bonded coarsely to each other, and, in the latter, places are seen further where cracks are generated around unit particles (Fig. 9*b*). In the case of a larger size of fine coke (9.35 μm), microcracks and pores are observed (Fig. 9*c*) even inside individual particles, in addition to the defects mentioned previously. The degree of generation of these defects is likely to be responsible for the reduction in the strength of the matrix.

Figure 10 shows the bonding condition at the interface between carbon fibers and matrix for V_f of short fibers = 30% and the treatment temperature being 600 and 1500°C. In the case of the treatment temperature 600°C, two differently characterized places are observed here and there, one where carbon fibers are well dispersed in the matrix, as shown in Fig. 10*a*-1, and the other where carbon fibers are congested because of bad permeability of the binder pitch as shown in Fig. 10*a*-2, and around these congested areas microcracks have often initiated. In the case of the treatment temperature 1500°C, the dispersion of carbon fibers in the matrix is similar to the case of the temperature 600°C, but microcracks (Fig. 10*b*-1) and macrocracks (Fig. 10*b*-2) are observed at interfaces of the congested carbon fibers presumably owing to shrinkage of the matrix. Thus, it is known that increase in V_f and subjecting to high-temperature treatment will increase generations of cracks in the case

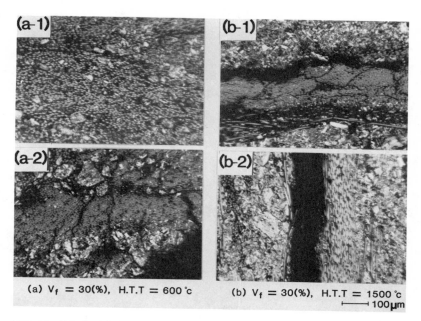

(a) V_f = 30(%), H.T.T = 600 °c   (b) V_f = 30(%), H.T.T = 1500 °c
⊢————⊣ 100μm

FIG. 10—*Polarized light microphotographs of a C/C composite cross section using short fiber.*

of short fibers. The degree of generation of these cracks is responsible for the reduction in the strength of the specimen.

Figure 11 shows the bonding condition at the interface between carbon fibers and matrix in the case of V_f of long fibers being 2.65%. For specimens with the pitch ratio of 1 and V_f of long fibers larger than 10%, and specimens with the pitch ratio of 0.4, the carbon fibers layered in the matrix remain in layers as they are, as shown in Fig. 11a. When V_f is less 6% with the pitch ratio 1, some places are seen where the carbon fibers remain in layers, but, in most of the sections, carbon fibers are observed well dispersed as shown in Fig. 11b. When carbon fibers form layers, the pitch impregnation in these layers is poor and often cause segregated layers with poor bonding at the interface between the carbon fiber layer and matrix, as shown in Fig. 11c, except some isolated areas where layers are well surrounded by matrix. Such segregated layers are found more often with higher treatment temperature in both blending ratio, 0.4 and 1.

From the fact that such segregated layers are generated always in positions where the fibers are not sufficiently dispersed in the matrix and remain in the form as layers, it is supposed that carbon fibers remaining in layers has caused directly this generation of segregated layers. In addition, when V_f is less than 6%, it is revealed that these layers are less generated by increasing

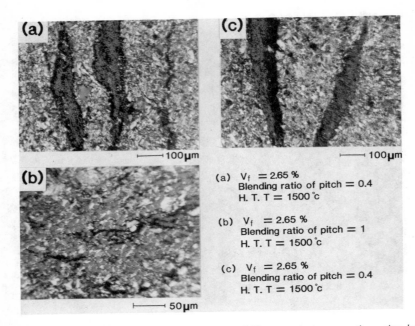

(a) V_f = 2.65 %
       Blending ratio of pitch = 0.4
       H. T. T = 1500 °c

(b) V_f = 2.65 %
       Blending ratio of pitch = 1
       H. T. T = 1500 °c

(c) V_f = 2.65 %
       Blending ratio of pitch = 0.4
       H. T. T = 1500 °c

FIG. 11—*Polarized light microphotographs of a C/C composite cross section using long fibers.*

the blending ratio of pitch, leading to a supposition that the increase in pitch mixing ratio has an effect on disperseness of carbon fibers in the matrix, and that this effect solely has contributed to manifesting the long fiber's reinforcement effect.

Accordingly the strength of a C/C composite with long fibers can presumably be enhanced more by increasing the value of V_f of the carbon fibers and by incorporating any technique which ensure that the fibers are dispersed in the matrix evenly.

When cut cloth is used, all fibers have remained in the matrix in the form as layers. The impregnation of pitch into these layers was poor. The bonding condition of carbon fibers to the matrix is observed in such that in the specimen heat-treated at 600°C the fibers are tightly surrounded by the matrix and, in the one treated at 1500°C, there are many segregated layers with poor bonding or many micro- and macrocracks along the fiber direction regardless of the blending ratio of pitch.

Conclusions

A fabrication of C/C composite materials using fine pulverized pitch coke to form their matrix was attempted, and the bending strength of the resultant composite materials were measured. By comparing these results with the data of ordinary carbon materials, feasibility to manufacture C/C composite materials by this process is examined. This study is not yet completed, but the experimental results hitherto are summarized as follows:

1. The C/C composite with short fibers has presumably the maximum bending strength at a certain V_f-value. The strength of the composite obtained by this experiment does not differ substantially from that of the matrix, but some mechanical improvement of the mixing performance of matrix material and fiber might provide an increased strength to a certain degree. Further distinct increase of the strength will not be expected as long as short fibers are used.

2. The C/C composite with long fibers has the bending strength 3 to 5 times as high as that of ordinary carbon material despite its low V_f-value of 2.65 to 6%. Further increase in V_f as well as improvement in the method for dispersing fibers evenly in the matrix might provide this composite material with a higher bending strength. It signifies that this development of C/C composite through this process has succeeded.

3. When cut cloth is used, the maximum bending strength three times as high as that of ordinary carbon materials is obtained, but there was raised problem that the permeability of pitch into the cloth layers in matrix is poor. A higher strength will not be expected as long as cut cloth is used.

4. The C/C composite obtained by this work present a quite different fracture behavior from ordinary carbon materials. Accordingly practical applications of these composite materials may spread to the fields where no particularly high strength is required but rather toughness is required.

Acknowledgment

We shall highly appreciate Shin-Nihon Seitetsu Chemical Co., Ltd. for offering coal-tar pitch as well as Nippon Carbon, Ltd. and Toho Besron Co., Ltd. for their carbon fibers.

References

[1] Carborundum Graphite Products Division "Carbitex," Sanborn, New York.
[2] Super-Temperature Corp., "RPG," Santa Fe Springs, California.
[3] Fitzer, E. and Karlische, K., 13th Biennial Conference on Carbon at Irvine, Extended Abstracts, 1977, p. 353.
[4] Fitzer, E., Hüttinger, W., and Manocha, L., *Extended Abstracts*, 14th Biennial Conference on Carbon at Pennsylvania, 1979, p. 240.
[5] Kawamura, K., Kimura, S., Yasuda, E., and Inagaki, M., *Tanso*, 1982, No. 109, p. 46.
[6] Iiya, K., *Handbook of "Huntai-Kogaku,"* Asakura-shoten, Japan, p. 65.

Shiro Kohara[1] and Norio Muto[1]

Fabrication of Silicon Carbide Fiber-Reinforced Aluminum Composites

REFERENCE: Kohara, S. and Muto, N., **"Fabrication of Silicon Carbide Fiber-Reinforced Aluminum Composites,"** *Recent Advances in Composites in the United States and Japan, ASTM STP 864*, J. R. Vinson and M. Taya, Eds., American Society for Testing and Materials, Philadelphia, 1985, pp. 456–464.

ABSTRACT: A fabrication method for silicon carbide (SiC) fiber-aluminum composites was investigated. An SiC fiber manufactured from an organosilicon polymer was used, and the combination of slurry impregnation and hot pressing was adopted in the present study. The method was improved so as to control the volume fraction of fibers in the fabricated composites. The composites with high fiber content were achieved by means of this method. During the fabrication process, degradation of the SiC fibers was observed. The addition of silicon to the aluminum matrix was effective in reducing the degradation. However, the complete prevention of the degradation of SiC fibers could not be achieved by silicon alloying.

KEY WORDS: composite materials, silicon carbide fibers, aluminum, fabrication, slurry, hot pressing

Silicon carbide (SiC) fibers synthesized from an organosilicon polymer is continuous and flexible and, in addition, has high tensile strength and excellent oxidation resistance [1–3]. These properties of the SiC fiber meet the requirements for the reinforcing fiber in metal matrix. To make a strong fiber-reinforced composite, a chemical reaction at the interface between fiber and matrix is necessary. This reaction brings about, to a limited extent, strong bonding between the fiber and matrix. However, excessive chemical reaction may result in degradation in the strength of the fibers, which would be consumed by the chemical reaction. The reinforcing fiber should be compatible with the matrix material. The compatibility of the SiC fibers with aluminum has been investigated [4]. It was shown that the SiC fibers were highly compatible with solid aluminum, but only fairly compatible with liquid aluminum. Various techniques for fabricating fiber-reinforced metal matrix

[1] Professor and research assistant, respectively, Institute of Interdisciplinary Research, Faculty of Engineering, University of Tokyo, Tokyo, Japan.

composites have been proposed. Among those techniques, a method combining slurry impregnation and hot pressing was adopted in the present study, since the slurry method was considered to be advantageous for continuous processing. However, the slurry method has the following disadvantages: (1) Control of the volume fraction of fibers in the fabricated composites is difficult; (2) residual pores between fibers cannot be eliminated completely, and (3) slurry is apt to fall off on handling. The consolidation of SiC fiber-aluminum composites by means of the slurry method has been reported by Kasai et al [5]. However, no countermeasures against those disadvantages in the slurry method were described in their report. In the present study, improvements were made to minimize those disadvantages in the slurry method and to apply this method to the continuous process.

The purpose of the present investigation is to determine the pertinent conditions for fabricating the SiC fiber-aluminum composites by the slurry impregnation and hot pressing method.

Experimental Procedure

SiC fibers manufactured by Nippon Carbon Co. Ltd. were used in the present study. A scanning electron micrograph of the SiC fibers is shown in Fig. 1. Tensile strength and elastic modulus of the fibers were determined by tension testing of the fiber specimens with a gage length of 20 mm. Various

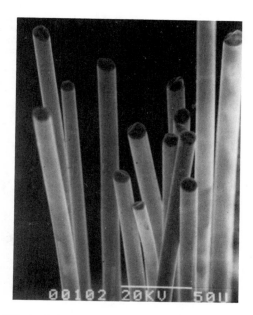

FIG. 1—*Scanning electron micrograph of SiC fiber.*

dispersion agents were tested for preparing aluminum powder slurry. Among the tested chemicals, sodium alginate was most effective as a dispersion agent for metal powder slurry. The slurry was prepared by mixing aluminum powder of 50% volume fraction with 1% sodium alginate solution while agitating. A yarn of 500 SiC filaments was immersed in the slurry and pulled up continuously through a nozzle with a small hole. The apparatus used for impregnating the yarn with slurry is schematically shown in Fig. 2. The impregnated yarn was dried in air and cut into pieces of settled length. The pieces were stacked in a graphite die and hot pressed in vacuum with pressures of 0.5 to 1 MPa at temperatures in the liquid state of aluminum: 953 to 973 K.

As the alternative method, some of impregnated yarns were heated by pulling up through a small furnace with the purpose of forming a kind of precursor before hot pressing. The handling of the impregnated yarn became easier by this precursor formation. Then, the precursors were cut into pieces of settled length and hot-pressed with the same conditions as before.

The microstructures and properties of the fabricated composites were examined. Tension testing of the composites was made using specimens with dimensions of 40 by 10 by 0.8 mm and aluminum cover skins on the grip portions.

Results

Figure 3 shows a Weibull plot of the strength of tested fiber specimens of more than 100. The measured mean tensile strength of the fiber was 2020 MPa (gage length: 20 mm), the mean elastic modulus 160 GPa, and the mean

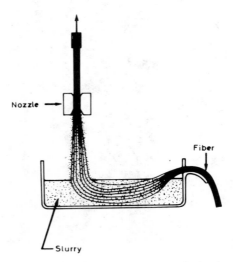

FIG. 2—*Apparatus for impregnating SiC yarn with aluminum powder.*

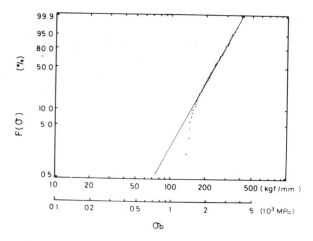

FIG. 3—*Weibull plot of tensile strength of SiC fiber.*

diameter 14.4 μm. The relationship between tensile strength and diameter of the fiber is shown in Fig. 4. The dependence of tensile strength on diameter of the fiber was recognized from this figure. Figure 5 shows cross sections of the impregnated yarn. It is shown that the yarn was well impregnated with aluminum powder and particles of the powder were uniformly distributed. The yarn was then hot-pressed at temperatures in the liquid phase of aluminum. The cross sections of the composite fabricated by hot pressing are shown in Fig. 6. No pores were observed in the sections, and the SiC fibers were distributed closely and uniformly.

The relationship between density and volume fraction of fibers is shown in Fig. 7. A dotted line in the figure represents the rule of mixture. All the measured values of density of the fabricated composites fell in the range of 98% to 100% of the calculated values. From these results, it can be said that this fabrication method is useful for obtaining sound composites with a high volume fraction of fibers. The relationship between tensile strength and volume fraction of fibers is shown in Fig. 8. A dotted line in the figure again represents the rule of mixture. Tensile strength increased with volume fraction of fibers at the region of low fiber content, but did not increase in the region of higher fiber content, and the values of tensile strength were considerably lower than those calculated by the rule of mixture.

Discussion

Little has been reported on the slurry method applied to the fabrication of composites [5,6]. One of the disadvantages of the slurry method is the difficulty in controlling volume fraction of fibers in the finally fabricated compos-

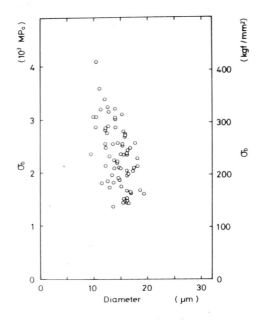

FIG. 4—*Relationship between tensile strength and diameter of SiC fibers.*

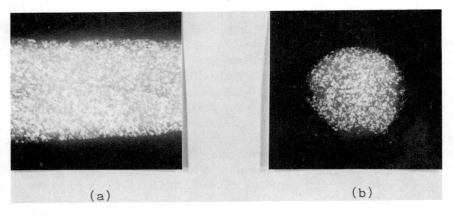

(a)   (b)

FIG. 5—*Cross sections of SiC yarn impregnated with aluminum powder: (a) longitudinal, (b) transverse.*

ites, and the other is the residual pores trapped between fibers. A small nozzle attached to the impregnation apparatus scrapes excess slurry from the surface of a yarn and yields intimate contact between fibers. Furthermore, a nozzle with a large hole gives a low volume fraction of fibers and one with a small hole gives a high volume fraction of fibers, so that the fiber content can be

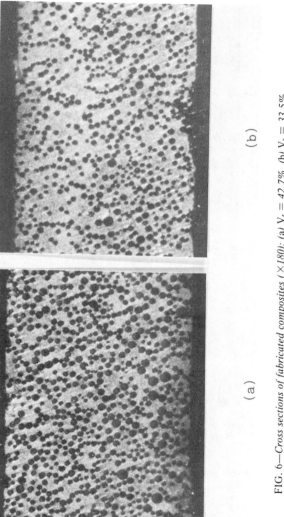

(b)

(a)

FIG. 6—Cross sections of fabricated composites (×180): (a) $V_f = 42.7\%$, (b) $V_f = 33.5\%$.

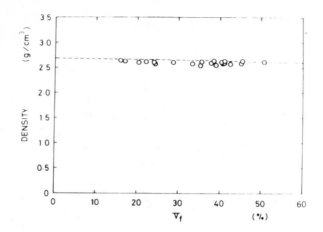

FIG. 7—*Relationship between density and volume fraction of fibers.*

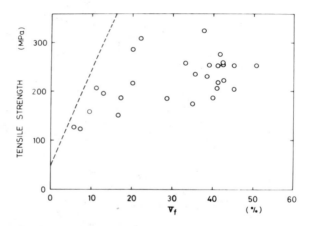

FIG. 8—*Relationship between tensile strength and volume fraction of fibers.*

roughly controlled by selecting a nozzle hole of adequate size. In addition, liquid-phase hot pressing is also effective for eliminating residual pores between fibers. Thus, the composites with a high volume fraction of fibers were obtained by the combination of the improved slurry method and liquid-phase hot pressing. Intermediate heating of the impregnated yarns before hot pressing made handling of the impregnated yarns easy, as previously reported by the authors [7]. However, the tensile strength of the fabricated composites was not improved very much by this treatment.

The values of tensile strength of the fabricated composites were lower than those calculated from the rule of mixture, one of the reasons being that degradation of the SiC fibers might have occurred during the fabrication process. So, the tensile strength of the fibers extracted from the fabricated composites by dissolving a matrix with sodium hydroxide solution was measured. Figure 9 shows the tensile strength and also the diameter of the fibers plotted against hot-pressing time, that is, the time held at 953 K during hot pressing. The data points for tensile strength represent the mean values determined from the Weibull plot of measured values as shown in Fig. 3. This graph shows that the tensile strength rapidly decreased with time, but the diameter remained about the same. It means that the properties of the SiC fibers may change essentially while the fibers are kept in contact with molten aluminum.

It has been reported that the compatibility of the SiC fibers with liquid aluminum could be improved by the addition of silicon to the aluminum matrix [4]. The effect of silicon in the aluminum matrix was investigated using two kinds of aluminum-silicon alloy powders: Al-1Si alloy and Al-10.5Si alloy. The former corresponds to the silicon content near the solubility limit of silicon in solid aluminum, and the latter the silicon content near the eutectic point. Hot pressing was carried out at 953 K for the former and at 873 K for the latter. The tensile strength of the SiC fibers extracted from the fabricated composites was plotted against hot-pressing time; this is also shown in Fig. 9. The effect of silicon addition in preventing degradation of the SiC fibers was only slight in Al-1Si alloy. In Al-10.5Si alloy, the effect was larger than in the case of Al-1Si alloy. However, the effect of silicon addition cannot be explained simply by the solubility limit of silicon in aluminum, since the temperature of hot pressing is lower in the case of Al-10.5Si alloy. Though the addition of silicon to aluminum had the effect of lessening degradation of the SiC fibers, it could not stop the degradation of the fibers. It may be necessary to protect the fibers from attack by molten aluminum, or to use the very-

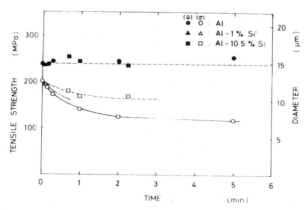

FIG. 9—*Changes in tensile strength and diameter with hot-pressing time.*

short-time fabrication method, when fabricating the SiC fiber-aluminum composites by liquid-phase processing.

Conclusions

A fabrication method for SiC fiber-aluminum composites was investigated and the following conclusions were obtained:

1. The method combining slurry impregnation and liquid-phase hot pressing was useful for fabricating the fiber-reinforced metal matrix composites with a high volume fraction of fibers.

2. The tensile strength of the SiC fiber-aluminum composites fabricated by liquid-phase hot pressing was lower than that expected from the rule of mixture. It may be attributed to the degradation of the SiC fibers due to the attack by molten aluminum.

3. The addition of silicon to aluminum was effective in reducing the degradation of the SiC fibers, but complete prevention of the degradation was not achieved.

References

[1] Yajima, S., Hayashi, J., Omori, M., and Okamura, K., *Nature*, Vol. 261, 1976, pp. 683–685.

[2] Yajima, S., Okamura, K., Hayashi, J., and Omori, M., *Journal of the American Ceramic Society*, Vol. 59, 1976, pp. 324–327.

[3] Yajima, S., Kayano, H., Okamura, M., Omori, M., Hayashi, J., Matsuzawa, T., and Akutsu, K., *American Ceramic Society Bulletin*, Vol. 55, 1976, p. 1065.

[4] Kohara, S. in *Proceedings*, Japan-U.S. Conference on Composite Materials, Tokyo, 1981, pp. 224–231.

[5] Kasai, Y., Saito, M., and Asada, C., *Journal of the Japan Society of Composite Materials*, Vol. 5, 1979, pp. 26–30, 56–61.

[6] Prewo, K. M. and Brennan, J. J., *Journal of Materials Science*, Vol. 17, 1982, pp. 1201–1206.

[7] Kohara, S. and Muto, N. in *Proceedings*, Fourth International Conference on Composite Materials, Vol. 2, Tokyo, 1982, pp. 1451–1455.

Ronald P. Nimmer,[1] *Kevorc A. Torossian,*[2] *and John S. Hickey*[1]

Fabrication and Spin Tests of Thick Laminated S2-Glass Flywheel Disks

REFERENCE: Nimmer, R. P., Torossian, K. A., and Hickey, J. S., **"Fabrication and Spin Tests of Thick Laminated S2-Glass Flywheel Disks,"** *Recent Advances in Composites in the United States and Japan, ASTM STP 864,* J. R. Vinson and M. Taya, Eds., American Society for Testing and Materials, Philadelphia, 1985, pp. 465–488.

ABSTRACT: Flywheels have been used to store and supply energy in a wide range of applications. In transportation-related applications, a primary design consideration is to store the maximum amount of energy per unit weight of the flywheel. Based upon this criterion, the high strength-to-density ratios of composite materials make them attractive materials for an efficient design. This investigation reports on a program to develop a composite flywheel consisting of a laminated S2-glass/epoxy central disk and a filament-wound graphite/epoxy outer ring. It is shown both analytically and experimentally that the presence of the outer ring significantly improves the energy density available in a simple laminated disk. Consideration is given to the subjects of fabrication, failure criteria, and experimental determination of ultimate speed and strength of prototype flywheels.

KEY WORDS: flywheels, composite, graphite, S2-glass, epoxy, failure criteria, first-ply failure, ultrasonic testing, spin testing

Although flywheels were commonplace in machinery even before the Industrial Revolution, it appears that they may still have new roles to play in our society today. Fluctuating costs and shortages of petroleum-based fuels have posed new constraints on energy use and created incentive for more efficient use of that energy. One concept under consideration applies a flywheel as an energy storage device in conjunction with either an internal combustion or battery-driven vehicle. Another potential application in a very different environment seems on the horizon as initial plans are made for earth-orbit space stations. In this case, solar energy will need to be stored for use when the station is out of direct sunlight. Flywheels are also under consideration for use in this application. With regard to efficiency in either of these uses, it appears that the substitution of composite materials for engineering metals in these

[1]General Electric Co., Corporate Research and Development, Schenectady, NY 12301.
[2]General Electric Co., Turbine Technology Laboratory, Schenectady, NY 12345.

465

mechanical devices offers advantages in operating efficiency which could be directly translated into additional energy savings. The work presented here reports the results of one program to enhance energy savings with effective use of composite materials.

During the late 1970s and early 1980s the Department of Energy sponsored research in the area of alternate energy use in ground transportation. Application of flywheel energy storage was one of the concepts developed with respect to this goal. In fact, significant reductions in fuel consumption have been demonstrated [1] when a flywheel is used in conjunction with a heat-engine in an automobile. Without a flywheel energy storage unit, the mileage of an ordinary vehicle decreases as the number of stops per mile increases. However, Ref 1 illustrates that the use of a flywheel can improve fuel efficiency by 210% over the Federal Urban Driving Cycle and 430% over the New York City Driving Cycle. These improvements are achieved by recovering braking energy and permitting the engine to perform in its most favorable load/speed range through load leveling. In addition, the flywheel can also offer significant advantages when used in conjunction with battery propulsion systems. Current lead-acid batteries limit electric vehicular multi-stop-and-go driving range and accelerating capability because of their inability to handle high power demands and still maintain high energy density. On the other hand, a hybrid flywheel/battery system can isolate the battery from the acceleration power demands as well as recover a substantial portion of the available braking energy. Recent laboratory tests [2] have indicated that a flywheel with only a 100 Watt-hour (Wh) usable energy storage capacity could increase the range of a 1665-kg (3700-lb) lead-acid battery electric vehicle by 30%.

As has been the case in many other applications of composite materials, the driving force behind the interest in composite flywheels is weight reduction and concomitant energy conservation. It can be easily shown [3] that the maximum energy which can be stored per mass of flywheel rotor is given by

$$\frac{W}{M} = k_s \left(\frac{\sigma_u}{\rho} \right) \tag{1}$$

where

$$W = \text{energy stored at failure,}$$
$$M = \text{mass of the flywheel,}$$
$$\sigma_u \text{ and } \rho = \text{ultimate strength and density of the material, respectively, and}$$
$$k_s = \text{form factor dependent upon flywheel design.}$$

Thus, for a given design, the highest energy density (W/M) will be attained with materials characterized by high strength-to-density ratios, giving rise to the attractiveness of composites. In general, although there are many bases for comparing flywheel designs, maximum energy density has become one of the most commonly used measures. While maximum energy densities of 46 Wh/kg (21 Wh/lb) have been reported for high-strength steel flywheels, composite materials have been looked to for energy densities between 65 and 85 Wh/kg (30 and 40 Wh/lb).

Several different composite flywheel designs have been pursued, including various types of filament-wound rings [4–10] as well as laminated disks, both flat [11–14] and contoured [14]. Reference 15 provides a general overview of recent work in the area of composite flywheel development. The specific work presented herein addresses a design which incorporates a filament-wound graphite ring and a laminated S2-glass disk assembled with an interference fit between the two components. Two such composite flywheels manufactured during this program are shown in Fig. 1. This particular approach combines the attractive design characteristics of dynamic stability with high energy density both on a per-unit swept-out volume and on a per-unit weight basis.

The idea of using both a solid disk and a filament-wound ring in one flywheel has been previously discussed by Gupta and Lewis [16]. Lustenader and Zorzi [17] suggested use of laminated glass/epoxy for the central disk as well as an interference fit between disk and ring to preclude separation of the components. Later, a more detailed design study as well as a survey of potential materials was reported by Nimmer in Ref 18. As a result, subsequent fabrication and testing focused on the use of laminated S2-glass/epoxy in the central disk and filament-wound graphite/epoxy in the outer ring. Suggested failure criteria and supporting nondestructive ultrasonic tests of the laminated central disks have also been discussed [19].

The encouraging results of the initial design analyses reported in Ref 18 were

FIG. 1—*Hybrid disk-ring flywheels.*

followed by a program to fabricate and test model flywheels, and the results of that program are described here. For the purpose of discussion, the work is divided into five general topics: fabrication of thick, laminated disks; failure criterion for these laminated disks; spin tests on laminated disks; manufacture and characterization of the filament-wound outer rings; and assembly and spin testing of complete hybrid disk-ring flywheels.

Fabrication and Characterization of Thick Laminated Flywheels

Although the difficulty of fabricating quality high-strength composite laminates increases as the thickness of the part increases, there are also advantages associated with the ability to manufacture thick flywheels. For example, for application to flywheel/electric bus systems, a flywheel package capable of storing as much as 12 kWh would be required. In order to store this amount of energy, a number of individual rotors would have to be joined—most likely with bonding techniques to avoid stress concentrations associated with bolt holes. For such an application, the ability to manufacture thick laminated disks would reduce the number of bonds required to assemble the system. This reduction in assembly operations could lead to lower costs. As a result, fabrication techniques for thick laminates were developed and 2 to 3.5-cm-thick (0.78 to 1.4-in.) disks were manufactured. These disks were substantially thicker than others reported in the literature [14,20].

Based upon the previous experience of the second author regarding the fabrication of thick structural components with fiber-reinforced composites, the hydroclaving procedure was chosen for the manufacture of the laminated central disks. In this process, the laminated prepregs are placed between two metal faceplates, vacuum bagged, and placed in a processing tank where they are exposed to a vacuum/pressure cycle. The standard cycle used to produce the laminates in this program began with 2 h at a vacuum pressure of 0.5 mm (0.02 in.) of mercury. After raising the tank temperature to 100°C (212°F), the heat/pressure transfer medium [hot asphalt at 160°C (320°F)] is introduced into the tank and the tank pressure is raised to 0.6894 MPa (100 psi) through introduction of nitrogen. Curing at this temperature and pressure lasts at least 10 h. It was felt that for the rather thick laminates to be manufactured in this program, the hydroclaving process would be more effective in producing uniformly cured, low-void-content products than standard compression molding techniques. On the other hand, the lower pressures during cure could lead to somewhat lower fiber-volume content. Dimensional uniformity was also a question.

Several steps were taken to establish the general characteristics and specific properties of the hydroclaved products. In order to accumulate data relevant to processing thick laminates, glass, resin, and void content were all measured using a burnout technique on sample laminates. The mass density of the glass used in these calculations was 2.52 g/cm^3 and the mass density of the resin was

1.21 g/cm³. The laminate thicknesses ranged from 2.0 mm (0.077 in.) to 35.5 mm (1.4 in.) and the diameters of the laminates ranged from 25 cm (10 in.) to 39.4 cm (15.5 in.).[3] Data spanning the entire scale of these dimensions are presented in Table 1. As can be seen from comparison of the data for the 2.0-mm and 25.0-mm-thick laminates, neither the void content nor the fiber content was significantly affected by the thicker laminate dimensions. In addition, measurements for each of the 30.5 and 39.5-cm-diameter (12 and 15.8-in.) laminates were taken at four different locations in the plane of the disk and are labeled 1, 2, 3, and 4 in Table 1. Excellent uniformity in the measured properties is to be noted. In the case of the 35.5-mm-thick (1.4-in.) laminate, specimens were also taken at three locations through the thickness at the center of the laminate and labeled A, B, and C in Table 1. Again, excellent uniformity is to be observed. With regard to dimensional uniformity, the measured thicknesses about the outside diameter of the 2.5-cm-thick (1-in.) laminates usually varied between 0.125 mm (5 mils) and 0.375 mm (15 mils) for the 25.4 and 30.5-cm-diameter (12 and 10-in.) disks. The thickness of the aluminum pressure platens was increased from 1.25 cm (0.5 in.) to 1.92 cm (0.75 in.) for the three 39.4-cm-diameter (15.5-in.) disks. It is felt that this contributed to the reduction in the variation in thickness of these 3.5-cm-thick (1.4-in.) disks to 0.05 mm (2 mils).

As another measure of the effectiveness of this fabrication cycle for thick laminates, Table 2 lists the expected composite properties described in the 3M information product sheet for the SP250-S2 unidirectional epoxy-glass prepreg used in this program. As can be seen, the recorded void content levels in the thick laminates are all below the value listed in Table 2. As expected, however, the fiber volume content of the thick laminates is somewhat less than the value of 53% suggested in Table 2.

In addition to the test data reported in Table 1, tension tests were carried out by Duke [21] at the Virginia Polytechnic Institute and State University (VPI and SU) to establish strengths and stiffnesses. Since Kulkarni [22] reported a dependence of strength upon uniaxial load direction with respect to test coupons of laminated graphite taken from material used to fabricate a laminated graphite flywheel, these tests were designed to provide similar information with respect to the laminated S2 glass used in this design. The results of these tests are summarized in Fig. 2, which presents laminate tensile strength data for loads applied in fiber as well as off-axis directions as a function of the angle between laminate fiber orientations. As can be seen in the figure, there is a strong dependence of tensile strength upon load orientation for the quasi-isotropic laminates with larger angles between fibers. As the angle between fibers, θ, is decreased, the tensile strength associated with a load in a fiber direction decreases substantially since the proportion of fibers in the load direction is decreasing. However, the off-axis strength reported in Fig. 2 in-

[3]Original measurements made in U.S. customary units.

TABLE 1—Measured laminate characteristics for various thicknesses.

Laminate No.	Dia. cm (in.)	Thickness cm (in.)		% Glass by Weight	% Glass by Volume	% Resin by Weight	Theoretical Specific Density	Measured Specific Density	% Voids	Glass Transition Temp °C
801B03	20.3 (8.0)	.20 (.077)		66.65	47.82	33.35	1.85	1.81	2.16	—
4	25.5 (10.0)	2.5 (1.0)		66.55	NA	33.45	1.8529	1.8226	1.633	120
31	30.5 (12.0)	2.5 (1.0)		66.33 (Avg.)	48.39 (Avg.)	33.67 (Avg.)	1.849 (Avg.)	1.812 (Avg.)	1.97 (Avg.)	117 (Avg.)
		Location	1	66.33	48.38	33.67	1.812	1.809	—	119
			2	66.34	48.35	33.64	1.850	1.811	2.1	117
			3	66.38	48.49	33.62	1.850	1.815	1.89	115
			4	66.29	48.36	33.71	1.849	1.813	1.92	117
33	39.4 (15.5)	3.56 (1.4)		67.94 (Avg.)	50.10 (Avg.)	32.06 (Avg.)	1.867 (Avg.)	1.832 (Avg.)	2.10 (Avg.)	113.4
		Location	1	67.77	50.04	32.23	1.870	1.835	1.871	115
			2	67.54	49.68	32.46	1.866	1.828	2.076	115
			3	67.60	49.83	32.40	1.832	1.832	1.900	114
			4	67.74	49.92	32.26	1.869	1.832	2.012	113
			A	68.32	50.50	31.68		1.837	2.12	114
			B	68.12	50.27	31.88		1.834	2.19	111
			C	68.20	50.36	31.80		1.835	2.19	113

39.4 cm (15.5 in.)

3.56 cm (1.4 in.)

TABLE 2—*Manufacturer's suggested composite*
properties for SP250-S2 glass/epoxy prepreg
(3M Co.)

Specific density	1.88
Thickness per layer	0.23 mm (0.009 in.)
Fiber volume	53%
Cured resin content	29% by weight
Void content	3% by volume

creases as θ decreases. Since the average stresses in a spinning disk will be applied in both fiber and off-axis directions at different circumferential locations, the off-axis strength results may provide more realistic insight into the ultimate strength of laminated flywheels.

It is interesting to note that the dependence of tensile strength upon load orientation and angle θ can be predicted qualitatively [23] using a so-called failed ply discount analysis [24]. Using this approach, the transverse and shear moduli in a laminae are set to zero when stress levels are indicative of matrix damage. The average laminate properties can then be recalculated and stress analysis repeated for higher loads. Ultimate failure is associated with failure in all the laminae, regardless of what that mode of failure may be. On the other hand, the general dependence of laminate strength upon load orientation and ply angle θ is not adequately reflected by the simple association of laminate failure with first fiber failure. In fact, Ref 25 shows that such an assumption actually leads to the prediction that off-axis strength is higher than strength in a fiber direction.

Analytical Expectations for Thick Laminated Flywheel Performance

Using the approach of failed ply discount, which qualitatively predicted the experimental results observed with respect to load axis dependence in tensile strength, a laminate analysis can be carried out to provide insight into the potential effects of laminae stacking sequence upon the flywheel's ultimate strength. The approach used in this analysis is explained in detail in Ref 19. Summarizing the approach, a plane-stress analysis is used to establish the state of average stress in the laminated disk. This average stress state is in turn used to calculate the state of stress in the individual laminae via laminate theory. With knowledge of the laminae stresses the rotational speed which causes the first ply failure in the disk can be calculated. Applying the approach of failed ply discount, the transverse and shear moduli of the failed lamina or laminae are then set to zero and a new set of average moduli calculated. With these new average moduli, the average state of stress due to an increment in the rotational speed of the flywheel can be calculated and these stresses in turn used to predict the new states of stress in individual laminae. Repeating this

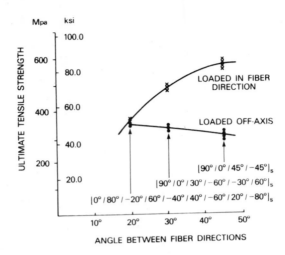

FIG. 2—*Laminate tensile strength as a function of angle between fiber directions.*

procedure, the rotational speed required to induce subsequent laminae failures can be calculated until all laminae have failed at all points in the plane of the disk. Setting transverse and shear moduli to zero in laminae where matrix damage is indicated by analysis is usually an overly severe degradation of the laminae properties and would be expected to yield a conservative prediction of failure.

Using this analytical approach, an effect of laminate stacking sequence does appear and is illustrated in Fig. 3, which describes the flywheel tip speed v_T necessary to damage all the laminae within a nondimensional radius $\bar{r}_D$. One possible failure criterion which is suggested by such an analysis is that ultimate

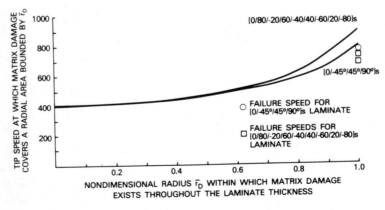

FIG. 3—*Comparison of lower-bound damage predictions for two laminated S2-glass/epoxy disks with different stacking sequences.*

failure occurs when all the laminae have been damaged throughout the disk. Based upon this criterion, initial damage would be expected to occur at the center of the disk and at a comparatively low tip speed with respect to ultimate failure. Using a maximum stress (or strain) criterion for individual lamina failure, it can be seen that this damage occurring near the center of the disk at low tip speeds is associated with transverse failure in individual laminae. In addition, the plane-stress analysis of a spinning disk indicates that the average stress state in the center of the disk is balanced biaxial. As a direct result of this state of average stress, laminate theory predicts two effects. First, all the laminae through the thickness of the disk will fail simultaneously in transverse tension, and second, the tip speed at which this initial failure occurs is independent of the laminating sequence provided it is quasi-isotropic in average stiffness. As can be seen in Fig. 3, the damage curves for two different laminating sequences remain extremely close until damage in the flywheel begins to extend into the outer region of the disk where the state of stress is dominated by the hoop stress. Although there are no "preferred strength directions" with respect to a state of balanced biaxial tension, there are preferred directions of loading for a uniaxial stress state as illustrated by the tension test results presented in Fig. 2. As a result, the analysis illustrated in Fig. 3 indicates that it will take higher tip speeds to damage the outside periphery of a flywheel with a $[0/80/-20/60/-40/40/-60/20/-80]_s$ laminating sequence than a flywheel with a $[0/-45/+45/90]_s$ sequence, since the former laminate has a higher off-axis strength than the latter. In both cases all laminae damage can be associated with transverse or shear failure on the basis of a maximum stress or strain criterion.

There is some experimental evidence that damage does occur first in the center of laminated flywheels and can be sustained without total failure. A complete discussion of this work can be found in Ref 19. Figure 4 illustrates results of ultrasonic velocity measurements presented in Ref 19 on S2-glass/epoxy flywheels before and after a spin test reaching a tip speed of 638 m/s (2093 ft/s). As can be seen from the figure, the longitudinal wave speeds through the thickness of the flywheel before spinning were uniform and in all cases faster than similar measurements in the wheel after testing to a peripheral speed of 638 m/s (2093 ft/s). In addition, the wave speeds in the wheel subjected to spin tests are slowest in the center of the disk with increasing velocities as the outside radius is approached. Reductions in wave speeds reflect degraded material moduli which are known to be associated with matrix damage in composites.

Spin Tests of Laminated Disks Without Rings

Five spin tests on laminated S2-glass epoxy disks were carried out. Four of the tests were conducted to destruction on flywheels similar to the one shown in Fig. 5 which were nominally 2.5 cm (1.0 in.) thick and fabricated by the General Electric Co. with the hydroclaving process previously described. A fifth,

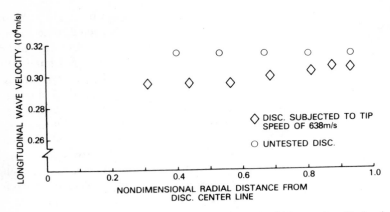

FIG. 4—*Comparison of longitudinal wave speeds through the thickness of an S2-glass/epoxy disk before and after spin tests.*

nondestructive test was conducted on a 1.0-cm-thick (0.4 in.) laminated fly-wheel built by the Lawrence Livermore Laboratory. This was the wheel on which nondestructive ultrasonic tests were conducted.

The spin tests for these laminated wheels were carried out in the General Electric Research and Development facility. The cylindrical containment tank for this facility has an internal diameter of 81.28 cm (32 in.) and is 60.96 cm (24 in.) deep. The tank is mounted in a concrete-reinforced pit below floor level. A 5-cm (2-in.) Barbour Stockwell air turbine and a 2.38-mm-diameter (3/32-in.) quill shaft were used to accelerate the flywheels. Shaft displacements were monitored throughout the test in order to identify critical speeds and keep track of the general dynamic state of the system. The turbine rpm record was displayed digitally, as well as recorded versus time on a strip chart. The vacuum level in the spin tank was also recorded on a strip chart during the test. Vacuum levels during a test were customarily 40 to 80 μm of mercury. The wheels were dynamically balanced before testing.

Table 3 summarizes the important information with respect to the five S2-glass laminated disk tests. The most important observation to be noted with respect to these tests is that all of the S2-glass disks failed at speeds of 762 m/s (2500 ft/s) and slower. Although Test No. 1 approaches a tip speed which is associated by analyses with failure in all laminas throughout the fly-wheel plane, all of the other ultimate test speeds were lower. Unfortunately, this limited amount of data cannot provide any clear indication of trend with respect to the effect of laminating sequence upon ultimate speed.

Finally, some observations with respect to failure events are called for. In both S2-glass Tests 1 and 3, a very rapid deceleration of the turbine took place at failure. Figure 6 presents the rpm time history for Test 1 and illustrates the rapid deceleration. Such extreme deceleration would be impossible to bring

FIG. 5—*Laminated, S2-glass/epoxy flywheel.*

about with the turbine air brake. In addition, there was no evidence of bearing seizure or foreign debris fouling the turbine in either case. One possible explanation for this event is an increase in the flywheel diameter during failure. In such a case, conservation of angular momentum would force the disk to decelerate as its moment of inertia increased. Although S2-glass Test 4 did not exhibit this rapid deceleration, the failure was associated with a very sudden increase in the measured shaft orbit. No shaft displacement of such size was observed at earlier stages of the test, and rotor dynamic analysis and experimental results reveal no critical speeds in this range. These observations all indicate that the ultimate speeds were associated with flywheel disk failures and not induced by the test system.

The debris in Tests 1, 3 and 4 was very fibrous in nature. Figure 7 illustrates the debris in the spin tank after Test 1. Figure 8 illustrates the largest piece of debris recovered. In addition to the fibrous debris in Tests 3 and 4, there were also some full-diameter pieces of debris approximately 0.3 to 0.6 cm (1/8 to 1/4 in.) thick. Figure 9 illustrates such a piece of debris from Test 3. Test 2 on the other hand was atypical in both speed and mode of failure. The failure was not only associated with a large shaft orbit, but also a sudden increase in spin pit vacuum pressure. Unlike earlier tests, the wheel continued to spin with ex-

TABLE 3—Test results for laminated S2-glass/epoxy flywheels.

Test No.	Material	Stacking Sequence	Outside Diameter, cm (in.)	Thickness, cm (in.)	Weight, kg (lb)	Rpm at Burst	Tip Speed at Burst, m/s (ft/s)	Energy Density at Burst, Wh/kg (Wh/lb)	Stress at Center at Burst, MPa (ksi)	Comments
1	S2-glass/ epoxy	(0/−45/45/90)	22.54 (8.8)	2.41 (0.95)	1.79 (3.94)	65 000	760 (2495)	40.1 (18.2)	427 (62)	drop in turbine rpm at failure; fibrous debris (General Electric Co. flywheel)
2	S2-glass/ epoxy	(0/−45/45/90)	22.54 (8.8)	2.54 (1.0)	1.81 (4.02)	37 500	439 (1440)	13.4 (6.02)	144 (21)	large orbit and increase in vacuum pressure occurred simultaneously prior to failure (General Electric Co. flywheel)
3	S2-glass/ epoxy	α-9	22.54 (8.8)	2.14 (0.842)	1.58 (3.49)	58 000	679 (2230)	31.9 (14.5)	345 (50)	drop in turbine rpm at failure; fibrous debris (General Electric Co. flywheel)
4	S2-glass/ epoxy	α-9	28.58 (11.25)	2.69 (1.06)	3.2 (7.05)	59 000	733 (2405)	37.22 (16.9)	397 (57.6)	large shaft orbit just prior to failure (General Electric Co. flywheel)
5	S2-glass	(0/90/30/−60/60/−30)	40.32 (15.78)	1.07 (0.42)	2.88 (6.35)	30 100[a]	638[a] (2094)	28.24[a] (12.82)	306.8[a] (44.5)	nondestructive test; wheel run to 30 100 rpm and acoustically tested before and after spin; differences observed (Lawrence Livermore Lab flywheel)

[a]Nondestructive test.

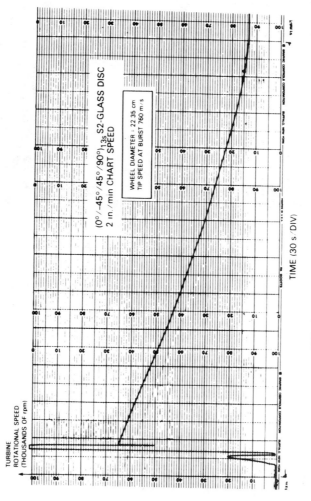

FIG. 6—*Rpm-time history of spin test No. 1.*

FIG. 7—*Debris in spin pit after Test 1.*

cessive orbit but without final failure. The large orbit eventually broke the shaft of the turbine as the wheel was being decelerated. The flywheel disk remained essentially in one piece but was completely delaminated as shown in Fig. 10. In addition, the speed at failure was substantially lower than in the other tests. The very sudden occurrence of the large shaft orbit without final failure in this case seems to suggest the possibility of material loss near the flywheel periphery resulting in substantial unbalance. The only other observation that can be made with respect to this test is that, in spite of the low tip speed at which failure occurred, the analysis discussed with respect to Fig. 3 does indicate the existence of transverse and shear damage over a substantial area of the disk.

Spin test results for laminated E-glass flywheels have also been reported by Hatch [12], Lustenader and Zorzi [17], and McGuire and Rabenhorst [20]. The laminated disks discussed in Refs 12 and 17 ranged in thickness from 0.41 to 2.54 cm (0.16 to 1 in.). In all cases the aluminum hub was attached to the laminated disk with bolts, necessitating holes through the laminates. The highest tip speed attained in these tests of laminated disks without external rings was 594 m/s (1950 ft/s). Reference 20 also reports spin test results with respect to laminated E-glass disks. There are two important differences in the configuration of the wheels tested in Ref 20. First, unlike Hatch's tests, the disks discussed in Ref 20 were attached to aluminum hubs with an elastomeric bond, eliminating stress concentrations. The bond was essentially identical to

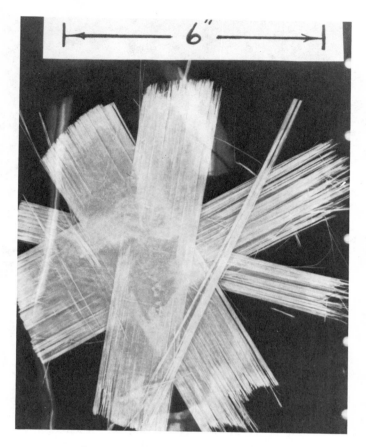

FIG. 8—*Largest piece of debris from Test 1.*

those used in the work reported here. Second, the wheels discussed in Ref *20* were thinner than any of those in Refs *17* or *20*, as well as those reported here, measuring less than 0.25 cm (0.1 in.) in thickness. Three tests reported in Ref *20* reached tip speeds of 794, 834, and 813 m/s (2606, 2735, and 2670 ft/s). The authors also report that when a 0.64-cm-diameter (0.25-in.) hole was drilled in the center of one of the 61-cm-diameter (24-in.) disks, the maximum tip speed reached by the wheel was only 590 m/s (1938 ft/s). This latter result is consistent with the results of Refs *12* and *17* for disks attached with bolt holes.

Although S2-glass/epoxy exhibits superior uniaxial strength with respect to E-glass/epoxy, consideration of the summary data in Table 3 indicates that all the thick S2-glass disks experienced ultimate failure before those reported in Ref *20* for thin E-glass/epoxy disks without holes. Several comments seem relevant to this observation. With respect to measured, unidirectional fiber

FIG. 9—*Full-diameter debris from Test 3.*

composite strengths, tests indicate that the S2-glass/epoxy used in the current flywheels can develop 1480 MPa (215 psi) [*21*] while unidirectional E-glass exhibits a strength of about 1100 MPa (160 ksi) [*26*]. Although this indicates a 34% increase in strength for the S2-glass relative to the E-glass, it may not be the most appropriate measure of performance in a spinning disk. The failure analyses suggested earlier make use of the off-axis strength of a laminate, and it was noted that at the average failure stresses for these laminates, the individual lamina stresses are not sufficient to cause fiber breakage. Laminate analyses suggest that the failures are associated with transverse and shear strengths. In such a case, off-axis strength tests might be a more sensible basis for comparison of expected material performance in a flywheel disk. Unfortunately these data are not available for the E-glass/epoxy system used in Ref *20*.

A second observation to be made with respect to a comparison of these spin tests is that the current tests were made with disks substantially thicker than the disks in Ref *20*. The careful material measurements of fiber-volume content, void content, and glass transition temperature all indicate appropriate levels of quality and cure in the thick laminates produced during this program. The only measured property that can be related to reduced strength would be the 48% fiber volume fraction, which is about 5% less than the expected value

FIG. 10—*Delaminated flywheel after spin Test 2.*

of 53% in thin laminates. However, one potential process-related problem which was not pursued is the existence of residual stresses in the disks. It would be expected that such stresses would be larger in the thicker disks and thus contribute to a reduction in performance. In light of the apparently undetermined cause for the lower failure speeds of these thick laminated disks, it would appear that the potential existence of significant residual stresses should be considered. As will be illustrated, improved laminated disk strength could substantially increase achievable energy densities in the hybrid flywheel.

Fabrication and Assembly of Filament-Wound Graphite/Epoxy Rings

The design concept of the flywheel shown in Fig. 1 is based upon the concept of improving the tip speed and energy density of the flywheel through addition of a graphite/epoxy ring. The rings for this program were filament wound using Thornel 300 graphite fiber and a Shell Epon 826 resin system. As in the case of the laminated disks, several of the rings were destructively tested to establish basic material properties. As the inside diameter of the finished rings increased from 20.3 to 36.8 cm (8.0 to 14.5 in.), the measured fiber volume content decreased from 63% to 58.75%. Void content was always less than 2%. The rings, which eventually were shrunk-fit around laminated disks, had 27.9 cm inside diameters (11.0 in.), 60.9% fiber volume fractions, and 0.49% void content. Ultrasonic measurement established the circumferential and radial stiffnesses of the ring at 127 GPa (18.4×10^6 psi) and 9.65 MPa (1.4×10^6 psi), respectively. Although a layer of bleeder cloth was applied to the outside diameter of the finished winding to absorb excess resin, photomicrographs still revealed a remaining resin layer. The rings were not machined.

In the case of the rings, an attempt was made to establish the magnitude of the residual longitudinal stresses in the ring. When cut annularly, the rings exhibited a tendency to close, indicating a residual moment producing tensile stresses at the inner radius and compressive stresses at the outside. Using circular, thin ring equations, the maximum residual stresses in the ring can be approximated by

$$\sigma_{max} = \frac{D_B - D_A}{4} \frac{Et}{R_m^2} \qquad (2)$$

where

$$
\begin{aligned}
D_B \text{ and } D_A &= \text{measured ring diameters before and after cutting,} \\
t &= \text{ring annular thickness,} \\
R_m &= \text{ring mean radius, and} \\
E &= \text{circumferential Young's modulus.}
\end{aligned}
$$

Based on such an approach, the maximum residual stress was estimated to be 69 MPa (10 ksi) at the inner radius of the ring.

For the thick annular rings the expected failure mode of the hybrid flywheel is delamination of the graphite ring due to tensile, radial stresses. In order to determine the radial strength of these rings, specimens measuring 6.25 by 2.54 by 0.25 cm (2.5 by 1.0 by 0.1 in.) were cut directly from a graphite ring and subjected to tension tests. The average ultimate tensile stress for six such tests was 22.3 MPa (3.2 ksi) but individual test results ranged from 14.47 to 27.51 MPa (2.1 to 3.9 ksi). As will be subsequently illustrated, these relatively low transverse strengths can have a significant effect on the efficiency of the design, and relevant issues such as resin strength and fiber/resin adhesion should be addressed to improve performance.

Assembly and Test of Hybrid Disk-Ring Flywheel

The laminated disks and filament-wound rings having been built and tested, the hybrid disk-ring flywheels shown in Fig. 1 were assembled and tested next. Using the results of the spin tests and tension tests previously discussed, a design curve for optimization of the flywheel's energy density through variation of the ring's annular thickness can be constructed, and is illustrated in Fig. 11.

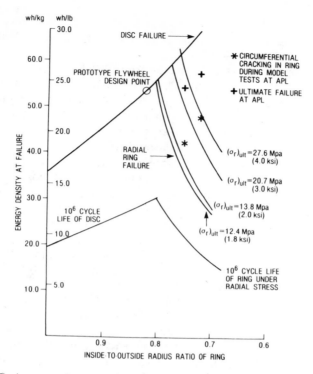

FIG. 11—*Design curves for maximizing energy density through variation of the ring annular thickness.*

Optimization of the hybrid flywheel is discussed more thoroughly in Ref *18*, but it can be seen from the figure that the mode of failure changes from laminated disk failure for high values of the ring's radius ratio to delamination of the filament-wound ring for lower radius ratios. The two flywheels built during this stage of the program were characterized by β-ratios associated with the expectation of failure through ring delamination. Since considerable scatter in radial ring strength was indicated by tension tests, failure curves for several levels of these strengths are included. Although Fig. 11 clearly indicates that the β-ratios for the flywheels were not optimum, they were seen as useful from the standpoint of further understanding the performance of the filament-wound rings for which strength data were limited. In addition, these first wheels also provided test data with respect to design of the wheel to preclude separation of the disk and ring.

Both flywheels shown in Fig. 1 were fabricated with a nondimensional interference fit of $\bar{\delta} = 0.002 \pm 2 \times 10^{-4}$ ($\bar{\delta} = \delta/R_o$) between the disk and ring. In both cases, the diameter of the laminated S2-glass disk was 27.94 cm (11.0 in.) and the nominal thickness 2.54 cm (1.0 in.). Wheel No. 1 had an outside diameter of 37.34 cm (14.7 in.) and an inside-to-outside radius ratio of $\beta = 0.75$. Wheel No. 2 had an outside diameter of 38.61 cm (15.2 in.) and an inside-to-outside radius ratio of $\beta = 0.72$.

The procedure followed in the manufacture of these flywheels was to manufacture the disk and the ring separately. The aluminum hub was then bonded to the disk. After bonding the hub, the outside diameter of the laminated disk was machined with the hub as the geometric center. The outside diameter of the disk was then machined to provide the appropriate interference fit. Both the outside diameter of the disk and the inside diameter of the ring were tapered to facilitate assembly. The taper on the rings was provided through the use of a tapered mandrel surface without machining. After machining, both disk and ring were exposed to cryogenic temperatures (liquid nitrogen) in order to accomplish the interference fit, and a small amount of pressure was applied with a hydraulic ram to complete the assembly. There was no observable evidence of damage due to assembly.

Both model wheels were tested at the Johns Hopkins Applied Physics Laboratory. In the case of Wheel 1, the test proceeded uneventfully to a tip speed of 780 m/s (2550 ft/s). In the vicinity of this tip speed, a significant increase in the vacuum pressure of the spin pit was measured. In spite of this effect, acceleration of the wheel was still possible. As the wheel continued to accelerate above 780 m/s (2550 ft/s), increasing vertical vibration was measured, reaching peaks of 8 g's. At 904 m/s (2970 ft/s) the flywheel pulled loose from the air turbine shaft. The maximum energy density reached in this test was 54 Wh/kg (24.5 Wh/lb). Post-test examination revealed an intact flywheel. A circumferential crack was visible midway between the inside and outside radii of the graphite/epoxy ring. Figure 12 illustrates the stress distributions expected in Flywheel 1 at 780 m/s (2550 ft/s). As can be seen, the transverse stress in the ring has reached levels near those associated with failure in the test coupons previously discussed.

Test results for Flywheel 2 were very similar. In addition to the standard measurements collected in the first test, a high-speed videotape recording (720 frames/s) was made during the second test. As in the previous test, the early stages of the test proceeded uneventfully. At a tip speed of 850 m/s (2785 ft/s) the appearance of a circumferential crack in the graphite/epoxy ring was visible on the videotape. The appearance of this crack was accompanied by a rise in the vacuum pressure of the test facility. Vertical vibrations also increased in magnitude above the tip speed of 850 m/s (2788 ft/s). At a tip speed of 931 m/s (3055 ft/s) the intact composite flywheel disk and ring separated from the aluminum hub at the elastomeric bond line. The maximum energy density reached in this test was 57 Wh/kg (26 Wh/lb). Again both the disk and the ring were intact when examined after completion of the test except for the circumferential crack in the ring and outside diameter damage to the ring incurred after the wheel separated from the hub.

The results of these tests bear out a number of analytical predictions. Per-

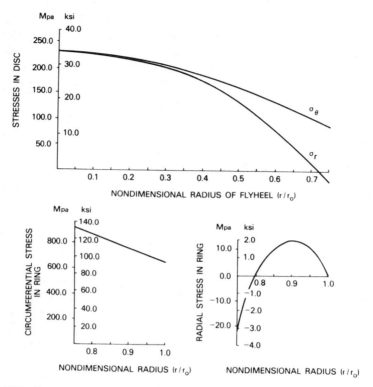

FIG. 12—*Stress distribution in hybrid flywheel No. 1 at 780 m/s (2550 ft/s).*

haps most important is the fact that the maximum energy density of a laminated flywheel can be substantially improved by the addition of a filament-wound graphite ring. The highest energy density previously attained with 2.54-cm-thick (1.0-in.) disks was improved 44% from 40 to 57 Wh/kg (18 to 26 Wh/lb) by the addition of a filament-wound graphite ring. Although Lustenader and Zorzi [17] reported improvement in energy density from 24 to 32.5 Wh/kg (11 to 14.75 Wh/lb) due to a Kevlar-49 ring on an E-glass disk, the energy densities achieved in these tests are still higher as predicted in Ref 18. In addition, it is evident that with a proper interference fit, separation of the disk and ring can be precluded. Finally, the failure mode observed in the test (circumferentially cracked ring) coincides with the predicted failure and indicates that thinner outside rings should produce improvements in the hybrid flywheel's performance. Both the initial and final failures observed in these tests are displayed in Fig. 11. As can be seen, in the first test the initial failure appears to have occurred at a radial stress of 15 MPa (2.2 ksi) while the second flywheel appears to have sustained a radial stress of 27 MPa (3.9 ksi) before delaminating. From the standpoint of containment, it is noteworthy that the flywheel continued to sustain load after its initial failure and that even after final failure, the wheels did not fragment. Two somewhat larger flywheels have been designed and built since these tests were completed and the designed radius ratio ($\beta = 0.82$) is also indicated in Fig. 11. As can be seen, the larger β-ratio should allow the flywheel to operate without ring delamination to ultimate speeds comparable to those observed in the Johns Hopkins tests.

Conclusions

It has been demonstrated that laminated disks up to 3.6 cm (1.4 in.) thick can be manufactured with fiber volume content between 48 and 50% and void content less than 1.8% for application to flywheel energy storage. Analysis indicates that initial damage to the matrix of such disks will occur at the center of the disk at comparatively low tip speeds. Although evidence of such damage has been observed in terms of reduced ultrasonic wave speeds for a flywheel tested above the threshold damage speed, the initiation of damage does not seem to be closely related to ultimate failure. Off-axis tension tests and ultimate failure analyses based upon laminate theory suggest that the stacking sequences with smaller angles between fiber directions in a unit cycle should lead to higher ultimate failure speeds, but too few spin test results are available to date to establish any trends. The ultimate tip speeds of the thick laminated disks were lower than expected compared with previously reported tests on thinner laminated disks.

Hybrid disk-ring flywheels composed of S2-glass/epoxy laminated disks and filament-wound graphite/epoxy rings assembled with an interference fit have been successfully fabricated. Initial spin tests substantiate analytical predictions of initial damage due to delamination of the outer ring. Final failure was due to increased system vibration in both tests. Improvements in ultimate energy densities of 44% have been demonstrated with respect to thick laminated disks without outer rings. Reduction of the inside-to-outside radius ratio of the graphite ring should allow ultimate energy densities of 55 Wh/kg (25 Wh/lb) and greater to be reached without delamination of the ring. This ultimate energy density is higher than reported energy densities for high-performance metal flywheels [27]. Improvement of the transverse strength of the filament-wound graphite/epoxy outer rings would improve performance to even higher energy densities. Improved strength for the laminated disks also seems to be a worthwhile goal of material and process development, since it would enhance performance of hybrid flywheels at higher β-ratios.

Acknowledgments

Initial development and testing of the hybrid flywheel was conducted under General Electric funding. The specific work reported in this paper was conducted under Department of Energy (DOE)/Lawrence Livermore Laboratory

Subcontract No. 2479309. The project monitor at Lawrence Livermore Laboratory was Dr. Satish Kulkarni. The spin tests at the Johns Hopkins Applied Physics Laboratory were conducted under the expert direction of Dr. David Rabenhorst under DOE/Lawrence Livermore Laboratory Contract 7825509, Section 3. The graphite/epoxy rings used in the hybrid flywheel were fabricated by Allan Hannibal at Lord Kinematics, Inc.

References

[1] Beachley, N. H. and Frank, A. A., "Final Report—Flywheel Energy Management Systems for Improving the Fuel Economy of Motor Vehicles," Report DOT/RSPA/DPB-50/7911, U.S. Department of Transportation, Washington, DC, Aug. 1979.

[2] Cornell, E. P. and Turnbull, F. G., "Mechanical Energy Storage Technology Application to Electric Vehicles," *Proceedings*, Contractors Review, Mechanical, Magnetic and Underground Energy Storage, Department of Energy, Washington, DC, Nov. 10–13, 1980.

[3] Lawson, L. J., "Design and Testing of High Energy Density Flywheels for Application to Flywheel/Heat Engine Hybrid Vehicle Drives," *Proceedings*, Intersociety Energy Conversion Energy Conference, Society of Automotive Engineers, 1971, pp. 1142–1150.

[4] Reedy, E. D. and Gerstle, F. P., Jr., "Design of Spoked-Rim Composite Flywheels," *Proceedings*, 1977 Flywheel Technology Symposium, U.S. Department of Energy Report CONF-771053, Washington, DC, March 1978.

[5] Knight, C. E., Jr., "Analysis of Deltawrap Flywheel Design," *Proceedings*, 1977 Flywheel Technology Symposium, U.S. Department of Energy Report CONF-771053, Washington, DC, March 1978.

[6] Knight, C. E., Jr., Kelley, J. J., Huddleston, R. R., and Pollard, R. E., "Development of 'Bandwrap' Flywheel," *Proceedings*, 1977 Flywheel Technology Symposium, U.S. Department of Energy Report CONF-771053, Washington, DC, March 1978.

[7] Satchwell, D. L., "High Energy Density Flywheel," *Proceedings*, Mechanical and Magnetic Energy Storage Contractors' Review Meeting, U.S. Department of Energy Report CONF-790854, Washington, DC, Dec. 1979.

[8] Younger, F. C., "A Composite Flywheel for Vehicle Use," *Proceedings*, Mechanical and Magnetic Energy Storage Contractors' Review Meeting, U.S. Department of Energy Report CONF-781046, Washington, DC, Oct. 1978.

[9] Hill, W. P., "Progress in Composite Flywheel Development," *Proceedings*, Mechanical and Magnetic Energy Storage Contractors' Review Meeting, U.S. Department of Energy Report CONF-781046, Washington, DC, Oct. 1978.

[10] Davis, D. C., "Advanced Composite Flywheel for Vehicle Application," *Proceedings*, Mechanical and Magnetic Energy Storage Contractors' Review Meeting, U.S. Department of Energy Report CONF-781046, Washington, DC, Oct. 1978.

[11] Rabenhorst, D. W. and Small, J. R., "Composite Flywheel Development Program: Final Report," Johns Hopkins Applied Physics Laboratory Report SDO-4616A, Baltimore, MD, April 1977.

[12] Hatch, B. D., "Alpha-Cross-Ply Composite Flywheel Development," *Proceedings*, 1977 Flywheel Technology Symposium, U.S. Department of Energy Report CONF-771053, Washington, DC, March 1978.

[13] Johnson, D. E. and Oplinger, D. W., "Failure Modes of Bi-directionally Reinforced Flywheels," *Proceedings*, 1977 Flywheel Technology Symposium, U.S. Department of Energy Report CONF-77053, Washington, DC, March 1978.

[14] Kulkarni, S. V., "Composite-Laminate Flywheel-Rotor Development Program," *Proceedings*, Mechanical and Magnetic Energy Storage Contractors' Review Meeting, U.S. Department of Energy Report CONF-790854, Washington, DC, Dec. 1979.

[15] Kulkarni, S. V., "Flywheel Rotor and Containment Technology Development Program of the U.S. Dept. of Energy," *Proceedings*, Third International Conference on Composite Materials, Paris, Aug. 25–29, 1980.

[16] Gupta, B. P. and Lewis, A. F., "Optimization of Hoop/Disk Composite Flywheel Rotor De-

signs," *Proceedings*, 1977 Flywheel Technology Symposium, U.S. Department of Energy Report CONF-77053, Washington, DC, March 1978.

[*17*] Lustenader, E. L. and Zorzi, E. S., "Status of the 'Alpha-Ply' Composite Flywheel Concept Development," *Proceedings*, Society for the Advancement of Material and Process Engineering Meeting, Anaheim, CA, May 3, 1978.

[*18*] Nimmer, R. P., "Parametric Design Analysis of a Hybrid Composite Flywheel Using a Laminated Central Disc and a Filament Wound Outer Ring," ASME Paper 80-DET-97, presented at the Design Engineering Technical Conference, Beverly Hills, CA, Sept. 28–Oct. 1, 1980.

[*19*] Nimmer, R. P., "Laminated Composite Flywheel Failure Analysis," *Proceedings*, Flywheel Technology Symposium, U.S. Department of Energy Report CONF-801022, Washington, DC, Oct. 1980.

[*20*] McGuire, D. P. and Rabenhorst, D. W., "Composite Flywheel Rotor/Hub Attachment Through Elastomeric Interlayers," *Proceedings*, 1977 Flywheel Technology Symposium, U.S. Department of Energy Report CONF-771053, Washington, DC, March 1978.

[*21*] Duke, J. C., "A Comparison of Quasi-Isotropic Fiber Reinforced Composite Laminates," Lawrence Livermore Laboratory Report UCRL-15225, Livermore, CA, Nov. 1979.

[*22*] Kulkarni, S. V., Stone, R. G., and Toland, R. H., "Prototype Development of an Optimized, Tapered-Thickness, Graphite/Epoxy Composite Flywheel," Lawrence Livermore Laboratory Report UCRL-52623, Livermore, CA, Nov. 1978.

[*23*] Nimmer, R. P., Torossian, K. A., and Hickey, J. S., "Laminated Composite Disc Flywheel Development; Second Interim Report," Lawrence Livermore Laboratory Report UCRL-15154, Livermore, CA, July 1979.

[*24*] Tsai, S. W. and Hahn, H. T., *Composite Materials Workbook*, U.S. Air Force Materials Laboratory Report No. AFML-TR-78-33, Wright-Patterson Air Force Base, Dayton, OH, March 1978, p. 195.

[*25*] McLaughlin, P. V., Jr., Dasgupta, A., and Chun, Y. W., "Composite Failure Analysis for Flywheel Design Applications," *Proceedings*, Flywheel Technology Symposium, U.S. Department of Energy Report CONF-801022, Washington, DC, Oct. 1980.

[*26*] "Scotchply[R] Reinforced Plastic Technical Data, Type 1002," 3M Company, St. Paul, MN, May 1969.

[*27*] Davis, D. and Hodson, D., "Rocketdyne's High-Energy-Storage Flywheel Module," *Proceedings*, 1977 Flywheel Technology Symposium, U.S. Department of Energy Report CONF-771053, Washington, DC, March 1978.

Yoshiro Minoda,[1] *Yoshiaki Sakatani,*[2] *Yasuhiro Yamaguchi,*[2]
Mamoru Niizeki,[3] *and Haruyoshi Saigoku*[3]

Toward Process Optimization by Monitoring the Electrical Properties During the Cure Cycle for CFRP

REFERENCE: Minoda, Y., Sakatani, Y., Yamaguchi, Y., Niizeki, M., and Saigoku, H., **"Toward Process Optimization by Monitoring the Electrical Properties During the Cure Cycle for CFRP,"** *Recent Advances in Composites in the United States and Japan, ASTM STP 864*, J. R. Vinson and M. Taya, Eds., American Society for Testing and Materials, Philadelphia, 1985, pp. 489–501.

ABSTRACT: Application of advanced composite materials, namely carbon fiber reinforced plastic (CFRP) (graphite/epoxy) to aerospace structures is currently expanding from simple components to complex-shape integrated structures. The objective of this study is to establish the feasibility of optimizing the cure cycle of integrated CFRP structure by in-process direct measurement of electrical properties of the article in an autoclave.

An instrumentation technique that included material and shape of electrodes and an insulator and bleeder system was investigated. Experimental studies using the above test setup attempted to determine correct pressurization timing under certain preprogrammed heating patterns thus resulting in cured CFRP laminates, which were practically voidless, with proper fiber volume fraction and thickness per ply.

KEY WORDS: carbon fiber reinforced plastic (graphite/epoxy), in-process monitoring, dielectrometry, electric resistance, pressurization timing

Applications of advanced composite materials, which have high specific strength and modulus, are being expanded to evermore complex-shape integrated structures.

The objective of this study is to establish the feasibility of optimizing the cure cycle of integrated carbon fiber reinforced plastics (graphite/epoxy) structure by in-process direct measurements of the electrical properties dur-

[1]R&D Institute of Metals and Composites for Future Industries, Tranomon Takagi Bldg., 1-7-2 Nishi-Shinbashi, Minato-ku, Tokyo 150, Japan.
[2]Nagoya Aircraft Works, Mitsubishi Heavy Industries, Ltd., 10 Oye-cho, Minato-ku, Nagoya 455, Japan.
[3]Aircraft Div., Fuji Heavy Industries, Ltd., 1-1-11 Yonan, Utsunomiya, Tochigi 320, Japan.

ing an autoclave cure to obtain high-quality structures at low cost. Conventional cure cycles use a fixed cure schedule with temperature monitoring, usually according to the prepreg manufacturer's instruction. But these fixed cure cycles do not precisely control the cure conditions of complex integrated structures of which the thickness is large and variable or which contain various prepreg properties and forms. Many authors have discussed the importance of in-process monitoring to control the cure cycle of resin matrix composites [1,2], and some experimental work to monitor the electrical properties such as dielectrometry has been reported by Lawless [3], and to monitor electrical conductivity by Brayden [4].

This paper describes the investigation of instrumental techniques to monitor the dielectric properties and electric resistance responses of the graphite/epoxy laminates during cure, and the results of experimental studies to obtain the optimum pressurization timing during cure which gives high-quality composites by both monitoring methods.

Dielectrometry

Experimental Procedure

Materials—The properties of the selected graphite/epoxy prepregs for the dielectrometry are given in Table 1a. This tape prepreg consists of conventional 177°C cure bleed type polyfunctional epoxy systems which show low viscosities and high resin flows before gelation in the standard cure cycle. For a one-step cure cycle, the prepreg manufacturer recommends pressurizing at the gel temperature. No recommendation is given for a two-step cure cycle.

TABLE 1a—*Properties of prepreg materials for dielectrometric monitoring.*

Type	Form	Prepreg	Laminate
Unidirectional tape (Mitsubishi Rayon Co. Pyrofil AS/410	341-mm width tape	resin content: 40% volatile content: 1% nominal thickness: 0.15 mm/ply	tensile strength: 1670 MPa tensile modulus: 137 000 MPa interlaminar shear strength: 107 MPa

TABLE 1b—*Properties of prepreg materials for resistance monitoring.*

Type	Form	Prepreg	Laminate
Graphite/epoxy woven fabric (3K-70-PW)	1-m-width roll	resin content: 42% volatile content: less than 2% nominal thickness: 0.185 mm/ply	tensile strength: 560 MPa tensile modulus: 68 300 MPa interlaminar shear strength: 70 MPa

Equipment—For the dielectrometry, AUDREY 204 (Tetrahedron Associates Inc.) was used to monitor the capacitance and dissipation factor changes of the prepreg laminate stack during hot-press and autoclave cure. Figure 1 shows the typical experimental apparatus setup for the dielectric monitor during autoclave cure. The dielectric range of the AUDREY 204 is capacitance from 0 to 500 pF with dissipation from 0 to 1.0. The frequency range is 100 to 1000 Hz from an internal oscillator.

Test Procedures—A series of experimental studies was performed to optimize the choice of electrodes, insulators, and bleeder systems for the monitoring of the dielectric changes. Then, three unidirectional 16-ply graphite/epoxy laminates were fabricated under the standard hot-press and autoclave cure cycles by applying pressure according to the monitored dielectric profiles. Each laminate was tested for thickness per ply, fiber volume content [ASTM Test Method for Fiber Content of Resin-Matrix Composites by Matrix Digestion (D 3171)], void content [ASTM Test Method for Void Content of Reinforced Plastics (D 2734)], and interlaminar shear strength [ASTM Test Method for Apparent Interlaminar Shear Strength of Parallel Fiber Composites by Short Beam Method (D 2344)].

Electrode Setup—For the various methods of fabrication, the electrode setups, including the electrode materials and shapes, insulator materials, and bleeder glass cloth ply numbers, were evaluated to obtain the proper dielectric profiles during cure with no defects on the laminate surfaces. Resultant typical electrode setup arrangements are shown in Fig. 2 for the thick laminate cure and the autoclave cure.

Optimization of Pressurization Timing—Typical dielectric profiles during the hot-press and autoclave cure are shown in Figs. 3 and 4. Three points of the pressurization timing (PT) were chosen for study in the above profiles (PT 1, PT 2, and PT 3).

PT 1 is the point where the dissipation shows the first peak and the capacitance begins to rise.

PT 2 is the point where the dissipation begins to rise to the second peak and the capacitance starts to fall.

PT 3 is the point where the dissipation shows a second peak and the capacitance levels off.

The resultant quality of the test laminates given in Table 2 indicates that PT 2 is the optimum point for the pressurization for both the hot-press one-step cure cycle and the autoclave two-step cure cycle. The laminates pressurized at PT 1 showed high fiber content (70%, 69%) but some voids (1.3%, 1.1%), and low thickness per ply (0.105 mm, 0.121 mm). The laminates pressurized at PT 2 showed no voids and high interlaminar shear strength (109.8 MPa in the autoclave-cured laminate). The laminates pressurized at PT 3 showed many voids (5.4%, 2.7%), high thickness per ply (0.147 mm, 0.140 mm) and resin-rich areas as shown in Fig. 5 typical cross sections.

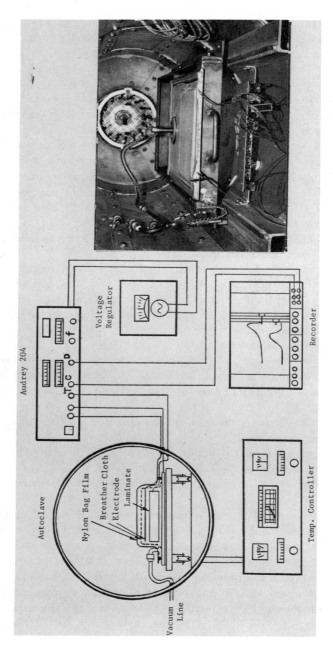

FIG. 1—*Experimental apparatus setup for the dielectric monitor during autoclave cure.*

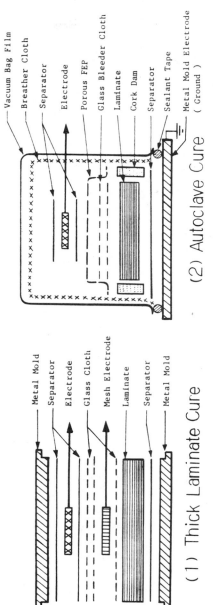

FIG. 2—Electrode setup for dielectrometry.

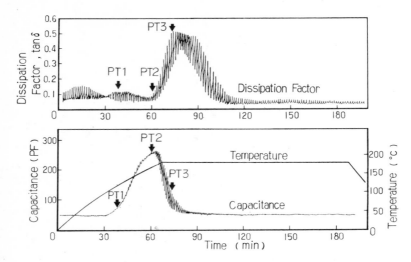

FIG. 3—*Dielectrometric profiles during hot-press cure for optimum pressurization timing (PT) (0.1 kHz ≦ frequency ≦ 1 kHz).*

Resistance Monitoring

Experimental Procedure

Experimental Apparatus—Figure 6 shows the experimental apparatus setup used for this study. The apparatus is a d-c circuit consisting of a d-c standard source, a constant resistance, an *X-Y* recorder for monitoring and plotting the voltage between the both ends of the constant resistance, and electrodes to be placed in the laminate. A method to place the electrodes on one side surface of the laminate was developed in consideration of future practical applications. The material of the electrodes is thin aluminum foil (0.05 mm thick), and the size is 25 by 25 mm. The electrodes were placed on the top surface of a laminate. One layer of perforated trifluoroethylene film and one layer of glass cloth were applied between the electrodes and the laminate to prevent direct contact of the electrodes with electrically conductive graphite fibers. The experiments were performed under the condition that the applied voltage to the circuit was 4.5 V constant and the constant resistance was 100 kΩ.

Standard Cure Cycle—The cure cycle was a currently used step cure cycle for bleed-type prepregs. The procedures began with a vacuum pressure of at least 560 mm Hg being applied to the bagged laminate, which had been loaded in an autoclave. Under the vacuum condition, the laminate was heated up to 120°C at 1.6°C/min, maintaining that temperature for one hour to bleed off excessive resin and air entrapped in the layers during the stacking operations. Next, a cure pressure of 6.0 kg/cm² (0.59 MPa) was applied at

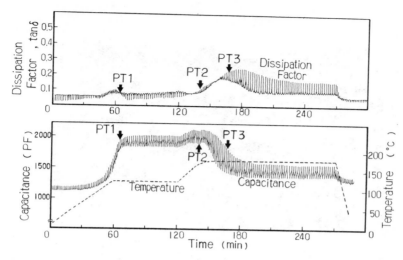

FIG. 4—*Dielectrometric profiles during autoclave cure for optimum pressurization timing (PT) (0.1 kHz ≤ frequency ≤ 1 kHz).*

TABLE 2—*Quality of laminates cured with dielectrometric monitoring.*

Procedure	Pressuri- zation Timing	Thickness, mm/ply	Density, 10^3 kg/m^3	Fiber Content, vol. %	Void Content, vol. %	Interlaminar Shear Strength, MPa
Hot press	PT1	0.105	1.62	70	1.3	...
One step	PT2	0.127	1.62	63	0	...
Cure cycle (Fig. 3)	PT3	0.147	1.50	62	5.4	...
Autoclave	PT1	0.121	1.61	69	1.1	51.0
Two step	PT2	0.133	1.66	68	0	109.8
Cure cycle (Fig. 4)	PT3	0.140	1.60	67	2.7	50.0

selected times during this stage. The vacuum pressure in the bag was vented to the atmosphere when the autoclave pressure reached 1.4 kg/cm^2 (0.14 MPa) during the course of the pressurization. After the completion of the pressurization, the laminate was heated up at 1.6°C/min to 177°C, then maintained for two hours at that temperature for the cure.

Standard Laminate—The laminate used in these experiments consists of 14 plies of bleed-type graphite/epoxy prepreg (Toray T-300/3601). The properties of the material are given in Table 1*b*. The cure temperature is 177°C, the cure time 2 h, and the time of pressurization recommended by the prepreg manufacturer is 30 min after the laminate reaches 120°C, the start of flow.

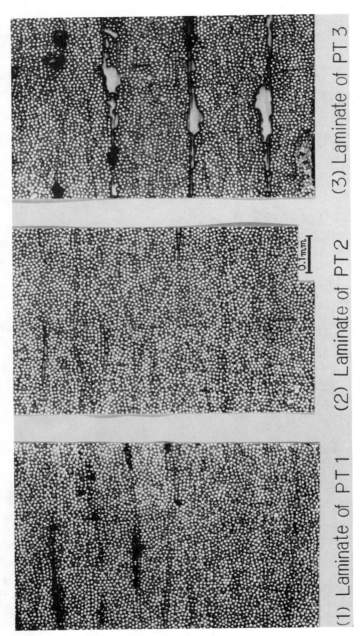

FIG. 5—*Typical cross sections of cured laminates (×112).*

(1) Laminate of PT 1   (2) Laminate of PT 2   (3) Laminate of PT 3

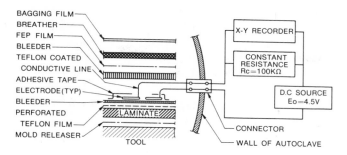

FIG. 6—*Experimental apparatus setup.*

For this experiment, the prepreg was exposed for 150 h to temperatures of 18 to 24°C after being removed from refrigeration, to simulate future practical usage. The size of the laminate was 150 by 200 mm.

Results and Discussion

Resistance Change Curve

Figure 7 shows the resistance change curve obtained in the experiments in which the standard laminates were cured with the standard cure cycle. During heat-up of the laminate, the voltage readings across the constant resistance increased. The voltage readings indicated a peak value close to 115°C, then they started to decrease. The voltage readings continued to decrease while the laminate temperature was held at 120°C.

Looking at the correlation between the heat-up rates and the voltage readings, the voltage readings increased in conjunction with the rising of the lami-

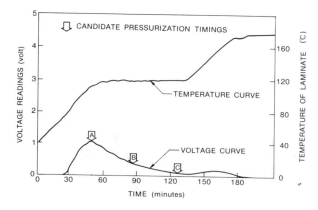

FIG. 7—*Candidate pressurization timings.*

nate temperature, and are related only to the laminate temperature itself but not to the heat-up rates of the following three cases: one with the heat-up rate of 1.6°C/min (standard), and others with heat-up rates of 1.1°C/min and 2.1°C/min.

Figure 8 shows the rate of voltage decrease when the laminate temperature was held at 120°C after a heat-up rate of 1.6°C/min. Excluding five minutes after the time when the voltage reading indicated a peak, the value of the voltage decreased linearly on a semilogarithmic graph.

Evaluation of Pressurization Timings

Three candidate points on the resistance curve were selected for evaluating the proper timing of pressurization in the standard cure cycle. These times are shown in Fig. 7. The times of pressurization were selected as follows:

- Point A—when the voltage reading indicates a peak.
- Point B—the point where the d-c resistance of the resin is 1.4 mΩ. The equation of translation voltage to d-c resistance is

$$Vc = Eo \cdot Rc/(Rs + Rc)$$

where

Vc = voltage reading across the constant resistance,
Eo = source voltage,
Rc = constant resistance, and
Rs = d-c resistance of the resin.

- Point C—time just before the second heat-up starts. Selected pressurization timings were evaluated for thickness per ply, interlaminar shear

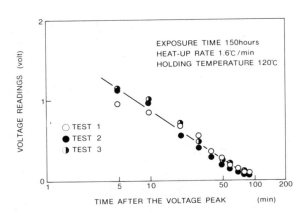

FIG. 8—*Voltage descent at the resin flow temperature.*

strength, fiber volume fraction, and void volume content of the cured laminates in the same manner as the dielectrometry's. Table 3 gives the results of the aforementioned tests.

The laminate pressurized at Point A had the lowest thickness among them, and the thickness of the laminates increased with increasing delay of pressurization. The laminate pressurized at Point B had the maximum interlaminar shear strength and the laminate pressurized at Point A had the lowest value. The laminate pressurized at Point A had the highest fiber volume fraction but the void volume content was 1.7 percent, which was the worst value obtained. The laminates pressurized at Points B and C indicated no void content; however, a few microscopic voids were observed in the laminate pressurized at Point C. Figure 9 shows photo micrographs of typical cross sections of the laminates pressurized at selected points.

Conclusions and Future Studies

1. Instrumental techniques including the electrode setup were established to monitor the electrical properties of the graphite/epoxy laminate during cure.

2. Both the dielectrometry and electrical resistance methods can be used to monitor autoclave cure of our candidate graphite/epoxy structures.

3. Experimental studies to obtain optimum pressurization timing which gives high-quality graphite/epoxy laminates were demonstrated in both monitoring methods.

4. Subsequent studies to clarify the relationship between the electrical properties and physiochemical properties such as viscosities and chemical reactions must be made.

5. Methods of applying monitoring to the autoclave cure of large complex-shape integrated structures will be explored. These include (a) the optimization of electrode setup on complex structures and (b) proper control of the cure cycle with the optimization of the pressurization timing from the monitored electrical properties.

TABLE 3—*Quality of laminate cured with resistance monitoring.*

Procedure	Pressuri- zation Timing	Thickness, mm/ply	Fiber Content, vol. %	Void Content, vol. %	Interlaminar Shear Strength, MPa
Autoclave	PT A	0.193	62.0	1.7	60.2
Two step	PT B	0.195	60.0	0	66.2
Cure cycle	PT C	0.198	59.7	0	60.0

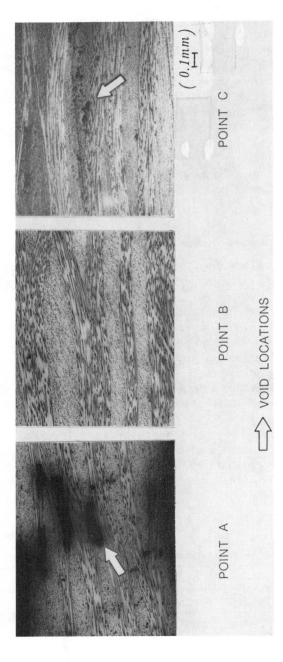

FIG. 9—*Microphotographs of cured laminates.*

Acknowledgments

This work was performed under the management of the Research and Development Institute of Metals and Composites for Future Industries as a part of the R&D project of Basic Technology for Future Industries sponsored by the Agency of Industrial Science and Technology, Ministry of International Trade and Industry, of Japan.

References

[1] Carpenter, J. F., "Instrumental Techniques for Developing Epoxy Cure Cycle" in *Proceedings*, National Symposium and Exhibition of the Society for the Advancement of Material and Process Engineering, Vol. 21, 1976, pp. 783–802.
[2] Yokota, M. J., "In-Process Controlled Curing of Resin Matrix Composites," *SAMPE Journal*, Society of Aerospace Material and Process Engineers, July/Aug., 1978, pp. 11–17.
[3] Lawless, G. W., "Dielectrometry as a Cure Monitoring Technique for Composite/Resin Processing" in *Proceedings*, ICCM-III (Third International Conference on Composite Materials), 1980, pp. 1585–1596.
[4] Brayden, T. H., "Laminate Conductivity and Events Occurring in the Cure Cycle," in *Proceedings*, 12th National Technical Conference of the Society for the Advancement of Material and Process Engineering, 1970.

Testing Methods

Akimasa Daimaru, [1] *Toshinobu Hata,* [1] *and Minoru Taya* [2]

Work of Fracture in Metal Matrix Composites

REFERENCE: Daimaru, A., Hata, T., and Taya, M., **"Work of Fracture in Metal Matrix Composites,"** *Recent Advances in Composites in the United States and Japan, ASTM STP 864,* J. R. Vinson and M. Taya, Eds., American Society for Testing and Materials, Philadelphia, 1985, pp. 505–521.

ABSTRACT: When one talks about "fracture toughness" of composites, one must be careful in distinguishing the energy release rate for the initiation of a crack (γ_I) and the energy absorbed during the fracture averaged over the whole history of fracture process (γ_F). Unlike metals, the value of the critical stress intensity factor K_{Ic} of composite was found not to reflect any basic material property. Instead, the measurement of γ_F seems to represent a meaningful toughness property of composite when the composite is a mixture of brittle and ductile phases, such as metal matrix composites (MMCs).

We have recently conducted experimental and analytical study on the value of γ_F of unidirectional fiber MMC and obtained a good agreement between the experiment and theory. In this paper, we report the experimental and theoretical results on a particulate MMC. The results of two different types of MMCs have revealed that three-point bending test of the MMC specimen with center notch can be used as a reasonable method to measure γ_F.

KEY WORDS: work of fracture, fracture toughness, metal matrix composites, three-point bending test

Metal matrix composites (MMCs) are superior to polymer matrix composites with respect to their performance under severe environments, particularly at elevated temperature. Though main emphasis has been placed upon the enhancement of the stiffness and strength of the composite, the higher toughness of the composite is equally important for some cases of application. However, these three properties often conflict with each other.

As to the measurement and prediction of the fracture toughness of MMCs, no attempt has been made to standardize the measurement of the fracture

[1] Research associate, Department of Mechanical and Aerospace Engineering, University of Delaware, Newark, Del. 19716; on leave from Honda R&D Co., Ltd., Wako R&D Center, 1-4-1 Chou, Wako-shi, Saitama 351, Japan.

[2] Associate professor, Department of Aerospace and Mechanical Engineering, University of Delaware, Newark, DE 19716.

505

toughness that is verifiable by a theory. In order to discuss the fracture toughness of MMCs, we must distinguish the type of reinforcement, that is, short or continuous fiber, since the mode of fracture process varies from continuous fiber to short fiber composites. Also the definition of the fracture toughness of composites must be clearly made. Most of the works on fracture mechanics [1–5] of composites in general have been on the analysis of the stress intensity or energy release rate of a continuous fiber composite except for those initiated by Kelly and his co-workers [6–9] who considered the micromechanical mode ahead of a crack in the inhomogeneous body, plastic deformation in the matrix, fiber debonding, and pullout.

When a crack progresses in a stop and go manner, the evaluation of the macroscopic fracture surface energy (γ_F) becomes more important than that of the crack initiation. This is indeed the case with MMCs which often consist of ductile metal and brittle fiber. The value of γ_F is defined as the energy absorbed during the fracture and averaged over the whole history of fracture process [10]. One of the simplest experimental methods to measure γ_F is the three-point bending test for a center-notched specimen. This method was first explored by Nakayama [11] for ceramic materials and modified later by Tattersall and Toppin [12] who used triangular notches at the midpoint. This method has been used by several researchers [13–14]. We [15] have recently used this method to measure γ_F of unidirectional metal matrix composites (carbon/aluminum) and also developed an analytical model to predict the value of γ_F. Our model is an extension of a two-dimensional model by Kelly and Cooper to a three-dimensional system, and it is applicable only to unidirectional fiber composites.

Unlike unidirectional metal matrix composites, an attempt to measure γ_F of short fiber (including particulate) MMCs has not been made. Kendall [16] proposed an empirical model to predict γ_F of a particulate reinforced polymer. This model is based on a kind of rule of mixtures approach in which the concept of "fracture surface fraction of particulate" is used to replace that of volume fraction. This model is supported by the fact that the observed fracture surface is not flat but quite wavy [17]. However, the observed fracture surface of particulate reinforced metal matrix composites is relatively flat. Thus the analytical model to predict γ_F of particulate MMCs must be developed independently of the polymer case.

In this paper, we will focus on the measurement of γ_F of a particulate MMC and also on the prediction of γ_F. By combining the previous results on unidirectional fiber MMC and the present ones on short fiber MMC, we then discuss the validity of the three-point bending test to see if the value of γ_F so measured reveals the fracture toughness of MMCs.

Since the detailed results of unidirectional MMC has been described in our recent paper [15], concise descriptions on the case of unidirectional MMC will be made wherever necessary in the following sections.

MMC Specimens

Two types of MMC materials were chosen in our study, unidirectional carbon fiber/6061-F aluminum and silicon carbide (SiC) particulate/6061-T6 aluminum. The former MMC materials fabricated by Materials Concepts, Inc., Columbus, Ohio, were supplied as a plate of dimensions 152.4 by 152.4 by 3.175 mm. This composite is made by hot-pressing precursors and 6061 aluminum foil where each precursor is already made by vacuum infiltration technique. No heat treatment was made for this composite. As a carbon fiber, Union Carbide made Thornel P55 (pitch) was used. Two different volume fractions of fiber, $V_f = 32$ and 48% were employed. The detailed informations on the composites as received (the micrographs of the cross section and the results of ultrasonic C-scan on these plates) have been described in our previous paper [15].

SiC particulate reinforced 6061 aluminum composite was supplied in plate form (thickness 5.08 and 10.16 mm) by DWA Composite Specialties Inc., Chatsworth, California. For this composite, two volume fractions of SiC particulate were used, $V_f = 20$, 30%. Also we prepared 6061-T6 aluminum without SiC particulate fabricated by the same manufacturing process as these composites. These materials are fabricated by a powder metallurgical technique followed by extrusion and rolling. Then, heat treatment (standard T6) was made. Typical views of SiC particulate/6061 aluminum composite are shown along the rolling direction (Fig. 1a) and its perpendicular direction (Fig. 1b). As seen from Fig. 1, the shape of the SiC particulate is more like a flake or cube with the average size being 6 to 12 μm. The ultrasonic C-scan (pulse echo 15 MHz 100 scan/in. scan) on this plate with ($V_f = 30\%$) is shown in Fig. 2 where practically no initial flaw was observed.

Typical stress-strain curves of the above MMCs are shown in Fig. 3a (unidirectional carbon fiber/6061 aluminum) and in Fig. 3b (V_f 30% SiC particulate of 6061 aluminum).

Three Point Bending Tests

In order to obtain the work of fracture of these MMC specimens, we have conducted a series of three-point bending tests. The specimen has triangular notches at the midpoint and is subjected to a load P at the midpoint as shown schematically in Fig. 4. A crack is initiated from the apex of the triangular ligament and propagates in a well-controlled manner resulting in a complete fracture of the ligament. Thus the load (P)-deflection (δ) curve is obtained. Typical P-δ curves of two different types of MMC are shown in Fig. 5a for V_f 32% carbon fiber/6061 aluminum and Fig. 5b for V_f 30% SiC particulate/6061 aluminum. The area underneath the P-δ curve is considered to be equal to the total energy absorbed during the whole fracture process (W_F). Then,

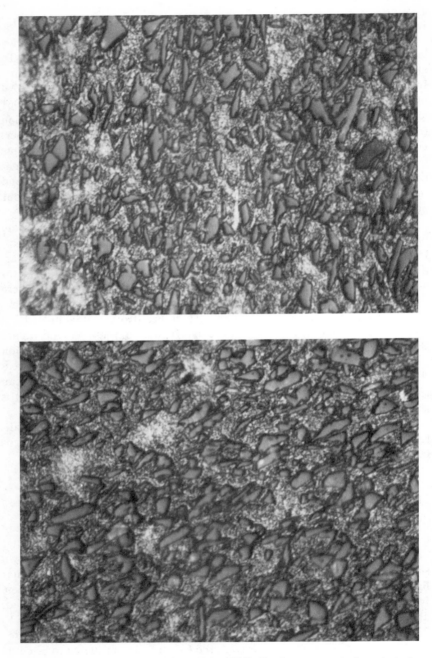

FIG. 1—*Typical section views of SiC particulate/6061 Aluminum composite by optical microscope with magnification 800 for* (a) *along the rolled direction (vertical direction)* (b) *along the section perpendicular to the rolled direction.*

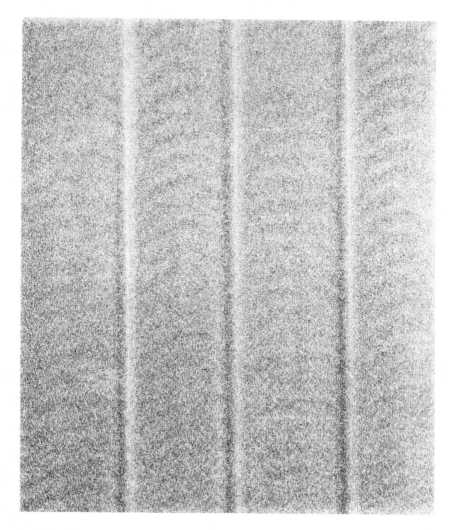

FIG. 2—*The result of ultra sonic C-scan of SiC particulate/6061 aluminum composite plate.*

the value of work of fracture (or macroscopic fracture surface energy) γ_F is calculated by $W_F/2A$ where A is the area of the triangular ligament. The crosshead speed was set at 2.5×10^{-5} (m/s). The span of the three-point bending specimen was 71.12 mm, and the dimension of the section was 10.16 by 2.54 mm for carbon fiber/aluminum composites, and four different section sizes were used for SiC particulate/aluminum composites, 10.16 by 3.175 mm, 10.16 by 5.08 mm, 10.16 by 10.16 mm, and 7.62 by 3.81 mm. The value of γ_F for these specimens are tabulated in Table 1. We have also measured γ_F

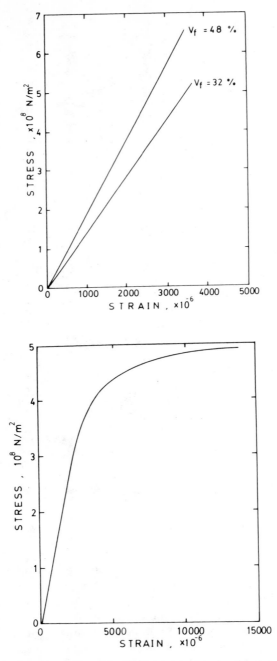

FIG. 3—*Stress-strain curves of MMC specimens* (a) *Unidirectional Carbon fiber/6061 Alumi-num,* $V_f = 32\%$ *and* 48%. (b) *SiC particulate/6061 Aluminum,* $V_f = 30\%$.

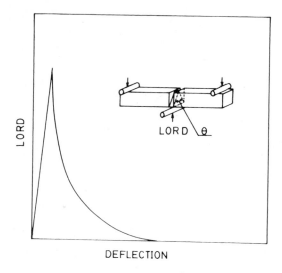

FIG. 4—*Three-point bending test.*

of 6061-T6 aluminum by the three-point bending test. In this case the specimen size was chosen as 71.12 by 10.16 by 2.54 mm for wrought aluminum and 71.12 by 10.16 by 5.08 mm for powder metallurgical aluminum so that the fracture can be completed (note that this three-point bending method is not suitable for ductile material because it is difficult to obtain the P-δ curve completely). The values of γ_F measured for 6061 aluminum-T6 are 57.2 KJ/m² for wrought aluminum and 16.9 KJ/m² for powder metallurgical aluminum which will be used to predict γ_F by a rule of mixtures model for reference.

Results and Discussion

The results of the three-point bending test for unidirectional carbon-fiber/6061 aluminum are plotted as open circles in Fig. 6 where open triangles with straight line denote the theoretical results, which will be described later. In Fig. 6, the result of SiC particulate/6061 aluminum with $V_f = 30\%$ and the cross-section geometry of 10.16 by 5.08 mm is plotted as a filled circle for the purpose of comparison. In the same figure, the value of γ_F based on a rule of mixtures is plotted as a dashed line, where $\gamma_F = V_m\gamma_m + V_f\gamma_f$ and where V denotes the volume fraction, γ is the work of fracture, and the subscripts m and f are for the matrix and reinforcement, respectively. The results for SiC particulate/6061 aluminum with the cross-section geometry of 10.16 by 5.08 mm are plotted as filled triangles for $V_f = 20$ and 30% in Fig. 7 where the

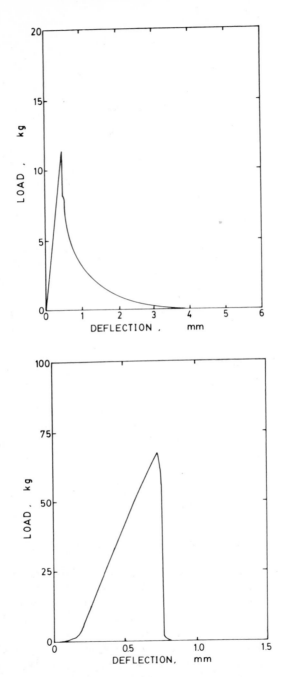

FIG. 5—*Typical P-δ curves of MMCs.* (a) V_f *32% Carbon fiber/6061 Al (the section size 10.16 × 2.54 mm)* (b) V_f *30% SiC particulate/6061 Al (the section size 10.16 × 5.08 mm).*

TABLE 1—*Work of fracture (γ_F) of MMC specimens.*

Type of Composites	V_f, %	Specimen Size, mm	Mean Value of γ_F, KJ m^{-2}	Number of Specimen Tested
Unidirectional carbon fiber/6061 aluminum composite	0	71.12 by 10.16 by 2.54	57.2	4
	32	71.12 by 10.16 by 2.54	6.05	6
	48	71.12 by 10.16 by 2.54	5.96	4
Particulate SiC/6061 aluminum composite	0	71.12 by 10.16 by 5.08	16.9	6
	20	71.12 by 10.16 by 5.08	3.13	8
		71.12 by 10.16 by 3.175	5.65	6
	30	71.12 by 10.16 by 5.08	4.32	6
		71.12 by 10.16 by 10.16	4.96	4
		71.12 by 7.62 by 3.81	5.29	4

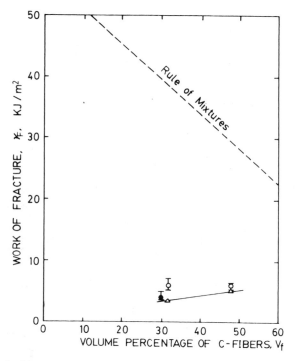

FIG. 6—*Work of fracture γ_F of unidirectional carbon fiber/6061 aluminum, versus V_f, experimental results (open circles) and theoretical ones, (open triangles connected by a straight line). The filled circle denotes the experimental result of V_f 30% SiC particulate/6061 Al. The dashed line is the result based on a rule of mixtures.*

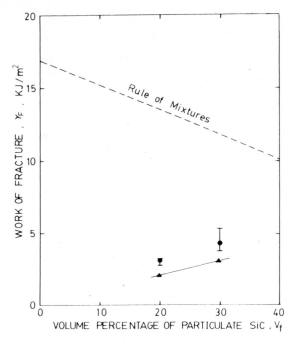

FIG. 7—*Work of fracture γ_F of SiC particulate/6061 aluminum versus V_f, experimental results (filled circles) and theoretical ones (filled triangles connected by a straight line). The dashed line is the result based on a rule of mixtures.*

filled triangles with straight line denote the theoretical results which will be discussed next.

A theoretical model to predict γ_F of unidirectional MMC has been recently proposed by the present authors [15]. This model takes account of three mechanisms which are operative in the fracture process of MMC: the elastic strain energy release rate, the plastic work along the matrix-fiber interface near the crack plane, and the fiber pullout. The formula to predict γ_F is given by

$$\gamma_F = \frac{1}{2}\left\{\frac{2\pi(1 - \nu^2)\bar{\sigma}^2 a_f}{3\bar{E}}(1 + \beta) + \alpha V_f(1 - \eta)^3 \frac{d\sigma_{fu}^2}{24\tau_y}\right\}$$

$$\beta = \frac{8(\sigma_{fu} - \sigma_f)\sigma_{fu}\bar{E}}{\pi\,\bar{\sigma}^2 E_m}\left[\frac{2}{5}\left(\frac{\bar{\sigma}}{\sigma_{fu}}\right)\sqrt{\frac{1}{2}\left(\frac{b_o}{a_f}\right)} - \frac{1}{2\pi}\left(\frac{d}{a_f}\right)\right]$$

(1)

The detailed derivation of Eq 1 is described in Ref *15*. In the derivation of Eq 1 we have used

$$\epsilon_0 = \frac{\sigma_m}{E_m} = \frac{\sigma_f}{E_f}$$

$$\nu = \nu_m = \nu_f \tag{2}$$

where

$\bar{E} = V_m E_m + V_f E_f,$
$E_i = $ Young modulus of the i-phase ($i = m$ and f),
$d = $ fiber diameter,
$\sigma_{fu} = $ fiber breaking stress,
$\tau_y = $ yield shear stress of the matrix,
$b_o = $ average spacing between fibers,
$\alpha = $ fraction of pulled-out fibers,
$\eta = $ ratio of the fiber breaking stress at a weak point to that of nonflawed
  fiber, and
$a_f = $ effective radius of the fractured section given by

$$a_f = \sqrt{\frac{wt}{\theta}} \tag{3}$$

In Eq 3, w and t denote the width and thickness of the specimen, and θ is the angle (in radian) of the apex of the triangle ligament (see Fig. 4). $\bar{\sigma}$ in Eq 1 is taken as the breaking stress of the composite and is related to the stresses in both phases, σ_m and σ_f by

$$\bar{\sigma} = V_m \sigma_m + V_f \sigma_f \tag{4}$$

where

$$\sigma_m = \bar{\sigma} E_m / \bar{E}$$

and thus σ_f is given by

$$\sigma_f = (\bar{\sigma} - V_m \sigma_m)/V_f \tag{5}$$

The data to compute γ_F for unidirectional carbon fiber/6061 aluminum are given in Table 2 except for α and η which are related to fiber pullout energy and tabulated in Ref 15. The values of γ so predicted (shown as open triangles connected by straight line) agree reasonably well with the experimental ones (shown as open circles) as evidenced by Fig. 6. The values predicted by a rule of mixtures (shown as a dashed line) far exceeds both the experimental and theoretical results. The scanning electron microscopy (SEM) pictures of the fractured section of unidirectional carbon fiber/6061 aluminum are shown in

TABLE 2—*Calculation data of unidirectional carbon/6061 aluminum composite.*

	Dimension	$V_v = 0.32$	$V_f = 0.48$
b_o	m	1.68×10^{-5}	1.37×10^{-5}
$\bar{E}$	N/m²	1.42×10^{11}	1.89×10^{11}
σ_m	N/m²	2.36×10^{8}	2.30×10^{8}
σ_f	N/m²	1.38×10^{9}	1.34×10^{9}
$\bar{\sigma}$	N/m²	6.02×10^{8}	7.63×10^{8}
$\nu_m = \nu_f$	1		0.3
E_m	N/m²		6.47×10^{10}
σ_{fu}	N/m²		2.06×10^{9}
d	m		1×10^{-5}
τ_y	N/m²		5.6×10^{7}
w	m		1.02×10^{-2}
t	m		2.54×10^{-3}
θ	radian		2.21
a_f	m		3.41×10^{-3}

fractured section of unidirectional carbon fiber/6061 aluminum are shown in Fig. 8a (at magnification 40), Fig. 8b (at magnification 640) for $V_f = 32\%$ and in Fig. 9a (at magnification 40), Fig. 9b (at magnification 640) for $V_f = 48\%$. It follows from Figs. 8 and 9 that the fracture surface is smooth inside each precursor, but not across the precursor-matrix interface due to the oxidated film which results in weak interface bonding.

As to the experimental results of SiC particulate/6061 aluminum the effect of specimen section size is not much (within ±15% range) as shown in Table 1. For a comparison, the result for the section size 10.16 by 5.08 mm is plotted in Fig. 6 as a filled circle and is found to be the same level as that of the unidirectional case. However, a theoretical model to predict γ_F of a particulate MMC should be developed differently from that of the unidirectional case.

According to the results of three-point bending tests there is some noticeable difference on the characteristic of P-δ curves between unidirectional MMCs and particulate MMCs as evidenced in Figs. 5a and b. The P-δ curve of unidirectional MMCs has two stages; the initiation (the curve going up hill) and propagation stage (the curve going down hill). On the other hand, particulate MMCs show the initiation stage but are followed by little propagation one. This fact suggests that the main contribution to γ_F in particulate MMCs is the elastic strain energy release rate corresponding to the first term in Eq 1 with $\beta = 0$. Thus, the formula to predict γ_F for particulate MMCs is given by

$$\gamma_F = \frac{\pi(1 - \nu^2)\,\bar{\sigma}^2 a_f}{3\bar{E}} \tag{6}$$

The data used to compute γ_F for SiC particulate/6061 aluminum are given in Table 3. The values of γ_F so predicted (indicated by filled triangles connected

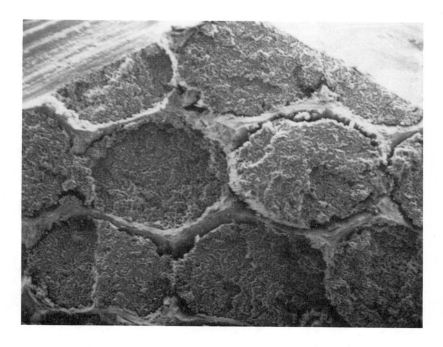

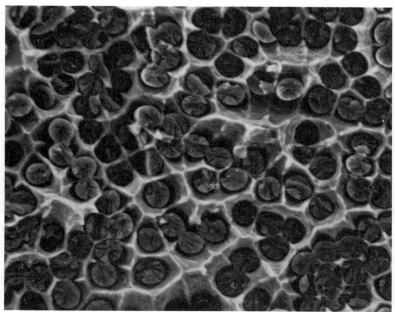

FIG. 8—*SEM pictures of the fractured section of the three-point bending test of* V_f *32% Carbon fiber/6061 Al specimen at the magnification of* (a) *40 and* (b) *640.*

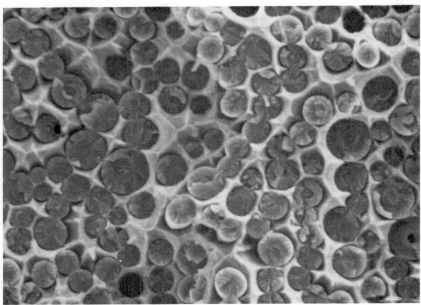

FIG. 9—*SEM pictures of the fractured section of the three-point bending test of* V_f *48% carbon fiber/6061 aluminum specimen at the magnification of* (a) *40 and* (b) *640.*

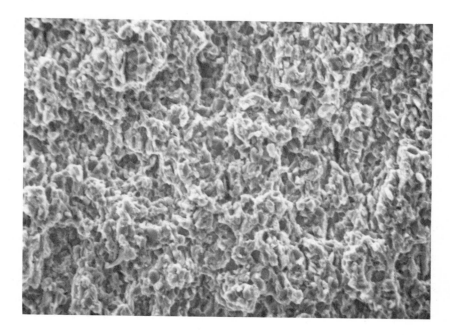

FIG. 10—*SEM pictures of the fractured section of the three-point bending test specimen of* V_f *30% SiC particulate/6061 Al composite with the section size 10.16 by 3.175 mm at the magnification of* (a) *320 and* (b) *2500.*

TABLE 3—*Calculation data of particulate SiC/6061 aluminum composites.*

Term	Dimension	$V_f = 20\%$	$V_f = 30\%$
$\bar{E}$	N/m^2	1.09×10^{11}	1.11×10^{11}
$\bar{\sigma}$	N/m^2	4.10×10^{8}	5.07×10^{8}
w	m	1.02×10^{-2}	
t	m	5.08×10^{-3}	
θ	radian	1.57	
a_f	m	5.73×10^{-3}	
$\nu_m = \nu_f$	1	0.3	

by a straight line) agree reasonably well with the experimental results (indicated by filled circles) as shown in Fig. 7. Typical SEM pictures of the fractured surface of particulate MMCs with $V_f = 30\%$ are shown in Fig. 10a (at magnification 320) and Fig. 10b (at magnification 2500). It follows from Fig. 10a that the fractured surface is reasonably flat, which supports the usage of Eq 6. It should be pointed up that a further experimental study remains to be made to confirm this fact and to check the effect of the size and geometry of the specimen.

Conclusions

A method of measuring the work of fracture (γ_F) of metal matrix composites is studied, and its theoretical verification is tried. It is found that the three-point bending test with center notched specimen is a simple yet dependable method to measure γ_F of MMC specimens regardless whether it is a continuous fiber system or a particulate system.

Acknowledgments

The authors are thankful for the financial support from Honda Research and Development and valuable comments from and discussions with Dr. H. Shimizu, DWA Composite Specialties, Inc., California, are also appreciated.

References

[1] Wright, W. A., Welch, D., and Jollay, J., "The Fracture of Boron Fibre-Reinforced 6061 Aluminum Alloy," *Journal of Materials Science*, Vol. 14, 1979, pp. 1218–1228.
[2] Cruse, T. A., "Tensile Strength of Notched Composites," *Journal of Composite Materials*, Vol. 7, 1973, pp. 218–229.
[3] Whitney, J. M. and Nuismer, R. J., "Stress Fracture Criteria for Laminated Composites Containing Stress Concentrations," *Journal of Composite Materials*, Vol. 8, 1974, pp. 253–265.

[4] Konish, H. J., Swedlow, J. L., and Cruse, A. T., "Fracture Phenomena in Advanced Fibre Composite Materials," *Journal of the American Institute of Aeronautics and Astronautics*, Vol. 11, 1973, pp. 40-43.

[5] Sih, G. C. and Chen, E. P., "Fracture Analysis of Unidirectional Composites," *Journal of Composite Materials*, Vol. 7, 1973, pp. 230-245.

[6] Cooper, G. A. and Kelly, A., "Tensile Properties of Fibre Reinforced Materials: Fracture Mechanics," *Journal of the Mechanics and Physics of Solids*, Vol. 15, 1967, pp. 279-297.

[7] Kelly, A., "Interface Effects and the Work of Fracture of Fiberous Composite" in *Proceedings*, Royal Society of London, Vol. A319, 1970, pp. 95-116.

[8] Cooper, G. A., "The Fracture Toughness of Composite Reinforced with Weakened Fibre," *Journal of Materials Science*, Vol. 5, 1970, pp. 645-654.

[9] Piggott, M. R., "Theoretical Estimation of Fracture Toughness of Fiberous Composites," *Journal of Materials Science*, Vol. 5, 1970, pp. 669-675.

[10] Phillips, D. C. and Tetelman, A. S., "The Fracture Toughness of Fibre Composites," *Composites*, Vol. 3, 1972, pp. 216-223.

[11] Nakayama, J., "Direct Measurement of Fracture Energies of Brittle Heterogeneous Materials," *American Ceramical Society Journal*, Vol. 48, 1965, pp. 583-587.

[12] Tattersall, H. G. and Tappin, G., "The Work of Fracture and its Measurement in Metals, Ceramics and Other Materials," *Journal of Materials Science*, Vol. 4, 1966, pp. 296-301.

[13] Gordon, J. E. and Jeronimidis, G., *Philosophical Transactions*, Royal Society of London, A294, 1980, pp. 545-550.

[14] Skinner, A., Koczakand, M. J. and Lawley, A., "Work of Fracture in Aluminum Metal-Matrix Composites," *Metallurgical Transactions*, Vol. 13A, 1982, pp. 289-297.

[15] Taya, M. and Daimaru, A., "Macroscopic Fracture Surface Energy of Unidirectional Metal Matrix Composites: Experiment and Theory," *Journal of Materials Science*, Vol. 18, 1983, pp. 3105-3116.

[16] Kendall, K., "Fracture of Particulate Filled Polymers," *British Polymer Journal*, Vol. 10, 1978, pp. 35-38.

[17] Andrews, E. H. and Walsh, A., "Rupture Propagation in Inhomogeneous Solids: An Electron Microscope Study of Rubber Containing Colloidal Carbon Black," in *Proceedings of the Physiological Society of London*, Vol. 72, 1958, pp. 42-48.

Akira Kobayashi[1] and Hiroshi Suemasu[1]

Liquid Crystal Film Visualization Approach to Fracture in Composites

REFERENCE: Kobayashi, A. and Suemasu, H., **"Liquid Crystal Film Visualization Approach to Fracture in Composites,"** *Recent Advances in Composites in the United States and Japan, ASTM STP 864*, J. R. Vinson and M. Taya, Eds., American Society for Testing and Materials, Philadelphia, 1985, pp. 522–531.

ABSTRACT: The cholesteric liquid crystals indicate a change of color in response to the temperature change by reflecting the specified wavelength of light due to the pitch alternation caused by the temperature change in the molecular layer of a liquid crystal, indicating red at lower temperature and blue at higher temperature.

Using these liquid crystals, heat evolution during rapid crack propagation can be estimated. However, rather much errors have been introduced due to poor response of liquid crystals.

In the present study, the heat evolution produced during rapid fracture in composites will be calculated through the *improved* dual-layer liquid crystal film method, which is to improve the response and to increase the accuracy in measurement of isothermal colored boundaries appeared in the liquid crystal film during fracture by obtaining two isothermals, T_1 and T_2, necessary for heat evolution calculation, at an earlier period of time after fracture, and by making the spacing between isothermals larger, leading toward the easier discrimination of colored boundaries, resulting in fewer errors. The results show that the time required to obtain two isothermals is reduced even by 50% for the glass-fiber composites. A videotape recorder system is used to record the colored isothermals, and velocity gages to measure the crack propagation velocity.

Thus obtained heat evolution is correlated with the crack propagation velocity.

KEY WORDS: composite, fracture, liquid crystal, visualization, heat evolution, crack propagation velocity, temperature, dual layer, running crack front, isothermals, glass-fiber composites, colored thermal boundary front

Liquid crystal was discovered by Friedrich Reinitzer in 1888. This is of optical-anisotropy liquid generating birefringence, and the cholesteric species indicate a change of color in response to the temperature change by reflecting the specified wavelength of light due to the pitch alternation caused by the

[1]Faculty of Engineering, Institute of Interdisciplinary Research, University of Tokyo, Komba-4-Chome-6-1, Meguro-ku, Tokyo, 153, Japan.

temperature change in the molecular layer of a liquid crystal, showing generally red at lower temperature and blue at higher temperature.

When a dynamic crack propagates in the solid, strong energy concentration
is observed at the running crack tip; hence the heat generation and subsequent heat conduction resulting in a temperature rise leading toward thermal
boundary front formation are expected. This was investigated theoretically by
Kambour and Barker [1], Rice and Levy [2], and Parvin [3], mainly on the
temperature rise.

For composites, a running crack has to advance with cutting fibers ahead
of a crack tip; therefore, it might be expected that much energy concentration
is required that results in heat generation and also in crack propagation velocity variation. The correlation between heat evolution and the crack propagation velocity can be investigated provided these quantities are measured during fracture. As to the crack velocity, this can be measured by gages.

Employing liquid crystals, heat evolution during rapid propagation can be
estimated. Experimental verification by this liquid crystal film technique so
far has been carried out by Fuller et al [4], Kobayashi and Suemasu [5,6],
and Kobayashi et al [7] for polymers and composites. However, many errors
have been introduced due to the poor response of liquid crystals. That is, the
crack propagation velocity $\dot{C}$ is very fast compared with the heat conduction
velocity; hence the change of color appearing in the liquid crystals cannot
follow the fracture phenomenon immediately. Actually, several time lags—
for example, 60 s after fracture for the glass-fiber composites (GFRP) [6]—
were required to observe two isothermals, T_1 and T_2, necessary for the heat
evolution calculation, the details of which are described later. Therefore,
these time lags must be made as short as possible in order to increase the
accuracy in estimation of heat evolution Q.

In the present research, the *improved* dual-layer liquid crystal film
method, which is to improve the response and accuracy in observing the isothermal colored boundaries appearing in the liquid crystal film, will be applied to glass-fiber composites to obtain the evolved heat during fracture.
Further details to increase the accuracy are described in what follows.

Experimental Procedures

Specimen

The glass-fiber polyester composites are from the Asahi Fiberglass Co. of
Japan (make, SLS-213C, 10-plied, 8-satin woven) of which the glass-fiber
content is 45 volume %. The size and the configuration are as shown in Fig. 1.
Also shown are the velocity gages and the thermocouple fixed on one side of a
specimen, and the liquid crystal film on the reverse side. The starting notch or
the initial crack, C_0 was machined and then a very fine cut was made by a
razor blade blow at the crack tip.

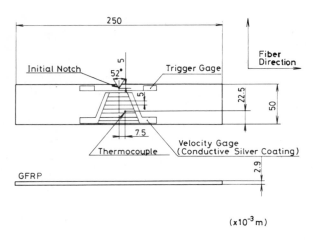

FIG. 1—*Specimen.*

Liquid Crystal Film Technique

Colored thermal boundary fronts appear in the liquid crystal film due to heat caused by the crack propagation following the crack front runs as shown in Fig. 2a: red at lower temperature and blue at higher temperature for cholesteric liquid crystal [5]. In the figure, T_1' is an isothermal for the colored thermal boundary front between green and blue (green/blue, in short), and T_2' red/colorless. However, these colored thermal boundary fronts spread slowly, gradually making mutual spacing between two isothermals, T_1' and T_2' ($T_1' \geq T_2'$), larger, and reaching the stage in which two isothermals can be discriminated after several time lags, $t = t_2$, say, 60 s after fracture for GFRP as shown in Fig. 2b [7]. As described later, these isothermals, T_1' and T_2', are necessary for calculation of heat evolution through the liquid crystal film visualization. In order to increase the accuracy in calculating evolved heat, the time lag, say $t = t_2$, should be made as short as possible.

So an *improved* dual-layer technique is developed. In view of the thermal boundary front formation, a boundary red/green, that is, yellow in principle, can be rather easily discriminated at an early period of colored thermal boundary front appearance, say, $t = t_1$ where $t_1 < t_2$. Therefore, formation of two different red/green boundary fronts almost simultaneously by providing dual-layer liquid crystal film as shown in Fig. 2c is tried. Actually, in the present experiment, two layers of liquid crystals, A and B, are placed as shown in Fig. 3. In the figure, the A-layer is composed of cholesterol oleyl carbonate and cholesterol nonanoate with a weight ratio of 3 to 2, and the B-layer is also composed of the same mixture with a weight ratio of 5 to 2. The colored thermal boundary fronts obtained are shown in Fig. 4. Two isothermals required for later calculation of heat evolution are, consequently, $T_1 = 24.2°C$ and $T_2 = 23.1°C$.

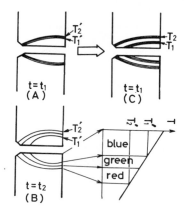

FIG. 2—*Thermal boundary front formation and discrimination of two isothermals* $(t_1 < t_2)$ $(T_1' > T_2')$ $(T_1 > T_2)$.

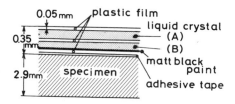

FIG. 3—*Dual-layer liquid crystal film concept.*

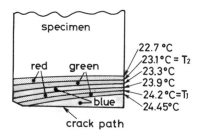

FIG. 4—*Obtained colored thermal boundary fronts in dual-layer concept.*

Velocity Gages

Velocity gages consist of a series of conducting wires, du Pont No. 4817 conductive silver coating material, placed on the composite specimen surface at prescribed intervals on the projected path of the crack and perpendicular to the direction of crack propagation as shown in Fig. 1. These wires form one leg of a bridge in the electronic circuit. When the wires are broken by the running crack, the corresponding time record is obtained from the trace on the oscilloscope, so that the average crack propagation velocity between two wires can be measured. Further details are given in Ref 8.

Measuring Block Diagram

The block diagram of data recording is shown in Fig. 5. As shown, the applied load is monitored by a load cell, the crack propagation velocity by velocity gages, the thermal boundary fronts due to change in color in the liquid crystal film by a video camera and also by an ordinary camera using ASA 400 color film. During the experiment, the illuminating lamps are employed to flash every 10 s. Special care is taken not to pour superfluous flash lamp heat on the specimen. A copper-constantan thermocouple of 0.1 mm diameter was placed on the velocity gage side as shown in Fig. 1. A loading apparatus of a conventional Instron-type tension tester was employed to achieve Mode I crack propagation with a crosshead speed of 5 mm/min.

Estimation of Heat Evolution

From the thermal boundary fronts obtained as shown in Fig. 4, the heat evolved during crack propagation can be calculated through the conventional heat conduction equation

$$Q = \rho c \sqrt{4\pi kt} \cdot \Delta T \cdot e^{y^2/4kt} \tag{1}$$

with the following assumptions, where $\Delta T = T_1 - T_2$:

1. One-dimensional heat flow.
2. No heat loss from the specimen surface.
3. Infinite specimen length.
4. No volume heat origin.

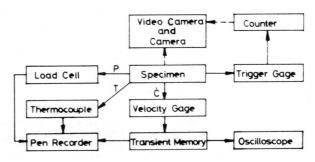

FIG. 5—*Measuring block diagram.*

5. Crack propagation velocity is fast enough compared with the heat conduction velocity.
6. The thermal boundary front is isothermal; red/green: $T_1 = 24.2°C$, red/green: $T_2 = 23.1°C$.
7. The square of the difference of heat evolution $(Q_1 - Q_2)^2$ is minimized, where Q_1 and Q_2 are the heat evolved on both crack faces per unit area of crack advance corresponding to the thermal boundary front T_1 and T_2, respectively. Actually, the obtained isothermals T_1 and T_2 show irregular wavy boundaries, so Assumption 7 is taken to minimize errors due to irregularity.

In the preceding:

Q = heat evolved on both crack faces per unit area of crack advance,
k = thermal diffusivity,
c = specific heat,
ρ = density,
t = time after fracture,
T = temperature, and
y = distance from the cracked plane.

Refer to Fig. 6 for coordinate information.
Actually, the following equation is used

$$Q = \rho c \sqrt{4\pi kt}\ e^{y_1^2/4kt}\ \frac{\displaystyle\int_0^b e^{y_2^2/4kt}\ (e^{y_2^2/4kt} - e^{y_1^2/4kt})\,dx}{\displaystyle\int_0^b (e^{y_2^2/4kt} - e^{y_1^2/4kt})^2\,dx}\ \Delta T \qquad (2)$$

where

b = specimen width = 50 mm,
$\rho c = 1.38 \times 10^6\,\mathrm{Jm^{-3}°C^{-1}}$, and
$k = 4.07 \times 10^{-7}\,\mathrm{m^2 s^{-1}}$.

Further details of equation (2) are given in the Appendix.

Experimental Results and Discussions

The average heat evolution obtained for seven specimens by the present dual-layer liquid crystal film technique as a function of a running crack front position 30 s after fracture is shown in Fig. 7, where x is the running crack front position. In Fig. 7 for two isothermals T_1 and T_2 obtained 30 s after fracture, we can measure y_1 and y_2 for any arbitrary x, so that heat evolution Q at the very x-position can be calculated by Eq 2.

First, it is emphasized that the present dual-layer technique makes it possi-

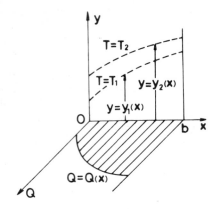

FIG. 6—*Coordinate information.*

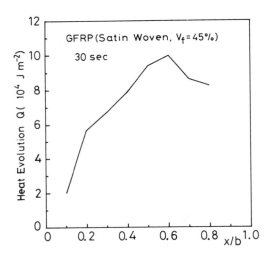

FIG. 7—*Heat evolution versus running crack front position.*

ble to observe two isothermals, T_1 and T_2, even 30 s after fracture, while it takes 60 s at least for the previous single-layer technique [7] to obtain discriminated colored isothermals of two thermal boundary fronts T_1 and T_2 necessary for heat evolution calculation. It is regretted that plain woven specimens were employed in the previous case [7], so that the comparison of heat evolution with the present dual-layer technique is rather meaningless as satin woven specimens are used in the present experiment. However, the heat evolution Q at the peak value is some 20% larger for the present case, if the difference due to the woven structure of GFRP specimens is just ignored.

Figure 8 shows the average crack propagation velocity for seven specimens obtained by velocity gages in terms of nondimensional running crack front

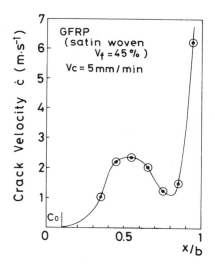

FIG. 8—*Crack propagation velocity versus running crack front position.*

position x/b, where b is the specimen width. In Fig. 8, black circles are the experimental values by velocity gages. Now in Figs. 7 and 8 it is noticed that both have peak values at around $x/b = 0.5 \sim 0.6$. Although the obtained crack propagation velocity profile is wavy, the heat evolution obtained by the present liquid crystal film technique, as shown in Fig. 7, is an increasing function of the crack propagation velocity as shown in Fig. 9. In Fig. 9, the shaded area means scatter, since the curves shown in Figs. 7 and 8 are those of averages for seven specimen data, respectively. No correlation is shown in Fig. 9 for the crack velocity of more than about 2.8 m/s, since the heat Q could not be calculated due to the difficulty of observing the two isothermals T_1 and T_2.

In Fig. 10 the temperature rise obtained at $x/b = 0.5$ 30 s after fracture, by the present dual-layer technique, is compared with the theoretical temperature rise through one-dimensional heat conduction theory. That is, the theoretical Q-value is calculated by using a heat conduction equation with $t = 30$ s, $y = 7.5$ mm (at the thermocouple position), and temperature rise ΔT measured by the thermocouple. Now using this Q-value, a theoretical curve of temperature rise $\sim y$ is obtained and is compared with the experimental temperature rises obtained by the dual-layer liquid crystal visualization technique at several y points at $x = 0.5\,b$, 30 s after fracture.

Conclusions

An *improved* dual-layer liquid crystal film visualization technique is proposed. By employing this improved technique, the time lag necessary to dis-

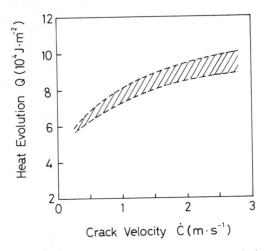

FIG. 9—*Heat evolution versus crack propagation velocity.*

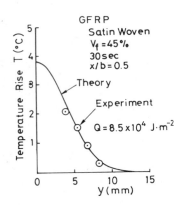

FIG. 10—*Comparison of theoretical and experimental temperature rise (dual-layer method for experimental verification).*

criminate two isothermals required to calculate the heat evolution is reduced by 50%, resulting in an increase in accuracy of heat calculated. Comparison with the theoretical temperature rise shows very good agreement with the present dual-layer technique, supporting a reasonable experimental liquid crystal film visualization technique, and correlations between heat evolution and crack propagation velocity during fracture in composites can be obtained.

Acknowledgments

The authors appreciate the assistance of Mr. Osamu Ichikawa, Mr. Hirotoshi Yasaka, and Miss Junko Nakashima during the course of this work. The research is partly supported by Government Subsidy for the Promotion of Scientific Research, Japan.

APPENDIX

Assuming the one-dimensional heat conduction, the heat evolution Q_1 is given for an isothermal T_1 by

$$Q_1(x) = \rho c \sqrt{4\pi kt} \cdot T_1 \cdot \exp(y_1^2/4kt) \tag{3}$$

Similarly for an isothermal T_2

$$Q_2(x) = \rho c \sqrt{4\pi kt} \cdot T_2 \cdot \exp(y_2^2/4kt) \tag{4}$$

Now in order to minimize errors due to irregular wavy isothermals obtained, we employ Assumption 7, that is

$$\frac{\partial}{\partial T_1} \int_0^b \{Q_1(x) - Q_2(x)\}^2 dx = 0 \tag{5}$$

from which we have

$$T_1 = \frac{\displaystyle\int_0^b \exp(y_2^2/4kt)\{\exp(y_2^2/4kt) - \exp(y_1^2/4kt)\}dx}{\displaystyle\int_0^b \{\exp(y_2^2/4kt) - \exp(y_1^2/4kt)\}^2 dx} \Delta T \tag{6}$$

where $\Delta T = T_1 - T_2$. Substitution of Eq 6 into Eq 3 leads to Eq 2.

References

[1] Kambour, R. P. and Barker, R. E., Journal of Polymer Science, Vol. A2, 1966, pp. 359–363.
[2] Rice, J. R. and Levy, N. in Physics and Plasticity, A. S. Aragon, Ed., MIT Press, Cambridge, Mass., 1969, pp. 277–293.
[3] Parvin, M., International Journal of Fracture, Vol. 15, No. 5, Oct. 1979, pp. 397–404.
[4] Fuller, K. N. G., Fox, P. G., and Field, J. E. in Proceedings of the Royal Society, London, Series A., No. 341, 1975, pp. 537–557.
[5] Kobayashi, A. and Suemasu, H. in Composite Materials; Mechanics, Mechanical Properties and Fabrication, K. Kawata and T. Akasaka, Eds., Japan Society for Composite Materials, Tokyo, 1981, pp. 339–346.
[6] Kobayashi, A. and Suemasu, H. in Progress in Science and Engineering of Composites, T. Hayashi, K. Kawata, and S. Umekawa, Eds., Proceedings, Fourth International Conference on Composite Materials, Japan Society for Composite Materials, Tokyo, 1982, pp. 633–640.
[7] Kobayashi, A., Munemura, M., Ohtani, N., and Suemasu, H., Journal of Applied Polymer Science, Vol. 27, 1982, pp. 3763–3768.
[8] Kobayashi, A., Ohtani, N., and Sato, T., Journal of Applied Polymer Science, Vol. 18, No. 6, June 1974, pp. 1625–1638.

Donald W. Oplinger,[1] Burton S. Parker,[1] Kanu R. Gandhi,[1] Roger Lamothe,[1] and Gary Foley[1]

On the Streamline Specimen for Tension Testing of Composite Materials

REFERENCE: Oplinger, D. W., Parker, B. S., Gandhi, K. R., Lamothe, R., and Foley, G., "On the Streamline Specimen for Tension Testing of Composite Materials," *Recent Advances in Composites in the United States and Japan, ASTM STP 864*, J. R. Vinson and M. Taya, Eds., American Society for Testing and Materials, Philadelphia, 1985, pp. 532–555.

ABSTRACT: Tensile and shear stress concentrations have been found in prior work to degrade the performance of tension test specimen designs in current use. In the particular case of tabbed specimens there is need for an alternative specimen design to avoid problems with tab failures and loss of tabs when testing in adverse environments. The streamline shape which is described here provides such an alternative. This paper discusses the theoretical background to the streamline shape, together with results of studies in which its performance is compared experimentally with other specimen types, especially tabbed specimens, in both static and fatigue testing environments.

KEY WORDS; tension test methods, fibrous composites, streamline shapes, hydraulic flow analogies

Tension test results in composite materials [1] are often adversely affected by deficiencies in the specimen geometry which produce localized tensile or shear stress peaks that lead to failure in objectionable locations and at stress levels below those which represent the true strength of the material. This paper addresses a class of tension specimen shapes which have the capability for minimizing the stress peaks responsible for adverse test results, thereby providing a strong tendency for failures to occur within the gage region of the specimen, as well as a reduction in the scatter observed in test results. This type of geometry, referred to as "streamline" shape, is of particular interest in providing an alternative to other specimen designs, especially those using bonded tabs, in cases when testing high-temperature, high-humidity, or other types of environment make tab retention difficult.

[1] Army Materials and Mechanics Research Center (AMMRC) Watertown, MA

The following discussion begins with some general observations on the behavior of typical tension test specimens in current use, followed by a discussion of the theoretical background to the streamline geometry concept, after which some experimental results are described which compare the performance of the streamline design with that of other specimen geometries, primarily tabbed specimens.[2]

General Observations of Tension Test Specimen Behavior

Figure 1 shows the tension specimen designs of interest here. The D3039 tabbed specimen is currently an industry standard for high-modulus materials. The D638 dogbone shape, which was originally introduced for plastics in general, has often been used for low-modulus composites. The last two designs in Fig. 1 are not in general use as yet, but are both of interest as alternatives to the tabbed specimen. The "bow tie" shape of Fig. 1 was introduced in 1969 [2]. Some of the stress analysis results which were reported in Ref 1 refer to large-taper-angle variations of the bow tie shape, as a means of drawing conclusions about the effect of taper angle.

Figure 2 summarizes the stress analysis results presented in Ref 1. All three specimen types considered there are subject to significant stress concentrations. In the case of the tabbed specimen these are quite severe, although a high degree of localization of these stresses together with the effect of clamping pressure by the test machine grips may help to reduce the extent of performance degradation caused by them.

In the case of the bow tie shape, the curves of peak stresses versus taper angle (Fig. 2b) are of interest. For small taper angles it can be argued that axial tensile stress is essentially uniform across the width of the specimen. In that case the tensile stress is given in terms of load by

$$\sigma_x \simeq \frac{P}{h(x)t}$$

where

$P = $ load,
$h(x) = $ specimen width, and
$t = $ thickness.

Making use of the equilibrium equation $\partial\sigma_x/\partial x = -\partial\tau_{xy}/\partial y$ gives the following for the shear stress distribution

$$\tau_{xy} \simeq \frac{yh_g}{h(x)^2}\frac{dh}{dx}\bar{\sigma} \qquad (1)$$

[2]Some of this material was presented at the Sixth Conference on Fibrous Composites in Structural Design (New Orleans, LA, Jan. 24-27, 1983). However, much of the material presented here is new, and this is the first instance in which it is presented in the open literature.

where

h_g = gage region width,
$\bar{\sigma}$ = gage region tensile stress, and
y = lateral coordinate.

The maximum value of shear stress at a given y obtained from Eq 1 occurs at $y = h$, and is given by

$$\tau_{xy\,edge} \simeq \frac{h_g}{h} h' \, \bar{\sigma} \qquad (2)$$

where $h' = dh/dx$. For the specimen as a whole the peak shear stress occurs where h'/h has its largest value. For the bow tie shape this occurs at the narrow (gage) end of the taper and is given by

$$\tau_{xy\,max} \simeq h' \, \bar{\sigma} \qquad (3)$$

while for the D638 shape it occurs at the wide (grip) end of the taper. Equation 3 leads to shear stress values which agree quite well with the finite-element results shown in Fig. 2b for various taper angles. For the D638 shape, Eq 2 slightly overestimates the maximum shear stress value because the horizontal boundary at the end of the grip region imposes a stress-free condition where Eq 2 requires a maximum. However, Eq 2 for general shapes and Eq 3 for the bow tie shape are quite useful in pointing out the geometric parameters which control the level of shear stress. In general, long slender shapes with gradual tapers are required to avoid shear failures in tension test specimens for organic matrix composite materials. For the bow tie shape in particular, Eq 3 indicates that the slope of the sides (that is, the tangent of the taper angle) should be less than the ratio of shear strength to tensile strength for the material being tested. For typical unidirectional materials this ratio is on the order of 1/20 to 1/50, corresponding to taper angles (that is, the inclination of each specimen boundary to the specimen axis) on the order of 1 to 3 deg.

Tensile stress concentrations are another aspect of the tension specimen design which is controlled by geometry. Figure 3 illustrates the relationship between tensile stress concentrations and breaking loads in D638 and bow tie-type specimens. Here the circles represent nominal gage region stresses at the loads causing specimen failure, while the exxes show maximum stresses predicted by finite-element analysis at the breaking loads. It is apparent that the maximum stresses all lie in the same range, while the nominal breaking stresses (which would normally be interpreted as material strength values) fall at considerably lower levels.

The reason for the occurrence of a local tensile stress peak in a tapered specimen is suggested in Fig. 4. This has to do with the fact that in the tapered region the shear stress which is zero along the symmetry axis rises monotonically along

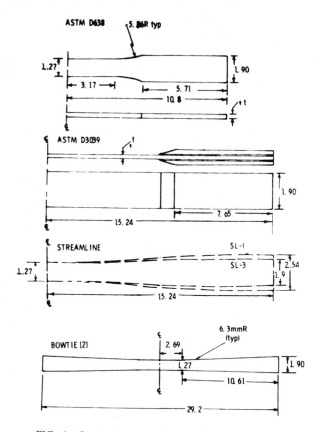

FIG. 1—*Specimen types of interest (dimensions in cm).*

the y-direction, in agreement with Eq 1, so that $\partial \hat{\tau}_{xy}/\partial y$ has the same sign throughout the tapered region; in the gage region, however, it is zero both at the symmetry axis and at the traction-free specimen boundary, forcing τ_{xy} to decrease with increasing y near the boundary, according to which the sign of $\tau_{xy,y}$ is opposite to what it is in the tapered region. This means that $\tau_{xy,y}$ goes through zero at some intermediate point along the x-direction, and because of the equilibrium equation cited earlier, $\sigma_{x,x}$ has to go through zero at the same point. It is found that this corresponds to a local *maximum* of σ_x versus x, and this will occur in any geometry in which the tapered region is terminated at a finite value of x. The main objective of the idealized version of the streamline shape discussed below is to provide a geometry for which the taper continuously approaches zero but never quite reaches it in a finite length, thus avoiding the condition associated with Fig. 4 which appears to be responsible for causing local stress peaks.

One additional consideration is that of breaks in the tab region for the case of

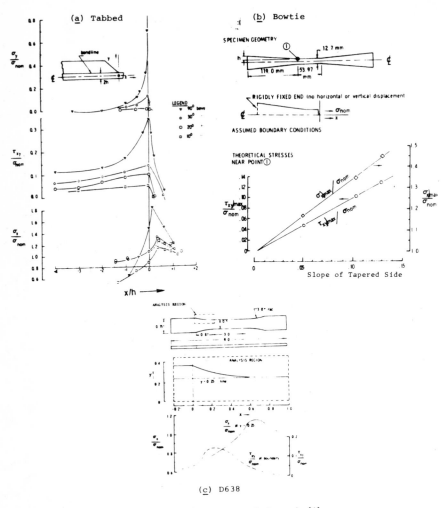

FIG. 2—*Previous stress analysis results* [1].

tabbed specimens. Figure 5 illustrates the importance of avoiding tab failures. (It is emphasized that these are failures within the tab region, not loss of tab bonds due to poor application procedures.) The data of Fig. 5 which correspond to tab breaks contribute strongly on the lower side of the distribution. In addition the coefficient of variation is considerably increased if the tab-break specimen results are retained. It should be noted that the ASTM Test Method for Tensile Properties of Fiber-Resin Composites (D 3039-76) which applies to tabbed specimens requires that no tab breaks occur within one specimen width of each tab, so that strict adherence to the specification would require dis-

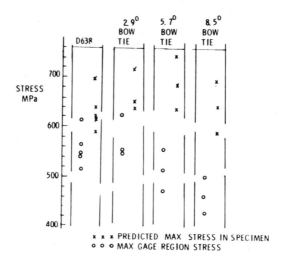

FIG. 3—*Effect of tensile stress concentrations on failure stresses for various specimen shapes.*

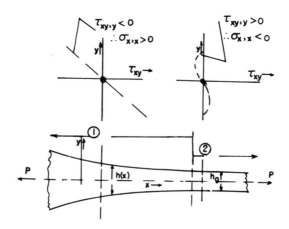

FIG. 4—*Origin of tensile stress concentrations.*

carding all of the results listed in Fig. 5 as "tab failures." This is shown even more dramatically in Fig. 6, which describes the distribution of failure locations in tabbed specimens consisting of woven aramid epoxies produced by several organizations in a cooperative study which was administered at AMMRC. Particularly in the case of high-temperature/high-humidity conditions, tab failures may be so severe a problem as to prevent tension test efforts from being completed.

The ideal specimen design is one which minimizes shear stresses and tensile stress peaks in the tapered region and avoids the inconveniences introduced by

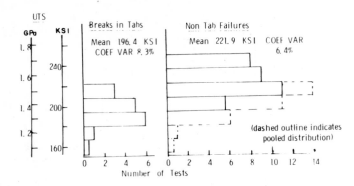

FIG. 5—*Effect of tab failures on test results: 14-ply SP250 S-glass (75°F, 50% relative humidity, 3-month exposure).*

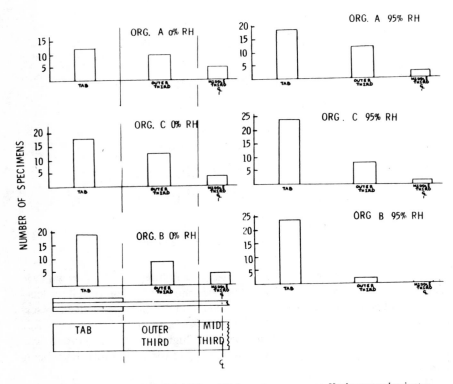

FIG. 6—*Failure location distribution in tabbed specimens—woven Kevlar epoxy laminates.*

bonded tabs, especially in cases where adverse environments make tab failures unavoidable or make tab retention difficult. The objective of this effort is to provide such a specimen design in the form of the streamline shape.

Theoretical Background to the Streamline Shape

The use of so-called "streamline" shapes has been in vogue for some time as a means of reducing stress concentrations in machine components [3–5]. According to the literature cited by Hetenyi [6], analogies between hydraulic flow and elastic stress fields have been used at least since 1925 to explore stress fields in solid components. Neuber [7] describes a class of geometries involving combinations of straight and circular boundary segments for which the analogy is exact, but for geometric shapes in general the analogies which have been introduced tend to be approximate. The main impetus for using such analogies being to develop boundary shapes which eliminate stress concentrations, those cited by Neuber are not necessarily of practical interest since they may not provide a beneficial effort of this kind. The approach used here was developed specifically for orthotropic bodies, although it appears also to apply to isotropic cases when the boundary geometry is appropriately restricted.

The two-dimensional elasticity equations in terms of displacements, for orthotropic materials, include

Equilibrium

$$\frac{\sigma_x}{\partial x} + \frac{\partial \tau_{xy}}{\partial y} = 0 \quad \frac{\partial \sigma_y}{\partial y} + \frac{\partial \tau_{xy}}{\partial x} \tag{4}$$

Constitutive Equations

$$\sigma_x = C_{11} u_{,x} + C_{12} v_{,y} \tag{5.1}$$

$$\sigma_y = C_{12} u_{,x} + C_{22} v_{,y} \tag{5.2}$$

$$\tau_{xy} = C_{66}(u_{,y} + v_{,x}) \tag{5.3}$$

where u and v are displacements in the x- and y-directions, respectively. These reduce to

$$C_{11} u_{,xx} + C_{66} u_{,yy} = -(C_{12} + C_{66}) v_{,xy} \tag{6.1}$$

$$C_{22} v_{,yy} + C_{66} v_{,xx} = -(C_{12} + C_{66}) u_{,xy} \tag{6.2}$$

For long gently tapered bodies (complying with the requirement for low taper angles to avoid shear failures) it is suspected that the terms in Eq 6 containing v can be ignored, corresponding to the assumption that $\partial v/\partial x$, the "vertical sliding" component of the shear strain, as well as the Poisson strain, $\partial v/\partial y$, appearing in Eq 5.1 are insignificant. (The validity of this approximation will be illustrated in the ensuing discussion of finite-element results.) As a result of this approximation it is noted that σ_x and τ_{xy} are given by

$$\sigma_x = C_{11} \frac{\partial u}{\partial x}; \qquad \tau_{xy} = C_{66} \frac{\partial u}{\partial y} \tag{7}$$

and Eqs 6 reduce to

$$C_{11} u_{,xx} + C_{66} u_{,yy} \cong 0 \tag{8}$$

In the following, the Airy's stress function, A, which is defined to relate to the stresses by

$$\sigma_x = \frac{\partial^2 A}{\partial y^2}; \qquad \sigma_x = \frac{\partial^2 A}{\partial x^2} \qquad \tau_{xy} = -\frac{\partial^2 A}{\partial x \partial y} \tag{9}$$

is introduced, together with the notation

$$x' = x; \qquad y' = \left(\frac{C_{66}}{C_{11}}\right)^{1/2} y \tag{10.1}$$

$$U = (C_{11} C_{66})^{1/2} u \tag{10.2}$$

$$V = \partial A/\partial y \tag{10.3}$$

It is found that combining Eqs 4, 7, 9, and 10 leads to

$$\frac{\partial U}{\partial x'} = \frac{\partial V}{\partial y'}; \qquad \frac{\partial U}{\partial y'} = -\frac{\partial V}{\partial x'} \tag{11}$$

from which it can be stated that the complex function W given by

$$W = U + iV \tag{12}$$

is an analytic function of z', where

$$z' = x' + iy' \tag{13}$$

and that the stress-free boundary in the z'-plane is one for which V is a constant. (It should be noted that the treatment here depends on assuming that along with the strain components involving the partial derivatives of v, σ_y is negligible.) In potential-flow problems the analogous situation is the case in which V represents a streamline, corresponding to a rigid wall across which no flow can take place. Solutions for potential flow in rigid-wall channels can thus be transformed into geometries for stress-free boundaries in the case of elasticity problem.

The situation shown in Fig. 7 is of interest here since it was found to give streamlines over part of the flow field which represent particularly favorable geometries for tension specimen shapes. The problem indicated in Fig. 7 is that of two-dimensional flow in a channel having a right-angle expansion. The textbook solution for this problem [8] leads to the following expressions: Setting

$$F_1 = -\frac{1}{\rho^2 - 1}[2\exp(\pi W/\bar{\sigma}_x w_h) - \rho^2 - 1] \qquad (14.1)$$

$$F_2 = \frac{1}{\rho^2 - 1}[\rho^2 + 1 - 2\rho^2\exp(-\pi W/\bar{\sigma}_x w_h)] \qquad (14.2)$$

$$c = w_h(C_{11}/C_{66})^{1/2} \qquad (14.3)$$

with

$$z' = \Omega(W) \qquad (15)$$

where

$$\Omega = \frac{c}{\pi}\left[\cosh^{-1}(F_1) - \frac{1}{\rho}\cosh^{-1}(F_2) - i\pi\left(1 - \frac{1}{\rho}\right)\right] \qquad (16)$$

and w_h is the half-width at the narrow end of the body, then setting V equal to a constant and varying U from $-\infty$ to $+\infty$, inserting values of W corresponding to these U and V into Eqs 14 and 16 and setting x' and y' to the real and imaginary parts of Ω corresponding to Eq 15, gives a streamline for the flow problem, or, correspondingly, a traction-free boundary for the stress problem. Within the present approximation, stresses of interest are generated from $V_{,x}$ and $V_{,y}$ as a result of Eq 10.3 together with Eq 9. These are obtained from the expressions

$$V_{,x} = \frac{i}{2}\left[\frac{1}{(d\Omega/dW)^*} - \frac{1}{(d\Omega/dW)}\right];$$

$$V_{,y'} = \frac{1}{2}\left[\frac{1}{(d\Omega/dW)^*} + \frac{1}{(d\Omega/dW)}\right] \qquad (17)$$

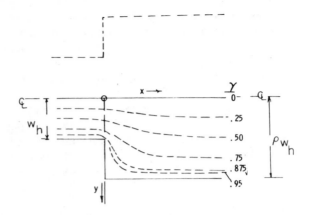

FIG. 7—*Flow in channel with contraction.*

in which the asterisk denotes complex conjugate of the expressions in parentheses. The derivative appearing in Eq 17 is found from the expression

$$\frac{d\Omega}{dW} = c\left[\frac{T-1}{T-\rho^2}\right]^{1/2} \tag{18.1}$$

where

$$T = \exp(\pi W/\bar{\sigma}_x w_h) \tag{18.2}$$

The second of Eq 17 is used to obtain $V_{,y}$ in order to evaluate σ_x by using the identity

$$V_{,y} = V_{,y'}(C_{66}/C_{11})^{1/2}$$

Figure 8 shows pertinent results obtained for an expansion ratio $\rho = 2.0$. The parameter associated with each curve is the streamline parameter γ which identifies the relative distance in the y-direction from the centerline to the streamline, at $x = +\infty$ or $-\infty$, where all streamlines are essentially horizontal, that is, $\gamma = y/w_h$. For γ equal to or less than 0.5 none of the curves in Fig. 8c has a peak in the vicinity of the taper, so that any curve for which $\gamma \leqq$ 0.5 gives a specimen boundary shape which will be free of tensile stress concentrations. There is evidence that $\gamma = 0.5$ gives this special behavior, that is, freedom from stress concentrations, for any streamline lying inside the $\gamma = 0.5$ streamline for any value of the expansion ratio. On the other hand, a value of γ somewhat less than 0.5 may be required to keep shear stresses within desired limits. In the practical designs which have been chosen the objective has been to keep the shear stress less than 0.05 times the gage region tensile stress.

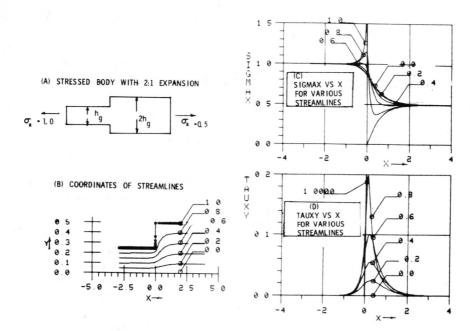

FIG. 8—*Stresses predicted by approximate theory for 2:1 expansion.*

Smaller values of γ tend to produce specimen designs which are longer than is desirable from the standpoint of economy. Similarly, large values of the expansion ratio ρ are desirable from the standpoint of focusing high stress values in the gage region, but tend to produce long specimen designs when the value of γ is adjusted to keep the shear stresses within bounds.

The validity of the approximate stress analysis is addressed in Fig. 9, which compares finite-element (FE) results with results of the approximate theory. The mesh for the FE solution (Fig. 9a) was generated using the streamline equations, Eqs 14 to 16, to allow element boundaries to be defined by lines of constant U and V. A program was written for use in an in-house FE program using eight-point isoparametric elements.

Comparison of stress results for three specimen shapes with the same expansion ratio are shown in Fig. 9b–d. In all cases the shear and axial tensile stresses obtained from the FE approach are almost indistinguishable from those given by the approximate theory, thus verifying the accuracy of the approximation. The FE results also demonstrate that the lateral tensile stresses are small enough to ignore.

These results also illustrate the effect of the streamline parameter on the length required for the specimen. Length is governed by the need for the stresses to settle out so that constant-width sections in both the grip region and the gage region can be provided. Practical specimen shapes (Fig. 10) have been selected

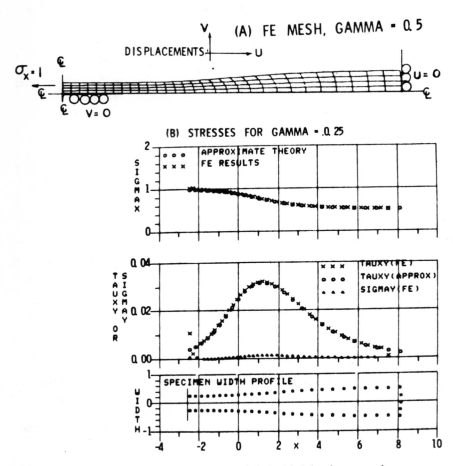

FIG. 9—*Comparison of approximate analysis with finite-element results.*

by using circular arcs to blend the theoretical shape to uniform-width segments in both regions. As seen in Fig. 9b, stresses for the case of $\gamma = 0.25$ settle out over the x-interval from -3 to $+8$, that is, a total length of 11 (all distances being relative to a half-width of 0.5 in the gage region), whereas for $\gamma = 0.50$ the distance for settling out is reduced to about 6. Noting that only half the length of an actual tension specimen is being considered here, for a gage width of 1.27 cm (0.5 in.) it is apparent that a specimen length of about 58 cm (25 in.) would be required to provide appropriate grip and gage regions for the case of $\gamma = 0.25$. In the compromise shapes shwon in Fig. 10, 30.5 cm (12 in.) has been used as a length selection. The effect of various departures from the ideal situation such as the lateral restraining effects of grips and the departure to the shape from ideality remains to be investigated. Experimental studies have been restricted to the two shapes of Fig. 10 corresponding to expansion ratios of 1.5 and 2.0.

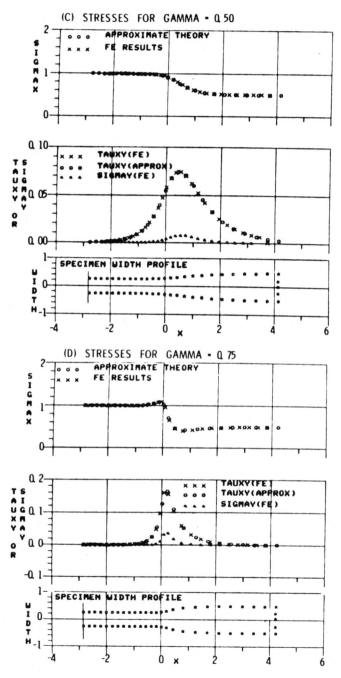

FIG. 9—*Continued*.

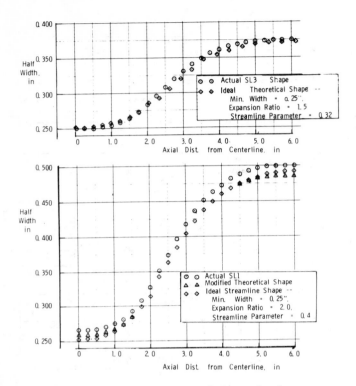

FIG. 10—*Theoretical versus practical streamline shapes.*

Experimental Studies

Specimen Machining

Conventional machining methods cannot be used to produce the streamline-type tension test specimen because of the continuous variation in the slope of the specimen profile. Although numerically controlled machining can be readily used, the aim of the study was to investigate methods which avoided the use of numerically controlled equipment because this is not available at some composite fabrication facilities. A method was developed based on the use of a converted nameplate engraver which was provided with an appropriate template, thus serving as a template-guided router. In addition, templates for the SL1 and SL3 shapes have been developed for an in-house Tensilkut[3] machine to establish that this type of commonly available equipment can successfully produce streamline specimens. The Tensilkut machine is quite easy to use, even by personnel with no special skills in machining.

The nameplate engraver, employed as a pantograph, provides a fast method

[3]Trade name, Tensilkut Corp., Waterbury, CT.

of specimen preparation, and several hundred streamline specimens have been produced with this equipment. Fiberglass/epoxy, Kelvar[4]/epoxy, graphite/epoxy, monolithic plastics, and even certain metal matrix composites have been machined with this device. Recent studies have also been conducted with the Tensilkut to determine the proper technique for obtaining results with that equipment that are equivalent to those obtained with the nameplate engraver. One negative feature of the latter is the fact that each of the various versions of the streamline shape requires a different "arm ratio" or multiplication factor between the template and specimen size. These settings vary continuously and cannot always be set accurately, resulting in unwanted variations in specimen dimensions. While the Tensilkut machine avoids this problem, as will be seen later it does not necessarily produce equivalent specimen quality to that of the pantograph, and studies are still being conducted to correct deficiencies in the technique of using the Tensilkut equipment.

Considerable research has been required to determine the optimum router bit for each type of material which has been tested under this effort. For both graphite/epoxy and glass/epoxy, router bits supplied by Starlite Industries have been found to be satisfactory, although fiberglass is not particularly difficult, and cutting bits which have been supplied by any of several manufacturers for the circuit board industry have been satisfactory. For graphite/epoxies which are more difficult than glass/epoxies, a fine-toothed bit (No. 403200) produced by Starlite under the generic name of Starbide[5] has been relatively satisfactory. As with other tapered tension specimen shapes on which work has been reported, inclined surfaces in graphite/epoxy laminates tend to be excessively flaw sensitive, and as with any highly brittle material, considerable care has to be exercised to avoid the degrading effects of surface flaws produced by the machining process. Figure 11 presents data which illustrate this point. Here two sets of data obtained with the SL3 specimen on $(0/90)_{4s}$ graphite/epoxy laminates are shown, along with data obtained on tabbed specimens. The triangular symbols represent specimens inadvertently machined with a relatively coarse-toothed cutter, while the remaining SL3 specimens, represented by circular symbols, were obtained with the fine-toothed cutter mentioned previously. The circles containing cross-strokes, it should be noted, were produced from specimens machined on the Tensilkut, while the remaining SL3 specimens were made on the pantograph, using the accepted cutter. It is obvious that the coarse cutter had a seriously degrading effect on the specimen quality.

Alumina (that is, FP[6]) aluminum metal matrix specimens, as well as FP magnesium and silicon carbide aluminum have been successfully prepared on the pantograph machine although tool life is a problem in such materials. Typically, the bit is advanced axially after one specimen is finished, so that

---

[4]Trade name, E. I. duPont de Nemours & Co., Inc., Wilmington, DE.
[5]Trade name, Starlite Industries, Rosemont, PA 19010.
[6]Trade name, E. I. duPont de Nemours & Co., Wilmington, DE.

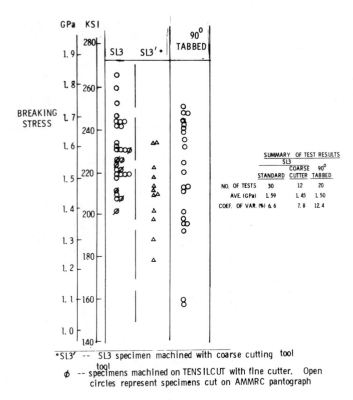

ø -- specimens machined on TENSILCUT with fine cutter. Open circles represent specimens cut on AMMRC pantograph

FIG. 11—*Test results for tabbed and streamline specimens: unidirectional T300 5208.*

subsequent specimens are subjected to unworn portions of the bit. Kevlar epoxies continue to be difficult to machine because of fraying of the specimen edges.

If numerical machining facilities are available, they can obviously be used to provide streamline specimen shapes. Test specimens as large as 120 cm (4 ft) in length have been made by numerical machining on very thick composite materials under consideration at AMMRC for application to Army bridging components.

Data Interpretation for Laminates of Varying Thickness

In in-house evaluation efforts on the streamline specimen, 60 by 120-cm (2 by 4-ft) laminate sheets are typically made up to provide specimens. Large thickness variations can occur in sheets of this size which can lead to misleading test results if not interpreted properly. Figure 12a shows the correlation of breaking stress with laminate thickness for a set of crossply graphite/epoxy SL3 specimens; here the stress values are calculated from a loaded area based

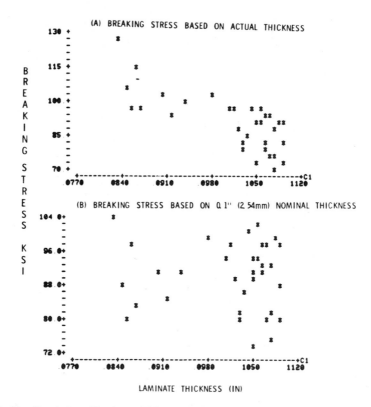

FIG. 12—*Correlation of laminate thickness and breaking strength, crossply T300 5208 graphite/epoxy, SL3 specimen.*

on the actual thickness of a given specimen. As a result a correlation of strength with specimen thickness is observed, while if a nominal thickness is assigned to the whole specimen lot as in Fig. 12*b*, no correlation with thickness is seen. In general, use of actual specimen thickness values should be avoided since thickness variations reflect only resin content changes and not changes in the area of load bearing material (that is, the fibers). Note that if the data of Fig. 12*a* are pooled as one population, the apparent scatter is much greater than in Fig. 12*b*. The former corresponds to a data range of 410 MPa (60 ksi) versus 230 MPa (34 ksi) in the latter. The scatter in the former case obviously includes the effect of thickness variations and does not reflect the true scatter in the strength of the material as the set of data based on nominal thickness does.

Static Test Results

Figure 13 as well as Fig. 11, discussed previously, gives comparisons of data obtained from streamline and tabbed specimens on unidirectional graphite

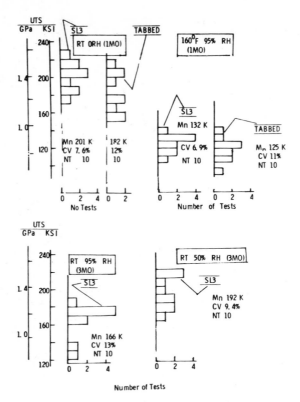

FIG. 13—*Tension test results on unidirectional S-glass/epoxy.*

and glass epoxies. In the case of Fig. 11, if attention is restricted to the well-machined specimens of the SL3 type (the open-circle data) the mean is considerably greater and the scatter less than that of the tabbed specimens. The same situation holds in Fig. 13 in the cases when both SL and tabbed specimens were used on similarly conditioned materials. In the case of Fig. 11 it is true that elimination of the two low points in the tab specimen data would reduce the scatter for that set, although the mean of the data would still be low compared with that of the SL3. However, no valid reason can be presently given for removing them, and until more data are obtained for this situation, the low points must be considered to be typical of the results that can be expected with tabbed specimens. In Fig. 13, the upper two sets of data for unidirectional S-glass epoxy show the SL3 to be again superior to the tabbed specimen. In this case two additional sets of data are included to illustrate that for some of the environmental conditioning situations to which this material was subjected, tab failures and losses of tabs became so severe a problem that the attempt to evaluate tabbed specimens for these cases had to be abandoned.

Results for crossply composites are shown on Fig. 14 and in Table 1. Figure 14 shows the distribution of specimen break locations for these tests, in the case of 0/90 deg graphite/epoxy. The ability of the SL3 specimen to focus breaks in the gage region for this material is clearly brought out here. Figure 15 shows a similar result for unidirectional graphite/epoxy materials, in this case pultruded materials which were 0.95 cm thick (0.375 in.). In the case of thinner unidirectional materials the focusing tendency of the streamline shape is not nearly so pronounced as it is for the specimens treated in Figs. 14 and 15.

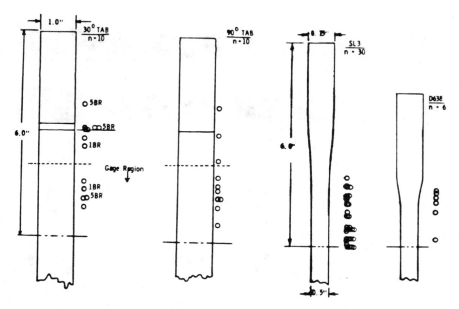

FIG. 14—*Specimen break locations in crossply T300 5208 graphite/epoxy tension tests.*

TABLE 1—*Test results on crossply composites.*

Material		3/4 SL1	SL1	SL3	D638	Tabbed		
						10-deg Bevel	30-deg Bevel	90-deg Bevel
$(0_2/90_2/0_3)_s$	No. of tests	19	29	43	25	10	8	9
SP250 E	Avg MPa	714	673	703	678	666	634	662
	(ksi)	(103.5)	(97.6)	(102.0)	(98.3)	(96.6)	(91.9)	(96.0)
Glass/epoxy	% Coefficient variation	4.1	3.9	3.2	5.3	4.5	4.9	4.8
$(0/90)_{45}$	No. of tests	9	...	43	6	...	9	10
T300-5208	Avg MPa	652	...	621	593	...	590	612
graphite/	(ksi)	(94.6)	...	(90.0)	(86.0)	...	(85.6)	(88.8)
epoxy	% Coefficient variation	8.3	...	8.5	3.6	...	6.6	5.2

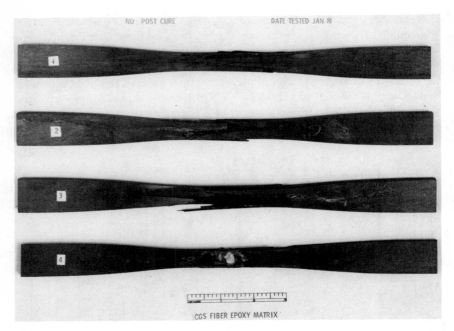

FIG. 15—*Specimen break characteristics in 0.9-cm-thick (0.375 in.) pultruded graphite/epoxy specimens.*

Data obtained with crossply graphite and glass/epoxies is given in Table 1. For the glass/epoxy materials the results obtained with the SL specimens were marginally better than those obtained with the tabbed specimens, although the difference is not great. Several specimen designs were treated here, including a 3/4-scale version of the SL1 (that is, all dimensions reduced by 25%) and tabbed specimens having three different tab angles. In terms of combined improvement in the mean of the data and reduction in the scatter, the SL3 specimen gave the best results for this $(0_{10}/90_4)$ crossply layup of SP250 Eglass. In the case of the tabbed specimens the results for the 10 and 90 deg beveled tabs were slightly better than those for 30 deg tabs, but somewhat less than those obtained for the SL3 specimens.

In the case of the crossply graphite/epoxy results obtained on bidirectionally reinforced T300 5208 with equal reinforcement in each direction, the results for the tabbed specimens look somewhat better, at least in terms of the scatter in the data. On the other hand, the SL specimens in this case represented an early stage development of machining techniques for the SL-type geometry, and it is quite likely that poor machining had a deleterious effect on the test results. The main advantage that can be claimed for the streamline shape in the case of crossply graphite/epoxy, at this point, is the ability to control failure locations as illustrated in Figs. 14 and 15.

Fatigue Test Results

Performance of the SL shape for fatigue testing is illustrated in Figs. 16 and 17. Figure 16 compares the use of D638 and SL specimens for generating S-N data on crossply glass/epoxies. Although the D638 is not normally used for fatigue testing, the lifetime scatter which it produced (about one decade) is not out of line with typical fatigue data for organic matrix composites. The lifetime scatter envelope for the SL, on the other hand, corresponded to only about one third of a decade. As shown in Fig. 17, the lack of heat generation in the SL specimen, even at cycling rates as high as 10 Hz, may account for the low scatter. Tabbed specimens are often used for fatigue testing of composites and are generally considered acceptable. However, the problems generated by tabs are exaggerated by the fatigue test environment, and the convenience of the SL specimen in avoiding such problems makes it especially attractive for fatigue testing. This is especially true in the case of Kevlar epoxy and metal matrix composites for which tab failures are even more of a problem than with other materials.

References 9-11 discuss other efforts in which streamline specimens have been used with good success in fatigue testing.

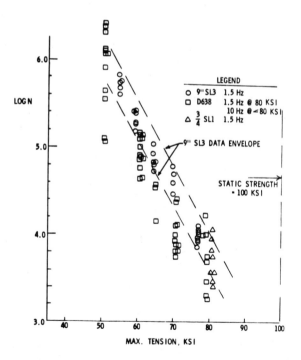

FIG. 16—*Fatigue test results in* $0_{10}/90_4$ *E-glass/epoxy: tension-tension,* R = 0.1, *various cycling rates.*

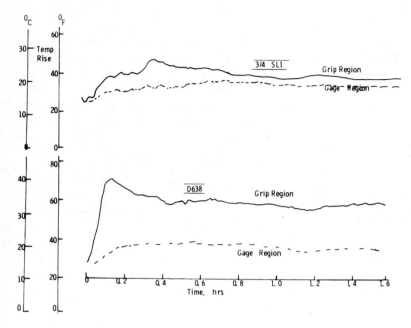

FIG. 17—*Comparison of specimen heating effects under cyclic loading.*

Conclusions

The streamline tensioin specimen shape appears to have great potential for tension testing of composite materials. Indeed it may well prove applicable to other types of material—ceramics and brittle plastics for example—where sensitivity to stress concentrations is excessive.

As a minimum, the streamline-type specimen offers a useful supplement to tabbed specimens where the latter are not suitable because of problems associated with testing at high temperature, high humidity, and fatigue testing.

Machining techniques for the streamline shape need continuing development. The most recently selected cutting tools appear to work well for unidirectional materials; there is a need to verify that they provide adequate performance for graphite/epoxy crossply composites. Except in the latter case, higher mean values, reduced data scatter, and better control of failure location are generally achieved with the streamline shape.

Acknowledgments

The finite-element program on which the results discussed in the theoretical section were obtained was provided by C. E. Freese.

A number of AMMRC materials testing engineers and students under the

Northeastern University Cooperative Program working at AMMRC provided considerable help and valuable suggestions regarding improvements in the streamline specimen and methods for machining it. In particular, P. Cavallaro developed computer programs for generating the data for use in machining Tensilkut templates. Bruce Andrews provided considerable guidance to students engaged in machining streamline specimens on in-house equipment. Among the students who helped to produce specimens were C. Cavallaro, M. Hochberg, R. Ivaskas, and A. Keturakis. Considerable help in data evaluation was provided by R. Pasternak and W. Bethany of AMMRC's Materials Property Branch.

Support in the preparation of this manuscript was provided by G. Zapata, P. Caira, and J. Considine. The first author also wishes to express his appreciation to his wife, Gloria Oplinger, for extensive support.

References

[1] Oplinger, D. W., Gandhi, K., and Parker, B. "Studies of Tension Test Specimens for Composite Material Testing," Army Materials and Mechanics Research Center Technical Report AMMRC TR 82-27, Watertown, MA, 1982.

[2] Dastin, S., Lubin, G. Munyak, J., and Slobodinski, A., "Mechanical Properties and Test Techniques for Reinforced Plastic Laminates" in *Composite Materials: Testing and Design, ASTM STP 460*, American Society for Testing Materials, Philadelphia, 1970, pp. 13–26.

[3] Lansard, R., "Filets without Stress Concentration" in *Proceedings*, SESA, Vol. 13, 1954, pp. 97–104.

[4] Peterson, R. E., *Stress Concentration Factors*, Wiley, New York, 1974, pp. 83–84.

[5] Heywood, B., *Designing by Photoelasticity*, 1st ed., Chapman and Hall, London, 1952.

[6] Hetenyi, M., "On Similarities Between Stress and Flow Patterns," *Journal of Applied Physics*, Vol. 12, 1941, pp. 592–595.

[7] Neuber, H., "Der eben Stomlinienspannungszustand mit Lastfreiem Rand," *Ingenieur Archiv*, Vol. 6, 1936, pp. 325–334.

[8] Walker, M., *The Schwartz-Christoffel Transformation and its Applications—A Simple Exposition*, Dover, New York, 1964, pp. 53–65.

[9] Roylance, M. E., "The Effect of Moisture on the Properties of an Aramid Epoxy Composite," Ph.D. thesis, Massachusetts Institute of Technology, Cambridge, MA, Sept. 1980.

[10] Foley, G., Roylance, M., and Houghton, W., "Life Prediction of Glass Fiber Composites Under Cyclic Loading" in *Proceedings*, International Conference on Advances in Life Prediction Methods, American Society of Mechanical Engineers, New York, 1983.

[11] Tsangarakis, N., et al, this publication, pp. 131–152.

Ronald P. Nimmer[1] *and Gerald G. Trantina*[2]

Design for Ultimate Strength with a Chopped Glass-Filled Phenolic Composite: Critical Flaw Sensitivity, Residual Stress, Fiber Orientation

REFERENCE: Nimmer, R. P. and Trantina, G. G., **"Design for Ultimate Strength with a Chopped Glass-Filled Phenolic Composite: Critical Flaw Sensitivity, Residual Stress, Fiber Orientation,"** *Recent Advances in Composites in the United States and Japan, ASTM STP 864*, J. R. Vinson and M. Taya, Eds., American Society for Testing and Materials, Philadelphia, 1985, pp. 556–582.

ABSTRACT: Because they retain strength and stiffness at temperatures as high as 180°C (356°F), injection molded, glass-filled phenolic composites have been used in structural applications. However, the strength of this material as measured with ASTM three-point flexure tests may be as much as 80% higher than the strength measured in tension tests. The objects of this investigation were to study the relative effects of fiber alignment, residual stress, and flaw distribution upon these measured values of strength and to assess the application of these data to the design process. There was no differential alignment of fibers observed between skin and core areas of the specimens which might lead to elevated flexural strengths. Residual compressive stresses between 7 MPa (1.0 ksi) and 14 MPa (2.0 ksi) were measured at the surface of the bend specimens. Finally, although the tensile strength data are well represented by a two-parameter Weibull model, there remains a 30% unexplained discrepancy between predicted and measured flexural strength.

KEY WORDS: ultimate strength, glass-filled phenolic, injection molded, short glass fibers, critical flaw sensitivity, Weibull statistical analysis, size effect, residual stresses, tension tests, flexure tests, fiber orientation

Many applications making use of chopped fiber composite materials require design with strength as a primary criterion. In such instances, three- and four-point bend tests are often used to define ultimate strength on the basis of experimental simplicity. However, it is often observed that the failure

[1]Mechanical engineer, General Electric Co., Corporate Research and Development, Schenectady, NY 12301.
[2]Manager, CAD Technology, General Electric Co., Plastics Technology Center, Pittsfield, MA.

limits defined by such bend tests are significantly higher than those measured in tension tests, leading to uncertainty in the appropriate definition of ultimate strength for the designer.

There are a number of mechanisms which could contribute to this observed discrepancy, including fiber orientation, residual stresses, and critical flaw distribution. Taggert et al [1] examine the standard ASTM Test Methods for Tensile Strength of Molded Electrical Insulating Materials (ASTM D 651-80) and Flexural Properties of Unreinforced and Reinforced Plastics and Electrical Insulating Materials (D 790-71), and discuss observed differences between Young's moduli and ultimate strength as measured by these two specimens. The primary focus of their work is fiber orientation effects. With respect to the flexure test, they point out that the chopped fibers are often oriented in the direction of the resin flow near free surfaces of the flexure specimen, thus giving rise to unrealistically high values of both Young's modulus and ultimate strength. However, even when proper account is made for this effect, predictions in Ref 1 for the ultimate failure of the flexure specimens based upon tension data are still considerably less than experimentally observed values. A second mechanism which could contribute to the higher apparent strength measured in flexure tests is a residual stress distribution. It is well known that the differential rates of cooling and solidification in molded plastics often create substantial compressive residual stresses near free surfaces and thus lead to higher failure loads. Distributions of this nature have been measured in a number of plastics [2,3].

Finally, the statistical distribution of inherent flaws in brittle materials has also been identified as a contributor to the observed difference in flexural and tensile strength. The Weibull statistical method [4] which provides the framework for this approach has been successfully applied to ceramic materials (see Ref 5 for example). More recently, its applicability has also been investigated with regard to composite materials. Bullock [6] reported significant success in applying a two-parameter Weibull model to explain differences in strengths measured by tension tests on strands of graphite fiber as well as unidirectional, composite, tension, and flexure specimens. On the other hand, Whitney and Knight [7] concluded that the results of their tests using tension specimens as well as three- and four-point flexure coupons made of unidirectional, graphite-epoxy composite did not correlate well with a two-parameter Weibull statistical failure model. Knight and Hahn [8] investigated the applicability of the technique with respect to randomly distributed, short fiber composites. Their results also exhibit some inconsistency in uniformly explaining these differences.

Although results in applying the two-parameter Weibull statistical model for failure have been mixed, its usefulness as a design tool is significant where it is shown to be applicable. In such cases, the observed differences in strength measured in tension and flexure tests are indicative of the variations of strength which could be expected in product components due to differences in physical size or induced stress distribution. Halpin et al [9] discuss techniques to account for these variations in engineering design within the framework of the Weibull statistical method, thus improving product reliability. The present investigation was undertaken to study the applicability of these techniques for design with injection molded, fiber-filled phenolic material. Such an investigation requires that the potential effects of flow-induced fiber orientation and residual stress also be carefully considered.

Specimens from both tension and bend tests were examined with an optical microscope to assess the degree of alignment induced in the test specimens. Bend tests were performed on full as well as sectioned bend specimens to quantify the effects of alignment upon observed Young's modulus. Regarding residual stresses, a method of successive material removal was applied to the bend specimens to define existing distributions in the flexural specimens. Finally, a two parameter Weibull model was applied to the tension test data and then used to predict the expected strength in flexure tests.

Material Description and Strength Data

In all the tests which are discussed, the material was a phenolic resin filled with glass and organic fibers as well as a smaller percentage of miscellaneous solid additives. By weight, the composite is 42% fiber and solid additive filled. Both the tension coupons as well as the bend specimens were injection molded.

The tension tests were conducted at room temperature using the 108-mm (4.32 in.) short, injection molded dogbone specimen specified in ASTM Method D 651-80 and illustrated in Fig. 1. The crosshead rate was 1.25 mm/min (0.05 in./min) in all tests, leading to a strain rate over the 0.95-cm (0.38 in.) gage section of $9.0 \times 10^{-5} \, s^{-1}$. The injection molded bend specimens were 12.7 cm (5 in.) long, 1.27 cm (0.5 in.) wide, and 0.64 cm (0.25 in.) thick. These specimens were loaded at room temperature in three-point bend configuration, also shown in Fig. 1, at crosshead rates of 2.54 cm/min (1.0 in./min), 0.254 cm/min (0.1 in./min), and 0.0254 cm/min (0.01 in./min). These crosshead rates translate to strain rates of $1.56 \times 10^{-3} \, s^{-1}$, $1.56 \times 10^{-4} \, s^{-1}$, and $1.56 \times 10^{-5} \, s^{-1}$, respectively.

Fifty tension test coupons were tested to failure. Although Ref 1 reports that tension tests using this coupon often result in grip failure due to fiber alignment effects, 42 of the 50 specimens failed in the gage section. An additional five specimens broke in the transition area and only three specimens broke in the grips. The average strength based on 47 specimens (gage and transition breaks) was 56.9 MPa (8250 psi) with a standard deviation of 4.5 MPa (650 psi). On the other hand, the average strength as measured from the three-point bend test was 105.7 MPa (15 350 psi) with a standard deviation of 4.0 MPa (575 psi) for the 2.54-cm/min crosshead rate and 104.0 (15 080 psi) with a standard deviation of 4.5 MPa (650 psi) for a crosshead rate of 0.254

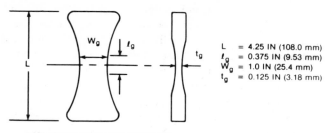

L   = 4.25 IN (108.0 mm)
t_g = 0.375 IN (9.53 mm)
W_g = 1.0 IN (25.4 mm)
t_g = 0.125 IN (3.18 mm)

ASTM D 651-80 TENSILE SPECIMEN

L = 5 IN (127.0 mm)
l = 4 IN (101.6 mm)
b = 0.5 IN (12.7 mm)
h = 0.25 IN (6.35 mm)
r = 0.1875 IN (4.76 mm)

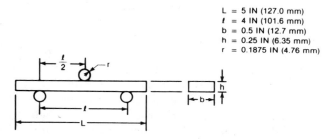

FIG. 1—*Tension and three-point bend specimens.*

cm/s. At a crosshead rate of 0.0254 cm/s, the average strength decreased to 96.8 MPa (14 040 psi) with a standard deviation of 5.4 MPa (775 psi). For the bend tests, the stress at failure was calculated as

$$\sigma_f = \frac{3P_f L}{2bd^2} \qquad (1)$$

where

P_f = failure load,
L = span between supports,
b = beam width, and
d = beam depth.

Fiber Orientation

The presence and importance of flow-induced orientation in specimens and components is well established. McCullough [*10*] has suggested an approach for quantifying this orientation and relating it to Young's modulus, shear modulus, and Poisson's ratio. Furthermore, in their discussion of ASTM tension and bend tests, Taggert et al [*1*] present photomicrographs taken from

the cross section of a bend specimen which clearly illustrate the pronounced variations in fiber alignment which can occur within molded parts. These photomicrographs clearly show fiber alignment in the direction of the specimen length (also the flow direction) near the free surfaces of the flexure bar, and fibers randomly oriented in the plane transverse to flow within the central region. The existence of such orientation will obviously invalidate the application of simple beam equations for the determination of Young's modulus and ultimate strength (such as Eq 1), as pointed out in Ref *1*. Therefore, a series of photomicrographs was taken of both tensile and flexural cross sections to qualitatively identify the presence of any such orientation in the specimens used for this investigation. In addition, a series of mechanical tests was performed to quantify the effect of spatially varying orientation in the flexure specimens.

The photomicrographs in Figs. 2a and 2b were taken from a cross section of the tension coupon's gage area with a line of view parallel to the load axis of the specimen. The small-diameter lighter color fibers are 14-μm-diameter glass fibers with original lengths prior to injection molding of approximately 3 mm (0.12 in.). The darker sections in the micrograph are organic fibers which seem to remain bundled in all the observed sections. Figure 2a is from a location immediately adjacent to the specimen's surface and Fig. 2b is from an interior location. Figures 3a and 3b, on the other hand, are tension specimen cross sections with a line of view perpendicular to the load axis. Figure 3a is a photomicrograph taken of an area adjacent to the surface and Fig. 3b of an interior area. As would be expected in a converging section, there is a definite alignment of flow. However, no apparent difference in the general orientation of fibers was observed through the thickness of a specimen.

Similar photomicrographical studies were made on the flexure specimens. As pointed out in Ref *1*, axial alignment of fibers near the surfaces of the beam significantly affects stiffness and strength measurements based upon homogeneous material assumptions. In a fashion similar to the tension coupons, Figs. 4a and 4b are photomicrographs of a beam cross section with line of view along the flexure coupon's length. Figure 4a represents material adjacent to a free surface and Fig 4b material from the specimen's center. Figures 5a and 5b also show material near a free surface and the interior respectively, but with the line of view perpendicular to the flow direction. Figures 4 and 5, as well as other photomicrographs which were studied, clearly show preferred fiber orientation along the length of the beam, but comparisons of material adjacent to a free surface with material from the interior show no obvious variations in alignment through the thickness as observed in Ref *1*.

A somewhat more quantitative comparison of fiber orientation at different locations in the flexural specimen can be obtained from measurements of the major axis of the elliptical fiber cross sections visible in Figs. 4 and 5. From simple trigonometry, glass fiber cross sections within 10% of circularity can be shown to have fiber axes oriented 10 deg or less to the line of sight in the

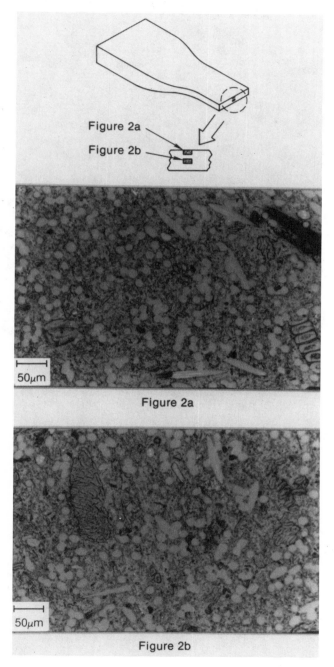

FIG. 2—*Photomicrographs of tension specimen cross section perpendicular to mold flow direction:* (a) *adjacent to free surface;* (b) *central region of specimen.*

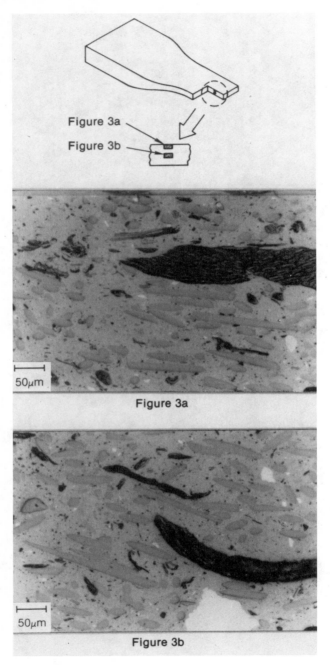

FIG. 3—*Photomicrographs of tension specimen cross section parallel to mold flow direction:* (a) *adjacent to free surface;* (b) *central region of specimen.*

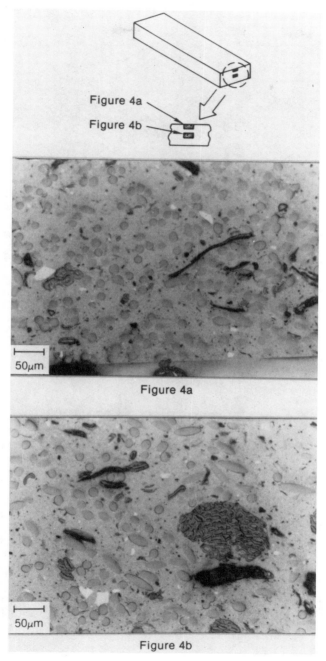

FIG. 4—*Photomicrographs of bend specimen cross section perpendicular to mold flow direction:* (a) *adjacent to free surface;* (b) *central region of specimen.*

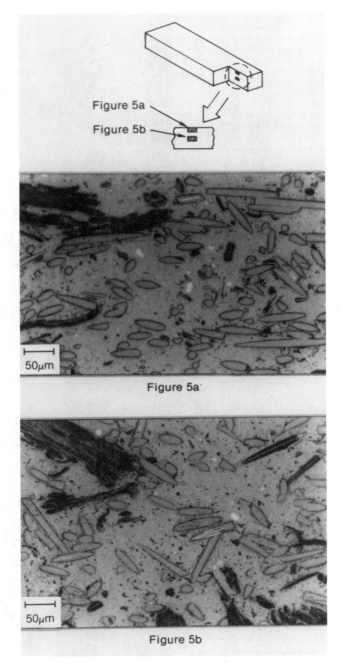

FIG. 5—*Photomicrographs of bend specimen cross section parallel to mold flow direction:* (a) *adjacent to free surface;* (b) *central region of specimen.*

photomicrograph. If the major axis of a cross section appearing as an ellipse is more than 100% larger than the near circular cross sections, the associated fiber is oriented at an angle of more than 60 deg to the line of sight. Using this technique with a limited number of photomicrographs of cross sections perpendicular to the flexural specimen's length indicates that 60% to 65% of the glass fibers in an area adjacent to a free surface are oriented within 10 deg of the axis of the beam. An additional 30% to 35% of the fibers are oriented between 10 and 60 deg to the beam's axis. Less than 5% of the fibers are oriented at an angle of more than 60 deg to the axis. Similar measurements made near the center of the beam showed only slightly different results. Here, approximately 50% of the fibers are within 10 deg of the beam axis and another 30% were between 10 and 60 deg of the axis.

As a final measure of the fiber orientation effect in the flexure specimens, flexural moduli were measured from ASTM molded specimens as well as 2.25 by 6.35 by 63.5-mm (0.09 by 0.25 by 2.5 in.) specimens cut from the original bars and shown in Fig. 6. If significant fiber orientation takes place in the boundary layer shown in Fig. 6, then the flexure specimen is inhomogeneous with different material properties in the core and the surface layer as noted in Ref *1*. For a three-point bend test, the relationship between load and displacement is

$$\delta = \frac{P l^3}{48(EI)_c} \qquad (2)$$

However, if the cross section of the beam is layered as in Fig. 6, then the quantity (EI) is a "composite" section modulus, as noted by the subscript c. In the case of the full ASTM bend specimen shown as Configuration 1 in Fig. 6, this section modulus can be written as

$$(EI)_1 = E_c \frac{1}{12} bh^3 \left\{ 1 + \left[\frac{E_s}{E_c} - 1 \right] \left[7\left(\frac{t_s}{h}\right) - 18\left(\frac{t_s}{h}\right)^2 \right. \right.$$

$$\left. \left. + 20\left(\frac{t_s}{h}\right)^3 - 8\left(\frac{t_s}{h}\right)^4 \right] \right\} \qquad (3)$$

where

E_c and E_s = Young's moduli of the core and skin layer, respectively,
$\quad t_s$ = skin layer thickness, and
$\quad h$ = full depth of the ASTM beam.

On the other hand for the smaller bend bar, machined from the full specimen and tested in Configuration 2 of Fig. 6, the section modulus is

$$(EI)_2 = E_c \frac{1}{12} (h) \left(\frac{b}{4}\right)^3 \left[1 + 2\left(\frac{E_s}{E_c} - 1\right)\left(\frac{t_s}{h}\right)\right] \qquad (4)$$

Now, if the load deflection relationships for Configurations 1 and 2 are measured and composite section moduli calculated from Eq 2, then an "effective" modulus can be calculated for each test configuration by dividing the right-hand side of Eqs 3 and 4 by the second moments of inertia of the respective sections. The ratio of the two effective moduli can then be expressed as

$$\frac{(E_{eff})_1}{(E_{eff})_2} = \left[1 + 7\left(\frac{E_s}{E_c} - 1\right)\left(\frac{t_s}{h}\right) - 18\left(\frac{E_s}{E_c} - 1\right)\left(\frac{t_s}{h}\right)^2 + 20\left(\frac{E_s}{E_c} - 1\right)\left(\frac{t_s}{h}\right)^3\right.$$

$$\left. - 8\left(\frac{E_s}{E_c} - 1\right)\left(\frac{t_s}{h}\right)^4\right] \div 1 + 2\left(\frac{E_s}{E_c} - 1\right)\left(\frac{t_s}{h}\right) \qquad (5)$$

This ratio of effective moduli as determined by test reflects the significance of the surface layer effect in the flexural bars and is dependent upon both the nondimensional surface layer thickness and the ratio of Young's moduli for the surface and core layers. If there is no surface layer orientation, the ratio of

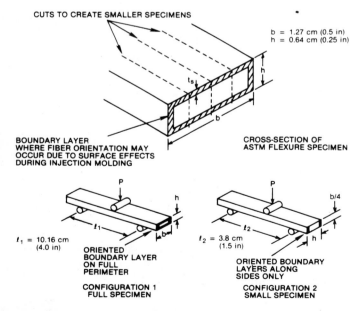

FIG. 6—*Flexure specimen cross sections and test configurations used to identify fiber orientation effects.*

effective moduli in Eq 5 must be 1.0. The presence of oriented fibers in a surface layer will cause the ratio in Eq 3 to be greater than 1.

Five specimens were tested in each configuration shown in Fig. 6, resulting in an average effective modulus of 10.0 GPa (1.47 × 10⁶ psi) for the full ASTM bar (Configuration 1) and an effective modulus of 9.51 GPa (1.38 × 10⁶ psi) for the smaller bar tested in Configuration 2. The standard deviation of these tests was 0.69 GPa (0.1 × 10⁶ psi). Figure 7 helps illustrate the significance of the 1.07 value for the effective modulus ratio as measured in these tests. Using Eq 5, each curve in Fig. 7 defines the relationship between the ratio of surface and core Young's moduli (E_s/E_c) and the nondimensional surface layer thickness (t_s/h) for a measured effective modulus ratio. Curves representing effective moduli ratios from 1.0 to 1.3 are included in the figure. As can be seen, an effective modulus ratio of 1.07 requires that the ratio E_s/E_c be less than 1.2 for most values of nondimensional layer thickness. The E_s/E_c ratio becomes significant in this case only if the oriented layer is less than 2% of the original beam depth or more than 40%. Based on these results in conjunction with observations from the photomicrograph studies, the effect of differential orientation of fibers at the surface and core of the specimens considered here appears to be minimal and is ignored in the residual stress and Weibull statistical studies which follow. For comparison, the point describing

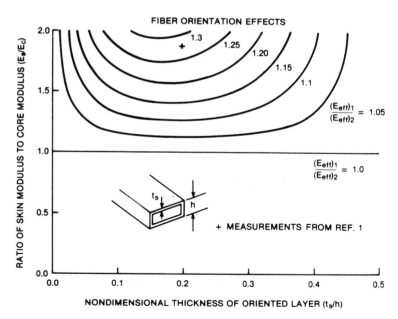

FIG. 7—*Allowable skin-to-core modulus ratios and nondimensional oriented layer thicknesses for observed ratios of effective moduli measured in Configurations 1 and 2.*

surface layer thickness and skin-to-core modulus ratio for the material studied in Ref *1* is also shown in Fig. 7. In this case, the authors observed a well-defined surface layer with a thickness 20% of the beam depth and a Young's modulus of 17.8 GPa (2.58 $\times$ 10⁶ psi) in the oriented layer compared to 9.58 GPa (1.39 $\times$ 10⁶ psi) in the core region. Had flexure tests as described here been applied to such a specimen, the ratio of effective moduli measured from the two tests would have been 1.27. Differences in material and processing parameters probably account for the difference in orientation observed in Ref *1* and the present investigation. Therefore, in the following discussions of residual stress and "size effect," it is assumed that differential fiber orientation across the specimens is negligible.

Residual Stresses

In addition to flow-induced fiber orientation, the molding process can also introduce residual stresses in parts and specimens. Stresses of this nature are usually large and compressive in the vicinity of the specimen surfaces with lower, tensile stresses in the interior. In a flexure test, the ultimate strength is calculated from the outer fiber stress at failure load. If failure in this brittle material is assumed to be due to tensile stress, then the compressive residual stresses at the surface would lead to erroneously high failure predictions.

In order to quantify this effect, the technique of successive removal of material coupled with measurement of the associated "bowing" of the rectangular flexural coupon was applied to the ASTM flexural coupons. This technique is described in detail by Treuting in Ref *11*. It is assumed that the material is homogeneous and that the residual stress distribution is independent of the length coordinate of the bar. Using Euler-Bernouli beam theory, the stress σ in a layer of material before removal is calculated as the sum of three components

$$\sigma = \sigma_a - \sigma_b - \sigma_c \tag{6}$$

where

$$\sigma_a = \frac{4}{3} \frac{Eh^2}{L^2} \frac{df}{dh} \tag{7}$$

$$\sigma_b = \frac{4Ehf}{L^2} \tag{8}$$

$$\sigma_c = \frac{1}{h} \int_h^{h_0} (\sigma_a - \sigma_b - \sigma_c)dh \tag{9}$$

In Eqs 6–9, f is the bow in the beam over the span L after the beam thickness has been reduced to thickness h and h_0 is the original thickness of the beam.

Measurements were taken on seven standard ASTM flexure specimens, 6.35 mm (¼ in.) thick, with 0.40 mm (¹/₆₄ in.) of thickness removed during each machining step until the specimen thickness was reduced to approximately 3.2 mm (⅛ in.). In order to account for the potential of asymmetry in the residual stress distribution while maintaining respectable accuracy in measurement, material was removed from the "gated" side of the molded specimen in four tests and the opposite side in the remaining three tests. The molding gate was used to define the side of the specimen.

Figure 8 illustrates the calculated residual stress distributions in the six specimens. As can be seen, these stresses are compressive at the edges of the specimen and tensile in the central region. Since the maximum bending stress occurs at the beam surface, the compressive residual stresses will raise the apparent strength of the bend specimens. The average nondimensional residual stress (σ_R/E) at the outer fiber, as measured in these six tests, was -0.09×10^{-2} with values ranging from a minimum of -0.0675×10^{-2} to a maximum of -0.1325×10^{-2}. Using Young's modulus of 10.1 GPa (1.47×10^6 psi) as measured earlier in the three-point bend tests of the full ASTM bars, the average outer fiber residual stress was 9.17 MPa (1330 psi) with minimum and maximum values of 6.82 MPa (990 psi) and 13.4 MPa (1950 psi), respectively. The two highest compressive residual stresses were measured on the

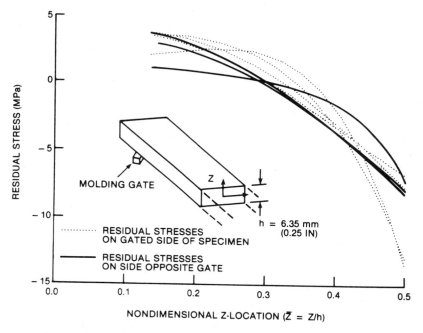

FIG. 8—*Residual stress distributions in injection-molded flexure coupons.*

"gated" side of the specimen. This is the side which experienced tensile stresses in all of the flexure tests conducted on the full ASTM bend bars.

With regard to measurement error, the nondimensional outer fiber residual stress can be simply expressed as

$$\frac{\sigma_R}{E} = \frac{4}{3} \frac{h^2}{L^2} \frac{\Delta f}{\Delta t} \tag{10}$$

Assuming that the thickness and bow measurements could be made to within 2.54×10^{-2} mm (0.001 in.), the average nondimensional residual stress will have a measurement error of $[(f_e/f) + (\Delta t_e/\Delta t)]$ where f is the measured bow after removing a layer of material Δt thick and f_e and Δt_e are the measurement errors in f and Δt, respectively. This, in turn, leads to an approximate error in the nondimensional residual stress of about $\pm 15\%$. In addition, the standard deviation of the Young's modulus as measured in flexure was 0.69 GPa (0.1×10^6 psi) for a mean value of 10.1 GPa (1.47×10^6 psi). Accounting for this variation, an approximate error bound on the outer fiber residual stress measured in these experiments is about $\pm 20\%$.

Critical Flaw Distribution Effects

Another mechanism which has often been successfully applied to explain differences in strengths as measured by tension and flexure tests of brittle materials is a statistical theory based upon a Weibull distribution of strength. Within the context of this theory, the observed discrepancy between strengths as measured by tension and flexure specimens can provide significant information regarding the effect of size and stress distribution upon failure in product components. Where it applies, this statistical approach provides an extensive methodology for determining component reliability based upon laboratory strength tests.

The fundamental basis for the Weibull failure model is the assumption that failure can be characterized by linear elastic fracture mechanics. In such a case, the failure stress, σ_f, can be expressed as

$$\sigma_f = \frac{K_c}{Y\sqrt{a}} \tag{11}$$

where

K_c = critical stress-intensity factor,
a = crack length, and
Y = factor dependent upon geometry.

Although K_c is a material property, there will be an inherent range of flaw sizes (a) in any set of specimens. This variation in flaw size leads to a corresponding variation in failure strength σ_f, as a result of Eq 11. The weakest-link model due to Weibull can be applied to characterize this distribution of strength and associated flaw size. In this model, the probability of failure, P, is expressed as

$$P = 1 - e^{-R} \tag{12}$$

where R is referred to as the risk of rupture and can be defined by either a surface or a volume integral

$$R_s = \int_S (\sigma/\sigma_0)^m \, dS = kS\left(\frac{\sigma_{max}}{\sigma_0}\right)^m \tag{13}$$

$$R_v = \int_V (\sigma/\sigma_0)^m \, dV = kV\left(\frac{\sigma_{max}}{\sigma_0}\right)^m \tag{14}$$

The former relation is based upon the assumption that the flaws associated with failure are distributed over the free surfaces of a component while the latter assumes distribution over the entire volume. The variable σ_{max} in Eqs 13 and 14 is the maximum stress in a specimen or component at failure. The exponent m is the Weibull modulus or shape factor and σ_0 is a normalizing constant. The Weibull modulus is a material property and is a measure of the scatter in the material's strength distribution. Smaller values of m are indicative of larger amounts of scatter in strength data.

The Weibull modulus m and reference stress σ_0 can be determined by substituting either Eq 13 or 14 into Eq 12 and rearranging into the form of

$$\ell n \, \ell n \, \frac{1}{(1 - P)} = m \, \ell n \, \sigma_{max} - (m \, \ell n \, \sigma_0 - \ell n \, kV) \tag{15}$$

In Eq 15, the probability of failure $P(\sigma_i)$ at or below a stress of $\sigma_{max} = \sigma_i$, can be defined from test data as

$$P(\sigma_i) = \frac{i - 0.5}{N} \tag{16}$$

where i is the ordering number of specimen failure (from lowest to highest strength) associated with σ_i and N is the total number of specimens. The experimental data defining P as a function of σ_i can then be least-squares fit to the linear relationship of Eq 15, thus defining m and σ_0.

Examination of Eqs 13 and 14 reveals that the probability of failure in a component at a given stress is dependent upon both the size (volume or surface area) and the stress distribution. Once the Weibull distribution relating stress σ_1 to probability of failure P has been defined for one set of test specimens, the stress, σ_2, associated with a given probability of failure in any other test specimen or component, can be determined using Eqs 13 or 14 in the form

$$(\sigma_2) = \sigma_1 \left(\frac{kV_1}{kV_2} \right)^{1/m} \tag{17a}$$

$$(\sigma_2) = \sigma_1 \left(\frac{kS_1}{kS_2} \right)^{1/m} \tag{17b}$$

Thus, by using Eqs 17a or 17b, the differences in size and stress distribution of test configurations can be statistically accounted for under the assumption of linear elastic fracture behavior.

The tension test specimen (illustrated in Fig. 1) and test technique described in ASTM D 651-80 were used to collect the statistical data for the Weibull model. Fifty specimens were tested to failure at a crosshead speed of 0.05 in./min (0.002 mm/min), resulting in test durations to failure of approximately 1.00 min. As can be seen from the typical load-displacement curve measured during these tests and displayed in Fig. 9, after the slack was taken up in the grips, the material response was quite linear. The data at the far left of Fig. 10 illustrate the distribution of strength observed in the tension tests. In addition, the least-squares fit of these data to a two-parameter Weibull model is also shown. The Weibull modulus, m, as determined from these data was $m = 15$ and the average net section stress at failure was 56.9 MPa (8250 psi). For a sample size of 50, Ref 12 indicates that the potential error in the statistically measured Weibull modulus is $\pm 20\%$. Although Ref 1 points out that results from finite-element analyses of this tension specimen show that the maximum stress is actually 8% higher than the net section stress, this difference proves to be negligible in the following calculations and has been neglected.

Having determined the Weibull modulus, m, from the tension tests, Eqs 17a and 17b can now be applied to predict the average strength and distribution for the three-point flexural specimens. For the tension specimens, the effective volume kV_T and the effective surface area kS_T are the actual volume and surface area of the gage section and are expressed simply as

$$kV_T = l_T b_T t_T \tag{18}$$

$$kS_T = 2l_T(b_T + t_T) \tag{19}$$

where l_T, b_T, and t_T are the length, width, and thickness of the gauge section, respectively. Because the three-point flexure test exhibits linear stress distri-

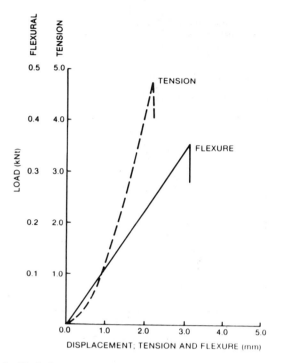

FIG. 9—*Typical load-displacement curves for tension and flexure tests.*

butions both along the length and through the thickness, it can be shown that the effective volume and surface area can be expressed as

$$kV_F = \frac{1}{2(m + 1)^2} \, l_F b_F h_F \tag{20}$$

$$kS_F = \frac{b_F l_F + (l_F h_F)/(m + 1)}{(m + 1)} \tag{21}$$

where

b_F and h_F = width and thickness of the specimen,
l_F = length between supports, and
m = Weibull modulus.

Although a residual stress distribution will alter these formulas, the resulting variation in predicted flexural strength was found to be less than 2%. For injection molded specimens tested without additional machining, one would expect flaws to be equally distributed over the entire volume of the specimen, thus making the volume model most appropriate.

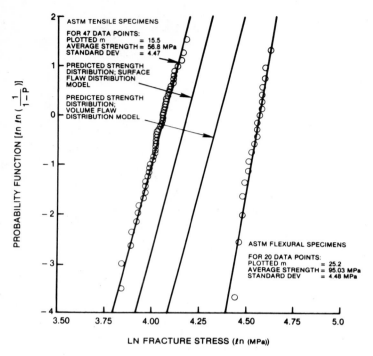

FIG. 10—*Measured and predicted Weibull distributions of strength for tension and flexure coupons.*

Table 1 illustrates the results of using the Weibull statistical data based upon the tension tests to predict the average strength of a three-point flexure specimen using both the surface and volume models. These predictions are compared with the average strength of 20 three-point flexure tests (ASTM D 790-71). For these tests, a crosshead rate of 2.54 mm/min (0.1 in./min) was applied, resulting in a time-to-failure of approximately one minute—a loading rate similar to the tension tests. The right-hand column of Table 1 reports the average flexural strength without correction for residual stresses. The adjacent column reports the average flexural strength assuming that an average residual compressive stress of 8.96 MPa (1300 psi) exists at the outer surface as previously measured in this investigation. As expected, the volume model for flaw distribution results in predictions closer to measured values from the flexure tests. However, the average flexural strength, corrected for residual stress, of 94.5 MPa (13.7 ksi) is still 29% higher than predictions based upon the tension test results. Figure 10 illustrates the distribution of measured flexural strength (including residual stress effects) compared with the distributions predicted from the statistical analysis of the tension data using both volume and surface models. In addition to the discrepancy in aver-

TABLE 1—*Comparison of measured and predicted average strengths.*

Observed Avg Tensile Strength	Predicted Avg 3-Point Flexural Strength; Surface Flaw Distribution Model	Predicted Avg 3-Point Flexural Strength; Volume Flaw Distribution Model	Observed Avg 3-Point Flexural Strength Corrected for Residual Stresses	Observed Avg 3-Point Flexural Strength Calculated from $\sigma_f = 3P_fL/2db^2$
56.9 MPa (8.2 ksi)	64.2 MPa (9.3 ksi)	73.6 Mpa (10.7 ksi)	95.45 MPa (13.7 ksi)	104.0 MPa (15.1 ksi)

age strengths, the flexure data exhibit less scatter than the tension data, which are also plotted in Fig. 10. Compared with a Weibull modulus of 15 for the tension tests, the modulus derived from fitting the flexure data to a two-parameter Weibull modulus is $m = 25$.

The average strength of brittle materials may be dependent upon the rate at which the test is conducted. If the rate of loading is slow, then subcritical crack growth may be possible. Since the material's statistical strength is related inversely to the distribution of crack sizes through fracture mechanics principles, subcritical crack growth during slow loading would be expected to lower average strength. Two additional sets of flexure tests were conducted to investigate this potential effect. The crosshead rates for these tests were 0.254 mm/min (0.01 in./min) and 25.4 mm/min (1.0 in./min), an order of magnitude slower and faster than the previous flexure tests. Figure 11 shows the average strengths of the three sets of flexure tests as a function of strain rate. Twenty standard ASTM flexure specimens were tested at each rate. The strengths reported in Fig. 12 include consideration of the average residual stress of 8.95 MPa (1300 psi). As can be seen from the figure, although the average strength at a strain rate of 1.5×10^{-4} (s^{-1}) was not significantly different from that at 1.5×10^{-3} (s^{-1}), there was a 7% reduction in average strength at the slowest strain rate, of 1.5×10^{-5} (s^{-1}). In addition, the Weibull modulus describing the data collected at a strain rate of 1.5×10^{-5} (s^{-1}) was $m = 17$ compared with $m = 25$ at 1.5×10^{-4} (s^{-1}) and $m = 29$ at 1.5×10^{-3}. Figure 12 illustrates the differences in Weibull strength distributions for the strain rates of 1.5×10^{-4} (s^{-1}) and 1.5×10^{-5} (s^{-1}).

With brittle materials, even careful machining can often introduce flaws which will have a substantial effect upon the average strength and scatter of strength data. To investigate this effect in the filled phenolic material under consideration here, standard ASTM flexural specimens were cut into smaller specimens as shown in Fig. 6. In the process, eight small flexure specimens nominally measuring 63.5 mm (2.5 in.) in length, 6.35 mm (0.25 in.) in width, and 2.05 mm (0.08 in.) in thickness were created from a single molded bar. Four of the smaller specimens cut in this matter then have one molded surface and one machined surface and four have two machined surfaces. From the group of specimens with two different surface types, 15 were tested

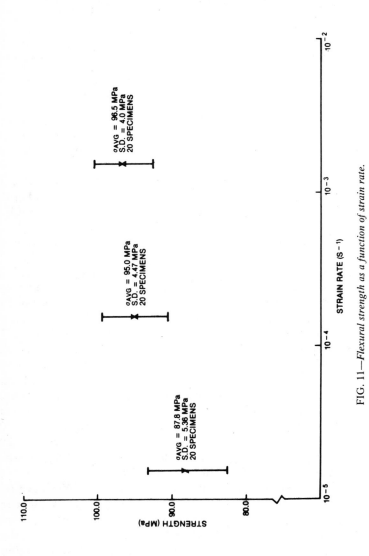

FIG. 11—*Flexural strength as a function of strain rate.*

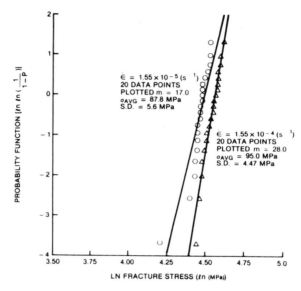

FIG. 12—*Measured distributions of strength for flexure tests at strain rates of 1.55 × 10⁻⁵ (s⁻¹) and 1.55 × 10⁻⁴ (s⁻¹).*

with the machined surface in tension and 15 were tested with the molded surface in tension. Forty specimens with two machined surfaces were also tested. In all cases the crosshead rate was 2.54 in./min (0.1 in./min), leading to failure in approximately 25 s. Figure 13 illustrates the marked difference in both average strength and Weibull modulus between the data collected when the specimens with one molded and one machined surface were tested with alternate sides in tension. The average strength with the molded surface in tension was 101.3 MPa (14.7 ksi) compared with 88.2 MPa (12.8 ksi) when the machined surface was in tension. Of perhaps even more significance is the fact that the calculated Weibull modulus of the strength data taken with the machined surface in tension was only 7.6 compared with 36.4 when the molded surface was in tension. This marked difference in the Weibull modulus and associated scatter in strength indicates that a very different and more critical distribution of flaws has been introduced in the specimens through the machining process. For high reliability, a component must be designed on the basis of the low-strength regime of the Weibull distribution. This requirement leads to substantial penalties for a machined material characterized by a Weibull modulus of 7.6. For comparison, Fig. 14 illustrates the strength distribution for the flexure tests performed on specimens with two machined surfaces. These data are very similar to the data in Fig. 13 for the machined surface in tension and reflect an average strength of 84.1 MPa (12.2 ksi) and a Weibull modulus of 5.3.

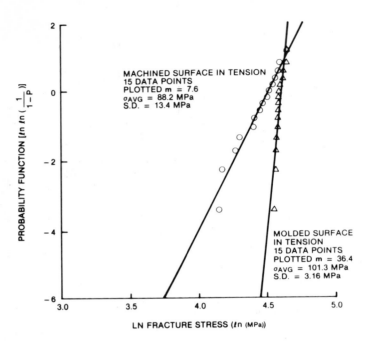

FIG. 13—*Measured distributions of strength for flexure tests with either molded or machined surfaces in tension.*

Discussion of Results

Figure 15 helps illustrate the importance of the Weibull statistical methodology to design with a material which is well characterized by such a model. In Fig. 15, the log of relative strength in a given component is plotted versus the log of relative effective volume or area parameter using Eqs 17a or 17b. The axes are nondimensionalized so that the relative strength is 1.0 for a relative effective area of 1.0. It can be seen from this figure that if the effective volume or area of a component is two orders of magnitude larger than that of the test specimen used to establish strength data, then for the same probability of failure with a Weibull modulus of 15, the maximum stress in the component should be only slightly more than 70% of the stress in the test specimen. If the Weibull modulus is 5, then the component stress should be approximately 40% of the test specimen stress for the same probability of failure.

For example, consider the rotating disk with central hole shown in Fig. 16 with a bore radius r_i of 2.54 cm (1.0 in.), an outer radius r_0 of 7.62 cm (3.0 in.), and a length L of 5.08 cm (2 in.). The hoop stress in such a disk can be expressed as

$$\sigma_\theta = \frac{3+\nu}{8} \rho(\omega r_0)^2 \left[1 + \left(\frac{r_i}{r_0}\right)^2 + \left(\frac{r_0}{r}\right)^2 \left(\frac{r_i}{r_0}\right)^2 - \frac{1+3\nu}{3+\nu} \left(\frac{r}{r_0}\right)^2 \right] \quad (22)$$

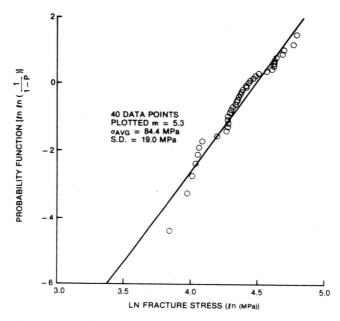

FIG. 14—*Measured strength distribution for flexure specimens with two machined surfaces.*

It can be shown that the effective volume of such a component is approximately

$$kV_D = \frac{\{-\sigma_\theta(r_i)/[d\sigma_\theta(r_i)/dr]\}\,2\pi r_0 L}{m + 2} \qquad (23)$$

Now, using the two-parameter Weibull model fitted to the observed strength distribution of a set of test specimens and described by Eq 12, a stress in a test specimen, σ_T, can be defined which is associated with a given probability of failure, P. Now Eq 17a can be applied to define the maximum stress in the rotating disk which is associated with the same probability of failure, leading to

$$\sigma_D = \sigma_T \left(\frac{kV_T}{kV_D}\right)^{1/m} \qquad (24)$$

For the disk described in Fig. 16, the effective volume is 10.32 cm³ (0.63 in.³). If the effective volume of the test specimen is 0.77 cm³ (0.047 in.³) (as in the case of the tension coupons used in this investigation) and the Weibull modulus is 15, then Eq 24 (or alternatively Fig. 15) indicates that the maximum stress in the disk, σ_D, should be 84% of σ_T to assure the same probability of

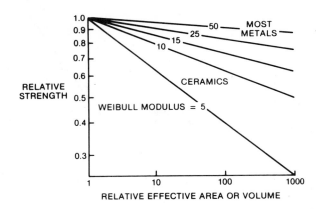

FIG. 15—*Relative strength as function of relative effective volume or surface area with Weibull modulus as a parameter.*

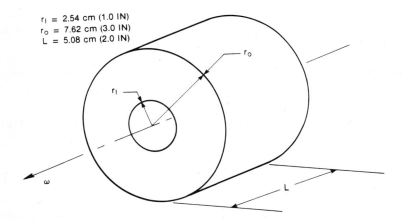

FIG. 16—*Rotating disk geometry.*

failure. For components with larger effective volumes, or material with a lower Weibull modulus, the design stress reduction would be increasingly severe. It should also be pointed out that for component geometries and loading conditions where closed-form expressions for effective volume are not available, the effective volume calculation has been coupled with finite-element analyses [13], thus extending the usefulness of this methodology where it is appropriate.

Conclusions

Application of ASTM Test Methods D 790-71 and D 651-80 to the injection molded, fiber filled phenolic material investigated here results in an average strength of 104.0 MPa (15 000 psi) from flexure tests and 56.9 MPa (8250 psi) from tension tests. Three mechanisms have been assessed with respect to their effects upon the observed discrepancy between tensile and flexural strength data: variation in fiber alignment across the flexural specimen's cross section, residual stresses in the flexural specimens, and flaw distribution effects as described by Weibull statistical theory. Although fiber alignment in the direction of the injection flow was clearly observed in both tension and flexure specimens, no enhanced alignment was observed near the specimen surfaces in comparison to their cores. Preferential fiber alignment near the surfaces of the bend specimens does not, therefore, appear to be a significant mechanism regarding the enhanced strength of three-point bend specimens in comparison with tensile specimens for the data examined in this investigation. On the other hand, compressive residual stresses ranging from 6.2 MPa (900 psi) to 13.1 MPa (1900 psi) were measured at the outer surfaces of the flexure specimens. The residual stresses in the core are smaller in magnitude and tensile in nature. The compressive stresses do serve to increase the apparent strength as measured by flexure tests. The geometry of the tension specimens precluded measurement of residual stresses by the material removal technique applied here. If they exist, however, the tensile stresses in the core would reduce the tension coupon's apparent strength. This effect would be less pronounced than in the flexure specimens, since the tensile residual stresses induced at the center of the specimens during cooling are smaller than the compressive stresses at the surfaces. Tests on a set of 50 tension specimens resulted in strength data which are well represented by a two-parameter Weibull model with a Weibull modulus of $m = 15$. If the flexural strength data are corrected downward by the average measured residual compressive stress of 8.96 MPa (1.3 ksi), an average strength from three-point flexural tests of 94.4 MPa (13.7 ksi) is obtained. On the other hand, if Weibull statistical theory is applied to the tension test data, then the predicted average strength for three-point flexural tests is 73.6 MPa (10.6 ksi). Consideration of tensile residual stresses in the tension specimens as well as expected errors in the measurement of residual stresses and determination of the Weibull modulus from a finite sample size might decrease this discrepancy. Differences in fiber volume and void content between the two specimens could also have an effect.

Additional tests were conducted using the three-point bend configuration to investigate load rate effects and machining effects. It appears that the strength data are sensitive to load rate consistent with the idea of subcritical crack growth of inherent flaws. In addition, if machined surfaces are loaded in tension, the average strength and Weibull modulus of the test data decrease from 101.3 MPa (14.7 ksi) to 88.2 MPa (12.8 ksi) and from 36 to 7, respectively. Machining, therefore, would have severe effects on the reliability of this fiber-filled phenolic material.

Although the Weibull statistical approach does not completely explain the discrepancy between tensile and flexural strengths measured in this investigation, it does seem to contribute significantly. If the Weibull modulus of 15 measured from tension tests is realistic, there would be substantial "size effects" upon strength which should be accounted for in component design. Unfortunately, application of this approach will be complicated by the effects of differences in fiber orientation between test specimens and components, forcing additional care in the choice of test specimens. It is clear, however, that strength data accumulated from ASTM D 790-71 flexure tests will be nonconservative for design on the basis of fiber orientation, residual stress, and Weibull flaw distribution effects.

Acknowledgment

The tension and flexure specimens used in this investigation were provided by the General Electric Plastics Business. The tension tests were conducted under the direction of M. D. Bertolucci and R. W. Smearing. The flexure testing and residual stress measurements were carried out by H. Moran at the General Electric Research and Development Center. S. A. Chamberlain and L. C. Miller both provided assistance in reducing and plotting the residual stress data. The authors would also like to acknowledge the interest and practical observations of Joe Williams, Ken Reynolds, and Gordon Osborne (General Electric D. C. Motor and Generator) which helped define this investigation.

References

[1] Taggert, D. G., Pipes, R. B., and Mosko, J. C., *Polymer Composites*, Vol. 1, No. 1, Sept. 1980, pp. 56-61.

[2] So, P. and Broutman, L. J. in *Proceedings*, 34th Annual Technical Conference, Atlantic City, NJ, Society of Plastics Engineers, April 1976, pp. 114-118.

[3] Choi, S. and Broutman, L. J., *Proceedings*, 41st Annual Technical Conference, Chicago, IL, Society of Plastics Engineers, May 1983, pp. 293-297.

[4] Weibull, W., *Journal of Applied Mechanics*, Vol. 18, Sept. 1951, pp. 293-297.

[5] Barnett, R. L. and McGuire, R. L., *American Ceramic Society Bulletin*, Vol. 45, No. 6, 1966, pp. 595-602.

[6] Bullock, R. E., *Journal of Composite Materials*, Vol. 8, April 1974, pp. 200-203.

[7] Whitney, J. M. and Knight M., "The Relationship Between Tensile Strength and Flexural Strength in Fiber Reinforced Composites," AFWAL Report TR-80-4104, Materials Laboratory, Air Force Wright Aeronautical Laboratories, Wright-Patterson AFB, Dayton, OH, Dec. 1980.

[8] Knight, M. and Hahn, H. J., *Journal of Composite Materials*, Vol. 9, Jan. 1975, pp. 77-89.

[9] Halpin, J. C., Jerina, K. L., and Johnson, J. A. in *Analysis of the Test Methods for High Modulus Fibers and Composites, ASTM STP 521*, American Society for Testing and Materials, Philadelphia, 1973, pp. 5-53.

[10] McCullough, R. L., *Treatise on Materials Science and Technology: Properties of Solid Polymeric Materials*, 10B, Academic Press, New York, 1977.

[11] Treuting, R. G., Lynch, J. J., Wishart, H. B., and Richards, D. G., *Residual Stress Measurements*, American Society for Metals, Cleveland, OH, 1951, p. 55.

[12] Jeryan, R. A., "Use of Statistics in Ceramic Design and Evaluation" in *Proceedings*, Fifth Army Materials Technology Conference, Newport, RI, March 21-25, 1977, Brook Hill Publishing Co., Chestnut Hill, MA, 1978, p. 42.

[13] Trantina, G. G. and deLorenzi, H. G., *Engineering for Power*, Vol. 99, No. 4, Oct. 1977, p. 559.

Elevated Temperature and Environmental Effects

Cecil G. Rhodes[1] and Robert A. Spurling[1]

Fiber-Matrix Reaction Zone Growth Kinetics in SiC-Reinforced Ti-6Al-4V as Studied by Transmission Electron Microscopy

REFERENCE: Rhodes, C. G. and Spurling, R. A., **"Fiber-Matrix Reaction Zone Growth Kinetics in SiC-Reinforced Ti-6Al-4V as Studied by Transmission Electron Microscopy,"** *Recent Advances in Composites in the United States and Japan, ASTM STP 864*, J. R. Vinson and M. Taya, Eds., American Society for Testing and Materials, Philadelphia, 1985, pp. 585–599.

ABSTRACT: Fiber-matrix reaction zone product formation and growth kinetics at 850 and 900°C in silicon carbide (SiC) reinforced Ti-6Al-4V have been studied by thin-foil transmission electron microscopy. The major reaction product is titanium carbide, which is sandwiched between a layer of Ti_5Si_3 and a fine-grained mixture of titanium carbide and Ti_5Si_3. The large volume fraction of titanium carbide compared to titanium silicide results from the presence of a carbon-rich layer at the surface of the fibers. Growth of the reaction zone and consumption of the fiber surface layer are nonuniform, with greater reaction occurring where the fiber contacts beta phase in the titanium matrix. During exposure at 850 and 900°C, the reaction products exhibit parabolic growth, with the titanium carbide growing at a faster rate than the titanium silicide. After consumption of the carbon-rich fiber surface layer, however, the titanium silicide growth rate increases over that of titanium carbide. Reaction zone growth rate and fiber surface layer consumption rate are faster at 900°C than at 850°C.

KEY WORDS: composite materials, titanium, silicon carbide, continuous fiber, transmission electron microscopy, reaction growth kinetics

Fiber reinforcement in a metal matrix can improve mechanical properties such as tensile strength and elastic modulus under uniaxial loading conditions [1]. Composites of this type can be consolidated by hot-pressing and fabricated into structural components by processes such as superplastic forming and diffusion bonding. Diffusion controlled reactions between fibers and metal matrix can occur during these elevated-temperature exposures, and the

---

[1] Rockwell International Science Center, Thousand Oaks, CA 91360.

presence of brittle reaction products thus formed can seriously degrade mechanical properties [1].

Various approaches have been taken to reduce the reaction kinetics between fiber and matrix; for example, coatings or boron fibers [1] and composition gradients on silicon carbide fibers [2]. However, for the development of effective diffusion barriers and the understanding of mechanisms by which reaction products degrade properties, the reaction product species, distribution, and growth rate must be known. Studies have been made of fiber-matrix reaction zones, generally by means of X-ray diffraction, electron microprobe, or scanning electron microscopy (SEM) [1,3]. Unfortunately, these techniques have spatial resolution limits (for identification purposes) on the order of 1 to 10 μm, and reaction zones narrower than this will go undetected by these techniques.

This paper describes the results of a study of fiber-matrix reaction zones in silicon carbide (SiC) reinforced Ti-6Al-4V using transmission electron microscopy (TEM), in which reaction products have been identified by X-ray microanalysis and convergent beam electron diffraction. These techniques have resolution limits on the order of 10 nm, an improvement in resolution of three orders of magnitude over microprobe or X-ray diffraction. In this work, reaction products in the "as-consolidated" condition have been identified and their growth behavior at 850 and 900°C has been followed.

Experimental Procedures

Materials

The composite used in this study was Ti-6Al-4V reinforced with fibers of silicon carbide (SCS-6); the SCS-6 fibers, which have a nonstoichiometric, carbon-rich mixture of carbon and silicon at the outer surface, are specially made for the reinforcement of titanium alloys [2]. The preform was consolidated by Amercom, Inc. and supplied to Rockwell International in the as-consolidated condition.

Thin-Foil Preparation

Electron microscopy thin-foil specimens were prepared from 0.5-mm (0.02 in.) transverse sections that were sliced from the bulk using a diamond cutoff wheel. Initial thinning was carried out by mechanical grinding of the 0.5-mm (0.02 in.) sections that were fastened to a specially designed polishing fixture by means of an acetone soluble wax. The fixture ensures flat and parallel surfaces as the specimen is ground through 70, 30, 15, and 9-μm diamond abrasive wheels on each surface successively to reach a final thickness of approximately 0.025 mm (0.001 in.). After cleaning in warm acetone, a 3-mm-diameter (0.12 in.) disk is cut from the thinned specimen. The disk is thinned to electron transparency in a Gatan ion milling apparatus. It typically takes from 6 to 10 h to perforate a 0.025-mm (0.001 in.) disk in the ion miller operating at 5 kV and 0.5 mA.

Electron Microscopy

Transmission electron microscopy was conducted with a Philips EM-400 electron microscope equipped with an X-ray detector and with scanning capability. The electron beam can be reduced to a minimum of 4-nm diameter for X-ray microanalysis of regions in the microstructure on the order of 10 nm. In this mode, chemical analysis by means of X-ray energy dispersive spectroscopy and structural analysis by means of microdiffraction can be accomplished on individual particles having dimensions as small as the aforementioned 10-nm range.

Heat Treatments

Heat treatments for studying growth kinetics at 850 and 900°C were performed in vacuum. Specimens were wrapped in tantalum foil, held in the vacuum furnace at temperature for 4, 8, 24, 48, or 100 h, and vacuum-cooled.

Results

As-Consolidated

The fiber-matrix interface as observed in a polished and etched cross section of the as-consolidated material is shown in Fig. 1. The figure reveals a measurable amount of reaction product that has formed during consolidation. The reaction zone appears to be about 1 μm thick and to have two distinct regions: a broad layer adjacent to the fiber having some internal structure and a narrower band adjacent to the titanium matrix (Fig. 1b).

Transmission electron microscopy of this interface reveals the details of the reaction zone (Fig. 2). The reaction zone is seen to consist of three layers rather than two as concluded from the SEM observation (Fig. 1). The very narrow (~ 70 nm) layer adjacent to the fiber surface could not be detected by observations of the polished and etched surfaces shown in Fig. 1. The bulk of the reaction zone is seen to be the intermediate layer, with the outermost layer comprising about one third of the total.

X-ray microanalysis indicates that silicon is concentrated in the very narrow layer adjacent to the fiber and in the outermost layer adjacent to the matrix (Fig. 3). The spectra of Fig. 3 show that neither aluminum nor vanadium is detected in the reaction products.

Microdiffraction of each of the three layers confirmed their identity. Phase identification was based on the comparison of interplanar spacings measured

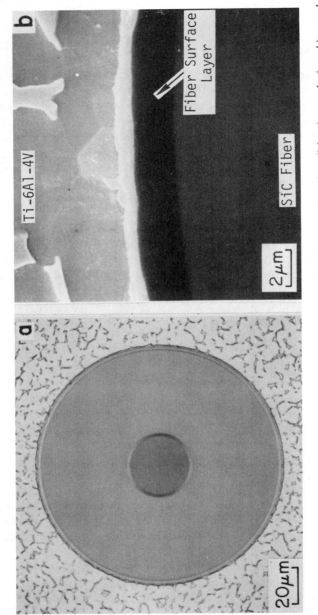

FIG. 1—*Polished and etched cross-section of silicon carbide reinforced Ti-6Al-4V, as consolidated: (a) light micrograph of graphite-cored fiber; (b) scanning electron micrograph of fiber/matrix interface.*

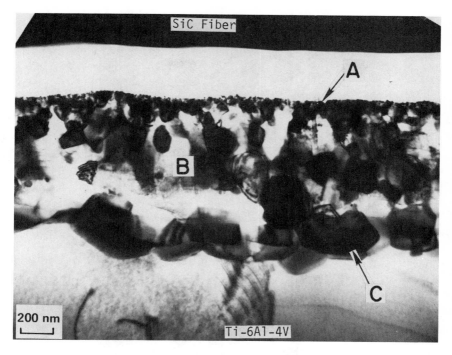

FIG. 2—*Thin-foil transmission electron micrograph of reaction zone between silicon carbide fiber and Ti-6Al-4V matrix showing inner layer, A, intermediate layer, B, and outer layer, C. Fiber has separated from reaction zone during final stages of thin-foil specimen preparation.*

from diffraction patterns with ASTM powder diffraction file data and on measured angles between various zone axes. The outermost layer was found to be Ti_5Si_3 (Fig. 4c), the intermediate layer was TiC (Fig. 4d), and the narrow inner layer was a mixture of TiC and Ti_5Si_3. Isolated particles of Ti_5Si_3 were also found dispersed through the TiC; these particles are the internal structure observed in the SEM image, Fig. 1b.

Effect of Heat Treatment

Extended exposure at 850 and 900°C results in growth of each of the three reaction layers. The total reaction zone grows at the expense of both the Ti-6Al-4V matrix and the fiber surface layer, with the latter being consumed in a nonuniform manner, resulting in an undulated surface, Fig. 5. At those places where the fiber contacts beta phase in the titanium matrix, the reaction zone consumes the fiber surface layer and grows more rapidly than where the fiber contacts alpha phase, as illustrated in Fig. 5. The result is a broad reaction zone and narrow surface layer where beta is contacted and a narrow reac-

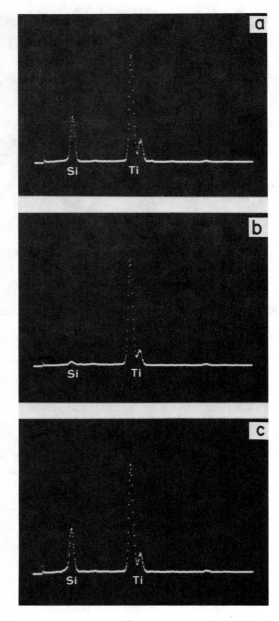

FIG. 3—*X-ray energy spectra from reaction zone components shown in Fig. 2; (*a*) inner layer,
(*b*) intermediate layer, (*c*) outer layer.*

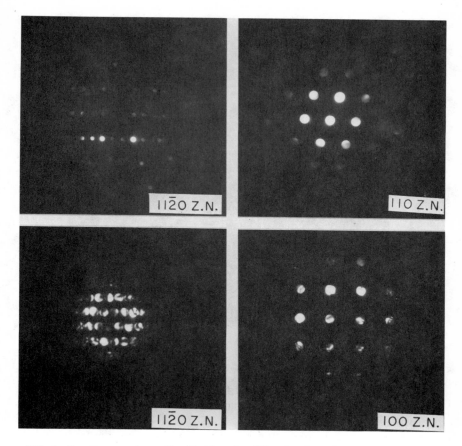

FIG. 4—*Reaction zone layers identified by microdiffraction. Inner layer contains Ti_5Si_3 (a) and TiC (b); outer layer is Ti_5Si_3 (c) and intermediate layer is TiC (d). Ti_5Si_3 has a hexagonal-close-packed crystal structure while TiC has a face-centered cubic structure.*

tion zone with a broad surface layer where alpha phase is contacted. At 850°C, the microstructure of Ti-6Al-4V consists of approximately 40% beta phase and 60% alpha phase, whereas at 900°C, the Ti-6Al-4V matrix contains roughly 50% alpha and beta phases [4]. During furnace cooling from these temperatures, the existing alpha phase grows at the expense of the beta phase until a mixture of about 90% alpha and 10% beta is reached. Hence, examination of the microstructure after heat treatment reveals where beta phase has contacted the fiber, but not the extent of beta phase at the reaction temperature.

Measurements have been made of total reaction zone width, individual reaction zone layer widths, and fiber surface layer width as a function of time and temperature. Because of the nonuniform growth of the reaction products

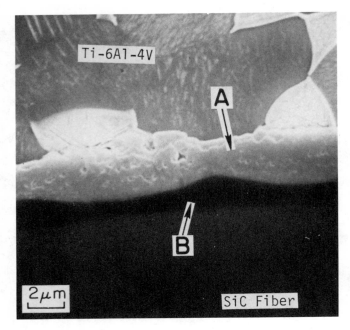

FIG. 5—*Scanning electron micrograph illustrating nonuniform thickness of reaction zone,* A, *and fiber surface layer,* B.

and nonuniform consumption of the fiber surface layer just described, the reported values are averages of a large number of measurements which consequently have fairly large error bars.

The 850°C data are presented in Figs. 6 and 7, where the fractional increases in thickness are plotted as a function of the square root of time. The linear relationship indicates parabolic growth of the reaction products and consumption of the fiber surface layer. Titanium carbide, which comprises the bulk of the reaction zone prior to heat treatment, grows at a faster rate than either the silicide layer or the inner mixed layer, thereby increasing the relative amount of carbide in the reaction zone with continued exposure at 850°C (Fig. 7).

The 900°C data are shown in Figs. 8 and 9, again with the fractional increases in thickness plotted as a function of square root of time. Comparison of Figs. 6 and 8 reveals that the reaction zone grows at a faster rate at 900°C than at 850°C and that the fiber surface layer is consumed at a faster rate at 900°C than at 850°C. Again, the titanium carbide layer grows more rapidly than the silicide and each of the three reaction zone layers grows at a higher rate at 900°C than at 850°C (compare Figs. 7 and 9).

It was observed that the reaction product growth rates appeared to change after the fiber surface layer had been completely consumed after extended

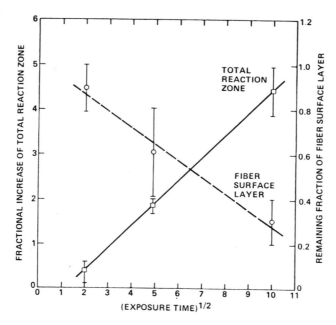

FIG. 6—*Fractional increase in total reaction zone thickness and remaining fraction of fiber surface layer as a function of square root of time at 850°C for silicon carbide (SCS-6) reinforced Ti-6Al-4V.*

exposure at these two temperatures. This phenomenon is illustrated in Fig. 10, where it can be seen that the titanium silicide adjacent to the fiber has grown to a greater thickness where the surface layer no longer exists. The growth rate of the silicide, then, appears to have increased to be greater than the titanium carbide growth rate after consumption of the fiber surface layer.

Discussion

Reaction Products

The observation that titanium carbide is the major component of the reaction zone in these SCS-6 reinforced titanium composites is in contrast to what others have reported for uncoated silicon carbide fibers in titanium or for silicon carbide coated boron (Borsic) fibers in titanium [1,5]. In those cases, the major constituent of the reaction zones was reported to be titanium silicide, $TiSi_2$ and Ti_5Si_3 in the former and Ti_5Si_3 in the latter.

Titanium carbide develops more extensively than titanium silicide in the SCS-6 composite as a direct result of the composition of the fiber surface. The fibers are grown with a carbon-rich nonstoichiometric composition at the outer surface and it is this surface layer that supplies carbon to the growth of

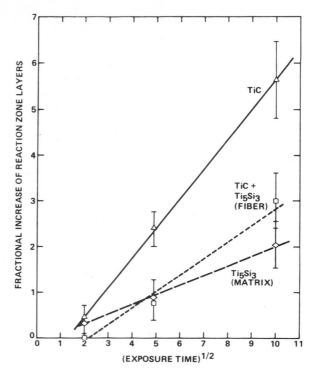

FIG. 7—*Fractional increase in individual reaction zone layer thicknesses as a function of square root of time at 850°C for silicon carbide (SCS-6) reinforced Ti-6Al-4V.*

titanium carbide. With more carbon than silicon available, titanium carbide will grow to a larger volume fraction than titanium silicide in the SCS-6 composites. However, once the fiber surface layer has been consumed and stoichiometric silicon carbide exposed for reaction with the matrix, prior results [1,5] suggest that titanium silicide should grow at a faster rate than the carbide. Such behavior was observed in the current work (Fig. 10).

The presence of Ti_5Si_3 at both the fiber-reaction zone interface and the matrix-reaction zone interface was unexpected inasmuch as others had reported that $TiSi_2$ formed adjacent to the silicon carbide fiber, with Ti_5Si_3 forming next to the matrix, in uncoated fiber-reinforced titanium [1]. In the absence of silicon-titanium diffusion couple experiments at 850 and 900°C, the reasons for Ti_5Si_3 forming at both interfaces can only be speculated upon. One possible explanation arises from the fact that the surface layer has a composition gradient (from carbon-rich at outer surface to stoichiometric silicon carbide internally) such that, as the layer is consumed, there is less silicon available for reaction. If titanium atoms diffuse readily through the existing titanium carbide, there will be a sufficient ratio of titanium and silicon atoms

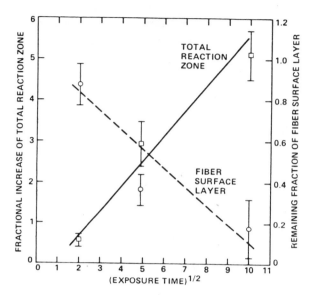

FIG. 8—*Fractional increase in total reaction zone thickness and remaining fraction of fiber surface layer as a function of square root of time at 900°C for silicon carbide (SCS-6) reinforced Ti-6Al-4V.*

for the growth of Ti_5Si_3 rather than the higher silicide, $TiSi_2$. Additional evidence for the formation of Ti_5Si_3 rather than $TiSi_2$ is found in the investigation of Borsic fiber reinforced Ti-6Al-4V where the only silicide present in the reaction zone was Ti_5Si_3 [5].

The nonuniform growth of the reaction zone and nonuniform consumption of the fiber surface layer are interrelated and appear to result from diffusivity and solubility of carbon in the alpha and beta phases of the titanium matrix. Carbon diffuses at a higher rate in beta phase than in alpha phase [6], hence the fiber surface layer would be expected to be consumed at a faster rate wherever it is in contact with beta phase. Figure 5 illustrates the rapid consumption of fiber surface layer adjacent to beta phase. In conjunction with its rapid diffusion rate, carbon has a lower solubility in beta phase than in alpha [6]. The lower solubility of carbon in beta leads to more precipitation of titanium carbide in the beta phase than in the alpha, as shown in Fig. 5.

There is extreme partitioning of alloying elements between alpha and beta phases in Ti-6Al-4V with the beta phase being lean in aluminum and rich in vanadium and the alpha phase being the converse. Other workers have shown that the addition of aluminum to the interface between silicon carbide fiber and titanium matrix reduces reaction kinetics [3]. Aluminum stabilizes the alpha phase and it is possible that the addition of aluminum to the interface promotes alpha phase formation and inhibits beta phase formation, thereby

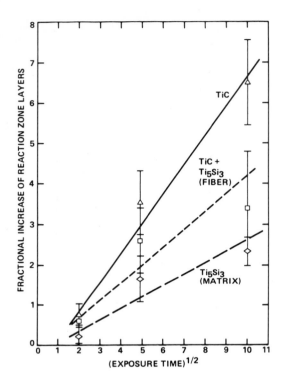

FIG. 9—*Fractional increase in individual reaction zone layer thicknesses as a function of square root of time at 900°C for silicon carbide (SCS-6) reinforced Ti-6Al-4V.*

reducing reaction zone formation kinetics. These results would suggest that, for a given temperature, an SCS-6 reinforced titanium alloy composite would exhibit increased reaction zone formation with increased volume fraction of beta phase.

Growth Kinetics

The parabolic growth relationship for the reaction zone products (see Figs. 6–9) is in agreement with the results of earlier workers [1]. The simple linear relationship between reaction zone width and square root of time, as observed in the current work, can be attributed to growth being controlled by diffusion through the reaction zone [1]. It is not clear from this study which atomic species is rate controlling, although the rapid growth of titanium carbide might indicate that carbon diffusivity through titanium carbide is the controlling factor.

The availability of carbon in the fiber surface layer is the key to the reaction zone growth kinetics. The titanium carbide grows at a faster rate than the

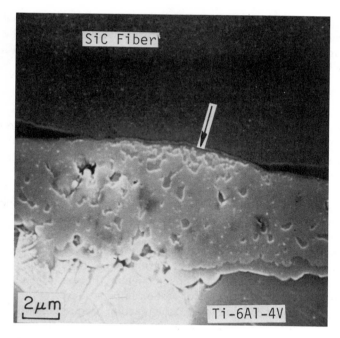

FIG. 10—*Scanning electron micrograph illustrating accelerated growth of* Ti_5Si_3 *in inner layer (*arrow*) after consumption of fiber surface layer.*

silicide apparently because there is more carbon than silicon in the fiber sur-face, for when the surface layer has been consumed, the silicide grows faster than the carbide (see Fig. 10). Such behavior points up the complexities in-volved in analyzing growth kinetics in a system that includes a composition gradient in one of the reacting components.

The curves through the data points in Figs. 6–9 do not extrapolate to zero. The offset (from zero) of the plots is not simply the result of preexisting reac-tants because the data are plotted as fractional increase in width, thereby eliminating any effects of reaction zone that is present prior to annealing. The apparent incubation period at both 850 and 900°C may be the result of a change in stoichiometry of the reaction products [7]; that is, the reaction zone products were formed initially at a consolidation temperature that is higher than the subsequent annealing temperatures. The incubation period, then, could be the time required for the various reaction products to achieve equi-librium stoichiometry at the lower temperatures.

The accelerated growth rates of each of the reaction zone products at 900°C compared with 850°C are not unexpected. The fact that no new compounds have formed at 900°C and that the relative growth rates are unchanged be-tween 850 and 900°C indicates that TiC and Ti_5Si_3 are the stable forms of their respective compounds at those temperatures. Because mechanical property degradation has been correlated with reaction zone width [1], the present results suggest that processing should be carried out at 850°C rather than 900°C or above.

Summary and Conclusions

This work has shown that the reaction zone that forms in the interface between SCS-6 fibers and Ti-6Al-4V matrix at 850 and 900°C consists of three layers of reaction products. The major reaction product is TiC, which is sandwiched between a layer of Ti_5Si_3 and a fine-grained mixture of TiC and Ti_5Si_3. The larger volume fraction of titanium carbide compared to titanium silicide results from the presence of a carbon-rich layer on the surface of the fibers, in contrast to silicon carbide coated boron fiber composites in which titanium silicide is the major reaction product.

Growth of the reaction zone and consumption of the fiber surface layer are nonuniform, with greater reaction occurring where the fiber contacts beta phase of the titanium matrix. This phenomenon appears to result from the high diffusivity and low solubility of carbon in beta phase.

During exposure at 850 and 900°C, the reaction products exhibit parabolic growth with the titanium carbide growing at a faster rate than either of the titanium silicide layers. After consumption of the carbon rich fiber surface layer, however, the titanium silicide growth rate increases over that of titanium carbide.

TiC and Ti_5Si_3 appear to be the stable forms of their respective compounds at 850 and 900°C. Because mechanical property degradation is a function of reaction zone width, the results of this work suggest that processing of SCS-6 reinforced titanium be carried out at 850°C rather than 900°C or higher.

Acknowledgments

We are pleased to acknowledge the experimental assistance of M. Calabrese and P. Q. Sauers. The composite material was supplied to us by T. E. Steelman of Rockwell International's North American Aircraft Operations Group. This report has been reviewed by the U.S. Department of Defense.

References

[1] Metcalfe, A. G., *Interfaces in Metal Matrix Composites*, Vol. 1, A. G. Metcalfe, Ed., Academic Press, New York, 1974, pp. 67-123.
[2] Cornie, J. A. in *Proceedings*, 4th Metal Matrix Composites Technology Conference, MMCIAC, Santa Barbara, CA, May 1982, pp. 30-1 through 30-9.
[3] Brewer, W. D. and Unnam, J. in *Mechanical Behavior of Metal-Matrix Composites*, J. E. Hack and M. F. Amateau, Eds., The Metallurgical Society of AIME, Warrendale, PA, 1983, pp. 39-50.

[4] Rhodes, C. G., Mitchell, M. R., and Chesnutt, J. C., "Fracture and Fatigue Characteristics in Titanium Alloys," Technical Report SC5227.1FR, Rockwell International Science Center, Thousand Oaks, CA, June 1982.

[5] Rhodes, C. G., "Study of Titanium Matrix Composites," Technical Report SC5288.1FR, Rockwell International Science Center, Thousand Oaks, CA, March 1981.

[6] Wagner, F. C., Bucur, E. J., and Steinberg, M. A., *Transactions*, American Society for Metals, Vol. 48, 1956, pp. 742–761.

[7] Klein, M. J., Reid, M. L., and Metcalfe, A. G., "Compatibility Studies for Viable Titanium Matrix Composites," Technical Report AFML-TR-69-242, Air Force Materials Laboratory, Wright-Patterson AFB, Ohio, 1969.

H. Thomas Hahn,[1] *D. G. Hwang,*[1] *and W. K. Chin*[2]

Effects of Vacuum and Temperature on Mechanical Properties of S2-Glass/Epoxy

REFERENCE: Hahn, H. T., Hwang, D. G., and Chin, W. K., **"Effects of Vacuum and Temperature on Mechanical Properties of S2-Glass/Epoxy Composite,"** *Recent Advances in Composites in the United States and Japan, ASTM STP 864*, J. R. Vinson and M. Taya, Eds., American Society for Testing and Materials, Philadelphia, 1985, pp. 600–618.

ABSTRACT: Mechanical properties of an S2-glass/epoxy composite have been characterized at room temperature and at 100°C in vacuum. The latter was to simulate the flywheel service environment. The epoxy was DER 332/Menthane Diamine (100/24.5), and the composite was filament-wound and cured 2 h at 150°C. It has been found that the composite continues to lose weight even after ~ 150 days of conditioning at 100°C in vacuum. The weight loss is accompanied by a change of color from translucent white to brown. The present composite shows less moisture absorption than most graphite/epoxy composites for aerospace applications. The 100°C/vacuum environment has only a minor effect on the mechanical properties, both static and fatigue. However, the moisture desorption in vacuum increases residual stresses and leads to ply cracking in laminates. The logarithmic fatigue life is linearly related to the fatigue stress, and the fatigue strength at 10^6 cycles is ~ 20% of the average static strength at room temperature and also at 100°C in vacuum. However, the corresponding fatigue strength at 100°C in air is below 20% of the average static strength. Whereas the static failure is brush-like, the fatigue failures at stresses lower than ~ 60% of the static strength are slivery. The failure mode at the elevated temperature tends to be more slivery than at room temperature. The composite is susceptible to modulus reduction in fatigue without any apparent longitudinal splitting. The maximum modulus reduction before failure increases with decreasing stress level.

KEY WORDS: composite materials, environmental effects, mechanical properties, fatigue, probability distribution, residual stresses

Composite materials have been found to be ideal for application in flywheel energy storage systems because of their high specific strength and nonhazardous failure modes [1,2]. The low mass density and high strength allow for the

[1]Professor of mechanical engineering and postdoctoral associate, respectively, Materials Research Laboratory, Washington University, St. Louis, MO 63130.
[2]Presently associate professor, National Ching-Hwa University, Taiwan, Republic of China.

storage of higher kinetic energy per unit mass through increased speed. Since high-performance composites are relatively new, however, their reliability in service environments has not fully been assessed yet. Therefore, the main objective of the present work was to investigate the behavior of a composite in a simulated flywheel service environment.

The design speed of a composite flywheel can be as high as 50 000 rpm, and hence a vacuum near 10^{-4} torr is required to avoid excessive aerodynamic heating. Even so, the temperature can easily reach ~100°C. However, the combined effect of elevated temperature and vacuum on mechanical properties has not yet been characterized except for an aramid fiber/epoxy composite [3].

The present paper is an extension of an earlier work [3]. The paper discusses the effects of elevated temperature and vacuum on static and fatigue properties of an S2-glass/epoxy composite. Also included are the elevated temperature stability in vacuum, the modulus reduction in fatigue, and the fatigue damage development.

Experimental Procedure

The composite studied was an S2-glass/epoxy filament-wound at the Lawrence Livermore National Laboratory (LLNL) in the form of 230 by 254-mm panels. The epoxy was DER 332/Menthane Diamine (100/24.5) cured 2 h at 150°C. The nominal fiber volume content was 65%. The actual fiber volume content of the panels varied from 63% to 66%.

Preconditioning was done in a vacuum oven at 100°C. Traveler specimens were used to monitor the weight change as a function of the period of preconditioning. A few specimens were preconditioned first at 75°C and then at 100°C. The ply cracking due to residual stresses was studied by conditioning 25-mm by 25-mm $(\pm 45)_{2s}$ specimens in a vacuum oven at 75°C.

An MTS and an Instron machine were used for mechanical tests. Tension and compression tests were performed in accordance with ASTM Test Methods for Tensile Properties of Fiber-Resin Composites (D 3039-76) and Compressive Properties of Unidirectional or Crossply Fiber-Resin Composites (D 3410-75), respectively. Shear tests were done according to the ASTM Recommended Practice for In-plane Shear Stress-Strain Response of Unidirectional Reinforced Plastics (D 3518-76).

All static tests at 100°C on the MTS machine were done in air by placing the specimen inside a glass/epoxy tube. The tube was 75 mm long by 75 mm in diameter. It had a heating pad inside and a fiberglass insulation on the outside. On the Instron machine, an Instron environmental chamber was used.

Fatigue tests were done at the stress ratio of 0.1 and the loading frequency of 4 Hz when the maximum stress was below 60% of the average ultimate tensile strength (UTS). At and above 60% UTS the frequency was reduced to 0.5 to 2 Hz. Only unidirectional specimens were tested in fatigue. A vacuum chamber designed at LLNL was used for fatigue tests at 100°C in vacuum [4].

Strain gages were used for the determination of the static properties. The modulus change in fatigue was studied by using an extensometer of 25.4-mm gage length. The specimen dimensions used are the same as those in Ref [4].

Results and Discussion

For ease of discussion, the following abbreviations are used throughout the paper:

LT: longitudinal tension
LC: longitudinal compression
TT: transverse tension
TC: transverse compression
 S: shear (longitudinal)
RT: room temperature in air
ET: elevated temperature ($=100°C$) in air
ETP: elevated temperature in air after preconditioning
ETV: elevated temperature in vacuum after preconditioning

Temperature Stability

Figure 1 shows the weight losses during preconditioning in vacuum. Some specimens were preconditioned first at 75°C and later at 100°C. For those specimens, weight loss stabilized at 75°C but increased again when temperature was increased at 100°C. Also, the surfaces changed color from translucent white to brown with the period of preconditioning. The composite is not believed to be stable at 100°C.

The effect of residual stresses on ply cracking in a $(\pm 45)_{2s}$ laminate can be seen in Fig. 2. The cracks were detected by spraying a red dye penetrant on the specimen surfaces. In all specimens, the crack density (crack number per unit length) increases as they are dried out. In the absence of moisture, the transverse ply stress in the laminate is estimated by [5]

$$\sigma_T = \frac{E_L(e^0 - e_L) + \nu_L E_T(e^0 - e_T)}{1 - \nu_L^2 E_T/E_L} \tag{1}$$

where

$$e^0 = \frac{E_L e_L + E_T(e_L + e_T)\nu_L + E_T e_T}{E_L + E_T(1 + 2\nu_L)} \tag{2}$$

Here E and ν are the Young's modulus and Poisson's ratio, respectively, and e is the curing strain. The subscripts L and T stand for longitudinal and transverse, respectively. Equation 2 gives the curing strain of the $(\pm 45)_{2s}$ laminate.

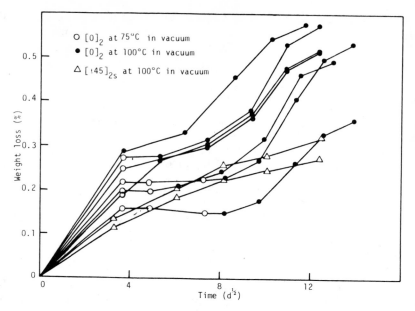

FIG. 1—*Weight losses during preconditioning.*

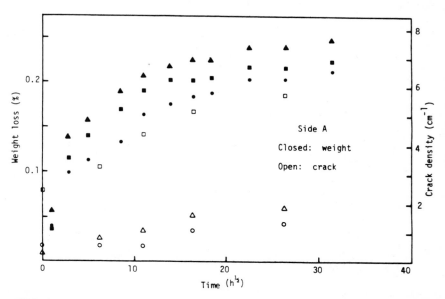

FIG. 2—*Increase of crack density after various periods of conditioning at 75°C in vacuum, side A.*

The RT moduli required in Eqs 1 and 2 are given in Tables 1 through 4. The thermal expansion coefficients of the present composite were found to be

$$\alpha_L = 5.52 \ \mu m/m/°C$$

$$\alpha_T = 23.34 \ \mu m/m/°C$$

Since the present composite was cured at 150°C, the resulting cure strains at RT (23°C) are

$$e_L = -701 \ \mu m/m$$

$$e_T = -2964 \ \mu m/m$$

The substitution of the curing strains and the moduli into Eqs 1 and 2 yields

$$\sigma_T = 34.2 \ MPa$$

The calculated residual stress is higher than the transverse tension strength at RT. Thus, the ply cracking in Fig. 2 is not unexpected. However, the largest number of cracks should have been in the specimens right after fabrication before any moisture absorption. The additional cracking shown in the figures indicates the time dependence of the ply cracking induced by residual stresses [6, 7] and the possibility of additional cure shrinkage at the elevated temperature.

TABLE 1—*Longitudinal tension properties and Weibull parameters for strength distributions.*

Days of preconditioning (vacuum at 100°C)	0	35~42	0
Test temperature, °C	RT	100	100
Fiber volume content, %	65	65	65
Failure stress			
avg, MPa (no. of specimens)	1983(14)	1782(10)	1788(10)
coefficient of variation, %	6.16	6.50	5.40
Failure strain			
avg, % (no. of specimens)	3.96(4)	3.86(4)	. . .
coefficient of variation, %	2.75	8.55	. . .
Modulus			
avg, GPa (no. of specimens)	56.41(5)	55.65(5)	. . .
coefficient of variation, %	3.40	3.40	. . .
Poisson's ratio			
avg, (no. of specimens)	0.242(4)	0.241(5)	. . .
coefficient of variation, %	1.77	2.23	. . .
Shape parameter	18.7	15.7	20.4
Characteristic strength, MPa	2039	1836	1835

TABLE 2—*Longitudinal compression and transverse tension properties.*

Longitudinal compression			
days of preconditioning (vacuum at 100°C)		0	60
test temperature, °C		RT	100
fiber volume content, %	65	64	64,65
(panel number)	(2A)	(2B)	(2B,2A)
failure stress			
avg, MPa (no. of specimens)	626(6)	798(5)	669(10)
coefficient of variation, %	5.96	5.29	10.5
Transverse tension			
fiber volume content, %	65	64	64,65
(panel number)	(2A)	(2B)	(2B,1B)
failure stress			
avg, MPa (no. of specimens)	25.3(5)	17(5)	29.8(9)
coefficient of variation, %	8.60	13.0	10.8
failure strain			
avg, % (no. of specimens)	0.14(2)	0.08(3)	0.18(4)
coefficient of variation, %			
modulus			
avg, GPa (no. of specimens)		19.2(5)	18.1(5)
coefficient of variation, %		4.97	5.08

TABLE 3—*Transverse compression properties.*

Days of preconditioning (vacuum at 100°C)	0	60
Test temperature, °C	RT	100
Fiber volume content, %	64,65	65,65
(panel number)	(2B,2A)	(2B,2A)
Failure stress		
avg, MPa (no. of specimens)	154(10)	130(11)
coefficient of variation, %	3.53	5.02

TABLE 4—*Longitudinal shear properties.*

Days of preconditioning (vacuum at 100°C)	0	56
Test temperature, °C	RT	100
Fiber volume content, %	65	65
Failure stress		
avg, MPa (no. of specimens)	70.4(5)	38.9(5)
coefficient of variation, %	6.32	3.83
Failure strain		
avg, % (no. of specimens)	>5	>5
coefficient of variation, %		
Modulus		
avg, GPa (no. of specimens)	8.89(5)	4.08(3)
coefficient of variation, %	3.95	

Longitudinal Tension Properties

The longitudinal tension properties at RT, ETP, and ET are summarized in Table 1. There is not much difference between the ETP and ET strengths. The elevated temperature of 100°C causes ~ 10% reduction in strength. However, the temperature has very little effect on the modulus.

The initial failure of the specimen is in the form of longitudinal splitting. The longitudinal cracks are initiated at one of the end tabs and grow toward the other end rather rapidly. Such sudden growth sometimes leads to the slipping of the extensometer knife edges. Therefore, an accurate measurement of failure strains by means of an extensometer was difficult.

The ligaments near the edges formed by the longitudinal splitting fail first, resulting in a sudden increase in strain. Thus the stress-strain relations show discontinuities near failure, Fig. 3. Sometimes fiber fractures at edges precede the longitudinal splitting. The fiber fractures soon grow into longitudinal splitting. In both cases, the longitudinal splitting reduces the net section, increasing the stress in the remaining section. Thus, the failure strains in Table 1 are greater than the corresponding failure stresses divided by the respective moduli. The final failure exhibits the typical brush-like failure mode. Most fibers are seen to fracture in the gage section.

Typical ETP stress-strain relations are compared with the RT relations in Fig. 3. The ETP environment has very little effect on the longitudinal tension properties. The ETP failure mode is similar to the RT failure mode.

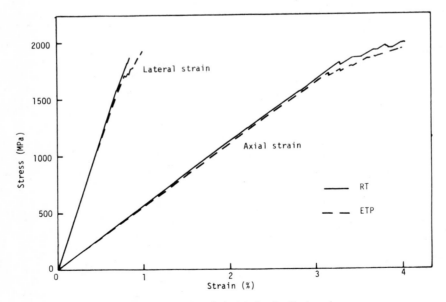

FIG. 3—*Stress-strain relations in longitudinal tension.*

The strength data were fit by a two-parameter Weibull distribution of the form

$$R_s(x) = \exp\left[-\left(\frac{x}{x_0} \right)^{\alpha_s} \right] \tag{3}$$

The strength shape parameter α_s and the characteristic strength x_0 were determined by the method of maximum likelihood [8], and the results are listed in Table 1.

Other Static Properties

The remaining static properties are listed in Tables 2 through 4. Both the LC and TT strengths at RT depend very much on the panels used. The panel with 65 volume % fibers yields a lower LC strength but a higher TT strength than does the other panel with 64 volume % fibers. However, both panels have the same strength in transverse compression. Also, the LC strength at ETP is the same for both panels.

From strain measurements of both surfaces of the specimen, the RT compression failure was found to be due to buckling. Thus, the LC strengths in Table 3 should be regarded as a lower bound. At 100°C after preconditioning, LC specimens fail either in buckling or in shear. In the latter case, failure occurs on a plane ~45 deg inclined to the loading. Yet there is no difference in strength between the two failure modes, Table 2. The shear failure mode was also observed in Kevlar 49/epoxy even at RT [4]. This failure mode is associated with fiber kinking on a microscopic scale.

In transverse tension, more ductile behavior is apparent on ETP. Although there is about 6% reduction in modulus, both failure strain and failure stress increase, Table 2, indicating the beneficial effect of increased resin flexibility.

The TC failure occurs on planes at ~45 deg to the loading. The TC strength shows a 16% reduction at ETP, Table 3. However, there is no difference between the RT and ETP failure modes.

The largest reductions at ETP observed in the shear strength and modulus, Table 4, are believed to be the result of ply cracking described in the preceding section. When the specimen is taken out of the oven and cooled down to RT, after having been preconditioned, the plies crack because of the high residual stresses. Thus the effective stiffness of the plies has already been reduced.

The ETP failure of $(\pm 45)_{2s}$ specimens is accompanied by a considerable necking of the specimen. The shear stress-strain relations are extremely nonlinear both at RT and ETP, Fig. 4. After the initial "yielding" the load transfer is mostly through the interlaminar surfaces. Note that the stress can still be increased ~50% after the initial "yielding."

The strength and modulus retention at ETP of the composite relative to the RT properties is compared with that of the epoxy in Fig. 5. Only those properties that are dominated by the matrix are shown. The epoxy data were taken

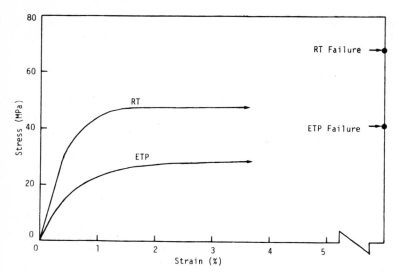

FIG. 4—*Shear stress-strain relations.*

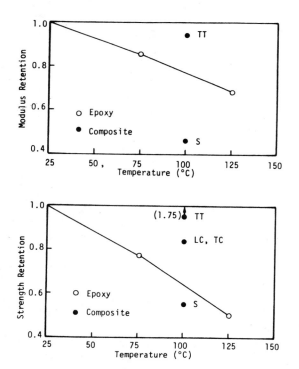

FIG. 5—*Comparison of property retention between composite and epoxy.*

from [9]. Except for the shear properties, the composite is less sensitive to the temperature than the epoxy. Note that the epoxy was not preconditioned.

Fatigue Properties

The *S-N* (stress versus number of cycles to failure) relations at three different test environments are shown in Fig. 6. In the figure, the maximum stress at each environment is normalized with respect to the average UTS at the corresponding environment. The numbers inside the parentheses denote the number of runout specimens. The ETV specimens were preconditioned for 61 to 139 days before fatigue testing.

Since the data in Fig. 6 indicates a linear relation between the normalized stress and the lifetime at every environmental condition, a linear regression analysis can be performed. To this end, an average lifetime is calculated at each stress level. Wherever Weibull distributions had already been determined, they were used instead of the actual data. The average lifetimes are then fit by an equation of the form

$$\frac{S}{\overline{X}} = a - b \log \overline{N} \tag{4}$$

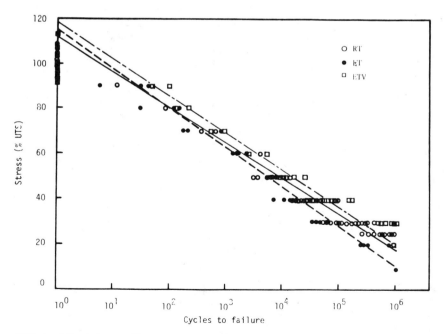

FIG. 6—*S-N relations at RT, ET, and ETV. Number of runout specimens: four at 25% UTS, two at 20% UTS at RT; three at 10% UTS at ET; ten at 30% UTS, one at 20% UTS at ETV.*

where $\bar{X}$ is the average ultimate tensile strength, and $\bar{N}$ the average fatigue life. The parameters a and b are listed in Table 5 together with the coefficients of correlation. The RT data at $0.25\bar{X}$ and the ETV data at $0.30\bar{X}$ are not included in the analysis since they deviate much from the linearity.

The environmental effects are shown in Fig. 6 on the basis of Eq 4. The ETV environment is seen to improve the normalized fatigue strengths while the ET environment is deleterious. Since both moisture and temperature are known to increase fatigue sensitivity, we may conclude that the observed improvement of the normalized fatigue strength is the result of the absence of moisture in vacuum.

The data at and below 50% UTS are fit by a Weibull distribution of the form

$$R_f(N) = \exp\left[-\left(\frac{N}{N_0}\right)^{\alpha_F}\right] \tag{5}$$

where $R_f(N)$ is the fraction of specimens surviving N cycles. The results for the fatigue shape parameter α_f and the characteristic lifetime N_0 are listed in Table 6 and are graphically shown in Fig. 7.

The RT shape parameter remains fairly constant at ~ 1.75 at stresses between 30 and 50% UTS; however, it increases to ~ 2.7 at 25% UTS. The change of the ETV shape parameter with the stress is not consistent. Yet the ETV parameter remains greater than the RT parameter at all stress levels.

The characteristic lifetime at ETV is longer than at RT; that is, the ETV environment improves the normalized fatigue strength although it reduces the static strength, as discussed earlier.

The fatigue failure sequences were monitored during testing. At all stress levels, longitudinal splitting was initiated at one of the end tabs, usually the lower moving grip, and preceded the final failure. The development of extensive longitudinal splitting could be taken as a sign of impending failure especially at low stress levels. A failure sequence at 70% UTS is shown in Fig. 8.

Whereas the static failure was brush-like, as mentioned earlier, the fatigue failures at stresses lower than $0.6\bar{X}$ were slivery. Above $0.6\bar{X}$, there was very little difference between the static and the fatigue failure mode. At ET and ETV, the failure modes tended to be more slivery than at RT. The same type of

TABLE 5—*Linear regression parameters for S-N relations.*

	a	b	Coefficient of Correlation
RT	1.115	0.154	0.9976
ETV	1.183	0.164	0.9953
ET	1.145	0.173	0.9967

TABLE 6—*Weibull parameters for fatigue life distributions.*

Fatigue stress, % UTS	25	30	40	50
RT—no. of specimens	15/4[b]	15	15	5
α_f	2.732	1.809	1.709	1.734
N_0	8.072×10^5	2.257×10^5	3.739×10^4	7.427×10^3
ETV[a]—no. of specimens	...	15/10[b]	15	15
α_f	...	3.046	1.943	2.893
N_0	...	1.339×10^6	8.620×10^4	1.404×10^4

[a]Specimens were preconditioned for 61 to 139 days.
[b]Number of runout specimens at 10^6 cycles.

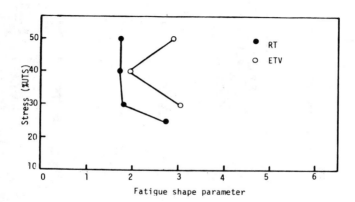

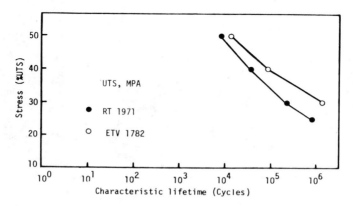

FIG. 7—*Weibull parameters for fatigue life distributions.*

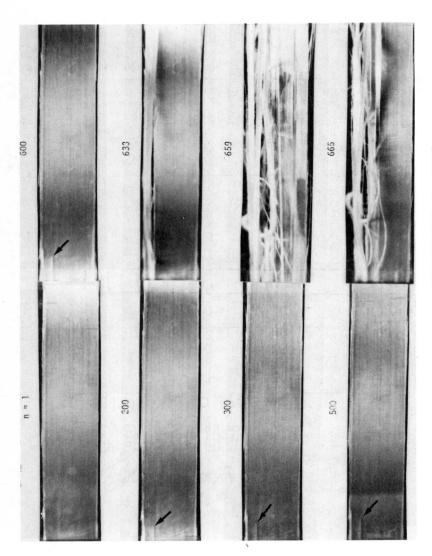

FIG. 8—*Fatigue failure process at 70% UTS, RT. Specimen 1lA-3.*

environmental effect was observed in static tension, as mentioned in the earlier section.

The strain energy density at the time of RT static failure is 35 MJ/m^3. At the fatigue stress of 60% UTS the strain energy density is reduced to 13 MJ/m^3. Since the energy to be dissipated upon failure is less in fatigue, the final failure will be less extensive.

To investigate the internal damage, a runout specimen tested at 25% UTS was examined at several cross sections on an optical microscope.

Figure 9 shows the damages at section A-A, which is cut slightly inside the tabs. The longitudinal splits observed outside the tabs are seen to continue into the tabs. It is to be noted that no fiber damage has been caused by the gripping.

Although the longitudinal splitting is mostly the result of matrix and interface failure, it involves fiber fracture as well to some extent, Fig. 10. However, the fractured fibers are quite limited to the region of splitting.

The lack of fiber damage was confirmed by burning the epoxy at 500°C of a runout specimen tested at 20% UTS and examining the fibers on an optical microscope. A few fiber breaks were observed along edges. However, no other fiber breaks could be detected on the surface.

The effect of the longitudinal splits on modulus was studied for a specimen (30B-B-8) tested at 40% UTS. The specimen did not show any splits except a little damage at edges up to 25 000 cycles. The modulus reduction at that point was ~5%. At 27 500 cycles a pair of parallel splits had already developed. The subsequent longitudinal splitting and the associated modulus change are schematically shown in Fig. 11. The longitudinal splitting does not reduce the modulus much unless it leads to the separation of a sliver from the specimen.

All the data on modulus change at RT are shown in Fig. 12. The maximum reduction in modulus before the full development of a sliver depends on the stress level: the higher the fatigue stress, the small the modulus reduction. For example, the maximum reduction is ~5% at 40% UTS, 7% at 30% UTS, and ~8% at 25% UTS. Therefore, it can be concluded that the global degradation before final failure, which is responsible for the modulus reduction, is more extensive at lower stresses. That is, the global degradation depends more on the number of cycles applied while the final failure is more stress dependent.

Conclusion

Mechanical properties of a filament-wound S2-glass/epoxy composite have been characterized in a flywheel service environment of 100°C and vacuum. The following conclusions derive from the present study.

1. Degradation of static properties at 100°C in vacuum is not significant. The temperature of 100°C has a more deleterious effect on the bulk epoxy than on the composite. One exception is the longitudinal shear property determined by using $(\pm 45)_{2s}$ laminate. The moisture desorption in vacuum increases

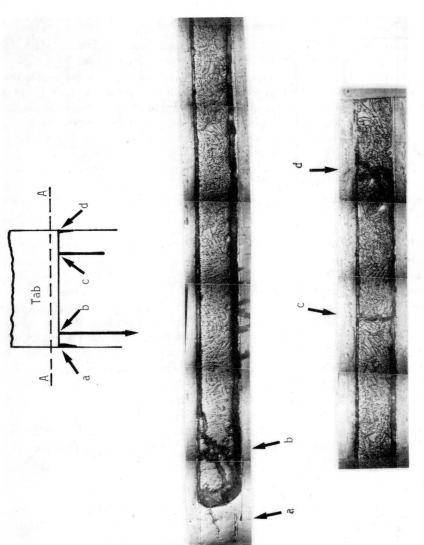

FIG. 9—*Photomicrograph of section within tabs.*

FIG. 10—Scanning electron micrograph of longitudinal splits.

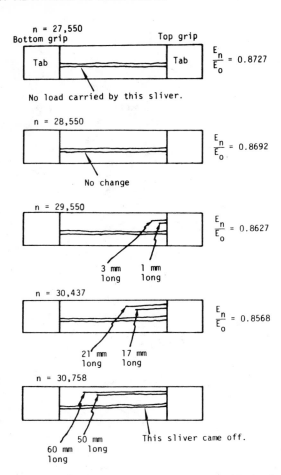

FIG. 11—*Growth of longitudinal splits and modulus change, RT, Specimen 30B-B-8.*

residual stresses which in turn leads to the cracking of plies. The ply cracking reduces the axial tension properties of the $(\pm 45)_{2s}$ laminate and hence the deduced shear properties.

2. The S2-glass/epoxy composite is very sensitive to fatigue. The fatigue strength at 10^6 cycles is only ~20% of the static strength at RT and at 100°C in vacuum.

3. The composite shows an excellent retention of fatigue strength at 100°C in vacuum in spite of its poor thermal stability.

4. The failure modes at stresses above 60% UTS are brush-like whereas those below 60% UTS are slivery. Failure at 100°C tends to be more slivery than bush-like.

5. The composite does not seem to be stable at 100°C in vacuum because it

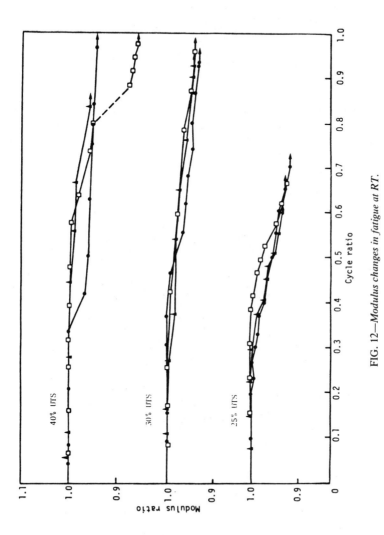

FIG. 12—*Modulus changes in fatigue at RT.*

continues to lose weight even after 150 days. A change of color to brown accompanies the weight loss.

6. The failure initiation is in the form of longitudinal splitting starting at the tab ends. Longitudinal splitting is more extensive in fatigue. The prevention of longitudinal splitting may prolong fatigue life.

7. The composite is susceptible to modulus reduction in fatigue without any apparent longitudinal splitting. The maximum modulus reduction before final failure increases with decreasing stress level.

8. Little damage is observed outside the regions of longitudinal splitting.

Acknowledgments

This paper is based on work supported by the Lawrence Livermore National Laboratory under Subcontract 6641009 as part of the Flywheel Program. The authors wish to thank T. T. Chiao, project engineer, for his guidance and helpful suggestions.

References

[1] *Proceedings*, 1975 Flywheel Technology Symposium, ERDA 76-85, ERDA and Lawrence Livermore Laboratory, Livermore, CA, 1975.
[2] *Proceedings*, 1977 Flywheel Technology Symposium, CONF-771053, Department of Energy, Washington, DC, 1978.
[3] Hahn, H. T. and Chin, W. K., "Mechanical Properties of an Aramid Fiber/Epoxy Composite for Flywheel Applications" in *Proceedings*, 37th Annual Conference, RP/CI, SPI, 1982, 24-C.
[4] Hahn, H. T., Hwang, D. G., Chin, W. K., and Lo, S. Y., "Mechanical Properties of a Filament-Wound S2-Glass/Epoxy Composite for Flywheel Applications," UCRL-15365, Lawrence Livermore National Laboratory, Livermore, CA, 1982.
[5] Tsai, S. W. and Hahn, H. T., *Introduction to Composite Materials*, Technomic Publishing Co., Westport, CT, 1980, p. 402.
[6] Hahn, H. T. and Hwang, D. G. in *Proceedings*, International Symposium on Engineering Science and Mechanics, H. M. Hsai and Y. L. Chou, Eds., National Cheng Kung University, Taiwan, Republic of China, 1981, p. 1229.
[7] Crossman, F. W., Mauri, R. E., and Warren, W. J., Hygrothermal Damage Mechanisms in Graphite-Epoxy Composites, NASA CR-3189, Dec. 1979.
[8] Cohen, A. C., *Technometrics*, Vol. 7, 1965, p. 579.
[9] Kolb, J. R., Chiu, I. L., and Newey, H. A., *Composites Technology Review*, Vol. 2, No. 3, 1980, p. 14.

Sokichi Umekawa,[1] Chi Hwan Lee,[2] Jo Yamamoto,[2] and Kenji Wakashima[2]

Effect of Coatings on Interfacial Reaction in Tungsten/Nickel and Tungsten/316L Composites

REFERENCE: Umekawa, S., Lee, C. H., Yamamoto, J., and Wakashima, K., **"Effect of Coatings on Interfacial Reaction in Tungsten/Nickel and Tungsten/316L Composites,"** *Recent Advances in Composites in the United States and Japan, ASTM STP 864,* J. R. Vinson and M. Taya, Eds., American Society for Testing and Materials, Philadelphia, 1985, pp. 619–631.

ABSTRACT: Effects of the fiber coating on interfacial phenomena in tungsten/nickel and tungsten/SUS316L (AISI 316L type stainless steel) composites were studied with special emphasis on the diffusion and the chemical reaction. A combination of tungsten fiber without doping and nickel matrix, in which recrystallization easily takes place at elevated temperature, was tested as a preliminary experiment. The combination is rather an extreme example with the recrystallization and interfacial reaction. Then, a composite with the tungsten fiber and 316L stainless steel matrix which would be more adoptable than pure nickel matrix for practical application was employed.

Oxides, carbides, and nitrides of such elements as aluminum, titanium, chromium, zirconium, and tantalum were examined as coating materials. The materials were selected on the basis of the chemical free energy of formation. The composites were exposed under the different kinds of thermal conditions. Effects of coating on the interfacial phenomena were evaluated by such characteristics of the composites as microstructure and mechanical behavior after the exposure.

Among all the coating materials tested, some of zirconium compounds were most effective in tungsten/nickel composites. Accordingly, zirconium dioxide and zirconium nitride coatings were mostly examined in a tungsten/316L stainless steel system. These coatings worked well as the reaction barrier at elevated temperature.

KEY WORDS: tungsten/nickel composites, tungsten/316L composites, interfacial reaction, fiber coating, coating materials, compounds, microstructure, mechanical properties

[1]Professor, Faculty of Science and Engineering, Department of Mechanical Engineering, Science University of Tokyo, Noda-shi, Chiba-ken, 278 Japan.

[2]Graduate student, Department of Materials Science and Engineering, associate, and professor, Research Laboratory of Precision Machinery and Electronics, Tokyo Institute of Technology, 4259 Nagatsuta, Midori-ku, Yokohama 227, Japan.

The major problem in a fiber-reinforced metal (FRM) for elevated-temperature usage is recrystallization of fiber as well as the chemical reaction between fiber and matrix. Strength and ductility of tungsten fiber obviously decrease due to the secondary recrystallization, especially for the equiaxial growth of new grains [1]. The recrystallization of tungsten fiber, which takes place at relatively low temperatures, depends on metallurgical and mechanical variables. Above all, diffusion of nickel and other specified elements in the matrix accelerates the recrystallization [2]. Utilizing a nickel-free matrix, iron-chromium-aluminum-yttrium (FeCrAlY) for example, was reported to be also effective in preventing the fiber from recrystallizing [3]. Fabrication of the FRM with a Fe-Cr-Al-Y-type matrix, however, seems to be rather complicated.

In the present work, the effects of the fiber coating on the interfacial phenomena were studied with special emphasis on the diffusion and the chemical reaction. A combination of tungsten fiber without doping and pure nickel matrix, in which recrystallization easily takes place, was tested as a preliminary experiment. The combination is rather an extreme example of the recrystallization and interfacial reaction. Next, a composite with the tungsten fiber and 316L stainless steel matrix which would be more adoptable than pure nickel for practical applications was employed. The coating materials were selected on the basis of the chemical-free energy of formation. Effects of the fiber coating on the interfacial reactions of composites were examined by metallography and mechanical testings.

Experimental Procedures

Materials

The fiber employed was commercial pure tungsten with a diameter of 500 μm. Two matrices were prepared from carbonyl nickel and commercial 316L stainless steel powder, 400 and 350 mesh, respectively. The chemical compositions of the fiber and the matrices are listed in Table 1. The combination of tungsten fiber without doping and nickel matrix, in which recrystallization of fiber and interfacial reaction would be most severe, was exercised to

TABLE 1—*Chemical compositions of tungsten fiber, carbonyl nickel, and 316L type stainless steel powder.*

Tungsten fiber	Fe	Mo	others	W				
Weight %	0.002	0.005	0.001	balance				
Carbonyl nickel	C		O		S	Fe	Ni	
Weight %	0.05 to 0.15		0.05 to 0.15		<0.001	<0.01	balance	
316L Stainless steel	Cr	Ni	Mo	Si	Mn	C	P	S	Fe
Weight %	17.19	12.93	2.21	0.72	0.23	0.017	0.014	0.006	balance

demonstrate the effectiveness of the coating onto the fiber. The 316L stainless steel matrix was selected for the fair compatibility with tungsten fibers and the better performance at elevated temperature.

Coatings

The straightened tungsten fibers of approximately 110 mm in length were cleaned by electropolishing in 2 percent NaOH solution. Oxides, carbides and nitrides of such metals as aluminum, titanium, chromium, and tantalum were examined for barrier coatings. The thickness of coating layers is summarized in Table 2. Coating on fiber was conducted by an activated reactive evaporation process using electron beam type physical vapor deposition techniques as previously presented [4]. X-ray diffraction revealed that most of coating substances obtained in the present work are compounds corresponding to the stoichiometric composition.

Fabrication and Thermal Exposure

Both coated and uncoated tungsten fibers were cut into lengths of 40 mm. The layers of unidirectionally arranged fibers were alternately embedded in layers of nickel matrix green of 1.5-mm thickness. The nickel matrix green thus piled up with the fibers was hot-pressed under a pressure of 35 MPa in an argon atmosphere for 7.2 ks. The sintering temperatures were 773 K for uncoated, 1173 K for Al_2O_3, Al_4C_3, TiC, Cr_2O_3, TaC, and 1373 K for ZrC and ZrO_2 coated fibers, respectively. For the composites with 316L stainless steel matrix, on the other hand, specimens were prepared by the following technique. Unidirectionally arranged layers of fibers were alternately settled in 316L stainless steel matrix powder. The powder thus composed with the fibers was compacted through two steps: (1) hot-pressed under a pressure of 27 MPa for 21.6 ks at 973 K and trimmed for the next step; (2) hot-pressed under a pressure of 35 MPa for 5.4 ks at 1373 K in an argon atmosphere. Both specimens for mechanical tests were machined to a size 40 mm long and 4 mm square. The fiber volume fractions were 11% for nickel matrix and 18% for 316L stainless steel matrix. Thermal exposure experiments were con-

TABLE 2—*Thickness of coating materials.*

Materials	Thickness, μm	Materials	Thickness, μm
ZrC	4	TiC	5
ZrO_2	4	Al_2O_3	7
ZrN	4	Cr_2O_3	3
TaC	1	Al_4C_3	1

ducted at 1423 K for the specimen with nickel matrix and at 1473 K for the specimen with 316 stainless steel matrix for some defined period of time up to 720 ks in an argon atmosphere.

Metallography

Metallographic observation and chemical analysis for microstructural variation were carried out by means of optical microscopy and electron microprobe X-ray analysis (EMX), respectively. Fracture surfaces of specimens were observed by scanning electron microscopy (SEM). Micro-Vickers hardness was measured related with the microstructure.

Results and Discussion

Interfacial Reaction

Tungsten/Nickel Composites—Various features of interfacial reaction in composites with and without coatings exposed at 1423 K for 720 ks are summarized in Fig. 1. For uncoated fiber FRM, interfacial reaction was not detected during fabrication which was processed at 773 K. It was found that the interfacial reaction is remarkable with an increase of exposure time, and a recrystallized zone of approximately 45 μm thick is formed at the circumferential region of the fiber after the exposure for 720 ks as shown in Fig. 1a. After hot-pressing at 1173 and 1373 K, no interfacial reaction was observed in most specimens with coated fibers. However, a little reaction accompanied by debonding of coating layer from fiber was detected in specimens with coatings such as Al_4C_3 and TaC. After exposure at 1423 K for 720 ks, most of the composite with coated fibers were observed to be effective against interfacial reaction. Little reaction was observed in the vicinity of interface of composites with ZrO_2 coated fibers as shown in Fig. 1b. Similar results were obtained in both FRM with ZrC and Al_2O_3 coated fibers. For TaC coatings, a thick reaction zone was found at the interface as shown in Fig. 1c. The reaction zone of 20 μm thick was observed in specimen with TiC coatings. For the Al_4C_3 coatings, recrystallization zone was found at the circumferential region of the fiber as presented in Fig. 1d. Effect of Cr_2O_3 coatings on the extent of recrystallization was almost similar to that of Al_4C_3 coatings. In the line profile as shown in Fig. 2, the ZrO_2 coating layer is very sound and none of diffusion between tungsten and nickel was detected

Tungsten/316L Stainless Steel Composites—For uncoated fiber, the reaction zone of about 8 μm thick was found in the specimen hot-pressed at 1373 K as shown in Fig. 3a. The reaction zone grew thicker with an increase of exposure time. After exposure at 1473 K for 720 ks, the reaction zone of approximately 70 μm thick was observed as shown in Fig. 3b. Here, the presence of cracks in the reaction zone presumably comes from the substantial

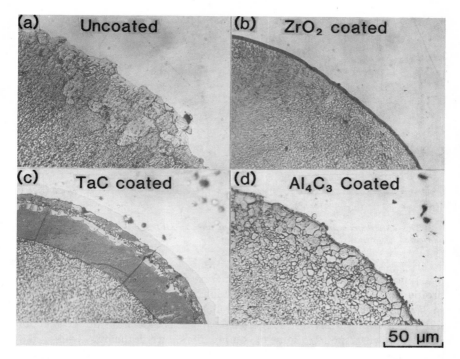

FIG. 1—*Microstructures in the vicinity of interface between fiber and matrix in tungsten/ nickel composites exposed at 1423 K for 720 ks. (a) Uncoated, (b) ZrO_2-coated, (c) TaC-coated and (d) Al_4C_3-coated fibers.*

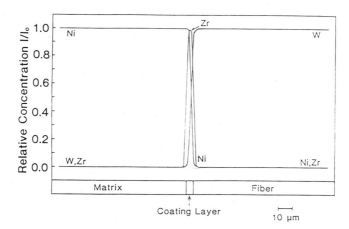

FIG. 2—*Line profile across interface between ZrO_2-coated fiber and nickel matrix, after exposure at 1423 K for 720 ks.*

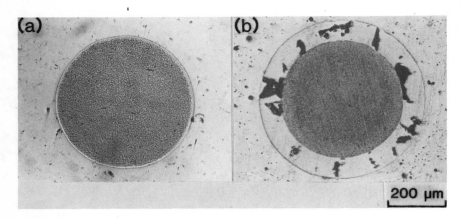

FIG. 3—*Microstructures of uncoated tungsten fiber in 316L stainless steel matrix:* (a) *as-fabricated,* (b) *exposed at 1473 K for 720 ks.*

difference of thermal expansion coefficients of the different phases. The line profile in Fig. 4 indicates that the reaction zone consists of several elements such as tungsten, iron, chromium, and nickel. The results of chemical analysis and microhardness test suggested that the composition of the reaction zone would correspond to some of compounds shown in the phase diagram of iron-tungsten system [5]. The relation between the thickness of the reaction zone and square root of the time was linear as shown in Fig. 5. Growth of the reaction zone, therefore, is expected to be a diffusion-controlled process. For

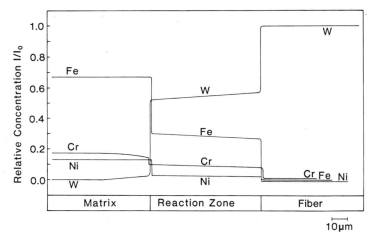

FIG. 4—*Line profile across interface between uncoated tungsten fiber and 316L stainless steel matrix, after exposure at 1473 K for 720 ks.*

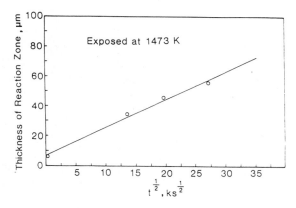

FIG. 5—*Thickness of reaction zone versus square root of exposure time at 1473 K.*

coated fiber in the tungsten/316L stainless steel system, such coating materials as ZrO_2 and ZrN were examined. In hot-pressing process at 1373 K and even after exposure at 1473 K, very little reaction was observed in the vicinity of interface between ZrO_2 coated fiber and matrix as shown in Figs. 6a and 6b. A similar result was obtained for in composites with ZrN-coated fibers. Electron backscatter and characteristic X-ray images of each elements in the vicinity of the ZrO_2 coatings were obtained by EMX as shown in Fig. 7. The presence of the coating layer between fiber and matrix can be detected in electron backscatter image as shown in Fig. 7a. In Fig. 7d, a concentrated zirconium layer is observed at the interface. Uniformly distributed patterns of each elements of tungsten, chromium, iron, and nickel were detectable as shown in Fig. 7b, c, e, and f. So far as the planar analysis is concerned, it

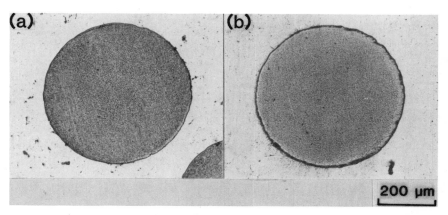

FIG. 6—*Microstructures of ZrO_2-coated fiber in 316L stainless steel matrix: (a) as-fabricated, (b) exposed at 1473 K for 720 ks.*

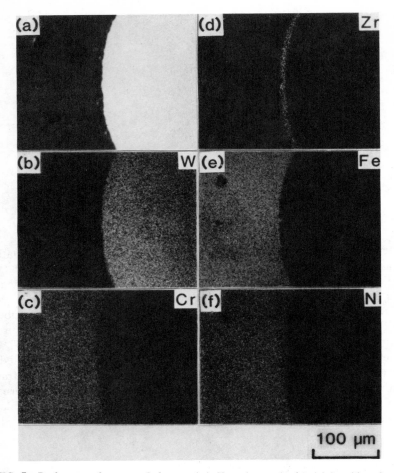

FIG. 7—*Backscatter-electron and characteristic X-ray images in the vicinity of interface between ZrO$_2$-coated tungsten fiber and 316L stainless steel matrix, after exposure at 1473 K for 720 ks.*

seems to be that neither of the elements of the matrix penetrates into the fiber, nor tungsten into the matrix. However, the line profile in Fig. 8 revealed that a small amount of chromium was detected in the ZrO$_2$ coating layer.

It can be concluded that coating materials such as ZrO$_2$ and ZrN play a role to prevent interfacial reactions between tungsten fiber and 316L stainless steel matrix.

Morphology and Mechanical Behaviors

Tungsten/Nickel Composites—The relation between the maximum load in bending test and time of exposure at 1423 K for composites both with and

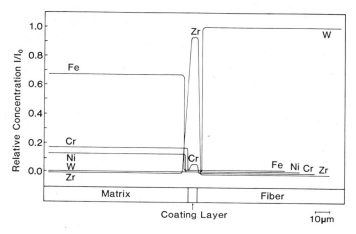

FIG. 8—*Line profile across interface between ZrO₂-coated fiber and 316L stainless steel matrix, after exposure at 1473 K for 720 ks.*

without coatings is shown in Fig. 9. Maximum load of composites with uncoated fibers was continuously decreased with the exposure time. The maximum load, however, was held with little change for composites with ZrO_2 and ZrC coated fibers for exposure times up to 720 ks. The load-deflection curves of the composites exposed at 1423 K for 720 K are shown in Fig. 10. The maximum load as well as deflection of coated fiber FRM is obviously greater

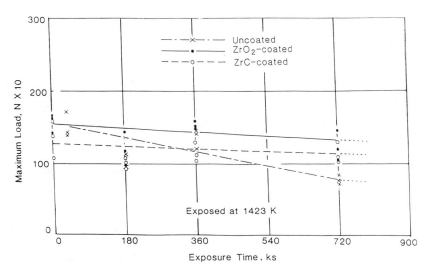

FIG. 9—*Maximum load in bending test at ambient temperature versus time of exposure at 1423 K in tungsten/nickel specimens.*

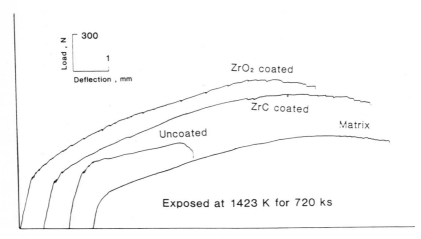

FIG. 10—*Load versus deflection curves of tungsten/nickel specimens in bending test at ambient temperature after exposure at 1423 K for 720 ks.*

than that of uncoated fiber FRM. The difference between the mechanical behaviors of these FRM as shown in Figs. 9 and 10 can be easily interpreted by comparing the morphologies of the fracture surfaces; an example is shown in Fig. 11. In uncoated FRM, the recrystallized zone of approximately 45 μm thickness was characterized by brittle failure as shown in Fig. 11a, and in ZrC coated fiber FRM, for example, no recrystallization or reaction were observed, as shown in Fig. 11b.

Tungsten/316L Stainless Steel Composites—Bending tests with specimens as-fabricated and exposed at 1473 K for 720 ks were conducted over the temperature range up to 773 K. In the as-fabricated condition, little difference between the maximum loads of specimens with coated and uncoated fibers was observed at elevated temperature, as shown in Fig. 12. In specimens exposed at 1473 K, the room temperature maximum load of composites with coated fibers was almost similar to that of composites with uncoated fibers as shown in Fig. 13. At elevated temperatures, the maximum load of composites with coated fibers, however, was considerably greater than that of composites with uncoated fibers. Specimens tested at 573 K are shown in Fig. 14. Specimens with coated fibers were bent without cracking, whereas specimens with uncoated fibers were broken accompanied by cracks. The difference of the maximum loads at elevated temperature between coated and uncoated fiber specimens as shown in Fig. 13 came from the difference of thickness of the reaction zone. This is again confirmed by comparing Fig. 13 with Fig. 12, in which similar values of maximum loads at elevated temperature for both specimens with and without coatings are shown. Impact tests with specimens both as-fabricated and exposed at 1473 K were carried out over the temperature range up to 1023 K. The results obtained by the impact test show an

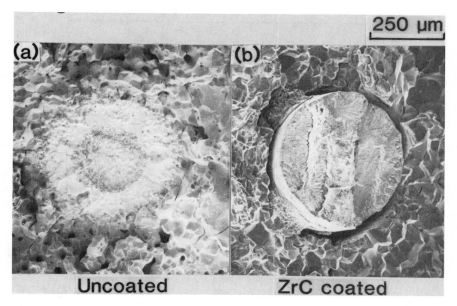

FIG. 11—*Fracture surfaces of tungsten/nickel composites in bending test at ambient temperature after exposure at 1423 K for 720 ks: (a) Uncoated fiber, (b) ZrC-coated fiber.*

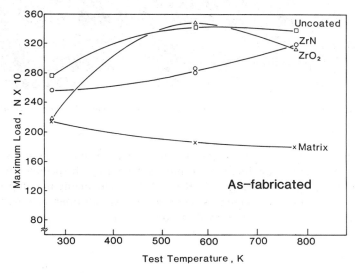

FIG. 12—*Maximum load in bending test versus temperature in as-fabricated tungsten/316L stainless steel specimens.*

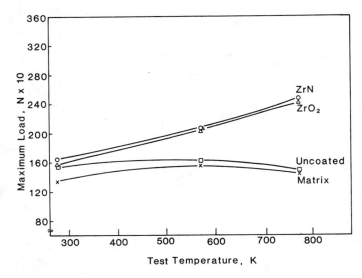

FIG. 13—*Maximum load in bending test versus temperature in specimens exposed at 1473 K for 720 ks.*

FIG. 14—*Appearances of tungsten/316L stainless steel specimens bending-tested at 573 K after exposure at 1473 K for 720 ks.*

essentially similar tendency for the coating effects to that of bending tests. However, the remarkable increase of impact strength with specimen was observed in the temperature range 473 K to 1023 K. The phenomenon presumably is attributed to the toughness increase of tungsten fiber accompanying an increase of ductility as described in previously [6].

Conclusions

The effect of coating to tungsten fiber on the interfacial reaction in tungsten/nickel and tungsten/316L stainless steel systems was examined with special emphasis on recrystallization of the fiber. Mechanical behaviors of the composites were also studied to evaluate the coating effect with relation to the interfacial reaction.

Oxides, carbides, and nitrides of such metals as aluminum, titanium, chromium, zirconium, and tantalum were exercised as coating materials in the tungsten/nickel system in which the reaction would take place severely. The coating materials were selected on the basis of low free energy of formation. Among all the coating materials tested, some of zirconium compounds were most effective in tungsten/nickel composites. Then, ZrO_2 and ZrN coatings were mostly examined in a tungsten/316L stainless steel system which is intended for practical applications at elevated temperature. Those coatings also worked well as the reaction barrier in that composite.

The effectiveness of coating onto the fiber results from a combination of intrinsic characteristics of the constituents, structural features of the coating layer, and the fabrication process of composites.

Acknowledgment

The authors wish to express their gratitude to Mr. T. Nakano and Mr. Y. Sugiura for both their experimental assistance and helpful discussions.

This study was supported partially by a Grant-in-Aid for Scientific Research from the Japan Ministry of Education, Science and Culture.

References

[1] Davis, G. L., *Metallurgia*, Vol. 57–58(10), 1958, pp. 177–184.
[2] Montelbano, T., Brett, J., Castleman, L., and Seigle, L., *Transactions*, Metallurgical Society of the American Institute for Mining, Metallurgical and Petroleum Engineers, Vol. 242, 1968, pp. 1973–1979.
[3] Essok, D. M., *Transactions*, Metallurgical Society of the American Institute for Mining, Metallurgical, and Petroleum Engineers, 1979, pp. 126–135.
[4] Lee, C. H., Yamamoto, J., and Umekawa, S. in *Proceedings*, Fourth International Conference on Composite Materials (ICCM-4), 1982, pp. 1289–1296.
[5] Hansen, M., *Constitution of Binary Alloys*, McGraw-Hill, New York, 1958, pp. 732–737.
[6] Allen, B. C., Maykuth, D. J., and Jaffee, R. I., *Journal of the Institute of Metallurgy*, Vol. 90, 1961, pp. 120–128.

Denise M. Aylor[1] and Robert M. Kain[2]

Assessing the Corrosion Resistance of Metal Matrix Composite Materials in Marine Environments

REFERENCE: Aylor, D. M. and Kain, R. M., **"Assessing the Corrosion Resistance of Metal Matrix Composite Materials in Marine Environments,"** *Recent Advances in Composites in the United States and Japan, ASTM STP 864,* J. R. Vinson and M. Taya, Eds., American Society for Testing and Materials, Philadelphia, 1985, pp. 632–647.

ABSTRACT: The corrosion resistance of graphite/aluminum (Gr/Al) and silicon carbon/aluminum (SiC/Al) metal matrix composite (MMC) materials and composite corrosion protection methods were investigated in marine environmental exposures ranging from 30 to 365 days. Accelerated corrosion of the Gr/Al resulted from galvanic interactions between the graphite fibers and the aluminum matrix. Corrosion of the SiC/Al was due to pitting, which was oriented at the silicon carbide-aluminum interfaces in the discontinuous forms of silicon carbide reinforcement. Differences in surface foil alloy and composite structure for the Gr/Al or type of silicon carbide reinforcement for the SiC/Al did not affect the materials' overall corrosion resistance. A variety of corrosion control coatings was found to be suitable for protection of the MMC, including organic and thermal-sprayed coatings.

KEY WORDS: metal matrix composites, graphite/aluminum, silicon carbide/aluminum, aluminum, coatings, marine corrosion

Select aluminum alloys have long been utilized in marine environments owing to their lightweight structure and generally good corrosion resistance. In particular, the 5000 series alloys and 6061 alloy aluminum have found successful application in seawater and in related marine atmospheres. The development of fiber-reinforced aluminum metal matrix composites has sparked interest in the use of these composites in marine environments due to the added strength, modulus, and stiffness properties over conventional aluminum alloys.

[1]Materials engineer, David W. Taylor Naval Ship Research and Development Center, Bethesda, MD 20084.
[2]Corrosion scientist, LaQue Center for Corrosion Technology, Inc., Wrightsville Beach, NC 28480.

The type of reinforcements to the aluminum matrix that were studied in this investigation included graphite (Gr) and silicon carbide (SiC) continuous fibers as well as discontinuous SiC in forms of particulate and whiskers. These reinforcements could promote additional corrosion concerns beside the pitting attack that is typical of some aluminum (Al) alloys. For example, the large potential difference between graphite and aluminum in the Gr/Al material invites accelerated galvanic corrosion upon exposure in a conductive electrolyte. In addition to the type of reinforcement in the composite, the variety of fabrication techniques employed could also add to increased corrosion at fiber-to-metal or metal-to-metal interfaces due to the structure of the composite.

The aim of this paper is to characterize the marine corrosion resistance of Gr/Al and SiC/Al metal matrix composites (MMC) through an understanding of their corrosion mechanisms. Changes in the corrosion resistance of the composites due to variations in surface foil alloy and composite structure fabrication for the Gr/Al MMC and different SiC forms (continuous fiber, whisker, or particulate) for the SiC/Al are also discussed. Finally, potential corrosion protection methods for Gr/Al and SiC/Al MMC are reviewed.

Experimental Test Program

Metal Matrix Composite Fabrication

Graphite/Aluminum—Gr/Al composites were all fabricated using the same basic process. In this process, tows (bundles) of continuous graphite fibers are formed into Gr/6061 Al composite wires by the titanium/boron (Ti/B) liquid metal infiltration process [1]. The wires are then stacked and hot-press diffusion bonded between aluminum alloy surface foils of Grades 6061 Al, 5056 Al, or 1100 Al. This forms a standard Gr/Al composite. An encapsulated composite has additional aluminum foils interwoven among the Gr/Al wires before the composite is bonded between the aluminum surface foils.

Silicon Carbide/Aluminum—SiC/Al composites included particulate (P), whisker (W), and continuous fiber (F) forms of silicon carbide in a 6061 Al alloy matrix. The SiC_p/Al panels were fabricated from atomized 6061 Al powder and 25 volume % SiC particles (1 to 10 μm size) blended together and hot-pressed between approximately 0.05-mm-thick (0.002 in.) 1100 Al alloy foils to form the desired shape. SiC_w/Al panels utilized 6061 Al powder and 20 weight % Grade F-9 silicon carbide whiskers (1 to 10 μm size). The blended powders were hot-pressed into two sheets, each 0.257 by 0.505 by 0.003 m (10.125 by 19.875 by 0.10 in.), and then saw-cut to the corrosion panel size.

SiC_F/Al composite panels were formed by successively stacking rows of 0.142-mm-diameter (0.0056 in.) SiC fibers (plasma-sprayed with 6061 Al) and 6061 Al alloys foils and diffusion bonding to yield a 2.5-mm-thick (0.1 in.) composite. The composite panels contained approximately 30 volume % silicon carbide fibers.

Surface Treatment Corrosion Protection Methods

Protection methods applied to the Gr/6061 Al MMC included sulfuric acid anodizing, chromate/phosphate conversion coating, electrodeposited aluminum/manganese, and electroless nickel. These coatings were applied to the aluminum foil surface by a variety of methods. The anodizing process formed a stable oxide coating on the aluminum surface by electrochemically polarizing the material in a sulfuric acid electrolyte and then by sealing in sodium dichromate. In contrast, the chromate/phosphate conversion coatings were formed on the composite surfaces by a chemical oxidation-reduction reaction which converted the natural aluminum oxide film to a chromate, a phosphate, or a complex oxide. Gr/Al panels with electrodeposited aluminum/manganese (Al/Mn) coatings utilized a very thin electroless nickel coating on their surfaces followed by plating of the Al/Mn. The electroless nickel coating involved no external current, but employed a chemical reducing agent to ultimately deposit nickel on the aluminum surface [2].

Coating systems applied to the SiC_w/Al panels consisted of sulfuric acid and chromic acid anodizing, various organic coatings, aluminum wire flame spray, and aluminum oxide plasma spray. Sulfuric acid and chromic acid anodizing was completed in standard anodizing tanks and then the panels were salt sealed. The surfaces of the organic coated panels were first prepared by sulfuric acid anodizing and sodium dichromate sealing. Epoxy primer and polyester or polyurethane topcoats were then applied to the panels employing conventional paint spraying techniques. The thermal sprayed panels were first grit-blasted with aluminum oxide and then either wire flame sprayed with 1100 Al or plasma sprayed with 99.5% Al_2O_3.

Marine Exposures

A complete identification of quantities of all composite and control specimens tested is included in Table 1. Test panel sizes varied among the composite and control specimens dependent on the quantity of material available. Exposed surface areas ranged from approximately 0.006 m² (9 in.²) to a maximum of 0.016 m² (25 in.²). Panel thicknesses varied between 0.002 m (0.075 in.) and 0.004 m (0.150 in.). In addition to the composite and control panels tested, Gr/6061 Al wires [each approximately 0.001 m (0.03 in.) diameter by 0.3 m (11 in.) long] were simultaneously exposed.

All specimens were cleaned with ethanol and air-dried prior to marine exposure. The test panels and wires were exposed in some or all of the following environments: (1) full immersion in quiescent, filtered natural seawater con-

TABLE 1—Panel exposures for Gr/Al and SiC/Al MMC.

Number of Panels Tested

Material / Environment:	25-m Atmospheric Lot		Splash/Spray		Filtered Seawater			
Test Duration (days):	60	180	60	180	30	60	90	180
Gr/6061 Al (encapsulated with 6061 Al foils)	…	2	…	2	…	1	1	1
Gr/6061 Al (standard with 5056 Al foils)	1	1	1	1	1	1	1	1
Gr/6061 Al (encapsulated with 5056 Al foils)	1	1	1	1	1	1	1	1
Gr/6061 Al (standard with 1100 Al foils)	1	1	1	1	1	…	…	1
Gr/6061 Al (encapsulated with 1100 Al foils)	…	2	…	2	…	…	…	2

Number of Panels Tested

Material / Environment:	25-m Atmospheric Lot				Splash/Spray		Alternate Tidal Immersion			Filtered Seawater			
Test Duration (days):	60	84	230	365	60	365	84	180	230	30	60	90	365
SiC$_p$/6061 Al	…	5	5	…	…	…	5	…	3	…	…	…	…
SiC$_w$/6061 Al	1	…	…	1	1	1	…	1	…	1	2	1	1
SiC$_F$/6061 Al	2	…	…	2	…	2	…	2	…	2	3	2	2
6061 Al controls	…	…	…	…	…	…	…	…	…	1	2	1	1

trolled at 30 ± 2°C (86 ±3.6°F), recirculated and completely refreshed six to seven times per day; (2) splash/spray exposure at a 30 cm (12 in.) test panel height above seawater; (3) alternate tidal immersion exposure; and (4) marine atmosphere exposure 25 m (80 ft) from the ocean. Exposure times ranged from 30 to 365 days. Panels were cleaned after test exposure by dipping in cold 30% HNO_3 and lightly brushing.

Results and Discussion

Prior exposure tests of Gr/Al composites in marine environments [3] identified the need for epoxy edge protection of the Gr/Al due to the problem of exfoliation at the edges when no protection was employed. The failure at the edges can be attributed to both the structure of the composite, which lends itself to disbonding at the wire to surface foil interface, and to a galvanic interaction between the graphite and the aluminum because of the exposed graphite fibers. Edge protection was applied to the SiC/Al composites but was later determined to be unnecessary due to the continuous matrix in the whisker and particulate SiC form and also because of the lack of accelerated galvanic interaction between the SiC and the Al.

Mass loss measurements made on the Gr/Al and SiC/Al panels were not accurate representations of the composites' corrosion due to the buildup of corrosion products and also the unknown quantity of water trapped in the epoxy edge protection which offset the mass loss. Thus, the corrosion analysis of all composites after exposure was limited to visual observations, metallography, and scanning electron microscopy.

Graphite/Aluminum

Corrosion Mechanisms—The corrosion mechanism for Gr/Al begins with pitting of the marine aluminum surface foils under low flow conditions. Previous testing [4] of Gr/6061 Al MMC and 6061-T6 Al control panels exposed for 30 days in various marine environments identified pit initiation on both material surfaces. Pitting progresses similarly on composite and control panels until the corrosion extends through the surface foils of the composite and provides marine exposure of the Gr/Al wires. Corrosion of the Gr/Al wires then comes into play. Figure 1 shows an as-received Gr/Al wire as well as wires after 30- and 70-day filtered seawater exposure. This figure highlights the buildup of aluminum corrosion products and the consequent increased wire diameter. The scanning electron micrograph in Fig. 2 pictures a Gr/Al wire exposed for 30 days in the splash/spray zone. The loss of aluminum from the wire with bare graphite fibers exposed is evident. This accelerated corrosion of the aluminum after such minimal exposure periods is predominately a result of the galvanic driving force of the Gr : Al couple. The Gr/Al wire cor-

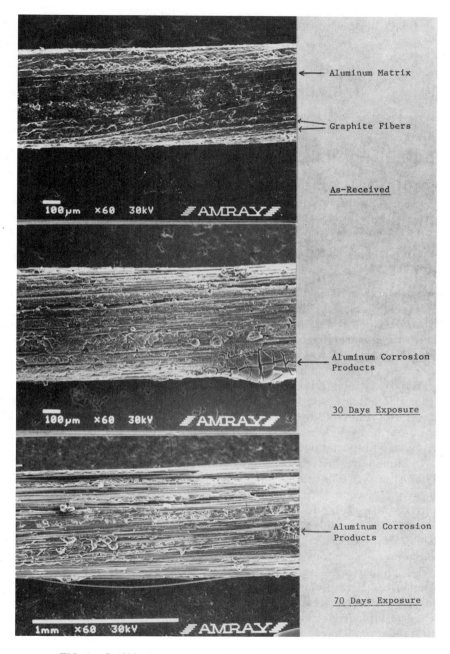

FIG. 1—*Gr/6061 Al wire before and after filtered seawater exposure.*

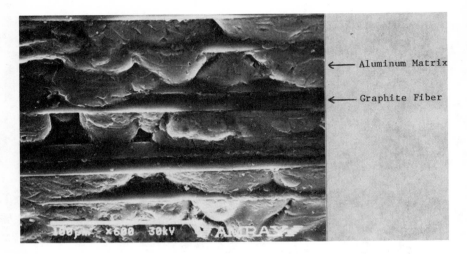

FIG. 2—*Gr/6061 Al wire after 30 days in splash/spray environment.*

rosion within a bonded composite panel causes volumetric expansion and eventual failure of the material.

Influence of Composite Structure—The corrosion resistance of the standard and encapsulated composites varied among the surface foil alloy tested. The 1100 and 5056 Al foil composites tested in splash/spray and marine atmosphere all exhibited light surface pitting through their 180-day exposures, with the exception of the encapsulated 1100 Al surface foil composite panel (Fig. 3). On this panel, degradation was due to edge exposure of the graphite fibers with resultant severe aluminum exfoliation and expulsion of the Gr/Al wires out of the side of the panel from the galvanic corrosion between the graphite and the aluminum.

In filtered seawater, the standard and encapsulated Gr/Al composites with 1100 Al surface foils showed no difference in corrosion behavior between the two types of composite structures. Both composite types experienced light surface pitting after 180-day exposure. The seawater-immersed composites with 5056 Al foils did exhibit more extensive pitting corrosion in the encapsulated form than in the standard form, but there was no penetration of the aluminum surface foils on the panels after their 180-day exposure periods.

To summarize, some slight differences in the corrosion behavior for the standard and encapsulated composites were observed, but no substantial corrosion resistance was gained by utilizing one form of composite over the other. These results agree with corrosion exposures completed by Vassilaros et al [3] on standard and encapsulated Gr/Al MMC with 6061 Al alloy foils.

Influence of Aluminum Alloy Surface Foil—The most accelerated attack of the Gr/Al composites, containing 6061, 5056, or 1100 Al alloy surface

FIG. 3—*Encapsulated Gr/Al panel with 1100 Al surface foils after 180-day 25-m (80 ft) atmospheric lot exposure.*

foils, was on those panels exposed in filtered seawater; the marine atmosphere and splash/spray environments were significantly less severe, with the former being slightly harsher. In general, no significant differences in corrosion attack were detected on the various surface foil alloy panels in marine atmospheric or splash/spray conditions after 180-day exposure.

The filtered seawater exposures revealed more accelerated corrosion on the Gr/Al MMC with 6061 Al foils than on the Gr/Al composites with 5056 Al or 1100 Al foils; the attack seen on the composite panels with 6061 Al foils was localized at the edges and progressed longitudinally along the wire path. These 6061 Al foil composite panels experienced blistering and exfoliation of the aluminum surface foils, resulting from exposure of the Gr:Al galvanic couple to the seawater. The 5056 Al foil panels exhibited substantial pitting corrosion (but not penetrating through the foils) on the encapsulated composites in seawater; the corrosion of the standard 5056 Al foil composites was limited to light surface pitting. Metal matrix composites with 1100 Al foils exposed in seawater experienced surface pitting similar to the standard 5056

Al foil panels. Figure 4 shows representative panels, highlighting the differences seen in seawater corrosion of the MMC.

Typically, marine aluminum alloys corrode in low-velocity seawater by pitting and the rate of pitting appears greater for 6061 Al than for 5056 and 1100 Al, which generally behave similarly [5]. For the Gr/Al composites tested, the differences in surface foil alloy or type of composite structure were not the major determinants of the composite corrosion resistance. The degradation of the encapsulated Gr/Al composites with 6061 Al or 1100 Al foils was not due to the rapid pitting of the surface foil alloy or the use of the encapsulated structure, but rather a result of accelerated galvanic corrosion from the Gr/Al couple exposed at the edges. Thus to effectively utilize Gr/Al composites in seawater, emphasis must be placed on preventing exposure of this galvanic couple.

Silicon Carbide/Aluminum

Corrosion Mechanisms—In general, the corrosion resistance of the SiC/Al MMC tested (particulate, whisker, and continuous fiber) was superior to the Gr/Al composites. This improvement is apparently due to the absence of any substantial galvanic driving force between the silicon carbide and the aluminum matrix. Observed corrosion on the SiC_p/Al panels was related to the corrosion resistance of the 1100 Al surface foils. The thinner 1100 Al foils utilized on the SiC_p/Al panels as compared to the Gr/Al composites with 1100 Al foils allowed the pitting to penetrate the surface foils within 230 days. After penetration of the surface foils, corrosion of the aluminum matrix continued and was concentrated around the silicon carbide particles. Figure 5a highlights the attack oriented around the SiC particles, showing a cross section of a SiC_p/Al panel after 230-day exposure in tidal immersion.

The mode of corrosion for the SiC_w/Al panels was also that of pitting. The size of the silicon carbide whiskers averaged smaller in diameter than the particulate, with the exception of a few larger-diameter whiskers scattered throughout the matrix. The scanning electron micrograph (SEM) included as Fig. 5b shows a cross section through a representative SiC_w/6061 Al panel after 60-day filtered seawater exposure. As in the SiC_p/Al panels, the pitting of the aluminum matrix was concentrated around the silicon carbide whiskers.

The aluminum pitting corrosion concentrated around the silicon carbide whiskers and particulate suggests that the crevices formed at these silicon carbide-aluminum interfaces may contribute to corrosion. Aluminum alloys are susceptible to crevice corrosion in seawater, and the corrosion usually takes the form of pitting [5]. Further analysis of these discontinuous SiC/Al composites is required to substantiate the validity of the crevice mechanism.

Degradation of the SiC_F/Al panels took the form of pitting of the 6061 Al surface foils that followed an intergranular path, which has been previously

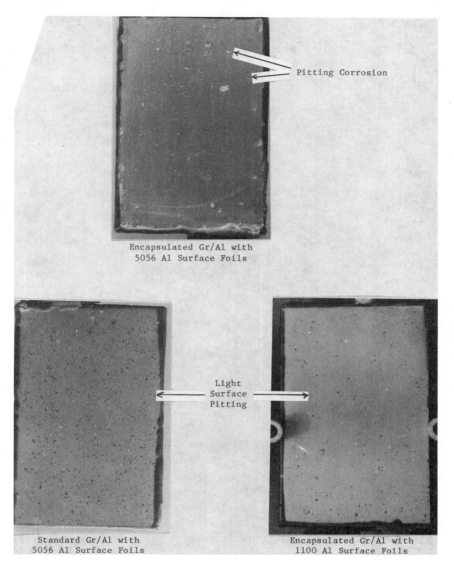

FIG. 4— *Various Gr/Al panels after 90-day exposure in filtered seawater.*

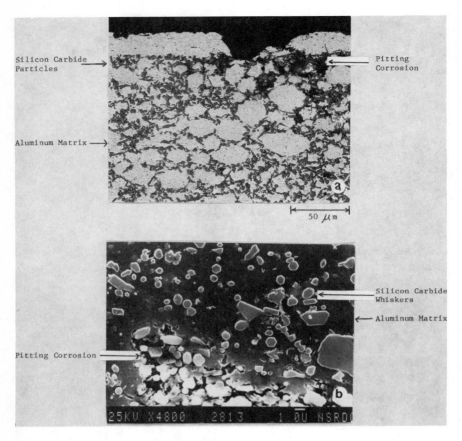

FIG. 5—*Cross sections through discontinuous SiC/Al panels: (a) SiC_p/6061 Al after 230-day tidal immersion exposure and (b) SiC_w/6061 Al after 60-day filtered seawater exposure.*

observed for 6061 Al in seawater [5]. Figure 6 shows evidence of the aluminum surface foil corrosion on a SiC_F/Al panel after exposure in filtered seawater for 60 days. Continuing exposure of the SiC_F/Al panels prevented the use of these composites for more detailed destructive metallographic analysis. The specimens which were inspected did not exhibit pitting through the aluminum foils and into the SiC/Al matrix. Thus, no assessment could be made as to whether the pitting was concentrated at the SiC-Al interfaces.

Silicon Carbide Reinforcement—In general, all three SiC reinforcements for the composite panels (continuous fiber, particulate, or whisker) retained good corrosion resistance through 365-day exposures. There were no catastrophic failures as with the Gr/Al panels, and the SiC/Al panels remained in test after the one-year period to generate longer-term data.

The most severe attack of the SiC_w/Al and SiC_F/Al composites was on

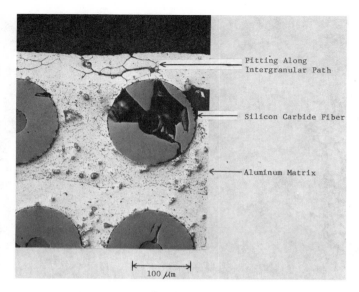

FIG. 6—*Cross-sectional view of SiC$_F$/6061 Al panel after exposure in filtered seawater for 60 days.*

those panels exposed in the filtered seawater environment. Figure 7 highlights the before-cleaning photographs of the composite and control panels after 60 days' seawater exposure, where the surface deposits are aluminum corrosion products from surface pitting. Behavior of the SiC/Al panels is similar, with the attack concentrating at the edges of the epoxy protection on the SiC$_w$/Al specimen and at more localized areas on the surface of the SiC$_F$/Al panel. The corresponding 6061 Al control panel shows superior corrosion resistance to the composites, exhibiting only a few select areas of attack on the surface.

Another comparison is made for the three types of SiC reinforced composites exposed in the 25-m (80 ft) atmospheric lot, which was the mildest environment for the SiC/Al panels. The pitting of the SiC$_w$/Al is slightly greater than that seen on the SiC$_F$/Al or SiC$_p$/Al panels after a maximum 365-day exposure, but overall the composite panels are similar (as in the filtered seawater environment).

A difference in composite corrosion behavior was detected between the SiC$_F$/Al and SiC$_w$/Al panels exposed in the splash/spray environment. Figure 8 shows more extensive aluminum corrosion products on the SiC$_w$/Al panel after 365 days in test, suggesting more severe surface pitting. The SiC$_w$/Al panels contained no aluminum surface foils as the silicon carbide continuous fiber reinforced composites did, and the pitting may have been slightly accelerated by the crevices created by the whiskers in the aluminum matrix. The degree of corrosion attack on the SiC/Al panels could not be assessed

6061 Al   SiC$_W$/6061 Al   SiC$_F$/6061 Al

FIG. 7—*SiC/Al and Al control panels after 60 days in filtered seawater.*

SiC$_W$/6061 Al   SiC$_F$/6061 Al

FIG. 8—*SiC/Al panels after one-year exposure in splash/spray zone.*

because the one-year exposure mark was an interim inspection and the corrosion products had to be preserved in order to obtain longer-term corrosion data.

Overall, visual observations of the SiC/Al composites exposed in a variety of marine environments for up to one-year periods revealed no substantial differences in the corrosion behavior between the types of silicon carbide reinforcement. Additional investigations of the corrosion behavior for the various SiC/Al composites will be initiated as longer-term exposures are completed.

Corrosion Protection Methods

A variety of surface treatments (both inorganic and organic) has been applied to Gr/Al and SiC/Al composites to identify adequate corrosion protection systems. The purpose of this section is to provide an indication of coatings that look promising after exposure in marine environments for a maximum of one year. Table 2 lists the coatings and thicknesses applied to the composite panels.

Of the coating systems evaluated on the Gr/Al composites, electroless nickel was the least successful. Severe composite degradation occurred (requiring removal of the panels after 70-day exposure in filtered seawater) due to galvanic interaction between the cathodic nickel coating and the anodic aluminum, exposed from pores in the nickel coating. Sulfuric acid anodizing provided good protection for the composites in all environments up through 180-day exposures. Thinning of the anodized layer with time was evident, and the composite panels were subsequently re-exposed to estimate the service life of this coating protection [4]. The chromate/phosphate conversion and the

TABLE 2—*Identification of coating systems applied to MMC.*

Material	Coating	Thickness, mm (mils)[a]
Gr/6061 Al (standard)	sulfuric acid anodizing	0.01 (0.5), 0.03 (1.0)
	chromate/phosphate conversion coating	. . .[b]
	electrodeposited aluminum/manganese	0.03 (1.0)
	electroless nickel	0.03 (1.0)
SiC$_w$/6061 Al	sulfuric acid anodizing	0.003 (0.1)
	chromic acid anodizing	0.003 (0.1)
	Al$_2$O$_3$ plasma spray	0.20 to 0.31 (8 to 12)
	epoxy primer with polyester or polyurethane topcoat	0.08 to 0.10 (3 to 4)
	1100 Al flame spray	0.20 (8), 0.38 (15)

[a]One mil equals 0.001 in.
[b]Coating weight approximately 70 mg/ft^2.

electrodeposited Al/Mn coatings experienced minimal degradation in the first year of exposure. Small surface pits were present on the Al/Mn coated panels and deeper, more localized pitting was seen on the conversion coated panels after 365-day exposure in filtered seawater; however, attack on these composites was not severe enough to warrant removal from the exposure environment.

Of the protection methods evaluated for the SiC_w/Al material, the coatings that exhibited the best corrosion resistance were the Al_2O_3 plasma spray, 1100 Al flame spray, and the organic epoxy coatings. After one-year test exposure, these coatings were unaffected by the environment, with the exception of ultraviolet effects from the marine atmosphere which dulled the brightly colored organic coatings. The anodized SiC_w/Al panels experienced pitting after only 30 days of filtered seawater exposure. SiC whiskers protruding from the aluminum matrix caused a nonuniform anodized surface and subsequent pitting of the composite. Anodizing would be a potential protection method for composites that contain aluminum foils on their surfaces, where the nonuniform surface would not be a concern.

Conclusions

1. The dominant corrosion mechanism for Gr/Al composites in marine environments is accelerated galvanic interaction between the aluminum and the graphite fibers. Suitable corrosion protection is required to effectively use Gr/Al in marine conditions.

2. Variations in the composite structure (standard versus encapsulated) or surface foil alloy (1100 versus 5056 versus 6061) do not significantly affect the corrosion resistance of the Gr/Al composites.

3. Pitting is the primary mode of attack for SiC/Al composites in marine environments. The pitting occurs at the silicon carbide/aluminum interfaces in the whisker and particulate SiC/Al composites and may be aggravated by a crevice situation at these interfaces.

4. The type of silicon carbide reinforcement (whisker versus particulate versus continuous fiber) utilized does not alter the overall corrosion behavior of the SiC/Al material.

5. Potential corrosion protection methods for aluminum matrix composites include organic coatings, thermal-sprayed coatings of aluminum and aluminum oxide, anodizing, electrodeposited Al/Mn, and chromate/phosphate conversion coatings. Of these methods, the first two are the most effective in the marine environment.

References

[1] Pfeifer, W. H. in *Hybrid and Select Metal Matrix Composites: A State of the Art Review*, W. J. Renton, Ed., American Institute of Aeronautics and Astronautics, New York, 1977, Chap. 6, pp. 159–252.

[2] Lyman, T., *Metals Handbook*, Vol. 2, *Heat Treating, Cleaning, and Finishing*, American Society for Metals, Metals Park, Ohio, 1984, pp. 443–445, and pp. 620–631.
[3] Vassilaros, M. G., Davis, D. A., and Steckel, G. L., "Marine Corrosion and Fatigue of Graphite Aluminum Composites" in *Proceedings*, Tri-Service Corrosion Conference, Colorado Springs, CO, Nov. 1980.
[4] Aylor, D. M., Ferrara, R. J., and Kain, R. M., "Marine Corrosion and Protection for Graphite/Aluminum Metal Matrix Composites," *Materials Performance*, Vol. 23, No. 7, July 1984.
[5] Godard, H. P. et al, *The Corrosion of Light Metals*, Wiley, New York, 1967, pp. 45–48, pp. 70–73, and pp. 125–141.

Jumpei Shioiri[1] and Katsuhiko Satoh[2]

Elevated-Temperature Internal Friction in an Oxide Dispersion-Strengthened Nickel-Chromium Alloy

REFERENCE: Shioiri, J. and Satoh, K., "Elevated-Temperature Internal Friction in an Oxide Dispersion-Strengthened Nickel-Chromium Alloy," *Recent Advances in Composites in the United States and Japan, ASTM STP 864*, J. R. Vinson and M. Taya, Eds., American Society for Testing and Materials, Philadelphia, 1985, pp. 648–658.

ABSTRACT: Internal friction in an yttrium oxide dispersion-strengthened nickel-chromium alloy (Inconel MA 754 alloy) was measured at temperatures ranging from 15 to 1200°C and at frequencies from 1 to 35 Hz. Measurements were made also for Nimonic 75 alloy, which has essentially the same chemical composition as the matrix of MA 754 alloy. In both alloys, the internal friction plotted against temperature is composed of a peak component and a background component. The activation energy of the grain boundary viscosity obtained from the peak component was 360 and 310 kJ/mol for MA 754 alloy and Nimonic 75 alloy, respectively. The background component was discussed in terms of the behavior of dislocations.

KEY WORDS: oxide dispersion-strengthened alloy, internal friction, high temperature

Because of their excellent load-carrying capacity at very high temperatures and because of their high melting temperatures compared with those of the current superalloys of more complicated chemical composition, the simple nickel-chromium alloys strengthened by oxide-dispersion are regarded as promising candidates for the next-generation materials of the combustors and vanes of gas turbine engines [1]. In the present work, the internal friction of an yttrium oxide dispersion-strengthened nickel-chromium alloy (Inconel MA 754 alloy) was measured. The internal friction is an important mechani-

[1]Professor, Department of Mechanical Engineering, College of Engineering, Hosei University, 3-7-2 Kajino-cho, Koganei-shi, Tokyo 184, Japan.
[2]Research associate, Department of Aeronautics, Faculty of Engineering, University of Tokyo, 7-3-1 Hongo, Bunkyo-ku, Tokyo 113, Japan.

cal property in itself, but, further, its temperature, frequency and strain amplitude-dependencies can provide important knowledge on the mechanism of deformation [2]. For comparison's sake, measurements were made also for Nimonic 75 alloy, which has essentially the same chemical composition as the matrix of MA 754 alloy.

Experimental Method

Measurements were made with a standard inverted torsion pendulum apparatus equipped with an infrared furnace. Figures 1 and 2 show the apparatus and a specimen, respectively. The test section (central parallel part) of the specimen is 25 mm in length and has a square cross section 3 mm in side length. The decay rate and frequency of the free oscillation of the torsion pendulum give, respectively, the internal friction and the elastic rigidity of the specimen. In this paper, as a measure of the internal friction, the "decrement" defined by $\Delta W/2W$ is used, where ΔW is the energy dissipated per

FIG. 1—*Inverted torsion pendulum apparatus equipped with an infrared furnace.*

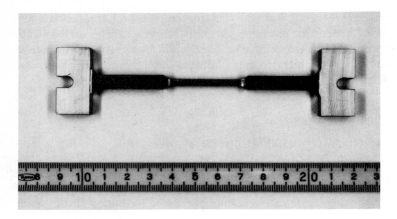

FIG. 2—*Specimen.*

cycle in the specimen and W is the strain energy stored in the specimen at the stroke end of the oscillation. In the torsion pendulum apparatus, the decrement can be obtained directly by measuring the logarithmic decrement of the free oscillation.

The tested materials were commercial products supplied by Inco Ltd. The chemical compositions are

MA   754: $Ni-0.06C-0.96Fe-0.001S-20.08Cr-0.32Al-0.42Ti$
$-0.39O-0.58Y_2O_3$

and

Nimonic 75: $Ni-0.10C-0.46Si-0.25Cu-4.12Fe-0.43Mn-19.53Cr$
$-0.35Ti-0.53Co-0.006S-2.4ppmPb$

The grain structure of MA 754 alloy used in this work was very similar to the standard structure characterized by the elongated grains and by the grain boundaries with irregularity, which was reported in Ref 3. The photomicroscopic structure of Nimonic 75 alloy was isotropic and the mean grain diameter was about 0.1 mm. Since the axis of the MA 754 specimen was taken in the direction of the hot working, the boundaries of the elongated grains could be active slip layers for torsional loading as far as their orientations were concerned.

Results and Discussion

Measurements were carried out at temperatures ranging from 15 to 1200°C and at frequencies from about 1 to 35 Hz. As illustrated in Fig. 3, in both of the alloys the measured internal friction plotted against temperature is composed of a peak component and a background component which increases monotonously with temperature. The Arrhenius plot of the temperature at which the peak appears against the pendulum frequency is shown in Fig. 4. The activation energies of the relaxation mechanism associated with the peak, which are determined from Fig. 4 for MA 754 and Nimonic 75, are 360 and 310 kJ/mol, respectively. Though the value of MA 754 is a little higher than that of Nimonic 75, the above values are around the activation energy for creep in Ni−20Cr alloy; for example, 368 kJ/mol [4] and 297 kJ/mol [5].

The above two components of the internal friction are usually seen together in the cases of the polycrystalline materials. Kê [6] ascribed the peak to the stress relaxation due to the viscous slip at the grain boundaries. Further, he pointed out [7] that the activation energy associated with the grain boundary slip is essentially identical to the activation energies for self-diffusion and for creep. Because of the correlation between the present results and the above finding by Kê, it may be reasonable to ascribe the peak of the internal friction observed both in MA 754 and Nimonic 75 to the grain boundary slip. However, it must be noted that in MA 754 there is a possibility that the diffusion around the dispersed particles also causes an internal friction peak [8–10].

Both the activation energy and the temperature (absolute) at which the peak appears in MA 754 are higher than those of Nimonic 75 by about 20%. The above correlation between the activation energy and the peak temperature may be explained in the following way. In general, the relaxation peak in a linear viscoelastic system appears when

$$\omega\tau = 1 \qquad\qquad (1)$$

where ω is the angular frequency and τ is the relaxation time. In the thermally assisted relaxation, τ can be given in the form of

$$\tau = A \exp(E/RT) \qquad\qquad (2)$$

where $\exp(E/RT)$ has the usual meaning and A is determined, in the present case, by the geometry of the grain structure and the pre-exponential factor in the grain boundary viscosity. Of course, A may differ considerably between MA 754 and Nimonic 75 owing to the peculiar grain structure in MA 754, but presumably the exponential factor will play a dominant role in τ. The above consideration can interpret the difference in the peak temperature in terms of the difference in the activation energy. The difference in the activation energy is presumably due to the difference in the chemical composition.

A large difference between the present two alloys is observed also in the background component of the internal friction which can be ascribed to the anelasticity of the grains themselves. Figures 5 and 6 show the differences in the frequency dependence and in the strain amplitude dependence, respec-

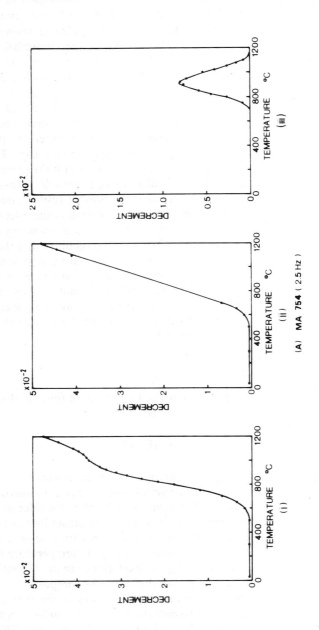

(A) MA 754 (2.5 Hz)

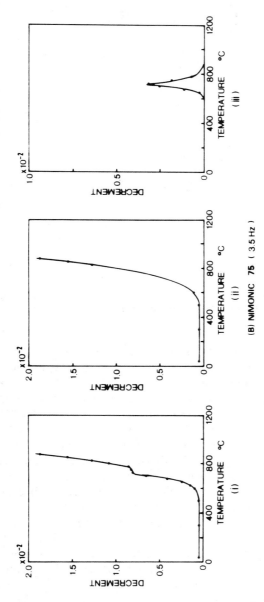

FIG. 3—*Internal friction plotted against temperature and its separation into components:* (i) *Measured internal friction.* (ii) *Background component.* (iii) *Peak component. (amplitude of maximum shear strain:* 3×10^{-5}).

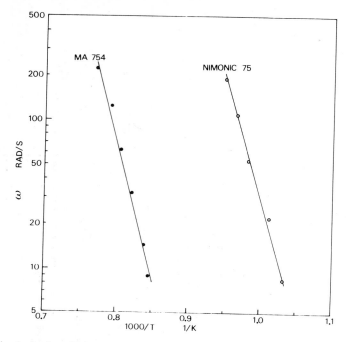

FIG. 4—*Arrhenius plot of the temperature at which the peak appears against the angular frequency of the pendulum (amplitude of maximum shear strain: 3 × 10⁻⁵).*

tively. As is seen in Fig. 5, the background components of MA 754 and Nimonic 75 begin to rise from about 600°C in a very similar manner, but eventually the rise in Nimonic 75 becomes steeper than in MA 754. The frequency dependence is also larger in Nimonic 75. Figure 6 shows that the strain amplitude at which the background component begins to rise in MA 754 is considerably higher than in Nimonic 75 especially at high temperatures.

The temperature dependence of the background component indicates that the internal friction is caused by the thermally assisted motion of the dislocations mobilized by the breakaway from the pinning points. The breakaway is also a thermally assisted process. When the temperature becomes high or when the frequency becomes low, the amplitude of the dislocation motion rate-controlled by the thermal activation will become large. In MA 754, when the amplitude of the dislocation motion becomes larger than the distance between the dispersed oxide particles, the effect of the restriction of the dislocation motion will become explicit. The above picture will explain the differences between the present two alloys in the temperature dependence and the strain amplitude dependence. The strain amplitude dependence shown in Fig. 6 indicates that the restriction of the dislocation motion by the dispersed particles, the main strengthening mechanism in MA 754, is effective and the

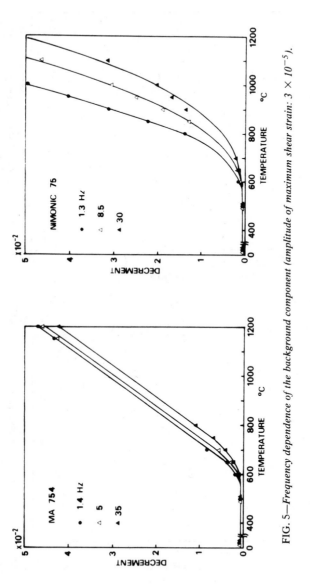

FIG. 5—*Frequency dependence of the background component (amplitude of maximum shear strain: 3×10^{-5}).*

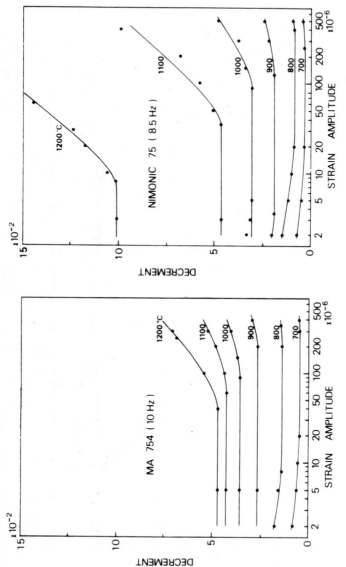

FIG. 6—Amplitude dependence of the background component.

increase in the internal friction with increasing temperature and strain amplitude is largely suppressed compared with Nimonic 75. However, when the
temperature becomes much higher or when the strain amplitude (accordingly, stress amplitude) becomes much larger, it becomes possible for dislocations to pass through the dispersed particles. At such high temperatures it
occurs by cross-slip or climb of dislocations thermally, that is, with the aid of
thermal activation, and at such large strain amplitudes it occurs by the bowing out of dislocations through the gaps between adjacent particles ather-
mally, that is, without the aid of thermal activation. Under these conditions,
the restriction of the dislocation motion by the dispersed particles is weakened and the internal friction will increase fairly remarkably with increasing
temperature and strain amplitude. In Fig. 6, the curve of 1200°C of MA 754
shows, at its large strain amplitude end, a sign of such an increase in the
internal friction.

The above discussion is of a qualitative nature, and in order to derive more
detailed knowledge on the strengthening mechanism of the oxide dispersion
alloy from the internal friction data, further theoretical work on the dislocation motion in the matrix with the dispersed particles seems necessary.

Conclusions

Internal friction measurements were made for MA 754 alloy (an yttrium
oxide dispersion-strengthened nickel-chromium alloy) and Nimonic 75 alloy
at temperatures ranging from 15 to 1200°C and at frequencies from 1 to
35 Hz. In both alloys, the measured internal friction plotted against temperature can be separated into a peak component and a background component.
The activation energies of the grain boundary viscosity determined from the
peak component were 360 and 310 kJ/mol for MA 754 alloy and Nimonic 75
alloy, respectively. Comparison between MA 754 alloy and Nimonic 75 alloy
in the temperature, frequency, and strain amplitude-dependencies of the
background component shows the effect of the restriction of the dislocation
motion by the dispersed oxide particles, which is the main strengthening
mechanism in MA 754 alloy. In order to derive more knowledge from the
internal friction data, further theoretical work is required.

References

[1] Signorelli, R. A., Glasgow, T. K., Halford, G. R. and Levin, S. R. in *Aeropropulsion 1979*,
    NASA CP-2092, National Aeronautics and Space Administration, Washington, DC, 1979,
    pp. 149–186.
[2] Zener, C., *Elasticity and Anelasticity of Metals*, The University of Chicago Press, Chicago,
    1948.
[3] Benjamin, J. S. and Larson, J. M., *Journal of Aircraft*, Vol. 14, No. 7, July 1979, pp. 613–
    623.

[4] Shahinian, P. and Achter, M. R., *Transactions*, American Society for Metals, Vol. 51, 1959, pp. 244-256.
[5] Sidey, D. and Wilshire, B., *Metal Science Journal*, Vol. 3, No. 2, March 1969, pp. 56-60.
[6] Kê, T. S., *Physical Review*, Vol. 72, No. 1, July 1947, pp. 41-46.
[7] Kê, T. S., *Physical Review*, Vol. 73, No. 3, Dec. 1948, pp. 267-268.
[8] Stobbs, W. M., *Philosophical Magazine*, Vol. 27, No. 5, May 1973, pp. 1073-1092.
[9] Koeller, R. C. and Raj, R., *Acta Metallurgica*, Vol. 26, No. 10, Oct. 1978, pp. 1551-1558.
[10] Monzen, R., Suzuki, K., Sato, A., and Mori, T., *Acta Metallurgica*, Vol. 31, No. 4, April 1983, pp. 519-524.

Thermomechanical Properties

Ronald D. Kriz[1] *and Hassel M. Ledbetter*[1]

Elastic Representation Surfaces of Unidirectional Graphite/Epoxy Composites

REFERENCE: Kriz, R. D. and Ledbetter, H. M., **"Elastic Representation Surfaces of Unidirectional Graphite/Epoxy Composites,"** *Recent Advances in Composites in the United States and Japan, ASTM STP 864*, J. R. Vinson and M. Taya, Eds., American Society for Testing and Materials, Philadelphia, 1985, pp. 661–675.

ABSTRACT: Unidirectional graphite/epoxy composites exhibit high elastic anisotropy and unusual geometrical features in their elastic-property polar diagrams. From the five-component transverse-isotropic elastic-stiffness tensor we compute and display representation surfaces for Young's modulus, torsional modulus, linear compressibility, and Poisson's ratios. Based on Christoffel-equation solutions, we describe some unusual elastic-wave-surface topological features. Musgrave considered in detail the differences between phase-velocity and group-velocity surfaces arising from high elastic anisotropy. For these composites, we find effects similar to, but more dramatic than, Musgrave's. Some new, unexpected results for graphite/epoxy include: a shearwave velocity that exceeds a longitudinal-wave velocity in the plane transverse to the fiber; a wave that changes polarization character from longitudinal to transverse as the propagation direction sweeps from the fiber axis to the perpendicular axis.

KEY WORDS: graphite/epoxy, acoustic properties, elastic properties, energy-flux deviation, group velocity, mode conversion

Graphite/epoxy composites with unidirectional fiber reinforcement exhibit transverse-isotropic elastic symmetry; thus, five independent elastic constants. In matrix form, one can represent the elastic-stiffness tensor as

[1]Materials research engineer and research metallurgist, respectively, Fracture and Deformation Division, Center for Materials Science, National Bureau of Standards, Boulder, CO 80303.

$$
[C_{ij}] = \begin{bmatrix} C_{11} & C_{12} & C_{13} & 0 & 0 & 0 \\ & C_{11} & C_{13} & 0 & 0 & 0 \\ & & C_{33} & 0 & 0 & 0 \\ & & & C_{44} & 0 & 0 \\ & & & & C_{44} & 0 \\ & & & & & (C_{11} - C_{12})/2 \end{bmatrix} \qquad (1)
$$

which is symmetrical, that is $C_{ji} = C_{ij}$. The elastic compliances, S_{ij}, relate inversely to the C_{ij}

$$
[S_{ij}] = [C_{ij}]^{-1} \qquad (2)
$$

In form, the S_{ij} matrix looks identical to the C_{ij} matrix except that $S_{66} = 2(S_{11} - S_{12})$.

While Eq 1 displays all the elastic-constant information as an "irreducible set," it displays directly very little of the wealth of practical information and interesting geometrical properties contained in C_{ij}.

For uniaxial graphite/epoxy, the present study considers many geometrical features of the elastic-constant tensor. For example: representation surfaces of Young's modulus, torsional modulus, linear compressibility, and Poisson's ratio. Following Musgrave [1], who studied materials such as zinc, we consider the velocity and slowness surfaces and find some remarkable features concerning the acoustic-energy flux. Following Ledbetter and Kriz [2], who focused on orthotropic materials, we find some peculiarities also in graphite/ epoxy. First, a shear velocity exceeds the longitudinal velocity in the plane perpendicular to the fibers. Second, a wave changes polarization character from longitudinal to transverse as the wave direction sweeps from the fiber axis to the perpendicular axis.

Calculations

Elastic Representation Surfaces

Representation of elastic properties by geometrical surfaces has obvious utility, as described by Nye [3]. Following a procedure outlined by Lekhnitskii [4], representation surfaces can be derived via the general elastic-compliance fourth-rank-tensor transformation

$$
S'_{ijkl} = a_{im} a_{jn} a_{kr} a_{ls} S_{mnrs} \qquad (3)
$$

Here we choose the x_3-axis parallel to the fibers of a unidirectional fiber-reinforced composite. We assume isotropy in the x_1-x_2 plane, transverse to the fibers. For hexagonal (transverse-isotropic) symmetry, transformations for all elastic constants are axially symmetric about x_3. Thus, elastic constants depend only on rotations, $+\theta$, away from x_3 toward the x_1-x_2 plane. Transformation equations for Young's modulus, $E_3(\theta)$, torsional modulus, $T_3(\theta)$, shear modulus, $G_{23}(\theta)$, linear compressibility, $K_3(\theta)$, and Poisson's ratio, $\nu_{32}(\theta)$, are in Voigt notation

$$E_3^{-1}(\theta) \equiv S'_{33}(\theta) = S_{11} + (2S_{13} + S_{44} - 2S_{11})x$$

$$+ (S_{11} + S_{33} - 2S_{13} - S_{44})x^2 \quad (4)$$

$$T_3^{-1}(\theta) \equiv [S'_{44}(\theta) + S'_{55}(\theta)]/2 = S_{44} + (S_{11} - S_{12} - S_{44}/2)(1 - x)$$

$$+ 2(S_{11} + S_{33} - 2S_{13} - S_{44})x(1 - x) \quad (5)$$

$$G_{23}^{-1}(\theta) \equiv S'_{44}(\theta) = S_{44} + 4(S_{11} + S_{33} - 2S_{13} - S_{44}/2)x(1 - x) \quad (6)$$

$$K_3(\theta) = S_{11} + S_{12} + S_{13} - (S_{11} - S_{33} + S_{12} - S_{13})x \quad (7)$$

$$\nu_{32}(\theta) \equiv -S'_{23}(\theta)/S'_{33}(\theta)$$

$$= E_3(\theta)[S_{13} + (S_{11} + S_{33} - 2S_{13} - S_{44})x(1 - x)] \quad (8)$$

where $x = \cos^2\theta$. We also consider the transformation of ν_{13} as shown in Fig. 1 where there are two rotations: α around x_2 followed by β around x'_1. Given the tilt angle α, we calculated ν_{13} in the x_2-x_3 plane. For this double rotation we find

$$\nu_{13}(\alpha, \beta) \equiv -S''_{13}(\alpha, \beta)/S'_{11}(\alpha) = -\{(S_{11} + S_{33} - S_{44})\cos^2\alpha \sin^2\alpha \cos^2\beta$$

$$+ S_{12}\cos^2\alpha \sin^2\beta + S_{13}[\cos^2\beta(\cos^4\alpha + \sin^4\alpha) + \sin^2\alpha \sin^2\beta]\}/S'_{11}(\alpha) \quad (9)$$

where

$$S'_{11}(\alpha) = S'_{33}(90 - \alpha) = E_3^{-1}(90 - \alpha) \quad (10)$$

When the tilt angle is zero β equals θ and Eq 9 reduces to

$$\nu_{13}(\theta) = -[S_{12}\sin^2\theta + S_{13}\cos^2\theta]/S'_{11}(\alpha) \quad (11)$$

Based on Eqs 4–11, we give polar diagrams, Figs. 2–8, that represent the unusual anisotropy of various elastic constants of unidirectional graphite/epoxy.

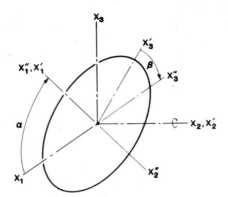

FIG. 1—*Coordinate transformations for ν_{13} (α, β). Uniaxial load along $x_I'' = x_1'$.*

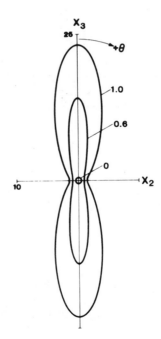

FIG. 2—*Polar diagram of Young's modulus E_3 for various fiber volume fractions, in GPa.*

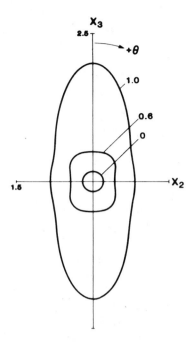

FIG. 3—*Polar diagram of torsional modulus* T_3 *for various fiber volume fractions, in GPa.*

Acoustic Representation Surfaces

The influence of elastic constants, calculated from Eqs 4–11, on stress-wave propagation is represented by polar diagrams of the eigenvalue solutions to Christoffel's equation [1]

$$(C_{ijkl}n_j n_k - \rho v^2 \delta_{il})p_i = 0 \qquad (12)$$

where

C_{ijkl} = fourth-rank elastic-stiffness tensor,
n_i = direction cosines of the wave vector measured from the principal material axes x_i,
ρ = the mass density,
δ_{il} = Kronecker delta,
v = phase velocity that propagates along n_i, and
p_i = particle-displacement direction cosines.

Solution of Eq 12 yields three eigenvalues, ρv^2, and the corresponding eigenvector, p_i, for each phase velocity. The orientation of the particle displacement with respect to the wave vector is denoted by subscripts (t, qt, ql) on

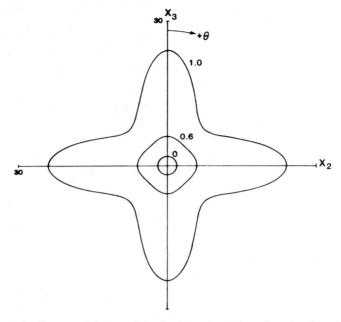

FIG. 4—*Polar diagram of shear modulus G_{23} for various fiber volume fractions, in GPa.*

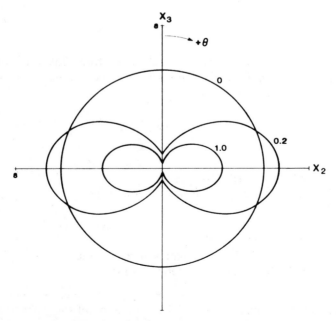

FIG. 5—*Polar diagram of linear compressibility K_3 for various fiber volume fractions, in GPa.*

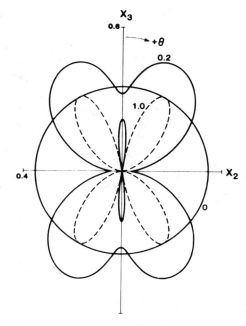

FIG. 6—*Polar diagram of Poisson's ratio* v_{32} *for various fiber volume fractions.*

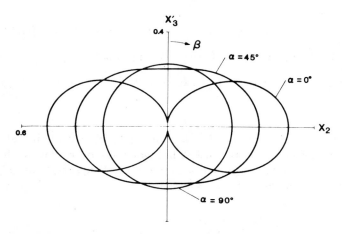

FIG. 7—*Polar diagram of Poisson's ratio* v_{13} *for various tilt angles,* α *(see Fig. 1), at a fiber volume fraction of 0.6.*

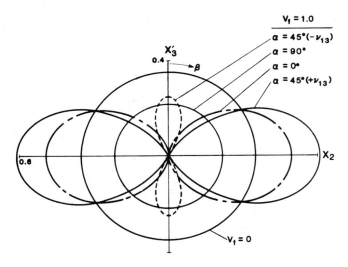

FIG. 8—*Polar diagram of Poisson's ratio v_{13}. Same as Fig. 7 with fiber volume fractions of 0.0 and 1.0.*

the phase velocities: t means that p_i is transverse to n_i ($n_i p_i = 0$), qt (quasi-transverse) means that the largest component of p_i is transverse to n_i ($0 < n_i p_i < 1\sqrt{2}$), and ql (quasi-longitudinal) means that the largest component of p_i is parallel to n_i ($1/\sqrt{2} < n_i p_i < 1$). Along principal material axes, pure transverse waves and pure longitudinal waves must exist for all wave phase velocities. The eigenvalue solutions, ρv^2, are the roots of the characteristic equation, written as the determinant of Eq 12

$$\begin{vmatrix} (C_{11}n_1^2 + C_{66}n_2^2 + C_{44}n_3^2) - \rho v^2 & (C_{12} + C_{66})n_1 n_2 & (C_{13} + C_{44})n_2 n_3 \\ (C_{12} + C_{66})n_1 n_2 & (C_{66}n_1^2 + C_{11}n_2^2 + C_{44}n_3^2) - \rho v^2 & (C_{13} + C_{44})n_2 n_3 \\ (C_{13} + C_{44})n_2 n_3 & (C_{13} + C_{44})n_2 n_3 & (C_{44}n_1^2 + C_{44}n_2^2 + C_{33}n_3^2) - \rho v^2 \end{vmatrix} = 0$$

$$(13)$$

Velocity surfaces are calculated from Eq 13 by considering all possible wave vector orientations. Because of hexagonal elastic symmetry, the velocity surfaces are also axially symmetrical around x_3. Thus, solutions to Eq 13 are calculated as functions of $+\theta$ from x_3

$$v_t(\theta) = [(C_{44}\cos^2\theta + C_{66}\sin^2\theta)/\rho]^{1/2} \tag{14}$$

$$v_{ql}(\theta) = [(C_{44} + C_{11}\sin^2\theta + C_{33}\cos^2\theta - \sqrt{C})/2\rho]^{1/2} \tag{15}$$

$$v_{qt}(\theta = [(C_{44} + C_{11}\sin^2\theta + C_{33}\cos^2\theta + \sqrt{C})/2\rho]^{1/2} \qquad (16)$$

where

$$C = [(C_{11} - C_{44})\sin^2\theta + (C_{44} - C_{33})\cos^2\theta]^2 + (C_{13} + C_{44})^2\sin^2 2\theta$$

One should note that subscripts t, qt, and ql represent only labels. In special cases, displacements t, qt, ql, and l may coexist on the same phase-velocity surface [2]. Thus, rarely, Eqs 15 and 16 yield polarizations different from the designated qt and ql subscripts. Equation 14 represents an exception where v_t is always a pure-transverse wave for all orientations.

Intersections of these velocity surfaces with the isotropic x_1-x_2 plane yield three concentric circles with pure-longitudinal or pure-transverse modes. Unlike isotropic materials, the two pure-transverse velocities calculated from Eqs 14 and 15 do not coincide in the isotropic x_1-x_2 plane where

$$v_t(90) = (C_{66}/\rho)^{1/2}; \qquad v_{qt}(90) = (C_{44}/\rho)^{1/2} \qquad (17)$$

The eigenvector solutions to Eq 12 are the particle displacements, $\{p_1:p_2:p_3\}$

$$p_1 = [n_1^2(C_{11} + C_{12})/2]^{1/2}/[\rho v^2 - (C_{11} - C_{12})/2 - C_{44}n_3^2] \qquad (18)$$

$$p_2 = [n_2^2(C_{11} + C_{12})/2]^{1/2}/[\rho v^2 + (n_2^2 - n_1^2)C_{11}/2 + C_{12}/2 - n_3^2 C_{44}] \qquad (19)$$

$$p_3 = [2n_3^2(C_{13} + C_{44})^2/(C_{11} + C_{12})]^{1/2}/\{\rho v^2 + (n_2^2 - n_1^2)C_{44}$$

$$+ n_3^2[2(C_{23} + C_{44})^2/(C_{11} + C_{12}) - C_{33}]\} \qquad (20)$$

The acoustic energy may propagate in a direction different from the wave vector. This direction of propagation is the energy-flux vector, and the corresponding velocity of propagation is the group velocity, v_g. For homogeneous plane waves, the energy-flux vector and v_g coincide [5]. For isotropic materials, energy-flux vector and group velocity coincide with the wave vector and with phase velocity. For anisotropic materials the energy-flux vector deviates from the wave vector by an angle, Δ. For plane homogeneous waves, the same deviation angle exists between phase and group velocities

$$v_g(\theta,\Delta) = v(\theta)/\cos \Delta \qquad (21)$$

Polar diagrams of group velocities can be calculated from Eq 21. We follow Musgrave [6] and plot v_g as a function of $\theta' = \theta + \Delta$. Due to the deviation of energy-flux, group-velocity surfaces appear distorted when compared with phase-velocity surfaces. Hence, polar diagrams of v_g are also called energy surfaces.

One calculates the deviation angle, Δ, from the direction cosines (n_i, L_i) of the wave vector and energy-flux vector, respectively

$$\Delta = \cos^{-1} n_i L_i \qquad (22)$$

For orthorhombic symmetry, Kriz [7] gave the direction cosines, L_i, of the energy-flux vector. For hexagonal symmetry, with transformations in the x_2-x_3 plane $(n_1 = 0)$, these become

$$L_1 = C_{66} p_1 p_2 n_2 + C_{44} p_1 p_3 n_3 \qquad (23)$$

$$L_2 = C_{66} p_1^2 n_2 + C_{22} p_2^2 n_2 + C_{23} p_2 p_3 n_3 + C_{44}(p_2 p_3 n_3 + p_3^2 n_2) \qquad (24)$$

$$L_3 = C_{22} p_3^2 n_3 + C_{33} p_2 p_3 n_2 + C_{44}(p_2^2 n_3 - p_2 p_3 n_2) \qquad (25)$$

Effects of fiber volume fraction on the effective C_{ij} were calculated using relationships derived by Datta, Ledbetter, and Kriz [8].

Results and Discussion

Polar diagrams shown in Figs. 2–8 were calculated using the elastic constants in Table 1. Matrix and graphite-fiber elastic constants were obtained from Ref 9. For comparison, Table 1 includes elastic constants for other polyacrylonitrile (PAN) graphite fibers.

The peculiar elastic behavior caused by the elastic anisotropy of the graphite fibers is shown in Figs. 2–8 at various fiber volume fractions. At zero volume fraction $(V_f = 0.0)$ the matrix is isotropic and spherical representation

TABLE 1—*Matrix and graphite fiber elastic properties in units of GPa, except v which is dimensionless*

		C_{33}	C_{11}	C_{12}	C_{13}	C_{44}	v_{32}	v_{12}
Matrix[a]	Ref 9	8.63	8.63	4.73	4.73	1.95	0.354	0.354
Fiber[b]	Ref 9	235	20.0	9.98	6.45	24.0	0.279	0.490
	Ref 14	240	20.4	9.40	10.5	24.0	0.35	0.497
	Ref 15	221	19.4	6.60[c]	5.8	20.3	0.23	0.32
Composite[d] $(V_f = 0.6)$		144	13.6	7.0	5.47	6.01	0.284	0.497

[a] Epoxy matrix composition is tetraglycidyl 4,4' diaminodiphenyl methane cured with diaminodiphenyl sulphone (TGDDM-DDS).
[b] Graphite fibers are pyrolized from polyacrylonitrile fibers.
[c] C_{12} is calculated from $C_{66} = (C_{11} - C_{12})/2$.
[d] Composite particles are calculated from equations derived in Ref 8.

surfaces (circular cross sections) are observed for all elastic constants. The largest orientational variations in elastic constants occur at $V_f = 1.0$. At intermediate V_f, we see that Poisson's ratios, linear compressibility, and Young's modulus are more affected by small changes in V_f. Torsional and shear modulus are affected less. Both Poisson's ratios (ν_{32} and ν_{13}) exhibit large negative values. For ν_{32} this is caused by variations in V_f and for ν_{13} this is caused by variations in tilt angle, α. In all cases, the Poisson's ratios satisfy the bounds derived by Christensen [10]

$$-1 < \nu_{12} < +1; \qquad \nu_{13} < (E_1/E_3)^{1/2} \qquad (26)$$

Elastic anisotropy exerts similar influences on the acoustic representation surfaces. Figures 9 and 10 demonstrate the influence of fiber volume fraction on phase-velocity and group-velocity surfaces. Large changes in flux-deviation angles, shown in Fig. 11, occur at low V_f. Hence, the cusps observed on the group-velocity curves in Fig. 10 result from changes in flux deviations.

Figure 12 shows surprising behavior of the particle displacement orientations at $V_f = 1.0$. For the v_{ql} surface the displacements in the x_2-x_3 plane undergo a transition for pure longitudinal along x_3 to pure shear along x_2. Consequently, in the x_1-x_2 plane one observes a shear wave moving faster than a longitudinal wave. While peculiar, similar properties were reported by Ledbetter and Kriz [2] for orthorhombic crystals. In orthorhombic crystals, this rare occurrence of $v_t > v_l$ leads to the interconnection of v_{ql} and v_{qt} surfaces into a single surface (also see Musgrave [11]). For hexagonal symmetry

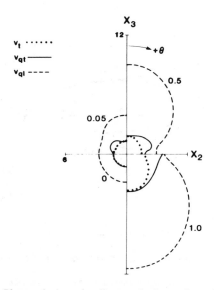

FIG. 9—*Phase-velocity polar diagram for four volume fractions.*

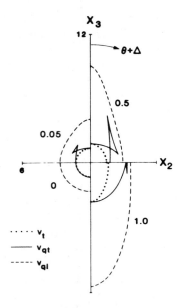

FIG. 10—*Group-velocity polar diagram corresponding to Fig. 9.*

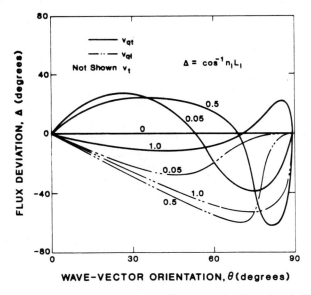

FIG. 11—*Flux-deviation directions for various fiber volume fractions. For clarity,* v_t *curves are omitted.*

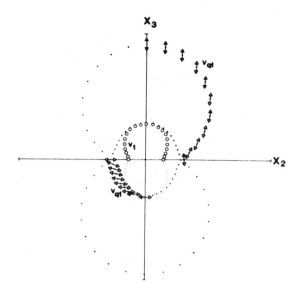

FIG. 12—*Polar diagram of displacement directions superimposed on the phase-velocity surfaces at* $V_f = 1.0$. *For clarity, only one quadrant shown.*

we observe $C_{44} = C_{22}$ when $V_f \cong 0.96$ and consequently $v_t = v_l$ in the x_1-x_2 plane. At volume fractions higher than 0.96 one observes a mode transition without the v_{ql} and v_{qt} surfaces interconnecting into a single surface.

This mode transition can be observed in Fig. 13. There the displacement deviations, $\delta = \cos^{-1} n_i p_i$, from the wave vector are shown versus θ. In Fig. 13 one observes that fiber volume fraction effects a smooth mode transition. The mode transition starts at $V_f = 0.3$ where v_{ql} transitions into v_{qt} at $\theta = 56°$. This transition does not continue into v_t; v_{qt} transitions back to v_{ql}. From $\theta = 56°$ to $\theta = 62°$ one observes a partial mode transition ($v_{qt} > _{ql}$). As V_f increases, this mode transition continues predictably; for $V_f > 0.96$, v_{qt} transitions into v_t. Thus, l, ql, qt, and t particle displacements are observed along a single surface.

While Hashin's composite-cylinder-assemblage model [12] can predict elastic constants for systems approaching $V_f = 1.0$, the geometric limit for hexagonal-packed circular fibers is $V_f = \pi/(2\sqrt{3})$ or 0.907, which is less than the volume fraction required for mode transition.

Such mode transitions can occur in composites at $V_f < 0.907$ when fiber constant $C_{44} \gg C_{22}$. Such a change in fiber elastic anisotropy could result from changing the fiber texture; but for fibers studied to date C_{44} exceeds C_{22} by only 5 to 20%.

The peculiar elastic anisotropy just discussed depends strongly on the fiber-

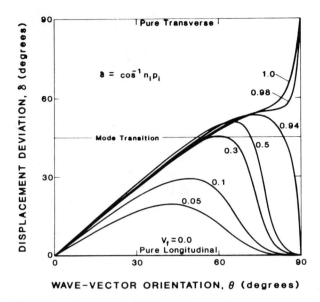

WAVE–VECTOR ORIENTATION, θ (degrees)

FIG. 13—*Displacement-direction deviations for* v_{ql}. *For higher fiber volume fractions, a mode transition occurs.*

elastic-constant values. The elastic constants for PAN graphite fibers shown in Figs. 2–8 at $V_f = 1.0$ were extrapolated by Kriz and Stinchcomb [9] using Hashin's composite-cylinder-assemblage model [12]. The correction $\beta_2 = K_{FTT}/(K_{FTT} + 2G_{FTT})$, noted by Hashin [13], changes the elastic constants predicted in Ref 9 less than 1% over the full range of fiber volume fractions. Thus, the fiber elastic constants in Table 1 are accurate and compare well with constants evaluated independently by Dean and Turner [14] and by Smith [15]. The well-developed radial structure of PAN graphite fibers observed by Diefendorf et al [16] is consistent with the transverse-isotropic symmetry reported by Kriz and Stinchcomb [9]. Fiber constants listed in Table 1 satisfy mechanical-stability conditions, that is Eq 1 must be positive-definite. The fiber Poisson's ratios also satisfy the bounds of Eq 26.

Conclusions

At low fiber volume fractions, Young's modulus E_3, linear compressibility K_3, and Poisson's ratios v_{32} and v_{13} are influenced most by the graphite-fiber anisotropy. Shear modulus G_{23} and torsional modulus T_3 are less affected by fiber anisotropy. Large negative values were observed for Poisson's ratios v_{13} and v_{32}, which were also influenced by fiber anisotropy. Poisson's ratio v_{32} is sensitive to small variations in fiber volume fraction. Poisson's ratio v_{13} is sensitive to rotations around the x_2-axis. Graphite-fiber anisotropy also affects both phase and group velocities, particularly the latter. In the x_1-x_2 plane the fiber constants $C_{44} > C_{22}$ cause a shear-wave speed to exceed a longitudinal wave speed; consequently the longitudinal wave along x_3 transitions into a pure shear wave in the x_1-x_2 plane. This mode transition depends on fiber volume fraction.

Acknowledgment

This work was supported partly by the Department of Energy, Office of Fusion Energy. During 1980-1982 Dr. Kriz was a National Research Council/National Bureau of Standards Postdoctoral Research Associate.

References

[1] Musgrave, M. J. P., *Crystal Acoustics*, Holden-Day, San Francisco, 1970, p. 83.

[2] Ledbetter, H. M. and Kriz, R. D., "Elastic-Wave Surfaces in Solids," *Physica Status Solidi (b)*, Vol. 11, 1982, pp. 475-480.

[3] Nye, J. F., *Physical Properties of Crystals*, Oxford U. P., London, 1957, pp. 143-148.

[4] Lekhnitskii, S. G., *Theory of Elasticity of an Anisotropic Body*, Holden-Day, San Francisco, 1963, p. 51.

[5] Hayes, M. and Musgrave, M. J. P., "On Energy Flux and Group Velocity," *Wave Motion*, Vol. 1, 1979, pp. 75-82.

[6] Musgrave, M. J. P., "On the Propagation of Elastic Waves in Aeolotropic Media II. Media of Hexagonal Symmetry" in *Proceedings*, Royal Society of London, Vol. 226, 1954, pp. 356-366.

[7] Kriz, R. D., "Mechanical Properties for Thick Fiber Reinforced Composite Materials Having Transversely Isotropic Fibers," Masters Thesis, College of Engineering, Virginia Polytechnic Institute and State University, Blacksburg, Va., 1976.

[8] Datta, S. K., Ledbetter, H. M., and Kriz, R. D., "Calculated Elastic Constants of Composites Containing Anisotropic Fibers," *International Journal of Solids and Structures*, Vol. 20, 1984, pp. 429-438.

[9] Kriz, R. D. and Stinchcomb, W. W., "Elastic Moduli of Transversely Isotropic Fibers and Their Composites," *Experimental Mechanics*, Vol. 19, 1979, pp. 41-49.

[10] Christensen, R. M., *Mechanics of Composite Materials*, John Wiley and Sons, 1979, p. 79.

[11] Musgrave, M. J. P., "On an Elastodynamic Classification of Orthorhombic Media" in *Proceedings*, Royal Society of London (*A*), Vol. 374, 1981, pp. 401-429.

[12] Hashin, Z., "On Elastic Behavior of Fiber Reinforced Materials of Arbitrary Transverse Phase Geometry," *Journal of the Mechanics and Physics of Solids*, Vol. 13, 1965, pp. 119-134.

[13] Hashin, Z., "Analysis of Properties of Fiber Composites with Anisotropic Constituents," *Journal of Applied Mechanics*, Vol. 46, 1979, pp. 543-550.

[14] Dean, G. D. and Turner, P., "The Elastic Properties of Carbon Fibers and Their Composites," *Composites*, Vol. 4, 1973, pp. 174-180.

[15] Smith, R. E., "Ultrasonic Elastic Constants of Carbon Fibers and Their Composites," *Journal of Applied Physics*, Vol. 43, 1972, pp. 2555-2561.

[16] Diefendorf, R. J., Riggs, D. M., and Sorensen, I. W., "The Relationships of Structure to Properties in Graphite Fibers," Technical Report No. AFML-TR-72-133, p. 65, Air Force Materials Laboratory, Wright-Patterson Air Force Base, Ohio.

Tsuyoshi Hayashi[1]

Shear Modulus of Epoxy Resin Under Compression

REFERENCE: Hayashi, T., **"Shear Modulus of Epoxy Resin Under Compression,"** *Recent Advances in Composites in the United States and Japan, ASTM STP 864*, J. R. Vinson and M. Taya, Eds., American Society for Testing and Materials, Philadelphia, 1985, pp. 676–684.

ABSTRACT: The matrix shear modulus of a unidirectionally fiber-reinforced composite material is one of the factors affecting its compressive shear mode fracture and it depends on the compressive stress.

In order to investigate the effect of compressive stress on the shear modulus, a special form of test specimen was designed such that compressive and torsional loads could be applied simultaneously. Test specimens were molded from epoxy resin for structural use.

Under every constant compressive load, specimens were subjected to torsional moment, and the corresponding strains were measured from which shear strains were calculated. Then the shear modulus was obtained as a function of compressive stress.

It was found that the shear modulus of epoxy resin decreased with increasing compressive stress in the range beyond the proportional limit, and the order of magnitude of its reduction was found.

KEY WORDS: composite material, matrix shear modulus, compressive stress, deterioration of shear modulus, epoxy resin

The matrix shear modulus of a unidirectionally fiber-reinforced composite is one of factors affecting its compressive shear mode fracture [1,2] for which the present author has proposed the theory of its compressive strength, based on the shear-instability concept and taking into consideration the deterioration of the shear modulus due to compression [2,3]. Recently Piggott and Harris [4] have studied the effect of compressive yield and fracture strengths of polyester and epoxy resins on the compressive fracture of those aligned fiber composites. And Kawabata has investigated the biaxial strength of epoxy resin [5]. But these papers have not dealt with the matrix shear modulus.

The matrix shear modulus depends upon the stress state, especially compressive stress, and its deterioration due to compression is expected. But such data have not yet been obtained.

[1]Professor emeritus, University of Tokyo; president Japan Plastic Inspection Association, 2-22-13 Yanagibashi, Taito-ku, Tokyo 111, Japan.

In the present study, as a typical matrix material epoxy resin, Epikote 828 with curing agent BF_3, which is widely used for structural composites, was used and its shear modulus was measured under various every constant compressive loads which were increased up to beyond the proportional limit.

In what follows, test specimen and loading apparatus, method of analysis, and test results are presented.

Test Specimens and Loading Apparatus

Test Specimens and Their Processing

Test specimens were made from Epikote 828 and curing agent BF_3 with weight ratio 100 to 3, by vacuum degassing for 20 min, and molded by using a cut-mold so as to give special symmetrical form as shown in Fig. 1. They were cured at 120°C for 3 h and aftercured at 180°C for 3 h. Heat deformation temperature of specimens, by the ASTM Test Method for Deflection Temperature of Plastics Under Flexural Load (D 648-82), was 190°C.

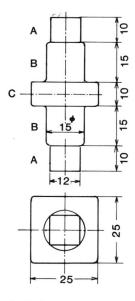

A=End portion

B=Cylindrical portion for
    strain measurement

C=Central portion for
    applying torsional moment

FIG. 1—*Test specimen and geometry.*

The specimens were formed and sized so that the residual stresses after curing and stress concentration would be as small as possible.

The end and central portions had square sections for applying torsion, and the inboard cylindrical portions were for measuring strains.

The tests were carried out in the International Organization for Standardization (ISO) standard atmosphere of 23°C and 50% relative humidity.

Fixture and Loading Apparatus

A test specimen was set in a fixture as shown in Fig. 2 which had a closed channel form attached to a lower loading platen of the testing machine. Each upper and lower flange of the channel had a square hole into which both end portions of the specimen, having square cross sections, were inserted.

A compressive load P was applied on the upper end of specimen. In order to apply torsion, a pulley having a square hole in the center was put into the central square-sectional portion of specimen, and torsional moment M was applied by two thin cables hung on the pulley whose diameter was D. Both cables were connected with a loading bar for applying variable weight W. Thus the torsional moment is $M = WD/2$. Axial and shear strains were measured by strain gages bonded on the cylindrical portions of specimen.

Figure 3 shows the referred axes and shear stress distribution.

Data Analysis

Denote the axial and circumferential directions of a specimen by L and T, directions making ± 45 deg with the specimen axis by x and y, and surface strains of the specimen referred to those directions by $(\epsilon_L, \epsilon_T, \gamma_{LT})$ and $(\epsilon_x, \epsilon_y, \gamma_{xy})$, respectively. Then we have the following relationships

$$\epsilon_L = \frac{1}{2}(\epsilon_x + \epsilon_y + \gamma_{xy})$$

$$\epsilon_T = \frac{1}{2}(\epsilon_x + \epsilon_y - \gamma_{xy})$$

$$\gamma_{LT} = \epsilon_y - \epsilon_x$$

ϵ_L, ϵ_T, and γ_{LT} are calculated from measured data of ϵ_x, ϵ_y, and γ_{xy}.

The compressive stress σ is P/A where $A = \pi d^2/4$, the sectional area of cylindrical portion of diameter d.

Twisting moment M applied from the pulley is transmitted to the upper and lower cylindrical portions of the specimen as M_1 and M_2. These moments cause the shear strains γ_{LT_1} and γ_{LT_2}, respectively. Assuming the linear shear

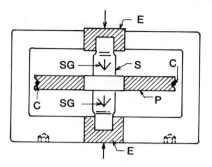

S=Test specimen
E=End cap
P=Pulley
SG=Strain gages
C=Cable

FIG. 2—*Fixture and loading system for test specimen.*

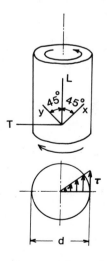

FIG. 3—*Referred axes and shear stress distribution.*

stress distribution on the circular cross sections of cylindrical portions, we have the following relationships

$$M_1 = \frac{\pi}{16} G d^3 \gamma_{LT_1} \quad \text{and} \quad M_2 = \frac{\pi}{16} G d^3 \gamma_{LT_2}$$

where G is the shear modulus of resin. Then

$$M = \frac{1}{2} D W \quad \text{and} \quad M = M_1 + M_2$$

where D is the pulley diameter and $W/2$ the cable tension. Thus we have

$$\frac{1}{2} D W = \frac{\pi}{16} G d^3 (\gamma_{LT_1} + \gamma_{LT_2})$$

Plotting W versus $(\gamma_{LT_1} + \gamma_{LT_2})$, we can find the required shear modulus G. Shear stress at the periphery of a circular section is given as

$$\tau = G \gamma_{LT} .$$

If the geometry of specimen and fixture is accurate, $M_1 = M_2$ and $\gamma_{LT_1} = \gamma_{LT_2}$.

From the data of W versus $(\gamma_{LT_1} + \gamma_{LT_2})$, we can obtain the shear modulus under various compression stresses.

If we put

$$M_0 = \frac{1}{2} M = \frac{1}{4} D W, \qquad \gamma_0 = \frac{1}{2} (\gamma_{LT_1} + \gamma_{LT_2})$$

we can obtain the usual expression

$$M_0 = \frac{\pi}{16} G d^3 \gamma_0$$

where M_0 and γ_0 mean the average moment and the average shear strain, respectively.

Test Results and Discussion

A compression test, a torsion test, and a combined compression-torsion test were carried out.

Compression Test

From compressive load P and measured strain, the initial elastic modulus $E = 3.23 \sim 3.36$ GPa, proportional limit $\sigma_P = 54.1 \sim 65.2$ MPa, and ultimate strength $\sigma_U = 87.7 \sim 111.7$ MPa were obtained. Figure 4 shows one of the stress-strain diagrams.

Torsion Test

Torsional moment was applied by varying the weight W from 0 to 294 $\sim$ 343 N with a step of 49 N (5 kgf). From measured strains ϵ_x and ϵ_y, the shear strain γ_{LT} on each cylindrical portion of the specimen was calculated. As mentioned before, the corresponding initial shear modulus G was found. Figure 5 shows an example for τ versus γ_{LT}.

From the torsion test we obtained shear modulus $G = 1215 \sim 1343$ MPa.

Compression-Torsion Test

Under a constant compressive load P, the torsional moment M was applied by varying the weight W from 0 to 300 $\sim$ 340 N (30 $\sim$ 35 kgf). For each compressive load, torsional moment was repeated by loading and unloading for one specimen. The reason why the same specimen was used for a serial test was to avoid any scatter that may have existed among specimens.

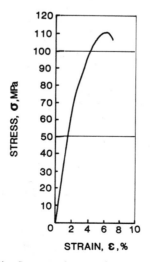

FIG. 4—*Stress-strain curve for compression.*

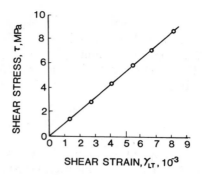

FIG. 5—*Shear stress-strain curve from torsion test.*

Figures 6 and 7 show test results for one of specimens. From these data the initial shear modulus G was obtained as a function of compressive stress.

Figure 8 shows the data for $G = G(\sigma)$.

These results show that the shear modulus of epoxy resin decreases with increasing compressive stress and also show the order of magnitude of its deterioration.

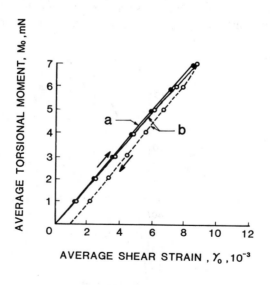

a = M_0 vs γ_0 for $\sigma = 16.17$ MPa
b = M_0 vs γ_0 for $\sigma = 32.34$ MPa

FIG. 6—*Average torsional moment versus average shear strain curves under constant compressive stresses.*

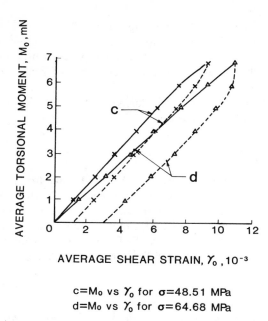

c=M_o vs γ_o for σ=48.51 MPa
d=M_o vs γ_o for σ=64.68 MPa

FIG. 7—*Average torsional moment versus average shear strain curves under constant compressive stresses.*

Conclusion

In order to clarify the shear behavior of matrix materials under compression, torsion and combined compression-torsion tests were carried out, using special-form specimens of an epoxy resin that was selected as a typical matrix material.

From test results it was confirmed that the shear modulus G of epoxy resin

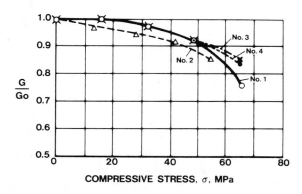

FIG. 8—*Retention of shear modulus versus compressive stress.*

(Epikote 828/BF$_3$) depends upon compressive stress and that the shear modulus decreases with increasing compressive stress. The order of magnitude of its reduction was also found. For compressive stress a little beyond the proportional limit, the shear modulus reduced by about 24% of its initial value for some specimens.

This result would be useful for clarifying the mechanism of shear mode fracture of aligned fiber composites due to compression.

Acknowledgment

The author would like to express his sincere thanks to Dr. N. Takashima, president, and Mr. T. Koike, technical director, of YUKA Shell Epoxy Co. for making the test specimens presented. He is also indebted to Mr. S. Tsutsumi, assistant of Chuo University, for making the loading apparatus, and to Mr. Y. Ohde, director, and Messrs. T. Takano and S. Kayama, staff, of Japan Plastic Inspection Association for performing the tests.

References

[1] Rosen, B. W. in *Fiber Composite Materials*, American Society for Metals, Metals Park, OH, 1965, Chapter 3, pp. 72-75.
[2] Hayashi, T., "On the Shear Instability of Structures Caused by Compressive Load," AIAA Paper No. 65-750, Joint Meeting of the American Institute of Aeronautics and Astronautics, the Royal Aeronautical Society, and the Japan Society for Aeronautical and Space Sciences, 1965.
[3] Hayashi, T., "Compressive Strength of Unidirectionally Fiber Reinforced Composites" in *Proceedings*, British Plastic Federation, Seventh International Reinforced Plastics Conference, Brighton, U.K., 1970.
[4] Piggott, M. R. and Harris, B., "Factors Affecting the Compression Strength of Aligned Fibre Composites" in *Proceedings*, Third International Conference on Composite Materials, ICCM-3, Paris, 1980, pp. 305-312.
[5] Kawabata, S., "Strength of Epoxy Resin under Multiaxial Stress Field" in *Proceedings*, Fourth International Conference on Composite Materials, ICCM-4, Tokyo, 1982, pp. 161-168.

Yoshihiro Takao[1]

Thermal Expansion Coefficients of Misoriented Short-Fiber Reinforced Composites

REFERENCE: Takao, Y., **"Thermal Expansion Coefficients of Misoriented Short-Fiber Reinforced Composites,"** *Recent Advances in Composites in the United States and Japan, ASTM STP 864*, J. R. Vinson and M. Taya, Eds., American Society for Testing and Materials, Philadelphia, 1985, pp. 685-699.

ABSTRACT: This paper examines the thermal expansion coefficients α_c of hybrid composites containing two- or three-dimensionally misoriented transverse-isotropic short fibers under the assumption that in the thermal expansion problem the transverse-isotropic fiber can be characterized by its longitudinal Young's modulus, Poisson's ratio, longitudinal thermal expansion coefficient, and transverse one—that is, isotropic mechanical properties and transverse-isotropic thermal ones. The analysis is based upon the Eshelby's equivalent inclusion method and average induced strain approach, which can take into account the interaction among all fibers at different orientations. It is therefore suitable for intraply hybrid systems. Numerical results are presented to demonstrate the effects of the distribution of fibers on α_c in the cases of carbon/glass/epoxy and carbon/Kevlar/epoxy composite. The effects of material constants on α_c are presented in detail.

KEY WORDS: thermal expansion, thermal strain, misoriented fiber, random distribution, hybrid composite, transverse-isotropic fiber, carbon, Kevlar, glass, aspect ratio

Discontinuous-fiber reinforced composites are attractive in their versatility in properties and relatively low fabrication cost. They consist of relatively short, variable length, and imperfectly aligned fibers distributed in matrix. The orientation of short fibers depends on the processing conditions employed and may vary from random to nearly aligned [1]. Thus it is imperative to take into account the effects of the bias in fiber orientation and variation in fiber aspect ratio on composite thermoelastic properties. Nowadays, the advanced composite containing anisotropic fibers such as carbon or Kevlar is one of the main concerns in the composite community. And the hybrid com-

[1]Associate professor, Research Institute for Applied Mechanics, Kyushu University, Kasuga, Fukuoka 816, Japan.

posite is, too, due to the freedom of tailoring composite properties and cost reduction by using lesser amounts of the expensive fibers [2]. It becomes necessary therefore to develop a theory which can be applied to the hybrid composites reinforced by anisotropic fibers such as carbon or Kevlar. The effective Young's modulus of misoriented short-fiber composites was well studied in the recent literature [3,4]. Thus we will study the thermal expansion coefficients α_c of the hybrid composites reinforced by misoriented anisotropic short fibers.

Even in the case of isotropic fibers, these effects have not been discussed except for the completely random distribution [5–7]. Chou and Nomura [5] and Craft and Christensen [6] used the so-called classical lamination analogy, where the interaction between all fibers at different orientation is not considered. The work by Takahashi et al [7] was based on the Eshelby's equivalent inclusion method [8–10].

The values of α_c for the composites reinforced by anisotropic fibers were obtained only for the cases of unidirectional ones [11–13]. Reference 13 used the theory based on the Eshelby's equivalent inclusion method for carbon/ aluminum composites.

In this paper we will develop the theory based on the Eshelby's equivalent inclusion method, which can take into account the interaction between all fibers. The present approach is particularly suitable for the advanced intraply hybrid composites, where there is no appropriate theory [2].

In the formulation, the general approach is described first followed by the restriction concerning the mechanical properties of fiber; that is, though the thermal anisotropy of fiber is included, the mechanical one is not considered in the present method, and it is shown that this restriction has practically no effect on α_c of our target composite system. Numerical results of α_c in the cases of carbon/glass/epoxy and carbon/Kevlar/epoxy composites are computed for both two- and three-dimensional distributions of fiber orientation with the parameter of distribution limit β. The effects of fiber distribution patterns (uniform or cosine type), aspect ratio, volume fraction, and material constants are presented, and the parameter having a strong effect on α_c is identified.

Formulation

Basic Equation

Our analytical model consists of N kinds of misoriented short fibers which are transverse-isotropic and embedded in an infinite isotropic matrix as shown in Fig. 1a. All fibers are modeled as prolate spheroids. Let the domain of the infinite body and fibers be denoted by D and Ω, respectively. Hence the domain of the matrix becomes $D - \Omega$. The domain Ω consists of N subdomains Ω_i, where the subscript i denotes the ith kind of fibers. Note that Ω_i can represent a particular inhomogeneity (fiber) or all inhomogeneities of type i.

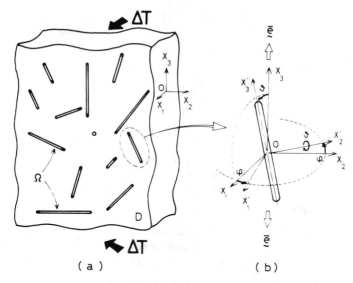

FIG. 1—*A calculation model.*

The elastic constants of the matrix and ith kind of fibers are denoted by $\mathbf{C}_m$ and $\mathbf{C}_{fi}$, respectively, and the thermal expansion coefficients of the matrix and ith kind of fibers are denoted by α_m and α_{fi}, respectively. The boldface symbols stand for tensorial quantity. The internal stress σ will be induced in the composite due to the mismatch between α_m and α_{fi} under the temperature change ΔT, and is a function of a position. The volume average of σ in $D - \Omega$ defines the volume averaged disturbance of strain in $D - \Omega$, $\bar{\mathbf{e}}$, due to the presence of all fibers as follows

$$\langle \sigma \rangle \left(\equiv \frac{1}{V_{D-\Omega}} \int_{D-\Omega} \sigma \, dV \right) = \mathbf{C}_m \bar{\mathbf{e}} \qquad (1)$$

where $\langle \ \rangle$ denotes the volume average of a quantity in $D - \Omega$ and $V_{D-\Omega}$ is the volume of matrix. Now a single fiber is introduced into the field where $\bar{\mathbf{e}}$ was applied. The orientation of fiber is defined by the angles θ and ϕ as shown in Fig. 1b. To apply the Eshelby's equivalent inclusion method to this single fiber the local coordinate system $x_1' x_2' x_3'$ is also adopted, where the x_3'-axis coincides with the fiber axis. In $D - \Omega$ and Ω_i, the thermal expansion $\alpha_m \Delta T$ and $\alpha_{fi} \Delta T$ are applied, respectively. Without losing generality we take $\Delta T = 1$. Then

$$\alpha_i^* = \alpha_{fi} - \alpha_m \qquad (2)$$

in Ω_i is related to σ in D and the Eshelby's equivalent inclusion method [8-10] yields

$$\sigma_i' = C_m'(\bar{e}' + e_i' - \alpha_i^{*'} - e_i^{*'}) \tag{3a}$$

$$\text{in } \Omega_i$$

$$= C_{fi}'(\bar{e}' + e_i' - \alpha_i^{*'}) \tag{3b}$$

and

$$e_i' = S_i'(\alpha_i^{*'} + e_i^{*'}) \tag{4}$$

where

σ_i' = disturbance of stress due to this single fiber,

S_i' = Eshelby's tensor (see Ref *10* in detail), and

$e^{*'}$ = fictitious strain called "eigen strain" or "transformation strain" and has some value in Ω_i but vanishes in $D - \Omega_i$.

Here the prime indicates tensorial quantity referred to the local coordinate system. Since the added single fiber can be regarded as any fiber in the composite, Eq 3 can hold for any inclusion in D. The fact that the volume average of σ in D vanishes [*14*] leads to

$$\bar{e} + \frac{1}{V_D} \sum_i^N \int_{\Omega_i} T_c(S_i' - I')(\alpha_i^{*'} + e_i^{*'})dV = 0 \tag{5}$$

where V_D, I, and T_c denote the volume of D, identity tensor, and transformation from the local to global coordinate system $x_1 x_2 x_3$, respectively. Here we state that $T_c C_m' T_c^{-1} = C_m$ and C_m is constant in D. T_c^{-1} denotes the transformation from the global to local coordinate system. From Eqs 3 and 4 we obtain

$$\alpha_i^{*'} + e_i^{*'} = \{(C_{fi}' - C_m')(S_i' - I') + C_{fi}'\}^{-1}\{C_{fi}'\alpha_i^{*'} - (C_{fi}' - C_m')T_c^{-1}\bar{e}\} \tag{6}$$

A substitution of Eq 6 into Eq 5 yields

$$\left(I - \frac{1}{V_D} \sum_i^N \int_{\Omega_i} T_c G_{1i}' T_c^{-1} dV\right)\bar{e} = -\frac{1}{V_D} \sum_i^N \int_{\Omega_i} T_c G_{2i}' \alpha_i^{*'} dV \tag{7}$$

where

$$G_{1i}' = (S_i' - I')\{(C_{fi}' - C_m')(S_i' - I') + C_{fi}'\}^{-1}(C_{fi}' - C_m')$$

$$G_{2i}' = (S_i' - I')\{(C_{fi}' - C_m')(S_i' - I') + C_{fi}'\}^{-1}C_{fi}' \tag{8}$$

The volume average of the induced strain in D due to α^* and e^*, γ_D is obtained as follows [*10*]

$$\gamma_D = \frac{1}{V_D} \sum_i^N \int_{\Omega_i} \alpha_i^* + \mathbf{e}_i^* dV \qquad (9)$$

Substituting Eq 6 into Eq 9 after the transformation to the global coordinate system, we finally obtain α_c as the sum of α_m and γ_D

$$\alpha_c = \alpha_m - \frac{1}{V_D} \sum_i^N \int_{\Omega_i} T_c \mathbf{G}'_{3i} T_c^{-1} dV \bar{\mathbf{e}} + \frac{1}{V_D} \sum_i^N \int_{\Omega_i} T_c \mathbf{G}'_{4i} \alpha_i^{*\prime} dV \quad (10)$$

where

$$\mathbf{G}'_{3i} = \{(\mathbf{C}'_{fi} - \mathbf{C}'_m)(\mathbf{S}'_i - \mathbf{I}') + \mathbf{C}'_{fi}\}^{-1}(\mathbf{C}'_{fi} - \mathbf{C}'_m)$$

$$\mathbf{G}'_{4i} = \{(\mathbf{C}'_{fi} - \mathbf{C}'_m)(\mathbf{S}'_i - \mathbf{I}') + \mathbf{C}'_{fi}\}^{-1}\mathbf{C}'_{fi}$$

$$(11)$$

Let the density function, volume of the single fiber, and volume fraction of the ith kind of fibers be denoted by $\rho_i(\theta)$ (see Fig. 1b; we assume ρ_i is the function of θ only), v_{fi}, and f_i, respectively. Then Eqs 7 and 10 are written as follows with the use of relations $dV = v_{fi}\rho_i(\theta) \sin \theta d\theta d\phi$ for the three-dimensional distribution of fiber orientation (3D) and $dV = v_{fi}\rho_i(\theta)d\theta$ for the two-dimensional one on the $x_2 x_3$-plane (2D)

$$\left\{ \mathbf{I} - \sum_i^N f_i \int_0^{\beta_i} \frac{1}{2\pi} \int_0^{2\pi} T_c \mathbf{G}'_{1i} T_c^{-1} d\phi \rho_i(\theta) \sin \theta d\theta \left(\int_0^{\beta_i} \rho_i(\theta) \sin \theta d\theta \right)^{-1} \right\} \bar{\mathbf{e}}$$

$$= - \sum_i^N f_i \int_0^{\beta_i} \frac{1}{2\pi} \int_0^{2\pi} T_c \mathbf{G}'_{2i} \alpha_i^{*\prime} d\phi \rho_i(\theta) \sin \theta d\theta \left(\int_0^{\beta_i} \rho_i(\theta) \sin \theta d\theta \right)^{-1}$$

$$\text{for 3D} \quad (7')$$

$$\left\{ \mathbf{I} - \sum_i^N f_i \int_0^{\beta_i} T_c \mathbf{G}'_{1i} T_c^{-1} \rho_i(\theta) d\theta \left(\int_0^{\beta_i} \rho_i(\theta) d\theta \right)^{-1} \right\} \bar{\mathbf{e}}$$

$$= - \sum_i^N f_i \int_0^{\beta_i} T_c \mathbf{G}'_{2i} \alpha_i^{*\prime} \rho_i(\theta) d\theta \left(\int_0^{\beta_i} \rho_i(\theta) d\theta \right)^{-1} \quad \text{for 2D} \quad (7'')$$

$$\alpha_c = \alpha_m - \sum_i^N f_i \int_0^{\beta_i} \frac{1}{2\pi} \int_0^{2\pi} T_c \mathbf{G}'_{3i} T_c^{-1} d\phi \rho_i(\theta) \sin \theta d\theta \left(\int_0^{\beta_i} \rho_i(\theta) \sin \theta d\theta \right)^{-1} \cdot \bar{\mathbf{e}}$$

$$+ \sum_i^N f_i \int_0^{\beta_i} \frac{1}{2\pi} \int_0^{2\pi} T_c \mathbf{G}'_{4i} \alpha_i^{*\prime} d\phi \rho_i(\theta) \sin \theta d\theta \left(\int_0^{\beta_i} \rho_i(\theta) \sin \theta d\theta \right)^{-1}$$

$$\text{for 3D} \quad (10')$$

$$\alpha_c = \alpha_m - \sum_i^N f_i \int_0^{\beta_i} T_c \mathbf{G}'_{3i} T_c^{-1} \rho_i(\theta) d\theta \left(\int_0^{\beta_i} \rho_i(\theta) d\theta \right)^{-1} \cdot \bar{e}$$

$$+ \sum_i^N f_i \int_0^{\beta_i} T_c \mathbf{G}'_{4i} \alpha_i^{*'} \rho_i(\theta) d\theta \left(\int_0^{\beta_i} \rho_i(\theta) d\theta \right)^{-1} \quad \text{for 2D} \quad (10'')$$

Here it is assumed that fibers are distributed within $0 \leq \theta \leq \beta_i$. Equation 7 (7' or 7″) represents, in general, six linear algebraic equations with six unknown components of $\bar{e}$ and provides a solution of $\bar{e}$, which is substituted into Eq 10 (10' or 10″) to obtain α_c. It is shown after some calculation that $\bar{e}$ has only two unknown components, namely, $\bar{e}_{11}$ ($= \bar{e}_{22}$) and $\bar{e}_{33}$, for 3D and three, namely, $\bar{e}_{11}, \bar{e}_{22}, \bar{e}_{33}$, for 2D with zero shear components.

Restriction on Mechanical Properties of Fiber

Two extreme cases of fiber volume fraction are considered here. First, at $f_i = 0$, α_c becomes α_m from Eq 10' or Eq 10″, which is reasonable. Next, at $f_N = 1, f_i = 0 \, (i \neq N)$, and $\alpha_N^{*'} = (\alpha^*, \alpha^*, \alpha^*, 0, 0, 0)$, γ_D should be $\alpha_N^{*'}$ for any β_N. Then from Eq 9

$$\frac{1}{V_D} \int_D \mathbf{e}_N^* dV = \mathbf{0} \tag{12}$$

The volume integral of Eq 3, after being transformed to the global coordinate system, leads to

$$\frac{1}{V_D} \mathbf{C}_m \int_D T_c(\bar{\mathbf{e}}' + (\mathbf{S}'_N - \mathbf{I}')\alpha_N^{*'} + (\mathbf{S}'_N - \mathbf{I}')\mathbf{e}_N^{*'}) dV$$

$$= \frac{1}{V_D} \int_D T_c \mathbf{C}'_{fN} T_c^{-1} T_c(\bar{\mathbf{e}}' + (\mathbf{S}'_N - \mathbf{I}')\alpha_N^{*'} + (\mathbf{S}'_N - \mathbf{I}')\mathbf{e}_N^{*'} + \mathbf{e}_N^{*'}) dV$$

$$\tag{13}$$

in this case, and Eq 5 is also rewritten as

$$\frac{1}{V_D} \int_D T_c(\bar{\mathbf{e}}' + (\mathbf{S}'_N - \mathbf{I}')\alpha_N^{*'} + (\mathbf{S}'_N - \mathbf{I}')\mathbf{e}_N^{*'}) dV = \mathbf{0} \tag{14}$$

When $\mathbf{C}'_{fN}$ represents the isotropic elastic constant or when all fibers are aligned along a direction, the value of $T_c \mathbf{C}'_{fN} T_c^{-1}$ becomes constant in D and comes out of the integral in Eq 13. Then Eq 12 is obtained from Eqs 13 and 14. In the case of anisotropic fiber the value of $T_c \mathbf{C}_{fN} T_c^{-1}$ is not the same for all fibers except for the case of aligned ones. A preliminary numerical calculation showed that Eq 12 does not hold for composites reinforced by misoriented fibers with anisotropic elastic constants.

Reference *13* shows that α_c of composites reinforced unidirectionally by transverse-isotropic carbon fibers depends mainly on an axial component C_{33} of stiffness of fiber, Young's modulus E_m, and Poisson's ratio ν_m of isotropic matrix among elastic constants of constituents. Then we assume that $\mathbf{C}'_{fN}$ represents the isotropic elastic constant. So, the present method takes into account the anisotropy of only thermal properties of fibers.

Numerical Results

Comparison with Previous Work

The main purpose of this study is to predict α_c for the composites reinforced by misoriented transverse-isotropic short carbon fibers. However, there is no available previous literature corresponding to the present work. Therefore the comparison will be made for two special cases, namely, unidirectional composites with transverse-isotropic fibers in their stiffness and thermal expansion coefficient [11,13], and composites reinforced by misoriented isotropic fibers [5,6]. The longitudinal α_{cL} and transverse components α_{cT} of α_c of unidirectional composites reinforced by isotropic fibers, which are computed by inputting the isotropic thermal expansion coefficient of fiber α_f into the present formulation, are plotted as a function of f for $\alpha_f = 27.0 \times 10^{-6}/°C$ and $-1.5 \times 10^{-6}/°C$ in Fig. 2a together with previous ones [11,13] reinforced by anisotropic fibers. This figure shows that the assumption of thermal isotropy is not appropriate. The values of α_c are nondimensionalized by the isotropic thermal expansion coefficient α_m. Present results of α_{cL} and α_{cT} with distribution limit $\beta = 0$ deg are plotted for Poisson's ratio of fiber $\nu_f = 0.17$ and 0.3 in Fig. 2b, where the thermal anisotropic properties are used, together with previous ones. It follows from Fig. 2 that the present method predicts practically the same results as the ones [11,13] which can consider both the mechanical and thermal anisotropy of fiber embedded in unidirectional composites and that $\nu_f = 0.17$ is appropriate. Material constants used in Fig. 2 are given by Table 1a for fibers with the same values of C_{33} and Table 1b for the epoxy matrix [1] except that Young's modulus of matrix $E_m = 3.43$ GN/m². Here, fiber volume fraction $f = 0.4$ and fiber aspect ratio $l/d = 100$. In the following figures carbon fiber is assumed to have a ν_f of 0.17. Figure 3 shows the comparison with other results based on classical lamination analogy for composites reinforced by completely random [density function $\rho(\theta) = $ constant and $\beta = 90$ deg] isotropic fibers. In 2D $\alpha_{c2} = \alpha_{c3}$ and in 3D $\alpha_{c1} = \alpha_{c2} = \alpha_{c3}$, where α_{ci} is the normal component of α_c along the x_i-direction. Present results of α_c/α_f in 2D are plotted in Fig. 3a for $l/d = \infty$, Young's moduli of fiber and matrix $E_{f(m)} = 86.2(2.8)$ GN/m²,

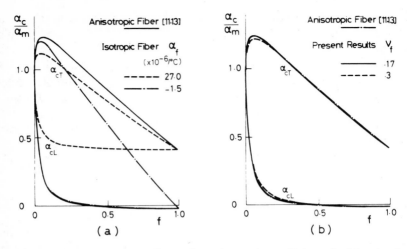

FIG. 2—*Effects of anisotropy of fiber on thermal expansion coefficients of unidirectional composites* α_c *with fiber volume fraction* f.

$\nu_{f(m)} = 0.2(0.35)$, and $\alpha_{f(m)} = 3(80) \times 10^{-6}/°C$ as a function of f together with previous theoretical and experimental results [6]. A present result of α_{c3} for $f = 0.34$ is smaller than the experimental one, which coincides with the fact that voids existing in practical specimens increase α_{c3} due to the reduced reinforcement by debonding originated from voids and larger thermal expansion of voids. Present results in 3D are plotted in Fig. 3b as a function of f for $l/d = 5$ and ∞ in the case of $E_{f(m)} = 72.4(2.8)$ GN/m², $\nu_{f(m)} = 0.2(0.35)$,

TABLE 1a—*Material constants of fiber in Fig. 2 [11]*

Material Properties	Components of Stiffness (GN/m²) Poisson's Ratio $\{\nu_f\}$					Thermal Expansion Coefficients ($\times 10^{-6}/°C$)	
	C_{11}	C_{33}	C_{44}	$C_{66}{}^e$	C_{13}	L^f	T^g
Aniso.[a] Mech.[c] Aniso. Ther.[d]	28.25	234.79	10.0	16.02	12.14	−1.5	27.0
Iso.[b] Mech.		234.79		{0.1}		−1.5	27.0
Aniso. Ther.		234.79		{0.17}		−1.5	27.0
Iso. Mech.		234.79		{0.17}		27.0	27.0
Iso. Ther.		234.79		{0.17}		−1.5	−1.5

[a]Aniso. = anisotropic (transverse-isotropic).
[b]Iso. = isotropic.
[c]Mech. = Mechanical properties.
[d]Ther. = thermal properties.
[e]$C_{66} = C_{11} - C_{12}$.
[f]L = longitudinal.
[g]T = transverse.

TABLE 1b—*Material constants of fiber and matrix used in present paper [1,11].*

Material	Mechanical Properties		Thermal Expansion Coefficients ($\times 10^{-6}/°C$)	
	Young's Modulus, E (GN/m^2)	Poisson's Ratio, ν	L^a	T^b
Carbon	250.0	0.17	−1.5	27.0
Kevlar	125.0	0.17	−2.0	59.0
Glass	76.0	0.2	4.9	(4.9)
Epoxy	3.5	0.38	66.0	(66.0)

^aL = longitudinal.
^bT = transverse.

and $\alpha_{f(m)} = 5(54) \times 10^{-6}/°C$ together with previous ones [5]. The effect of l/d is pointed out clearly by the present methods.

Thermal Expansion of Carbon/Glass/Epoxy and Carbon/Kevlar/Epoxy Composites

Figures 4 and 5 shows α_c of carbon/glass/epoxy intraply hybrid composites as a function of β ($\equiv \beta_i$) for both 2D and 3D. Material constants are given in Table 1b. The effects of the type of distribution in fiber orientation on α_c/α^*

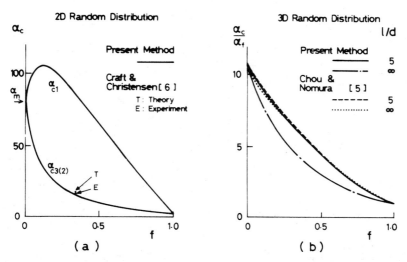

FIG. 3—*Thermal expansion coefficients of composites α_c reinforced by isotropic fiber with completely random fiber distribution as a function of* f: *uniform type of fiber distribution and distribution limit β = 90 deg; (a) $E_{f(m)}$ = 86.2(2.8) GN/m^2, $\nu_{f(m)}$ = 0.2(0.35), $\alpha_{f(m)}$ = 3(80) $\times 10^{-6}/°C$, l/d = ∞; (b) $E_{f(m)}$ = 72.4(2.8) GN/m^2, $\nu_{f(m)}$ = 0.2(0.35), $\alpha_{f(m)}$ = 5(54) $\times 10^{-6}/°C$.*

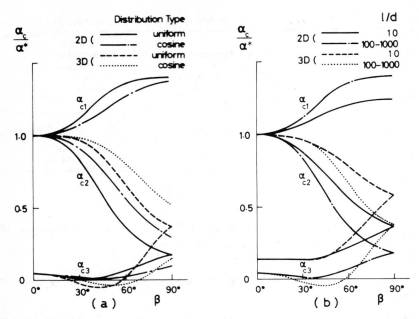

FIG. 4—*Effects of* (a) *distribution type* ($l/d = 100$, $f_G = f_C = 0.2$) *and* (b) *aspect ratio* l/d *(uniform type, $f_G = f_C = 0.2$) on thermal expansion coefficients of carbon/glass/epoxy composites α_c with distribution limit β.*

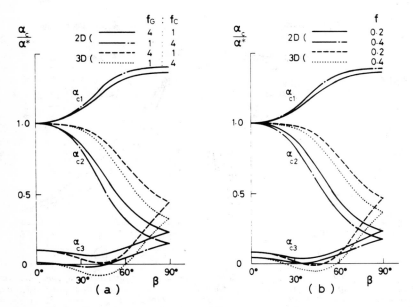

FIG. 5—*Effects of* (a) *volume fraction ratio of glass to carbon* $f_G : f_C$ ($f_G + f_C = 0.4$) *and* (b) *total volume fraction* f ($f_G : f_C = 1$) *on α_c with β; uniform type, $l/d = 100$.*

are shown in Fig. 4a, where $l/d = 100$, $f = 0.4$, the ratio of glass fiber volume fraction to carbon $f_G:f_C = 1$, and α^* denotes α_{c1} at $\beta = 0$ deg. The value of α_c for the cosine type of distribution $[\rho_i(\theta) = \cos \theta]$ is obtained roughly by shifting α_c for the uniform type $[\rho_i(\theta) = \text{constant}]$ to the right side as expected. Then α_{c2} is higher for the cosine type than the uniform type. And α_{c3} of the cosine type is higher at large values of β and lower at small values of β due to the fact that there is a minimum for α_{c3}. This phenomenon was pointed out in the case of the bidirectional laminate of carbon/epoxy with experimental results [12,15]. The present results show that it is more significant in 3D than in 2D. At $\beta = 90$ deg, α_{c3} for 3D is higher than 2D as expected. The effect of aspect ratio l/d on α_c/α^* is shown in Fig. 4b for $f = 0.4$ and $f_G:f_C = 1$. In Fig. 4 and following figures we consider the uniform type of distribution. The value of α^* for $l/d = 10$ is different from the one for $l/d = 100 - 1000$ as shown in Table 2. The effects of (a) volume fraction ratio of glass to carbon $f_G:f_C$ with $f = 0.4$ and (b) total volume fraction f itself with $f_G:f_C = 1$ on α_c/α^* for $l/d = 100$ in Fig. 5 are small compared with those in Fig. 4.

We have computed the values of α_c by changing the material parameters of the matrix and fiber within the practical range to assess the degree of the effects of the parameters, and they are shown in Figs. 6–8 as a function of β for both 2D and 3D. In this calculation $l/d = 100$, $f = 0.4$, $f_G:f_C = 1$, and the data given by Table 1b are used if not mentioned otherwise. Material properties of glass are kept constant. It follows from Fig. 6 that the effects of thermal expansion coefficients of matrix α_m and carbon-type fiber α_{fL} and α_{fT} on α_c/α^* are small within the calculated range. It should be noted that the effects on α_c itself are large especially for the case of α_m as shown in Table 2. This is also pointed out for isotropic fibers [16]. The effects of mechanical properties of the matrix [(a) Young's modulus E_m and (b) Poisson's ratio ν_m] and carbon-type fiber [(a) Young's modulus E_f and (b) Poisson's ratio ν_f] on α_c/α^* are shown in Figs. 7 and 8, respectively. The variation of E_m has strong effects and the variation of ν_f has negligible effects. Figure 9 shows α_c/α^* of carbon/Kevlar/epoxy intraply hybrid composites as a function of β for $l/d = 100$ and $f = 0.4$ with the parameter being volume fraction ratio of Kevlar to

TABLE 2—α^* in Figs. 4-9 ($\times 10^{-6}/°C$).

Fig. No.	4a	4b	5a	5b	6a	6b
------------	56.84	54.91	60.03	71.64	*	56.58
....	*[a]	*	53.22	*	157.3	56.93

Fig. No.	6c	7a	7b	8a	8b	9
------------	51.58	57.04	52.0	*	56.9	73.42
....	64.09	56.30	59.47	57.35	56.77	64.89

[a] Asterisk denotes 56.84.

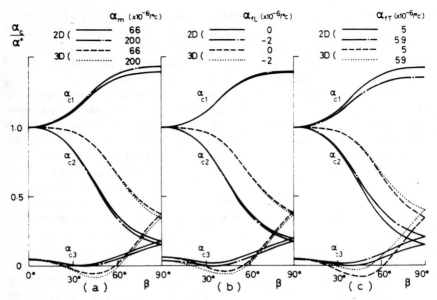

FIG. 6—*Effects of thermal expansion coefficients of (a) matrix* α_m, *(b) carbon-type fiber: longitudinal component* α_{fL}, *and (c) transverse component* α_{fT} *on* α_c *with* β; *uniform type,* $l/d = 100$, $f_G = f_C = 0.2$.

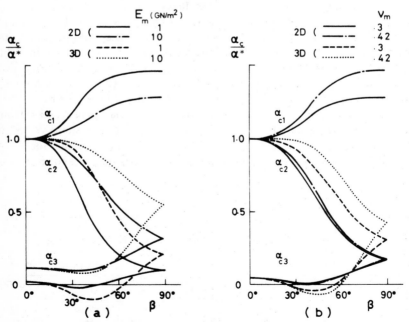

FIG. 7—*Effects of (a) Young's modulus* E_m *and (b) Poisson's ratio* ν_m *of matrix on* α_c *with* β; *uniform type,* $l/d = 100$, $f_G = f_C = 0.2$.

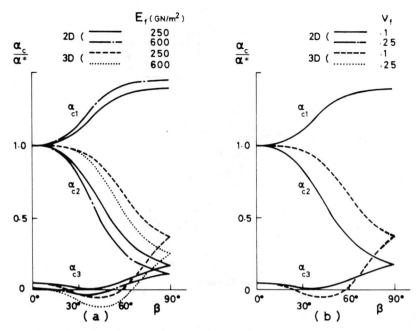

FIG. 8—*Effects of (a) Young's modulus* E_f *and (b) Poisson's ratio* ν_f *of carbon-type fiber on* α_c *with* β; *uniform type,* $l/d = 100$, $f_G = f_C = 0.2$.

carbon $f_K : f_C$. Material constants are given by Table 1b; we assume that Poisson's ratio of Kevlar is also 0.17. It follows from Figs. 4–9 that the effects of the variation of parameters on α_c / α^* are divided into four groups within the practical calculated range: strong effects from the distribution-type, l/d, and E_m; medium from volume fraction, ν_m, and E_f; small from thermal expansion coefficients of each phase; and negligible from ν_f.

Conclusion

1. The theory to predict the thermal expansion coefficients α_c of hybrid composites reinforced by misoriented anisotropic (transverse-isotropic) short fibers is developed under the assumption that transverse-isotropy of elastic constants of fiber has no significant effect on α_c and that the transverse-isotropic fiber can be characterized by longitudinal Young's modulus, Poisson's ratio, and longitudinal and transverse thermal expansion coefficients. The present theory is based on the Eshelby's equivalent inclusion method and average induced strain approach and takes into account the interaction among all fibers. It is therefore appropriate for intraply hybrid composites.

2. The effects of the distribution limit of fiber on α_c are obtained for cases

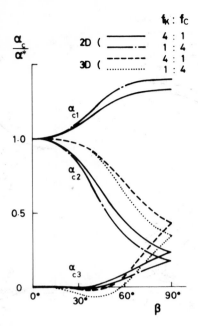

FIG. 9—*Thermal expansion coefficients of carbon/Kevlar/epoxy composites α_c with the parameter of volume fraction ratio of Kevlar to carbon $f_K : f_C$ as a function of β; uniform type, $1/d = 100$, $f (\equiv f_K + f_C) = 0.4$.*

of two- or three-dimensional distribution of fiber orientation with both uniform type and cosine type fiber density functions.

3. Carbon/glass/epoxy and carbon/Kevlar/epoxy composites with misoriented short fibers are studied and it is pointed out that the variation of aspect ratio, Young's modulus of matrix, and distribution type of fiber have significant effects on the nondimensional thermal expansion coefficients of composites α_c / α^*.

References

[1] Hull, D., *An Introduction to Composite Materials*, Cambridge University Press, U.K., 1981.
[2] *Hybrid and Select Metal Matrix Composites: A State-of-the-Art Review*, W. J. Renton, Ed., American Institute of Aeronautics and Astronautics, 1977.
[3] Takao, Y., Chou, T. W., and Taya, M., "Effective Longitudinal Young's Modulus of Misoriented Short Fiber Composites," *Journal of Applied Mechanics*, Vol. 49, 1982, pp. 536–540.
[4] Takao, Y., Taya, M., and Chou, T. W., "Effective Longitudinal Young's Modulus of Two-Dimensionally Misoriented Short Fiber Composites" in *Proceedings*, Fourth International Conference on Composite Materials (ICCM-IV), T. Hayashi, K. Kawata, and S. Umekawa, Eds., Tokyo, 1982; Japan Society for Composite Materials, 1982, pp. 1091–1098.
[5] Chou, T. W. and Nomura, S., "Fiber Orientation Effects on the Thermoelastic Properties

of Short-Fiber Composites," *Fiber Science and Technology*, Vol. 14, 1980-1981, pp. 279–291.

[6] Craft, W. J. and Christensen, R. M., "Coefficient of Thermal Expansion for Composites With Randomly Oriented Fibers," *Journal of Composite Materials*, Vol. 15, 1981, pp. 2-20.

[7] Takahashi, K., Harakawa, K., and Sasaki, T., "Analysis of Thermal Expansion Coefficients of Particle-Filled Polymers," *Journal of Composite Materials (Supplement)*, Vol. 14, 1980, pp. 144-159.

[8] Wakashima, K., Otsuka, M., and Umekawa, S., "Thermal Expansions of Heterogeneous Solids Containing Aligned Ellipsoidal Inclusions," *Journal of Composite Materials*, Vol. 8, 1974, pp. 391-404.

[9] Eshelby, J. D., "The Determination of the Elastic Field of an Ellipsoidal Inclusion, and Related Problems" in *Proceedings of the Royal Society of London*, Series A, Vol. 241, 1957, pp. 376-396.

[10] Mura, T., *Micromechanics of Defects in Solids*, Martinus Nijhoff Publishers, The Hague, The Netherlands, 1982.

[11] Ishikawa, T., Koyama, K., and Kobayashi, S., "Thermal Expansion Coefficients of Unidirectional Composites," *Journal of Composite Materials*, Vol. 12, 1978, pp. 153-168.

[12] Uemura, M., Iyama, H., and Yamaguchi, Y., "Thermal Residual Stresses in Filament-Wound Carbon-Fiber-Reinforced Composites," *Journal of Thermal Stresses*, Vol. 2, 1979, pp. 394-412.

[13] Takao, Y. and Taya, M., "Thermal Expansion Coefficients and Thermal Stresses in an Aligned Short Fiber Composite With Application to a Short Carbon Fiber/Aluminum," submitted for publication.

[14] Mori, T. and Tanaka, K., "Average Stress in Matrix and Average Elastic Energy of Materials With Misfitting Inclusions," *Acta Metallurgica*, Vol. 21, 1973, pp. 571-574.

[15] Rogers, K. F. et al, "The Thermal Expansion of Carbon Fiber-Reinforced Plastics: Part 1 The Influence of Fiber Type and Orientation," *Journal of Materials Science*, Vol. 12, 1977, pp. 718-734.

[16] Schapery, R. A., "Thermal Expansion Coefficients of Composite Materials Based on Energy Principles," *Journal of Composite Materials*, Vol. 2, 1968, pp. 380-404.

Kozo Kawata,[1] *Shozo Hashimoto,*[2] *Nobuo Takeda,*[3] *and Shozo Sekino*[4]

On High-Velocity Brittleness and Ductility of Dual-Phase Steel and Some Hybrid Fiber Reinforced Plastics

REFERENCE: Kawata, K., Hashimoto, S., Takeda, N., and Sekino, S., **"On High-Velocity Brittleness and Ductility of Dual-Phase Steel and Some Hybrid Fiber Reinforced Plastics,"** *Recent Advances in Composites in the United States and Japan, ASTM STP 864,* J. R. Vinson and M. Taya, Eds., American Society for Testing and Materials, Philadelphia, 1985, pp. 700–711.

ABSTRACT: First, the one-bar method developed by the present authors for the characterization of materials in high-velocity uniaxial tension up to breaking is stated. Second, the results of high-velocity tension ranging up to 1.4×10^3 s^{-1} on a dual-phase steel (shown in Table 1), pure iron A and B (shown in Table 2), and carbon/glass hybrid composites (shown in Table 3) are stated, comparing with the data of static tension. The dual-phase steel shows the tendency of high-velocity ductility, differing from the pure irons. Three types of carbon fiber/glass fiber hybrid composites tested show the tendencies of high-velocity ductility with varying increasing rate according to the hybrid ratio, in contrast to the fact that carbon fiber reinforced plastic is strain-rate insensitive or high-velocity brittle.

KEY WORDS: high-velocity tension, impact, high-velocity ductility, high-velocity brittleness, dual-phase steel, pure iron, hybrid composite, test method, evaluation

[1]Department of Mechanical Engineering, Faculty of Science and Technology, Science University of Tokyo, Yamazaki, Noda, 278 Japan.
[2]Institute of Interdisciplinary Research, Faculty of Engineering, University of Tokyo, Komaba, Meguro-ku, Tokyo, 153 Japan.
[3]Japan Atomic Energy Institute, Takasaki Radiation Chemistry Establishment, Watanuki-cho, Takasaki, 370-12 Japan.
[4]Fundamental Research Laboratories, Nippon Steel Corp., 1818 Ida, Nakahara-ku, Kawasaki, 211 Japan.

In the advanced fields of engineering, high-velocity deformation by high-velocity loading occurs frequently. For the design of impact-resistant structure, stress-strain diagrams at high strain rates are needed. The stress-strain diagrams of materials measured by quasi-static testing machines are ordinarily the ones at the order of strain rate of 10^{-3} s^{-1}. The one bar method was developed by the present authors for obtaining tensile stress-strain diagrams up to breaking at the strain rate range up to 10^3 or 10^4 s^{-1} [1,2]. Also, new concepts of high-velocity brittleness and high-velocity ductility of materials [3,4] were shown by the present authors.

In this paper, the data of mechanical properties in high-velocity tension at about 10^3 s^{-1} are reported for three groups of materials. The first material is a dual-phase steel. The need to lower automobile vehicle weights has resulted in the development of steels with increased specific strength. The high-strength microalloyed steels thus developed suffer from lower formability compared with conventional low-carbon steels. So, dual-phase steels were developed to satisfy both of these two conflicting requirements of high strength and improved formability. For the analysis of high-speed forming and impact-resistant structure design, the mechanical properties in high-velocity tension should be clarified. Data on pure irons showing typical results of clear high-velocity brittleness of body-centered cubic lattice (bcc) metals are reported for comparison. The good properties in high-velocity tension of the dual-phase steels seem to be based upon the properties as composite metal material. The third material reported is a carbon fiber/glass fiber (CF/GF) hybrid composite. In earlier papers [5,6,2], data on high-velocity ductility of glass fiber reinforced plastic (GFRP) and high-velocity brittleness or insensitivity of carbon fiber reinforced plastic (CFRP) were reported by the present authors. It is a quite interesting problem how the hybrids behave for high-velocity tension.

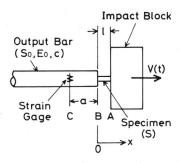

FIG. 1—*Principle of "one-bar method" for material characterization in high-velocity tension.*

Procedure

Equipment and Its Principle

The testing system consists of a hammer, an impact block, a specimen and an output bar (Fig. 1). This system is included in the block-to-bar type and the authors designate it the "one bar method" since this method uses only one bar and new analysis formulas. This system has the features of exact response without apparent vibrational effect such as are seen for short load cell causing multiple passages of reflected wave and of giving enough data up to breaking in tension. When the impact block is given an impact by the hammer, the specimen is deformed and the strain $\epsilon_g(t)$ of the output bar is measured at a distance a from the impact end. The stress $\sigma(t)$ and strain $\epsilon(t)$ of the specimen are calculated by the one-dimensional stress wave theory as follows [1]

$$\left.\begin{aligned} \epsilon(t) &= \frac{1}{l} \int_0^t \left[V(\tau) - c\epsilon_g \left(\tau + \frac{a}{c} \right) \right] d\tau \\[2em] \sigma(t) &= \left(\frac{S_0}{S} \right) E_0 \epsilon_g \left(t + \frac{a}{c} \right) \end{aligned}\right\} \tag{1}$$

where

$$
\begin{aligned}
l \text{ and } S &= \text{length and cross-sectional area of specimen,} \\
S_0, E_0, \text{ and } c &= \text{cross-sectional area, Young's modulus, and elastic longi-} \\
&\quad\;\; \text{tudinal wave velocity of output bar, respectively, and} \\
V(t) &= \text{velocity of impact block, measured by an electro-optical} \\
&\quad\;\; \text{extensometer.}
\end{aligned}
$$

According to the one-bar method, two kinds of high-velocity tension testing machine were constructed. In the first type, a rotating disk is used for impacting the hammer. The maximum peripheral speed of the disk is about 100 m/s. In the second type, a pendulum is used for impacting the impact block, as a Charpy or Izod type. The hammer speed at the impact point is about 3.5 m/s.

Specimen Preparation

The specimen size for the above-mentioned high-velocity tension testing machines is shown in Fig. 2. Detailed data of tested materials are given in the "Experimental Results and Discussion" section.

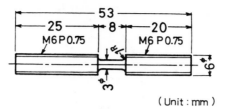

(Unit : mm)

FIG. 2—*Specimen size.*

Procedure

In a practical experiment, a calculating system for dynamic stress-strain diagrams is utilized. When the hammer impacts the impact block, the transient output bar strain $\epsilon_g(t)$ is recorded in a digital transient wave memory. Combining it with the impact block velocity $V(t)$, a digital computer calculates dynamic stress $\sigma(t)$ and strain $\epsilon(t)$ using Eq 1 and also absorbed energy per unit volume $E_{ab}(t)$.

Experimental Results and Discussion

Dual-Phase Steel

Material for the investigation is a 0.12% carbon steel and was prepared as follows: annealed at 800°C for 10 min, oil quenched in 50°C oil bath, and then air cooled. The chemical analysis is given in Table 1. The grain size of each phase was as follows. The grain size of ferrite was ASTM No. 6 and that of martensite islands was smaller by several numbers in ASTM No. than ferrite. A typical dynamic stress-strain diagram is shown in Fig. 3, comparing with the static one. Peak stress σ_p, total elongation ϵ_t, and absorbed energy per unit volume E_{ab} are higher than the corresponding values in static tension. These tendencies certify that this dual phase steel is of high quality for high-speed forming and shock-resistant structure. These properties show a remarkable contrast to the corresponding properties of pure irons shown later.

TABLE 1—*Chemical analysis of dual-phase steel, weight %.*

C	Si	Mn	P	S	Al	N
0.12	1.36	1.60	0.002	0.005	0.017	0.0015

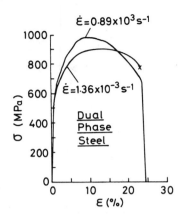

FIG. 3—*Stress-strain diagrams of a dual-phase steel for dynamic and static tension.*

Pure Irons A and B

Materials for the investigation were prepared as follows: Pure iron A (pure iron designated as A):   Thick plate of 130 mm in thickness heated at 1200°C for 1 h was rolled to 15 mm in thickness and air cooled. Pure iron B: Material A was wet-hydrogen treated at 600°C for 375 h, dry-hydrogen treated at 600°C for 32 h, and annealed in vacuum at 600°C for 135 h. We can calculate mean distance of diffusion ($\sqrt{Dt}$, D:-diffusion coefficient of hydrogen in steel, t:-time) of hydrogen, for $D = 2 \times 10^{-4}$ cm^2/s at 600°C. Sample size is 10 mmϕ. We can see that hydrogen goes out in about 20 min. 135-h baking is quite enough. The chemical analyses are given in Table 2. In the B material, the contents of carbon and nitrogen were less than 3 ppm, the limit of chemical analysis. Micro Vickers hardness (200 g) was 68 to 60 and grain size was ASTM No. 2 to No. 3 which was coarser by number 1 than that of pure iron A. The dynamic stress-strain diagrams are shown in Figs. 4 and 5 in comparison with the static ones. In both materials A and B, the increasing of peak stress σ_p and the decreasing of total elongation ϵ_t in dynamic tension were observed. Especially, strain-rate dependence was quite remarkable in material B. For example, total elongation decreases from about 59% to about 2% for strain rate increasing ratio of about one million times. Strain-rate dependence of σ_p,

TABLE 2—*Chemical analysis of pure irons A and B, weight %.*

Material	C	Si	Mn	P	S	N	O
A	0.002	0.01	0.01	0.003	0.0004	0.002	0.0077
B	3 ppm	0.01	0.01	0.003	0.0004	3 ppm	

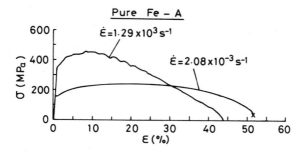

FIG. 4—*Stress-strain diagrams of pure iron A for dynamic and static tension.*

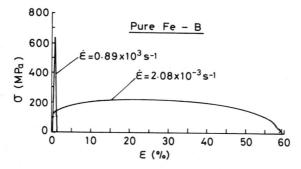

FIG. 5—*Stress-strain diagrams of pure iron B for dynamic and static tension.*

ϵ_t, and E_{ab} of pure irons A and B is clearly different from the corresponding values of dual-phase steel, as shown in Fig. 6.

High-velocity brittleness of pure irons A and B seems to be significant as bcc metal materials. In pure iron B of the highest purity, the microstructure shows the occurrence of grain boundary cracking; see Fig. 7. This seems the reason for high-velocity brittleness of this material. The microstructure of this material deformed in high-velocity tension shows that when a slipband comes to motion, deformation is concentrated in this band and so the numbers of actuated slipbands are smaller than in low-velocity tension. When a slipband reaches the grain boundary, new slipbands must be actuated in the neighboring grain to maintain successive deformation. But, in iron, the potential slipband number is rather limited and some time will be needed to actuate the new slipbands. Under this condition in high-velocity tension, it will be difficult to actuate new slipbands necessary to maintain high-velocity plastic deformation in the neighboring grain in time and in accordance with external high-velocity deformation. When the grain boundary strength is not suffi-

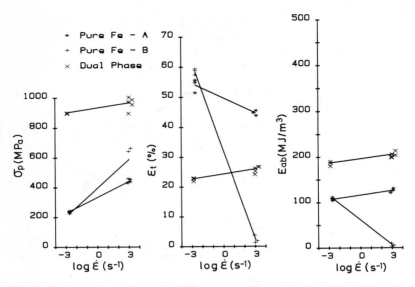

FIG. 6—*Strain-rate dependence of peak stress σ_p, total elongation ϵ_t, and absorbed energy per unit volume E_{ab} in pure irons A and B and dual-phase steel.*

cient to withstand the stress concentration that has thus occurred, cracking occurs at grain boundaries.

For dual-phase steel, in the first place, grain size is much smaller than for pure irons. The grain boundary is stronger because of carbon and nitrogen [7], and especially because martensite islands aligned along ferrite grain boundaries play the role as a reinforcement, thus preventing the local necking of each ferrite grain. In dual-phase steel, necking elongation is larger than expected from uniform elongation. These differences may be the reason why dual-phase steel shows better elongation even in high-velocity tension.

CF/GF Hybrid FRP

Three types of carbon fiber/glass fiber hybrid composite were prepared as shown in Table 3. Test specimens (Fig. 2) were taken from these cured plates with the longitudinal axis which coincided with the direction of fill. Some examples of dynamic stress-strain diagrams are shown in Figs. 8–10 in comparison with the statical ones. The ratio of longitudinal GF/CF is 2:1 in Hybrid 101, and 1:2 in Hybrid 102 and Hybrid 103. The difference between 102 and 103 is shown in Table 3. The cloths of 101 and 102 are plain woven and 103 is satin woven. The volumetric contents (%) of CF/GF (fill) in the hybrids were approximately 10.6/21.2, 20.8/10.4, and 28.7/14.4 for 101, 102, and 103, respectively. The fact that in 101, of which ratio of GF is comparatively high, the clear difference in the form of a stress-strain diagram seen between

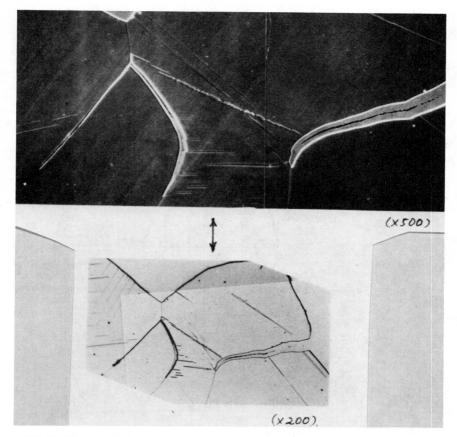

(×500)

(×200)

FIG. 7—*Micrograph of section structure of fracture surface in wet hydrogen treated pure iron B under high-velocity tension* ($\dot{\epsilon} = 1.4 \times 10^3$/s).

TABLE 3—*Composition of CF/GF hybrids.*

Hybrid	101	102	103
Toray Cloth	5101	5104	1306
	T-300*-1000[a]		T-300*-3000[b]
Fill ⎰ Torayca (No/25 mm)	17.7	25	43
⎱ Glass ECE 225 1/2 (No/25 mm)	35.4	12.5	22
Warp Glass ECE 225 1/0 (No/25 mm)	30	22.5	10
Tensile strength of T-300* MPa		3138	3040
		(320 kgf/mm²)	(310 kgf/mm²)
Tensile strength of Toray		78	
epoxy matrix, MPa	...	(8 kgf/mm²)	...

[a]1000 filaments.
[b]3000 filaments.

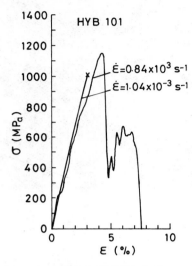

FIG. 8—*Stress-strain diagrams of HYB 1 CF/GF hybrid composite for dynamic and static tension.*

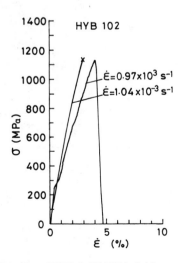

FIG. 9—*Stress-strain diagrams of HYB 2 CF/GF hybrid composite for dynamic and static tension.*

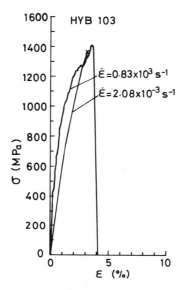

FIG. 10—*Stress-strain diagrams of HYB 3 CF/GF hybrid composite for dynamic and static tension.*

dynamic and static ones is suggestive. Anyway, the strain-rate dependence of σ_p, ϵ_t, and E_{ab} is positive (Fig. 11). In pure CFRP, high velocity brittleness or insensitiveness is known [5,6,2]. On the contrary, in pure GFRP, remarkably high-velocity ductility is found, and the fracture surface in dynamic tension is clearly different from that of CFRP [5,6,2]. As seen from the above-mentioned data, the hybrid composition with glass fiber is effective in improving the mechanical properties of CFRP in high-velocity tension. This improvement seems to be based upon the high-velocity ductility of GFRP.

The total tendencies of all materials investigated in this paper are summarized in Table 4. The detailed experimental data are shown in Table 5 with experimental scatter. The scatter is small, and obtained S-S curves in dynamic tension have enough reproducibility.

Conclusion

The "one-bar method" for material characterization in high-velocity tension is a very excellent one. It can give dynamic stress-strain diagrams up to breaking without disturbance. Many new results for the understanding of high velocity deformation of materials, such as a dual-phase steel, pure irons, and CF/GF hybrids, were obtained.

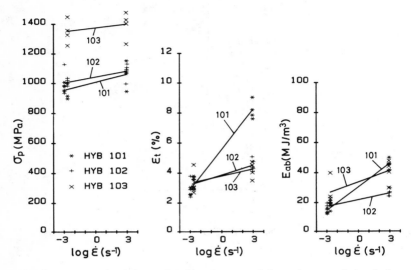

FIG. 11—*Strain-rate dependence of peak stress* σ_p, *total elongation* ϵ_t, *and absorbed energy per unit volume* E_{ab} *in HYB 1, 2, and 3, hybrid composites.*

TABLE 4—*Strain-rate dependence of tensile mechanical properties in a dual-phase steel, pure irons A and B and three CF/GF hybrid composites.*

Material	$\dfrac{\sigma_{pd}}{\sigma_{ps}}$	$\dfrac{\epsilon_{td}}{\epsilon_{ts}}$	$\dfrac{E_{abd}}{E_{abs}}$
Dual phase	1.07	1.15	1.10
Pure Fe-A	1.84	0.83	1.17
Pure Fe-B	2.59	0.04	0.06
HYB 101	1.11	2.69	2.90
HYB 102	1.08	1.38	1.46
HYB 103	1.03	1.25	1.53

d: $\dot{\epsilon} \cong 10^3 \ (S^{-1})$,
s: $\dot{\epsilon} \cong 10^{-3} \ (S^{-1})$,

TABLE 5—*Experimental data on σ_p, ϵ_t and E_{ab}.*

Material		$\dot{\epsilon}$ (1/s)	N	σ_p (MPa)		ϵ_t (%)		E_{ab} (MJ/m³)	
				M	Σ	M	Σ	M	Σ
Dual-phase	S	1.36×10^{-3}	3	887	2.5	22.8	0.59	184	5.1
steel	D	1.42×10^3	5	949	40.8	26.1	1.10	202	5.9
Pure iron	S	2.08×10^{-3}	3	235	2.3	54.1	2.24	106	2.6
A	D	1.08×10^3	3	433	12.0	44.7	0.82	125	4.6
Pure iron	S	2.08×10^{-3}	3	223	5.1	58.6	1.07	108	2.4
B	D	1.04×10^3	3	578	111	2.3	1.28	6.4	3.3
	S	1.56×10^{-3}	6	939	44.4	3.1	0.51	15.5	2.7
HYB 101	D	0.85×10^3	4	1040	84.3	8.2	0.62	44.9	3.3
	S	1.56×10^{-3}	7	986	63.2	3.3	0.54	17.6	3.1
HYB 102	D	0.88×10^3	5	1060	50.3	4.5	0.40	25.7	2.3
	S	2.08×10^{-3}	4	1230	78.4	3.4	0.83	26.5	9.1
HYB 103	D	0.80×10^3	4	1370	89.1	4.3	0.55	40.4	7.6

NOTES

$\dot{\epsilon}$ (1/s) = strain rate (*S*: static, *D*: dynamic).
σ_p (MPa) = peak stress.
ϵ_t (%) = total breaking strain.
E_{ab} (MJ/m³) = absorbed energy per unit volume.
M = mean.
Σ = standard deviation.
N = number of specimen.
S = quasi-static.
D = dynamic.

References

[1] Kawata, K., Hashimoto, S., Kurokawa, K., and Kanayama, N., "A New Testing Method of the Characterization of Materials in High Velocity Tension" in *Mechanical Properties at High Rates of Strain 1979* (Conference Series No. 47), J. Harding, Ed., Institute of Physics, Bristol and London, 1979, pp. 71–80.

[2] Kawata, K., Hashimoto, S., and Takeda, N. in *Proceedings*, 13th Congress of the International Council of the Aeronautical Sciences, Seattle, Aug. 1982, pp. 826–836.

[3] Kawata, K., Hashimoto, S., and Kurokawa, K., "Analysis of High Velocity Tension of Bars of Finite Length of BCC and FCC Metals with Their Own Constitutive Equations" in *High Velocity Deformation of Solids* (*Proceedings* of IUTAM Symposium, Tokyo, 1977), Springer-Verlag, Berlin, 1979, pp. 1–15.

[4] Kawata, K., *Theoretical and Applied Mechanics* (*Proceedings* of 15th International Congress on Theoretical and Applied Mechanics, Toronto, 1980), North-Holland, Amsterdam, 1980, pp. 307–317.

[5] Kawata, K., Hondo, A., Hashimoto, S., Takeda, N., and Chung, H. L., "Dynamic Behaviour Analysis of Composite Materials" in *Proceedings*, Japan-U.S. Conference on Composite Materials, Tokyo, 1981; Japan Society for Composite Materials, Tokyo, 1981, pp. 2–11.

[6] Kawata, K., Hashimoto, S., and Takeda, N., "Mechanical Behaviours in High-Velocity Tension of Composites" in *Proceedings*, Fourth International Conference on Composite Materials (ICCM-IV) Oct., 1982, Japan Society for Composite Materials, Tokyo, 1982, pp. 829–836.

[7] Matsui, H. and Kimura, H., *Journal of the Japan Institute of Metals*, Vol. 47, 1983, p. 294.

Summary

Summary

The 44 papers included in this publication cover various aspects of thermo-mechanical and physical behavior of composites which have been studied over the past two years both in the United States and Japan. For convenience, the aspects covered are grouped into nine subject areas: Fracture, Fatigue, Stress Analysis, Dynamic Behavior, Design, Fabrication Methods, Testing Methods, Elevated Temperature and Environmental Effects, and Thermomechanical Properties. Brief summaries of the nine subject areas will be given.

Fracture

The fracture process of various composite systems is investigated. Although laminated composites were focused on in most papers, new types of composites were also examined; for example, helical fiber reinforced ceramics, and steel fiber reinforced polymer cement composites. Fukuda discussed a model to calculate the stress concentration factor of the nearest fibers adjacent to a broken fiber in an unidirectional hybrid composite. Drzal and Rich studied the effect of various surface treatments of fiber on the interfacial shear strength and found that some surface finishes resulted in a brittle zone between fiber and matrix, thus increasing the interfacial shear strength. Fracture toughness by three-point bending test was measured on a helical tantalum fiber/silicon carbide composite by Kagawa et al and on steel fiber/ polymer cement composite by Horiguchi. Failure mechanism in laminated composites were studied by focusing on transverse ply failure by Kister et al, on a curved free-edge by Klang and Heyer, while the initial and post failure in the stress-strain curve of the composite was studied by Ohira and Uda.

Fatigue

Fatigue damage was focused on from various viewpoints, causes of damage, and measurement methods. Tsangarakis et al investigated the tension-tension fatigue damage of a fiber-reinforced fiber/aluminum composite and found that fabrication defects reduced the fatigue limit, but the fiber imperfection and cyclic frequence did not. Techniques measuring fatigue damage were discussed on a notched laminated composite by Kress and Stinchcomb and also by Cantrell et al. Kress and Stinchcomb attempted to correlate "damage states" to the other mechanical properties and Cantrell et al proposed a new ultrasonic measurement technique which picks up only "real damage" in composites but not pseudo-damage such as surface irregularity. Harris and Morris reported on the experimental results on the compression-compression fatigue of a notched laminated composite and found that its fatigue life is a strong function of stacking sequence or specimen thickness.

Stress Analysis

Stress analysis is grouped into two areas: analytical and numerical techniques. Mura and Taya solved the thermal residual problem in a short fiber metal matrix composite by using the concept of Somigliana dislocations and Eshelby's equivalent inclusion. Rehfield et al calculated analytically the stress field in a tabbed compression specimen to identify the possible failure point in the specimen. A similar attempt was made by Cunningham et al on the stress field in and around the end tabs of a composite specimen, and they found that the best tab design to reduce the stress concentration and to yield consistent experimental results has a 10-deg tab angle and no cutoff thickness. The Saint-Venant effect in composite plate was studied numerically by Okumura et al, who found that the rate of attenuation of nonuniform local stress distribution is smaller in composite than in isotropic plate. Akasaka and Asano developed a numerical method based on an extremum principle of total potential energy and computed the stress and deformation of a sandwich panel having curved faceplates under pressure loading. The numerical results on the strains and deformations were compared with the experiment, resulting in good agreement between them.

Dynamic Behavior

Various aspects of dynamic behavior of composite were focused on: acoustic damping, wave propagation, impact, and buckling. Acoustic damping in composite was studied on stiffened graphite/epoxy panel subjected to also static shear load by Soovere and on ferrite-resin composites by Yamauchi and Emoto. Soovere studied the effect of a high random acoustic load together with static shear load near initial buckling on the dynamic response of integrated stiffened graphite/epoxy panels both experimentally and analytically and found that the damping in noncritical modes can increase by as much as a factor of four at initial buckling. Yamauchi and Emoto found in their experiment on ferrite-resin composite that the factor affecting the damping coefficient (DC) most is the thickness of metal coating on the composite plate, and the value of DC was not influenced much by the frequency. The mechanism and behavior of low-velocity impact on composites were studied by Chamis and Sinclair and also by Ross et al. Chamis and Sinclair studied impact resistance of unidirectional composite by use of a Charpy tester with the aim of studying the energy-absorbing mechanisms. They also examined several analytical models, that is, mechanistic, statistical, and transient finite-element-method models. The last model was attempted on a laminated composite in detail by Ross et al, who compared the numerical results by the 3-D finite element method with the experiments and obtained good agreement between them. The buckling behavior of viscoelastic composite columns was studied analytically by Wilson and Vinson, who used a quasi-static approach and found that the viscoelastic effects are significant in reducing the critical buckling load by 10 to 20%.

Design

Design was discussed in two aspects: optimum design of the composite itself in selecting the best combination of fiber type with its alignment and matrix material, and the design of composite structures made of various composite materials, most often laminated composites. Miki reported a design method for a laminated composite with required flexural stiffness. In Miki's method, three parameters—fiber type, matrix type, and ply angle—are determined at the end of optimization process. Morita et al obtained two optimum compositions of fibers in hybrid composites used for the magnetic powder core of the magnetic wedges in an induction motor. The main objective in their work was to enhance the flexural strength by keeping a certain level of permeability. Ko and Pastore studied the design of a 3-D fabric composite in order to obtain the accurate prediction of its strength. The analytical model is based on a maximum failure criterion and the predicted value agreed well with the experiment. Kawashima et al investigated experimentally the optimum design of three composite structures that are to be used for space structural components. The evaluation of these composite structures was made by subjecting them to acoustic noise environment, thermal cycling, and sinusoidal vibration.

Fabrication Methods

Fabrication conditions on various composites were reported, aiming at improving the mechanical properties of a composite. Cho and Okura studied processing conditions on a carbon-carbon composite where the matrix is pulverized coke, and three types of fiber were examined, that is, short, long fibers, and cloth. The evaluation of the c/c composite fabricated was made by measuring the bending strength. They found that the composite with long fibers has the best bending strength. The method fabricating silicon carbide fiber/aluminum composite by slurry impregnation and liquid phase hot pressing was discussed by Kohara and Muto, who found that as the volume fraction of fibers increases, a departure of the strength from the value of a rule of mixture is enhanced and addition of silicon to the aluminum matrix is effective in reducing the degradation of silicon carbide fibers. Nimmer et al examined the fabrication and design criteria for two types of composite flywheels and also reported some fabrication methods for thicker flywheels by hydroclaving process. Minoda et al discussed the curing condition of an integrated graphite/epoxy laminated composite and proposed a new testing method, in-process direct measurement of electrical properties of articles in an autoclave in order to better monitor the cure cycles of the composite.

Testing Methods

Further efforts in improving testing methods were made by several researchers. Daimaru et al measured the work of fracture (γ_F) of continuous and short fiber metal matrix composites by a three-point bending test. The specimen has a center notch with a triangle ligament. The analytical model proposed by Daimaru et al predicted the experimental values of γ_F very well. Oplinger et al examined various types of tension specimens both experimentally and numerically. They concluded that the tension specimen with a streamline shape is quite promising for laminated composites and ceramics, and it is a good alternative one to tabbed specimens when the tabbed specimens cannot be used under a more stringent environment. Nimmer and Trantina discussed two testing methods to measure the strength of composites, the ASTM three-point flexural test and ordinary tension test, and concluded that the strength measured by the ASTM flexural test overestimates the true strength. Kobayashi and Suemasu studied the dynamic crack propagation in a glass fiber composite by measuring the crack velocity and the temperature increase due to the energy dissipation at a crack tip. They reported an improved dual-layer liquid crystal film method to record the temperature distribution.

Elevated Temperature and Environmental Effects

Metal matrix composites are often used at high temperatures. Thus, three papers studied the microscopic behavior of metal matrix composites at high temperatures, and the other two papers discussed the environmental effects on composites. Rhodes and Spurling studied the growth kinetics of the reaction zone products in SiC/Ti-6Al-4V composites and found that the major product is titanium carbide, among others, and its growth obeys a diffusion type law, that is, the thickness is proportional to (time)$^{1/2}$. In order to avoid reaction products, Umekawa et al used coated fibers for two types of metal matrix composites, tungsten/nickel and tungsten/SUS316, and concluded that zirconium compounds were effective as a coating material for these composites. Internal friction measurements were made on oxide dispersion-strengthened nickel-chromium (Ni-Cr) alloys at various temperatures and frequencies by Shioiri and Satoh, and the activation energy of the grain boundary viscosity was obtained from the peak component and will provide useful data for the analysis of the creep behavior of Ni-Cr alloys. Aylor and Kain investigated the corrosion resistance of several aluminum-based metal matrix composites in a marine environment. The metal matrix composites investigated are continuous and short silicon carbide fiber/aluminum and continuous graphite/aluminum composites. Since all the composites used were found to show excessive corrosion, some suitable coatings were recommended. Hahn et al studied the behavior of a composite flywheel in vacuum and at 100°C and found that the static properties of the composite were not degraded significantly and that the fatigue strength was not reduced at all after exposure to vacuum and 100°C.

Thermomechanical Properties

In order to use composite as a structural component, the thermomechanical properties of the composite must be found either experimentally or analytically. One of the most accurate methods to determine the elastic constants is the wave propagation method. This was studied by Kritz and Ledbetter on unidirectional graphite fiber/epoxy composites. Additional findings in their work were that shear wave velocity exceeds longitudinal wave velocity in the plane transverse to the fiber axis. In the standard analysis of the mechanical properties of composites, the shear modulus (G) of the matrix has been always assumed to be constant. Hayashi challenged this assumption and showed experimentally in a combined shear and compression test that G becomes a function of the compressive stress.

Takao proposed a new model to predict the thermal expansion coefficients of a hybrid short fiber composite with fibers being misoriented two- or three-dimensionally. Unlike the conventional approach based on laminate theory, this model accounts directly for the interaction between misoriented short fibers by use of Eshelby's method. Although most of the studies on the mechanical properties of the composite are focused on the static properties, a knowledge of the mechanical properties at high strain-rate becomes equally important under some circumstances. Kawata et al obtained such data experimentally on dual-phase steel and graphite/glass/epoxy hybrid composites at high strain rates. They found that dual-phase steel showed considerable ductility at high-velocity impact compared with the matrix pure iron, and graphite/glass/epoxy composites also showed considerable ductility at high-velocity impact despite the fact that usually graphite/epoxy composites are brittle under high velocity impact.

J. R. Vinson

University of Delaware, Newark, DE 19716; conference chairman and coeditor.

M. Taya

University of Delaware, Newark, DE 19716; conference secretary and coeditor.

Index